Fundamentals of Queueing Theory

AF864964

WILEY SERIES IN PROBABILITY AND STATISTICS

Established by WALTER A. SHEWHART and SAMUEL S. WILKS

Editors: *David J. Balding, Noel A. C. Cressie, Garrett M. Fitzmaurice, Iain M. Johnstone, Geert Molenberghs, David W. Scott, Adrian F. M. Smith, Ruey S. Tsay, Sanford Weisberg*
Editors Emeriti: *Vic Barnett, J. Stuart Hunter, Jozef L. Teugels*

A complete list of the titles in this series appears at the end of this volume.

Fundamentals of Queueing Theory

Fourth Edition

Donald Gross
George Mason University
Fairfax, Virginia

John F. Shortle
George Mason University
Fairfax, Virginia

James M. Thompson
Freddie Mac Corporation
McLean, Virginia

Carl M. Harris

WILEY

FUNDAMENTALS OF QUEUEING THEORY
Fourth Edition

Copyright © 2008 by John Wiley & Sons, Inc. All rights reserved.

Authorized reprint by Wiley India (P.) Ltd., 4435-36/7, Ansari Road, Daryaganj, New Delhi 110002.

All rights reserved. AUTHORIZED REPRINT OF THE EDITION PUBLISHED BY JOHN WILEY & SONS, INC. No part of this book may be reproduced in any form without the written permission of the publisher.

Limits of Liability/Disclaimer of Warranty: The publisher and the author make no representation or warranties with respect to the accuracy or completeness of the contents of this work and specifically disclaim all warranties, including without limitation warranties of fitness for a particular purpose. No warranty may be created or extended by sales or promotional materials. The advice and strategies contained herein may not be suitable for every situation. This work is sold with the understanding that the publisher is not engaged in rendering legal, accounting, or other professional services. If professional assistance is required, the services of a competent professional person should be sought. Neither the publisher nor the author shall be liable for damages arising here from. The fact that an organization or Website if referred to in this work as a citation and/or a potential source of further information does not mean that the author or the publisher endorses the information the organization or Website may provide or recommendations it may make. Further, readers should be aware that Internet Websites listed in this work may have changed or disappeared between when this work was written and when it is read.

Wiley also publishes its books in a variety of electronic formats. Some content that appears in print may not be available in electronic books. For more information about Wiley products, visit our website at: www.wiley.com.

Reprint: 2014

Printed at: Print India Press, Sahibabad, UP

ISBN: 978-81-265-3406-7

Carl Harris, 1940–2000

This book is dedicated to the memory of Carl M. Harris. Carl and I first entertained the idea of a queueing text back in 1968 and collaborated on the first three editions. We were friends and colleagues from that time until his untimely death from a heart attack while exercising at a local gym, one month after his 60th birthday.

Carl was the BDM International Professor of Operations Research and the founding chair of the Systems Engineering and Operations Research Department for the Volgenau School of Information Technology and Engineering at George Mason University, Fairfax, Virginia. In 1999, he was awarded the Institute for Operations Research and the Management Sciences (INFORMS) Kimball Medal in recognition of distinguished service to the Operations Research profession and the Operations Research Society of America (INFORMS' predecessor). He was the society's 39th president. Carl's research interests were in the areas of applied probability and statistics, particularly queueing theory and stochastic processes. He authored or co-authored about 80 scholarly papers (on many of which I was fortunate enough to have worked with him.) In addition to the first 3 editions of this book, he co-authored with Saul I. Gass, *The Encyclopedia of Operations Research and Management Science.*

His warmth, collegiality, and friendship are sorely missed.

DONALD GROSS

CONTENTS

PREFACE

The changes in this fourth edition reflect the feedback from numerous students, teachers, and colleagues since the third edition came out ten years ago. Almost all the material from the third edition is kept in the fourth, however, with a fair amount of editing and reorganization.

Chapter 2 contains a new section on choosing the number of servers and a new subsection on computational issues of the Erlang B formula. The chapter now begins with a section on birth–death processes, which was the old Section 1.10 from the previous edition. Chapter 3 is substantially edited and contains a new section on retrial queues. Chapter 5 contains an expanded discussion of the level crossing method developed by Percy Brill. Chapter 7 is now split into two separate chapters: Chapter 7, Bounds and Approximations, and Chapter 8, Numerical Techniques and Simulation. Chapter 7 includes a new section on network approximations, and Chapter 8 includes a new section on numerical inversion of transforms.

Two appendices are added back to this edition, one on transforms and generating functions and the other on differential and difference equations (by popular request). The appendix on the QtsPlus software is completely rewritten to reflect the changes and expansion made to the software. Also, a subsection on how to use the software is added to Chapter 1. Finally, many more examples and problems are added. In this edition we do not denote which problems are solvable on the computer — we leave that up to the discretion of the student and/or instructor. We also do not include here a table on suggested text material for various course lengths (quarter, semester, etc.). Again, we believe that the instructor is the best one to decide.

For errata, updates, and other information about the text and associated QtsPlus software, see the text website:

`<http://mason.gmu.edu/~jshortle/fqt4th.html>`.

Donald Gross
John F. Shortle
James M. Thompson

Fairfax, Virginia
March 2008

ACKNOWLEDGMENTS

We are grateful for the assistance given to us over the years by many professional colleagues and students, whose numerous comments and suggestions have been so helpful in improving our work. We particularly thank Prof. Andrew Ross, Eastern Michigan University, Prof. Percy Brill, University of Windsor, and Prof. John Mullen, New Mexico State University, for their very detailed comments, many of which are reflected in this edition. We also appreciate the help and encouragement of our research colleagues, Dr. Martin Fischer and Dr. Denise Masi of Noblis (formerly Mitretek Systems) and Prof. Brian Mark of George Mason University.

With heartfelt thanks, we extend special appreciation once more to our families for their unlimited and continuing encouragement and to all the people at John Wiley & Sons who have been wonderfully supportive of us through these four editions. We also appreciate the support of the Volgenau School of Information Technology and Engineering and the Department of Systems Engineering and Operations Research at George Mason University.

D. G.
J. F. S.
J. M. T.

CHAPTER 1

INTRODUCTION

All of us have experienced the annoyance of having to wait in line. Unfortunately, this phenomenon continues to be common in congested, urbanized, "high-tech" societies. We wait in line in our cars in traffic jams or at toll booths; we wait on hold for an operator to pick up our telephone calls; we wait in line at supermarkets to check out; we wait in line at fast-food restaurants; and we wait in line at banks and post offices. We, as customers, do not generally like these waits, and the managers of the establishments at which we wait also do not like us to wait, since it may cost them business. Why then is there waiting?

The answer is simple: There is more demand for service than there is facility for service available. Why is this so? There may be many reasons; for example, there may be a shortage of available servers, it may be infeasible economically for a business to provide the level of service necessary to prevent waiting, or there may be a space limit to the amount of service that can be provided. Generally these limitations can be removed with the expenditure of capital, and to know how much service should then be made available, one would need to know answers to such questions as, "How long must a customer wait?" and "How many people will form in the line?" Queueing theory attempts (and in many cases succeeds) to answer these questions through detailed mathematical analysis. The word "queue" is in more common usage

Fundamentals of Queueing Theory, Fourth Edition.
By D. Gross, J. F. Shortle, J. M. Thompson, and C. M. Harris
Copyright © 2008 John Wiley & Sons, Inc.

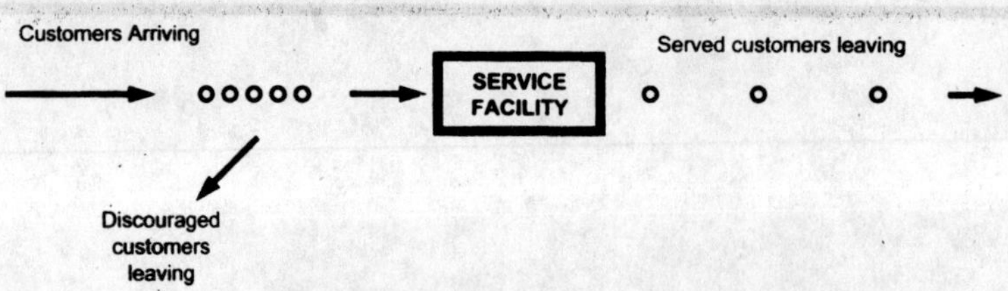

Figure 1.1 A typical queueing process.

in Great Britain and other countries than in the United States, but it is rapidly gaining acceptance in this country, although it must be admitted that it is just as displeasing to spend time in a queue as in a waiting line.

1.1 Description of the Queueing Problem

A queueing system can be described as customers arriving for service, waiting for service if it is not immediate, and if having waited for service, leaving the system after being served. The term "customer" is used in a general sense and does not imply necessarily a human customer. For example, a customer could be a ball bearing waiting to be polished, an airplane waiting in line to take off, or a computer program waiting to be run. Such a basic system can be schematically shown as in Figure 1.1. Although any queueing system may be diagrammed in this manner, it should be clear that a reasonably accurate representation of such a system would require a detailed characterization of the underlying processes.

Queueing theory was developed to provide models to predict the behavior of systems that attempt to provide service for randomly arising demands; not unnaturally, then, the earliest problems studied where those of telephone traffic congestion. The pioneer investigator was the Danish mathematician A. K. Erlang, who, in 1909, published "The Theory of Probabilities and Telephone Conversations." In later works he observed that a telephone system was generally characterized by either (1) Poisson input, exponential holding (service) times, and multiple channels (servers), or (2) Poisson input, constant holding times, and a single channel. Erlang was also responsible for the notion of stationary equilibrium, for the introduction of the so-called balance-of-state equations, and for the first consideration of the optimization of a queueing system.

Work on the application of the theory to telephony continued after Erlang. In 1927, E. C. Molina published his paper "Application of the Theory of Probability to Telephone Trunking Problems," which was followed one year later by Thornton Fry's book *Probability and Its Engineering Uses*, which expanded much of Erlang's earlier work. In the early 1930s, Felix Pollaczek did some further pioneering work on Poisson input, arbitrary output, and single- and multiple-channel problems. Additional work was done at that time in Russia by Kolmogorov and Khintchine, in France by Crommelin, and in Sweden by Palm. The work in queueing theory picked up momentum rather slowly in its early days, but accelerated in the 1950s, and there has been a great deal of work in the area since then.

There are many valuable applications of the theory, most of which have been well documented in the literature of probability, operations research, management science, and industrial engineering. Some examples are traffic flow (vehicles, aircraft, people, communications), scheduling (patients in hospitals, jobs on machines, programs on a computer), and facility design (banks, post offices, amusement parks, fast-food restaurants).

Queueing theory originated as a very practical subject, but much of the literature up to the middle 1980s was of little direct practical value. However, queueing theorists have once again become concerned about the application of the sophisticated theory that has largely arisen since the close of World War II. The emphasis in the literature on the exact solution of queueing problems with clever mathematical tricks is now becoming secondary to model building and the direct use of these techniques in management decisionmaking. Most real problems do not correspond exactly to a mathematical model, and increasing attention is being paid to complex computational analysis, approximate solutions, sensitivity analyses, and the like. The development of the practice of queueing theory must not be restricted by a lack of closed-form solutions, and problem solvers must be able to put the developed theory to good use. These points should be kept in mind by the reader, and we attempt to illustrate them whenever possible throughout this text.

1.2 Characteristics of Queueing Processes

In most cases, six basic characteristics of queueing processes provide an adequate description of a queueing system: (1) arrival pattern of customers, (2) service pattern of servers, (3) queue discipline, (4) system capacity, (5) number of service channels, and (6) number of service stages.

1.2.1 Arrival Pattern of Customers

In usual queueing situations, the process of arrivals is stochastic, and it is thus necessary to know the probability distribution describing the times between successive customer arrivals (interarrival times). It is also necessary to know whether customers can arrive simultaneously (batch or bulk arrivals), and if so, the probability distribution describing the size of the batch.

It is also necessary to know the reaction of a customer upon entering the system. A customer may decide to wait no matter how long the queue becomes, or, on the other hand, if the queue is too long, the customer may decide not to enter the system. If a customer decides not to enter the queue upon arrival, the customer is said to have *balked*. A customer may enter the queue, but after a time lose patience and decide to leave. In this case, the customer is said to have *reneged*. In the event that there are two or more parallel waiting lines, customers may switch from one to another, that is, *jockey* for position. These three situations are all examples of queues with *impatient customers*.

One final factor to be considered regarding the arrival pattern is the manner in which the pattern changes with time. An arrival pattern that does not change with time (i.e., the probability distribution describing the input process is time-independent) is called a *stationary* arrival pattern. One that is not time-independent is called *nonstationary*.

1.2.2 Service Patterns

Much of the previous discussion concerning the arrival pattern is appropriate in discussing service. Most importantly, a probability distribution is needed to describe the sequence of customer service times. Service may also be single or batch. One generally thinks of one customer being served at a time by a given server, but there are many situations where customers may be served simultaneously by the same server, such as a computer with parallel processing, sightseers on a guided tour, or people boarding a train.

The service process may depend on the number of customers waiting for service. A server may work faster if the queue is building up or, on the contrary, may get flustered and become less efficient. The situation in which service depends on the number of customers waiting is referred to as *state-dependent* service. Although this term was not used in discussing arrival patterns, the problems of customer impatience can be looked upon as ones of state-dependent arrivals, since the arrival behavior depends on the amount of congestion in the system.

Service, like arrivals, can be stationary or nonstationary with respect to time. For example, learning may take place, so that service becomes more efficient as experience is gained. The dependence on time is not to be confused with dependence on state. The former does not depend on the number of customers in the system, but rather on how long it has been in operation. The latter does not depend on how long the system has been in operation, but only on the state of the system at a given time, that is, on how many customers are currently in the system. Of course, a queueing system can be both nonstationary and state-dependent.

Even if the service rate is high, it is very likely that some customers will be delayed by waiting in the line. In general, customers arrive and depart at irregular intervals; hence the queue length will assume no definitive pattern unless arrivals and service are deterministic. Thus it follows that a probability distribution for queue lengths will be the result of two separate processes—arrivals and services—which are generally, though not universally, assumed mutually independent.

1.2.3 Queue Discipline

Queue discipline refers to the manner in which customers are selected for service when a queue has formed. The most common discipline that can be observed in everyday life is first come, first served (FCFS). However, this is certainly not the only possible queue discipline. Some others in common usage are last come, first

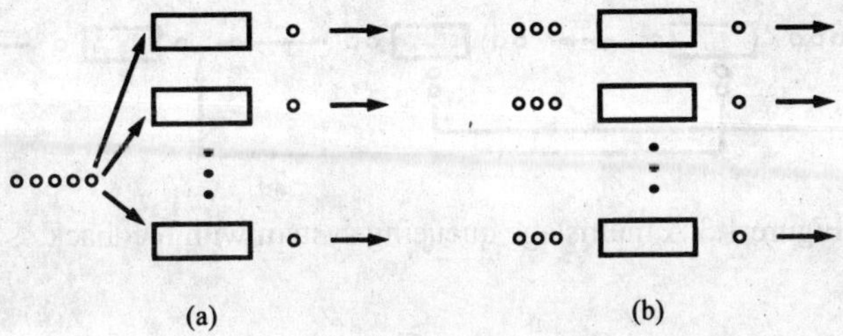

Figure 1.2 Multichannel queueing systems

served (LCFS), which is applicable to many inventory systems when there is no obsolescence of stored units, as it is easier to reach the nearest items, which are the last in; selection for service in random order independent of the time of arrival to the queue (RSS); and a variety of *priority* schemes, where customers are given priorities upon entering the system, the ones with higher priorities to be selected for service ahead of those with lower priorities, regardless of their time of arrival to the system.

There are two general situations in priority disciplines. In the first, which is called *preemptive*, the customer with the highest priority is allowed to enter service immediately even if a customer with lower priority is already in service when the higher-priority customer enters the system; that is, the lower-priority customer in service is preempted, its service stopped, to be resumed again after the higher-priority customer is served. There are two possible additional variations: the preempted customer's service when resumed can either continue from the point of preemption or start anew. In the second general priority situation, called the *nonpreemptive* case, the highest-priority customer goes to the head of the queue but cannot get into service until the customer presently in service is completed, even though this customer has a lower priority.

1.2.4 System Capacity

In some queueing processes there is a physical limitation to the amount of waiting room, so that when the line reaches a certain length, no further customers are allowed to enter until space becomes available as the result of a service completion. These are referred to as finite queueing situations; that is, there is a finite limit to the maximum system size. A queue with limited waiting room can be viewed as one with forced balking where a customer is forced to balk if it arrives when the queue size is at its limit. This is a simple case, since it is known exactly under what circumstances arriving customers must balk.

1.2.5 Number of Service Channels

As we shortly explain in more detail, it is generally preferable to design multiserver queueing systems to be fed by a single line. Thus, when we specify the number of service channels, we are typically referring to the number of parallel service stations

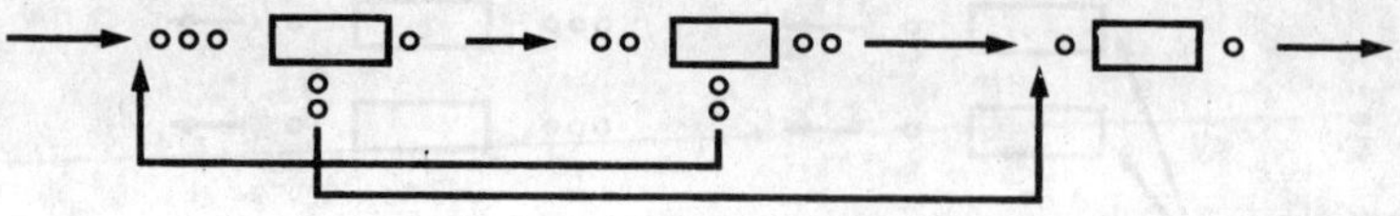

Figure 1.3 A multistage queueing system with feedback.

that can serve customers simultaneously. Figure 1.1 depicts an illustrative single-channel system, while Figure 1.2 shows two variations of multichannel systems. The two multichannel systems differ in that the first has a single queue, while the second allows a queue for each channel. A hair-styling salon with many chairs is an example of the first type of multichannel system (assuming no customer is waiting for any particular stylist), while a supermarket or fast-food restaurant might fit the second description. It is generally assumed that the service mechanisms of parallel channels operate independently of each other.

1.2.6 Stages of Service

A queueing system may have only a single stage of service, as in the hair-styling salon, or it may have several stages. An example of a multistage queueing system would be a physical examination procedure, where each patient must proceed through several stages, such as medical history; ear, nose, and throat examination; blood tests; electrocardiogram; eye examination; and so on. In some multistage queueing processes recycling or feedback may occur. Recycling is common in manufacturing processes, where quality control inspections are performed after certain stages, and parts that do not meet quality standards are sent back for reprocessing. Similarly, a telecommunications network may process messages through a randomly selected sequence of nodes, with the possibility that some messages will require rerouting on occasion through the same stage. A multistage queueing system with some feedback is depicted in Figure 1.3.

The six characteristics of queueing systems discussed in this section are generally sufficient to completely describe a process under study. Clearly, a wide variety of queueing systems can be encountered. Before performing any mathematical analyses, however, it is absolutely necessary to describe adequately the process being modeled. Knowledge of the basic six characteristics is essential in this task.

It is extremely important to use the correct model or at least the model that best describes the real situation being studied. A great deal of thought is often required in this *model selection procedure*. For example, let us reconsider the supermarket mentioned previously. Suppose there are c checkout counters. If customers choose a checkout counter on a purely random basis (without regard to the queue length in front of each counter) and never switch lines (no jockeying), then we truly have c independent single-channel models. If, on the other hand, there is a single waiting line and when a checker becomes idle, the customer at the head of the line (or with the lowest number if numbers are given out) enters service, we have a c-channel model. Neither, of course, is generally the case in most supermarkets. What usually happens is that queues form in front of each counter, but new customers enter the queue that is the shortest (or has shopping carts that are lightly loaded). Also, there is a great deal of jockeying between lines. Now the question becomes which choice of models (c independent single channels or a single c-channel) is more appropriate. If there were complete jockeying, the single c-channel model would be quite appropriate, since even though in reality there are c lines, there is little difference, when jockeying is present, between these two cases. This is so because no servers will be idle as long as customers are waiting for service, which would not be the case with c truly independent single channels. As jockeying is rather easy to accomplish in supermarkets, the c-channel model will be more appropriate and realistic than the c-single-channels model, which one might have been tempted to choose initially prior to giving much thought to the process. Thus it is important not to jump to hasty conclusions but to select carefully the most appropriate model.

1.3 Notation

As a shorthand for describing queueing processes, a notation has evolved, due for the most part to Kendall (1953), which is now rather standard throughout the queueing literature. A queueing process is described by a series of symbols and slashes such as $A/B/X/Y/Z$, where A indicates in some way the interarrival-time distribution, B the service pattern as described by the probability distribution for service time, X the number of parallel service channels, Y the restriction on system capacity, and Z the queue discipline (Appendix 1 contains a dictionary of symbols used throughout this text). Some standard symbols for these characteristics are presented in Table 1.1. For example, the notation $M/D/2/\infty/\text{FCFS}$ indicates a queueing process with exponential interarrival times, deterministic service times, two parallel servers, no restriction on the maximum number allowed in the system, and first-come, first-served queue discipline.

In many situations only the first three symbols are used. Current practice is to omit the service-capacity symbol if no restriction is imposed ($Y = \infty$) and to omit the queue discipline if it is first come, first served ($Z = \text{FCFS}$). Thus $M/D/2$ would be a queueing system with exponential input, deterministic service, two servers, no limit on system capacity, and first-come, first-served discipline.

The symbols in Table 1.1 are, for the most part, self-explanatory; however, a few require further comment. The symbol G represents a general probability distribution; that is, no assumption is made as to the precise form of the distribution. Results in these cases are applicable to any probability distribution. These general-time distributions, however, are required to represent independent and identically distributed random variables.

It may also appear strange that the symbol M is used for exponential. The use of the symbol E, as one might expect, would be too easily confused with E_k, which is used for the type-k Erlang distribution (a gamma with an integer shape parameter). So M is used instead; it stands for the Markovian or memoryless property of the exponential, which is developed in some detail in Section 1.9.

Table 1.1 Queueing Notation $A/B/X/Y/Z$

Characteristic	Symbol	Explanation
Interarrival-time distribution (A) / Service-time distribution (B)	M	Exponential
	D	Deterministic
	E_k	Erlang type $k(k = 1, 2, \ldots)$
	H_k	Mixture of k exponentials
	PH	Phase type
	G	General
# of parallel servers (X)	$1, 2, \ldots, \infty$	
Max. system capacity (Y)	$1, 2, \ldots, \infty$	
Queue discipline (Z)	FCFS	First come, first served
	LCFS	Last come, first served
	RSS	Random selection for service
	PR	Priority
	GD	General discipline

The reader may have noticed that the list of symbols is not complete. For example, there is no indication of a symbol to represent bulk arrivals, to represent series queues, to denote any state dependence, and so on. If a suitable notation does exist for any previously unmentioned model, it is indicated when that particular model is brought up in the text. However, there still remain models for which no symbolism has either been developed or accepted as standard, and this is generally true for those models less frequently analyzed in the literature.

1.4 Measuring System Performance

Up to now the concentration has been on the physical description of queueing processes. What, then, might one like to know about the effectiveness of a queueing system? Generally there are three types of system responses of interest: (1) some measure of the waiting time that a typical customer might be forced to endure; (2) an indication of the manner in which customers may accumulate; and (3) a measure of the idle time of the servers. Since most queueing systems have stochastic elements, these measures are often random variables and their probability distributions, or at the very least their expected values, are desired.

There are two types of customer waiting times, the time a customer spends in the queue and the total time a customer spends in the system (queue plus service). Depending on the system being studied, one may be of more interest than the other. For example, if we are studying an amusement park, it is the time waiting in the queue that makes the customer unhappy. On the other hand, if we are dealing with machines that require repair, then it is the total down time (queue wait plus repair time) that we wish to keep as small as possible. Correspondingly, there are two customer accumulation measures as well: the number of customers in the queue and the total number of customers in the system. The former would be of interest if we desire to determine a design for waiting space (say, the number of seats to have for customers waiting in a hair-styling salon), while the latter may be of interest for knowing how many of our machines may be unavailable for use. Idle-service measures can include the percentage of time any particular server may be idle, or the time the entire system is devoid of customers.

The task of the queueing analyst is generally one of two things. He or she is either to determine the values of appropriate measures of effectiveness for a given process, or to design an "optimal" (according to some criterion) system. To do the former, one must relate waiting delays, queue lengths, and such to the given properties of the input stream and the service procedures. On the other hand, for the design of a system the analyst might want to balance customer waiting time against the idle time of servers according to some inherent cost structure. If the costs of waiting and idle service can be obtained directly, they can be used to determine the optimum number of channels to maintain and the service rates at which to operate these channels. Also, to design the waiting facility it is necessary to have information regarding the possible size of the queue to plan for waiting room. There may also be a space cost that should be considered along with customer-waiting and idle-server costs to obtain the optimal system design. In any case, the analyst will strive to solve this problem by analytical means; however, if these fail, he or she must resort to simulation. Ultimately, the issue generally comes down to a trade-off of better customer service versus the expense of providing more service capability, that is, determining the increase in investment of service for a corresponding decrease in customer delay.

1.5 Some General Results

We present some general results and relationships for $G/G/1$ and $G/G/c$ queues in this section, prior to specific model development. These results will prove useful in many of the following sections and chapters, as well as providing some insight at this early stage.

Denoting the average rate of customers entering the queueing system as λ and the average rate of serving customers as μ, a measure of traffic congestion for c-server systems is $\rho \equiv \lambda/c\mu$ (often called *traffic intensity*). When $\rho > 1$ ($\lambda > c\mu$), the average number of arrivals into the system exceeds the maximum average service rate of the system, and we would expect, as time goes on, the queue to get bigger and bigger, unless, at some point, customers were not allowed to join. If we are interested in steady-state conditions (the state of the system after it has been in operation a long time), when $\rho > 1$, the queue size never settles down (assuming customers are not prevented from entering the system) and there is no steady state. It turns out that for steady-state results to exist, ρ must be strictly less than 1 (again, assuming no denial of customer entry). When $\rho = 1$, unless arrivals and service are deterministic and perfectly scheduled, no steady state exists, since randomness will prevent the queue from ever emptying out and allowing the servers to catch up, thus causing the queue to grow without bound. Therefore, if one knows the average arrival rate and average service rate, the minimum number of parallel servers required to guarantee a steady-state solution can be calculated immediately by finding the smallest c such that $\lambda/c\mu < 1$.

What we most often desire in solving queueing models is to find the probability distribution for the total number of customers in the system at time t, $N(t)$, which is made up of those waiting in queue, $N_q(t)$, plus those in service, $N_s(t)$. Let $p_n(t) = \Pr\{N(t) = n\}$, and $p_n = \Pr\{N = n\}$ in the steady state. Considering c-server queues in steady state, two expected-value measures of major interest are the mean number in the system,

$$L = \mathrm{E}[N] = \sum_{n=0}^{\infty} np_n,$$

and the expected number in queue,

$$L_q = \mathrm{E}[N_q] = \sum_{n=c+1}^{\infty} (n-c)p_n.$$

1.5.1 Little's Formulas

One of the most powerful relationships in queueing theory was developed by John D. C. Little in the early 1960s (see Little, 1961, for the original proof—a host of papers refining the proof followed in the ensuing decades). Little related the steady-state mean system sizes to the steady-state average customer waiting times as follows. Letting T_q represent the time a customer (transaction) spends waiting in the queue prior to entering service and T represent the total time a customer spends in the system ($T = T_q + S$, where S is the service time, and T, T_q, and S are random variables), two often used measures of system performance with respect to customers are $W_q = \mathrm{E}[T_q]$ and $W = \mathrm{E}[T]$, the mean waiting time in queue and the mean waiting time in the system, respectively. Little's formulas are

$$L = \lambda W \tag{1.1a}$$

and

$$L_q = \lambda W_q. \tag{1.1b}$$

Thus it is necessary to find only one of the four expected-value measures, in view of Little's formulas and the fact that $\mathrm{E}[T] = \mathrm{E}[T_q] + \mathrm{E}[S]$, or, equivalently, $W = W_q + 1/\mu$, where μ, as before, is the mean service rate.

Although the following does not constitute a proof, we illustrate the concept of Little's formulas by considering a *sample path* of one *busy period* (time from when

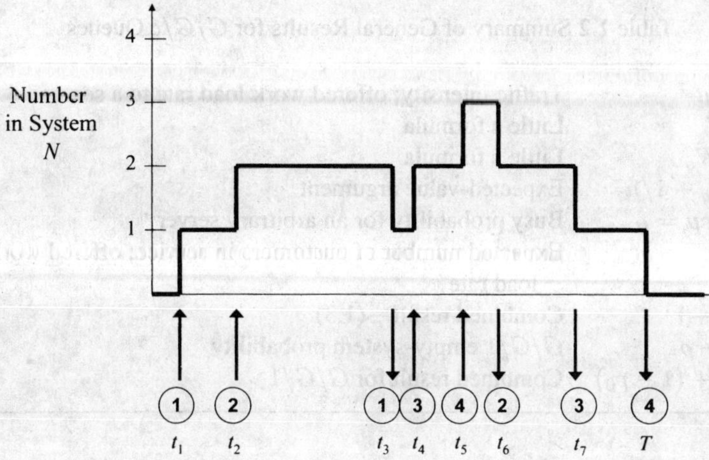

Figure 1.4 Busy-period sample path.

a customer enters an empty system until it next empties out again). Consider the illustration in Figure 1.4, where the number of customers (say, N_c) that arrive over the time period $(0, T)$ is 4.

The calculations for L and W are

$$\begin{aligned} L &= [1(t_2 - t_1) + 2(t_3 - t_2) + 1(t_4 - t_3) + 2(t_5 - t_4) \\ &\quad + 3(t_6 - t_5) + 2(t_7 - t_6) + 1(T - t_7)]/T \\ &= (\text{area under curve})/T \\ &= (T + t_7 + t_6 - t_5 - t_4 + t_3 - t_2 - t_1)/T \end{aligned} \tag{1.2a}$$

and

$$\begin{aligned} W &= [(t_3 - t_1) + (t_6 - t_2) + (t_7 - t_4) + (T - t_5)]/4 \\ &= (T + t_7 + t_6 - t_5 - t_4 + t_3 - t_2 - t_1)/4 \\ &= (\text{area under curve})/N_c. \end{aligned} \tag{1.2b}$$

Thus we see from (1.2a) and (1.2b) that the area under curve is $LT = WN_c$, which yields $L = WN_c/T$. The fraction N_c/T is the number of customers arriving over the time T and is, for this period, the arrival rate λ, so that $L = \lambda W$. A similar argument would hold for a picture of the number in the queue N_q over the period (0, T), yielding $L_q = \lambda W_q$. While this is not a proof (since it needs to be shown that these relationships hold in the limit over many busy periods as time goes to infinity), one can see the idea behind the relationships.

An interesting result that can be derived from Little's formulas [(1.1a) and (1.1b)] and the relation between W and W_q is

$$L - L_q = \lambda(W - W_q) = \lambda(1/\mu) = \lambda/\mu. \tag{1.3}$$

Table 1.2 Summary of General Results for $G/G/c$ Queues

$\rho = \lambda/c\mu$	Traffic intensity; offered work load rate to a server
$L = \lambda W$	Little's formula
$L_q = \lambda W_q$	Little's formula
$W = W_q + 1/\mu$	Expected-value argument
$p_b = \lambda/c\mu = \rho$	Busy probability for an arbitrary server
$r = \lambda/\mu$	Expected number of customers in service; offered work load rate
$L = L_q + r$	Combined result—(1.3)
$p_0 = 1 - \rho$	$G/G/1$ empty-system probability
$L = L_q + (1 - p_0)$	Combined result for $G/G/1$

But $L - L_q = \mathrm{E}[N] - \mathrm{E}[N_q] = \mathrm{E}[N - N_q] = \mathrm{E}[N_s]$, so that the expected number of customers in service in the steady state is λ/μ, which we will denote by r. Note for a single-server system that $r = \rho$ and it also follows from simple algebra that

$$L - L_q = \sum_{n=1}^{\infty} np_n - \sum_{n=1}^{\infty}(n-1)p_n = \sum_{n=1}^{\infty} p_n = 1 - p_0.$$

From this, we can easily derive the probability that any given server is busy in a multiserver system in the steady state. We denote this probability by p_b. Since we have just shown that the expected number present in service at any instant in the steady state is r, it follows from the symmetry of the c servers that the expected number present at one server is r/c. Then, by a simple expected-value argument, we can show that $p_b = \rho$, since

$$r/c = \rho = 0 \cdot (1 - p_b) + 1 \cdot p_b.$$

For a single-server queue ($G/G/1$), the probability of the system being idle ($N = 0$) is the same as the probability of a server being idle. Thus $p_0 = 1 - p_b$ in this case, and $p_0 = 1 - \rho = 1 - r = 1 - \lambda/\mu$. The quantity $r = \lambda/\mu$, the expected number of customers in service, has another interesting connotation. It is sometimes also referred to as the *offered load*, since, on average, each customer requires $1/\mu$ time units of service and the average number of customers arriving per unit time is λ, so that the product $\lambda(1/\mu)$ is the amount of work arriving to the system per unit time. Dividing this by the number of servers c (which yields ρ) gives the average amount of work coming to each server per unit time.

Table 1.2 summarizes the results of this section.

1.6 Simple Data Bookkeeping for Queues

At this point, it might be useful to use a table format to show how the random events of arrivals and service completions interact for a sample single-server system

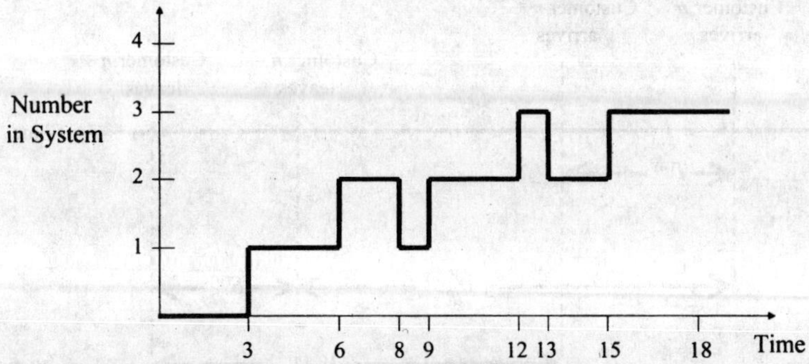

Figure 1.5 Sample path for queueing process.

to form a queue. In the following, we begin at time 0 with a first arrival and then update the system state when events (arrivals or departures) occur—thus the name *event-oriented bookkeeping* is used for this sort of table.

Consider the elementary case of a constant rate of arrivals to a single channel that possesses a constant service rate. (Figure 1.5 is an illustration of this with interarrival times of 3 and serve times of 5.) These regularly spaced arrivals are to be served first come, first served (FCFS). Let it also be assumed that at time $t = 0$ there are no customers waiting and that the channel is empty. Let λ be defined as the number of arrivals per unit time, and $1/\lambda$ then will be the constant time between successive arrivals. The particular unit of time (minutes, hours, etc.) is up to the choice of the analyst. However, consistency must be adhered to once the unit is chosen so that the same basic unit is used throughout the analysis. Similarly, if μ is to be the rate of service in terms of completions per unit time when the server is busy, then $1/\mu$ is the constant service time. We would like to calculate the number in the system at an arbitrary time t, say, $n(t)$, and the time the nth arriving customer must wait in the queue to obtain service, say, $W_q^{(n)}$. From these, it then becomes easy to compute the major measures of effectiveness. Under the assumption that as soon as a service is completed another is begun, the number in the system (including the customer in service) at time t is determined by the equation

$$\begin{aligned} n(t) = \{&\text{number of arrivals in } (0, t]\} \\ &- \{\text{number of services completed in } (0, t]\}. \end{aligned} \tag{1.4}$$

It should be pointed out that there are usually three waiting times of interest—the time spent by the nth customer waiting for service (or line delay), which we write here as $W_q^{(n)}$; the time the nth customer spent in the system, which we shall call $W^{(n)}$; and what is called the virtual line wait $V(t)$, namely, the wait a fictitious arrival would have to endure if it arrived at time t. The reader is cautioned that various authors are not consistent and each of these quantities is sometimes referred to simply as the waiting time.

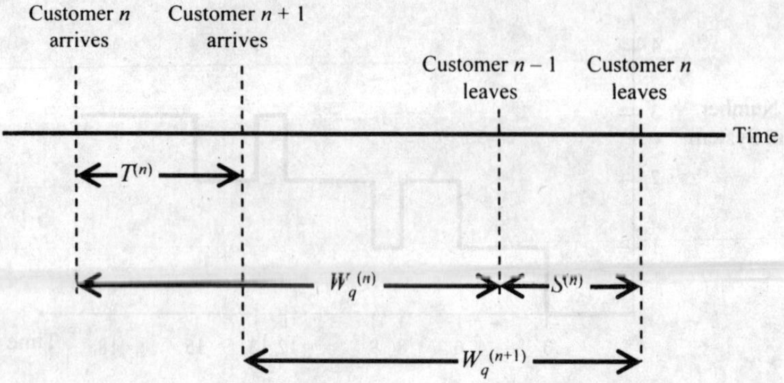

Figure 1.6 Successive $G/G/1$ waiting times.

To find the waiting times in queue until service begins, we observe that the line waits $W_q^{(n)}$ and $W_q^{(n+1)}$ of two successive customers in *any* single-server queue (deterministic or otherwise) are related by the simple recurrence relation

$$W_q^{(n+1)} = \begin{cases} W_q^{(n)} + S^{(n)} - T^{(n)} & (W_q^{(n)} + S^{(n)} - T^{(n)} > 0), \\ 0 & (W_q^{(n)} + S^{(n)} - T^{(n)} \leq 0), \end{cases} \tag{1.5}$$

where $S^{(n)}$ is the service time of the nth customer and $T^{(n)}$ is the interarrival time between the nth and $(n+1)$st customers. This can be seen by a simple diagram as shown in Figure 1.6. (This is an important general relation that is also utilized in later portions of the text.)

Bookkeeping has to do with updating the system status when events occur, recording items of interest, and calculating measures of effectiveness. Event-oriented bookkeeping updates the system state only when events (arrivals or departures) occur. Since there is not necessarily an event every basic time unit, in next-event bookkeeping the master clock is increased by a variable amount each time, rather than a fixed amount as it would be in time-oriented bookkeeping. The event-oriented approach will be illustrated here by an example, using the arrival and service data given in Table 1.3.

We see from simple averaging calculations for columns (5) and (6) in Table 1.4 that the mean line delay of the 12 customers was $40/12 = \frac{10}{3}$, while their mean system waiting time turned out to be $70/12 = \frac{35}{6}$. Furthermore, we observe that we can estimate the mean arrival rate as $\frac{12}{31}$ customers per unit time, since there were 12 arrivals over the 31-time-unit observation horizon. Thus the application of Little's law to these numbers tells us that the average system size L over the full time horizon was

$$L = \lambda W = \frac{70/12}{31/12} = \frac{70}{31}.$$

The mean queue size can be computed similarly.

Table 1.3 Input Data

i	1	2	3	4	5	6	7	8	9	10	11	12
Interarrival time between customers $i+1$ and i	2	1	3	1	1	4	2	5	1	4	2	—
Customer i service time	1	3	6	2	1	1	4	2	5	1	1	3

Table 1.4 Event-Oriented Bookkeeping

(1) Master Clock Time	(2) Arrival/ Departure Customer i	(3) Time Arrival i Enters Service	(4) Time Arrival i Leaves Service	(5) Time in Queue	(6) Time in System	(7) No. in Queue Just After Master Clock Time	(8) No. in System Just After Master Clock Time
0	1-A	0	1	0	1	0	1
1	1-D					0	0
2	2-A	2	5	0	3	0	1
3	3-A	5	11	2	8	1	2
5	2-D					0	1
6	4-A	11	13	5	7	1	2
7	5-A	13	14	6	7	2	3
8	6-A	14	15	6	7	3	4
11	3-D					2	3
12	7-A	15	19	3	7	3	4
13	4-D					2	3
14	8-A;5-D	19	21	5	7	2	3
15	6-D					1	2
19	9-A;7-D	21	26	2	7	1	2
20	10-A	26	27	6	7	2	3
21	8-D					1	2
24	11-A	27	28	3	4	2	3
26	12-A;9-D	28	31	2	5	2	3
27	10-D					1	2
28	11-D					0	1
31	12-D					0	0

1.7 Poisson Process and the Exponential Distribution

The most common stochastic queueing models assume that interarrival times and service times obey the exponential distribution or, equivalently, that the arrival rate and service rate follow a Poisson distribution. In this section we will derive the Poisson distribution and show that assuming the number of occurrences in some time interval to be a Poisson random variable is equivalent to assuming the time between successive occurrences to be an exponentially distributed random variable.

We consider an arrival counting process $\{N(t), t \geq 0\}$, where $N(t)$ denotes the total number of arrivals up to time t, with $N(0) = 0$, and which satisfies the following three assumptions:

(i) The probability that an arrival occurs between time t and time $t + \Delta t$ is equal to $\lambda \Delta t + o(\Delta t)$. We write this as Pr{arrival occurs between t and $t + \Delta t\} = \lambda \Delta t + o(\Delta t)$, where λ is a constant independent of $N(t)$, Δt is an incremental element, and $o(\Delta t)$ denotes a quantity that becomes negligible when compared to Δt as $\Delta t \to 0$; that is,

$$\lim_{\Delta t \to 0} \frac{o(\Delta t)}{\Delta t} = 0.$$

(ii) $\Pr\{\text{more than one arrival between } t \text{ and } t + \Delta t\} = o(\Delta t)$.

(iii) The numbers of arrivals in nonoverlapping intervals are statistically independent; that is, the process has independent increments.

We wish to calculate $p_n(t)$, the probability of n arrivals in a time interval of length t, n being an integer ≥ 0. We will do this by first developing differential-difference equations for the arrival process. For $n \geq 1$ we have

$$\begin{aligned} p_n(t + \Delta t) = {} & \Pr\{n \text{ arrivals in } t \text{ and none in } \Delta t\} \\ & + \Pr\{n - 1 \text{ arrivals in } t \text{ and one in } \Delta t\} \\ & + \Pr\{n - 2 \text{ arrivals in } t \text{ and two in } \Delta t\} + \cdots \\ & + \Pr\{\text{no arrivals in } t \text{ and } n \text{ in } \Delta t\}. \end{aligned} \tag{1.6}$$

Using assumptions i, ii, and iii, (1.6) becomes

$$p_n(t + \Delta t) = p_n(t)[1 - \lambda\, \Delta t - o(\Delta t)] + p_{n-1}(t)[\lambda\, \Delta t + o(\Delta t)] + o(\Delta t), \tag{1.7}$$

where the last term, $o(\Delta t)$, represents the terms $\Pr\{n - j$ arrivals in t and j in Δt; $2 \leq j \leq n\}$.

For the case $n = 0$, we have

$$p_0(t + \Delta t) = p_0(t)[1 - \lambda\, \Delta t - o(\Delta t)]. \tag{1.8}$$

Rewriting (1.7) and (1.8) and combining all $o(\Delta t)$ terms, we have

$$p_0(t + \Delta t) - p_0(t) = -\lambda\, \Delta t p_0(t) + o(\Delta t) \tag{1.9}$$

and

$$p_n(t+\Delta t) - p_n(t) = -\lambda\,\Delta t p_n(t) + \lambda\,\Delta t p_{n-1}(t) + o(\Delta t) \quad (n \geq 1). \tag{1.10}$$

We divide (1.9) and (1.10) by Δt, take the limit as $\Delta t \to 0$, and obtain the differential-difference equations

$$\lim_{\Delta t \to 0}\left[\frac{p_0(t+\Delta t) - p_0(t)}{\Delta t} = -\lambda p_0(t) + \frac{o(\Delta t)}{\Delta t}\right],$$

$$\lim_{\Delta t \to 0}\left[\frac{p_n(t+\Delta t) - p_n(t)}{\Delta t} = -\lambda p_n(t) + \lambda p_{n-1}(t) + \frac{o(\Delta t)}{\Delta t}\right] \quad (n \geq 1),$$

which reduce to

$$\frac{dp_0(t)}{dt} = -\lambda p_0(t) \tag{1.11}$$

and

$$\frac{dp_n(t)}{dt} = -\lambda p_n(t) + \lambda p_{n-1}(t) \quad (n \geq 1). \tag{1.12}$$

We now have an infinite set of linear, first-order ordinary differential equations to solve. Equation (1.11) clearly has the general solution $p_0(t) = Ce^{-\lambda t}$, where the constant C is easily determined to be equal to 1, since $p_0(0) = 1$. Next, let $n = 1$ in (1.12), and we find that

$$\frac{dp_1(t)}{dt} = -\lambda p_1(t) + \lambda p_0(t),$$

or

$$\frac{dp_1(t)}{dt} + \lambda p_1(t) = \lambda p_0(t) = \lambda e^{-\lambda t}.$$

The solution to this equation is

$$p_1(t) = Ce^{-\lambda t} + \lambda t e^{-\lambda t}.$$

Use of the boundary condition $p_n(0) = 0$ for all $n > 0$ yields $C = 0$ and gives

$$p_1(t) = \lambda t e^{-\lambda t}.$$

Continuing sequentially to $n = 2, 3, \ldots$ in (1.12) and proceeding similarly, we find that

$$p_2(t) = \frac{(\lambda t)^2}{2} e^{-\lambda t}, \qquad p_3(t) = \frac{(\lambda t)^3}{3!} e^{-\lambda t}, \qquad \ldots \tag{1.13}$$

From (1.13), we conjecture that the general formula is

$$p_n(t) = \frac{(\lambda t)^n}{n!} e^{-\lambda t}. \tag{1.14}$$

It is left as an exercise (Problem 1.13) to use mathematical induction to verify (1.14), which is the well-known formula for a Poisson probability distribution with mean

λt. Thus if we consider the random variable defined as the number of arrivals to a queueing system by time t, this random variable has the Poisson distribution given by (1.14) with a mean of λt arrivals, or a mean arrival rate (arrivals per unit time) of λ.

Poisson processes have a number of interesting additional properties. One of the most important is that the numbers of occurrences in intervals of equal width are identically distributed (stationary increments). In particular, for $t > s$, the difference $N(t) - N(s)$ is identically distributed as $N(t+h) - N(s+h)$, with probability function

$$p_n(t-s) = \frac{[\lambda(t-s)]^n}{n!} e^{-\lambda(t-s)}.$$

This can easily be seen by the following argument. Since the Poisson has independent increments (assumption iii), there is no loss of generality if $N(s)$ and $N(s+h)$ are assumed to be zero. Then if the Poisson derivation is carried out for both $N(t)$ and $N(t+h)$ under assumptions i, ii, and iii, the foregoing formula results for each (see Problem 1.16).

We now show that if the arrival process is Poisson, an associated random variable defined as the time between successive arrivals (interarrival time) follows the exponential distribution. Let T be the random variable "time between successive arrivals"; then

$$\Pr\{T \geq t\} = \Pr\{\text{no arrivals in time } t\} = p_0(t) = e^{-\lambda t}.$$

Therefore we see that the cumulative distribution function of T can be written as

$$A(t) = \Pr\{T \leq t\} = 1 - e^{-\lambda t},$$

with corresponding density function

$$a(t) = \frac{dA(t)}{dt} = \lambda e^{-\lambda t}.$$

Thus T has the exponential distribution with mean $1/\lambda$. We would intuitively expect the *mean* time between arrivals to be $1/\lambda$ if the *mean* arrival rate is λ. Our analysis substantiates this. It can also be shown that if the interarrival times are independent and have the same exponential distribution, then the arrival rate follows the Poisson distribution, and a proof of this assertion follows.

To begin, let the cumulative distribution function (CDF) of the arrival counting process, $\Pr\{N(t) \leq n\}$, be denoted by $P_n(t)$. Then it follows that

$$\begin{aligned} p_n(t) &= \Pr\{N(t) = n\} \\ &= P_n(t) - P_{n-1}(t). \end{aligned}$$

But

$$P_n(t) = \Pr\{(\text{sum of } n+1 \text{ interarrival times}) > t\}.$$

However, the sum of independent and identically distributed exponential random variables has an Erlang distribution (which is a special type of gamma distribution); hence

$$P_n(t) = \int_t^\infty \frac{\lambda(\lambda x)^n}{n!} e^{-\lambda x}\, dx. \qquad (1.15)$$

The transformation of variables $u = x - t$ gives

$$\begin{aligned} P_n(t) &= \int_0^\infty \frac{\lambda^{n+1}(u+t)^n}{n!} e^{-\lambda t} e^{-\lambda u}\, du \\ &= \int_0^\infty \frac{\lambda^{n+1} e^{-\lambda t} e^{-\lambda u}}{n!} \sum_{i=0}^{n} u^{n-i} t^i \frac{n!}{(n-i)!i!}\, du, \end{aligned}$$

from the binomial theorem. The summation and integral may be switched to give

$$P_n(t) = \sum_{i=0}^{n} \frac{\lambda^{n+1} e^{-\lambda t} t^i}{(n-i)!i!} \int_0^\infty e^{-\lambda u} u^{n-i}\, du.$$

But the integral in the above equation is essentially the well-known gamma function and equals $(n-i)!/\lambda^{n-i+1}$. So

$$P_n(t) = \sum_{i=0}^{n} \frac{(\lambda t)^i e^{-\lambda t}}{i!},$$

which is clearly recognizable as the CDF of the Poisson process.

The Poisson-exponential arrival process derived here is sometimes referred to as completely random arrivals. Although the reader might think that completely random would allude to some sort of haphazard arrival process or a uniform distribution for interarrival times, when encountered in queueing literature it specifically refers to the Poisson-arrival-rate-exponential-interarrival-time pattern. This can be explained in light of the following characteristic of a Poisson process. Given that k arrivals have occurred in an interval $[0, T]$, the k times $\tau_1 < \tau_2 < \cdots < \tau_k$ at which the arrivals occurred are distributed as the order statistics of k uniform random variables on $[0, T]$. Note that it is not the interarrival times, but rather the times at which the arrivals occurred, that are uniformly distributed. This can be proved as follows.

The differential element of the conditional probability density may be written as

$$\begin{aligned} f_\tau(t|k)\, dt &\equiv f(t_1, t_2, \ldots, t_k | k \text{ arrivals in } [0,T])\, dt_1\, dt_2 \cdots dt_k \\ &\approx \Pr\{t_1 \le \tau_1 \le t_1 + dt_1, \ldots, t_k \le \tau_k \le t_k + dt_k | k \text{ arrivals in } [0,T]\}. \end{aligned}$$

Using the definition of conditional probability gives

$$\begin{aligned} &f_\tau(t|k)\, dt \\ &= \frac{\Pr\{t_1 \le \tau_1 \le t_1 + dt_1, \ldots, t_k \le \tau_k \le t_k + dt_k \text{ and } k \text{ arrivals in } [0,T]\}}{\Pr\{k \text{ arrivals in } [0,T]\}} \end{aligned}$$

The numerator of the right-hand side above can be found by making direct use of the Poisson probability function and its properties, since we wish to find the probability that exactly one event occurs in each of the k time intervals $(t_i,\ t_i + dt_i)$, and no events occur elsewhere, that is, in $T - dt_1 - dt_2 - \cdots - dt_k$. Therefore, since the probability of k occurrences in a time t is Poisson, we have

$$f_\tau(t|k)\,dt \approx \frac{\lambda\,dt_1 e^{-\lambda dt_1}\,\lambda\,dt_2 e^{-\lambda dt_2}\cdots\lambda\,dt_k e^{-\lambda dt_k} e^{-\lambda(T-dt_1-dt_2-\cdots-dt_k)}}{(\lambda T)^k\,e^{-\lambda T}/k!}$$

$$= \frac{k!}{T^k}dt_1\,dt_2\cdots dt_k.$$

Hence

$$f_\tau(t_1, t_2, \ldots, t_k|k \text{ arrivals in } [0,T]) = \frac{k!}{T^k}, \tag{1.16}$$

which is identical to the joint density of the order statistics of k random variables uniform on $[0, T]$.

One important consequence of the uniform property of the Poisson process is that the outcomes of random observations of a stochastic process $X(t)$ have the same probabilities as if the scans were taken at Poisson-selected points. When $X(t)$ is a queue, this property is called *PASTA*, for "Poisson arrivals see time averages."

Making similar assumptions to those above for arrivals, one could utilize the same type of process to describe the service pattern. If we change the three assumptions in the beginning of this section slightly by using the word service instead of arrival and condition the probability statements by requiring the system to be nonempty, we obtain a Poisson service rate and an exponential service-time distribution for describing the service pattern. In the following section, we prove an important property of the exponential distribution which aids in a relatively simple analysis of queueing problems when arrival and service patterns exhibit the Poisson–exponential characteristics as derived in this section.

1.8 Markovian Property of the Exponential Distribution

We will now prove the Markovian or memoryless property of the exponential distribution. To explain this property in words, suppose service times are exponentially distributed. This property states that the probability that a customer currently in service has t units of remaining service is independent of how long it has already been in service. Thus we wish to prove that

$$\Pr\{T \le t_1 | T \ge t_0\} = \Pr\{0 \le T \le t_1 - t_0\}. \tag{1.17}$$

The proof is relatively straightforward and proceeds as follows. From the definition of conditional probability we have

$$\begin{aligned}\Pr\{T \le t_1 | T \ge t_0\} &= \frac{\Pr\{(T \le t_1) \text{ and } (T \ge t_0)\}}{\Pr\{T \ge t_0\}} \\ &= \frac{e^{-\lambda t_0} - e^{-\lambda t_1}}{e^{-\lambda t_0}} \\ &= 1 - e^{-\lambda(t_1 - t_0)} = \Pr\{0 \le T \le t_1 - t_0\}.\end{aligned}$$

It is also true that the exponential distribution is the only continuous distribution that exhibits this memoryless property. (The only other distribution to exhibit this property is the geometric, which is the discrete analog of the exponential.) The proof of this assertion rests on the fact that the only continuous function solution of the equation

$$g(s+t) = g(s) + g(t)$$

is the linear form

$$g(y) = Cy, \tag{1.18}$$

where C is an arbitrary constant. This rather intuitive result turns out not to be a trivial matter to prove. However, the proof is well documented in the literature (e.g., see Parzen, 1962), and, for the purposes of this text, this additional detail is not necessary. Then, under the assumption of this result, we proceed as follows. We wish to show that if the memorylessness of (1.17) holds, then it follows that

$$\Pr\{T \le t\} = F(t) = 1 - e^{Ct}.$$

Now subtract both sides of (1.17) from 1 and denote the complementary CDF by $\tilde{F}$. Thus

$$\tilde{F}(t_1 | T \ge t_0) = \tilde{F}(t_1 - t_0). \tag{1.19}$$

From the laws of conditional probability, we can write (1.19) as

$$\frac{\tilde{F}(t_1 \text{ and } T \ge t_0)}{\tilde{F}(t_0)} = \frac{\tilde{F}(t_1)}{\tilde{F}(t_0)} = \tilde{F}(t_1 - t_0),$$

or

$$\tilde{F}(t_1) = \tilde{F}(t_0)\tilde{F}(t_1 - t_0).$$

Letting $t = t_1 - t_0$ yields

$$\tilde{F}(t + t_0) = \tilde{F}(t_0)\tilde{F}(t),$$

and when natural logarithms are taken of both sides, it is found that

$$\ln \tilde{F}(t + t_0) = \ln \tilde{F}(t_0) + \ln \tilde{F}(t).$$

It thus follows from (1.18) that

$$\ln \tilde{F}(t) = Ct, \quad \text{or} \quad \tilde{F}(t) = e^{Ct}.$$

Thus $F(t) = 1 - e^{Ct}$.

There are many possible and well-known generalizations of the Poisson process, most of which have rather obvious applications to queues and are taken up in greater detail later in the text. The most obvious of the generalizations is a truncation of the infinite domain, that is, omission of some of the nonnegative integers from the range of possible values. This is done whenever the removed values are either theoretically meaningless or practically unobservable. An example of this occurs in the $M/M/c/c$ queue and gives rise to *Erlang's loss* or the *Erlang-B formula.* The only change to be made here, with caution, is the rescaling of the respective probabilities because the Poisson terms no longer sum to one.

Another generalization arises if we go back to the axiomatic derivation and no longer permit λ to be a constant independent of time. If instead the functional relationship is denoted by $\lambda(t)$, then the probability of one occurrence in a small time increment is rewritten as $\lambda(t)\Delta t + o(\Delta t)$, and it turns out that the resulting distribution of the counting process is the so-called nonhomogeneous Poisson given by

$$p_n(t) = e^{-m(t)} \frac{[m(t)]^n}{n!}, \quad m(t) = \int_0^t \lambda(s)\, ds \qquad (n \geq 0).$$

A third, and very common, generalization occurs when one relaxes the Poisson assumption that more than one occurrence in Δt has probability $o(\Delta t)$. Instead, let

$$\Pr\{i \text{ occurrences in } (t, t + \Delta t)\} = \lambda_i\, \Delta t + o(\Delta t) \qquad (i = 1, 2, \ldots, n)$$

with

$$\sum_{i=1}^{n} \lambda_i = \lambda.$$

It should be immediately clear now that this is equivalent to allowing the event of i simultaneous occurrences in Δt with probability $\lambda_i \Delta t + o(\Delta t)$, and each individual stream of occurrences of the same batch size (i) itself forms a Poisson process. If these substreams are denoted by $N_i(t)$, then it should also be clear that the total process is $N(t) = \sum_i iN_i(t)$, with probability function

$$\begin{aligned} p_n(t) &= \Pr\{n \text{ occurrences in } [0, t]\} \\ &= \sum_{i=0}^{n} e^{-\lambda t} \frac{(\lambda t)^i}{i!} c_n^{(i)} \qquad (c_0^{(0)} \equiv 1), \end{aligned}$$

where $c_n^{(i)}$ is the probability that i occurrences give a grand total of n (i.e., the probability associated with the i-fold convolution of the batch-size probabilities $\{\lambda_i/\lambda\}$). The process $N(t)$ is known as the *multiple Poisson* and clearly also has the stationary and independent-increment properties.

The foregoing probability function $p_n(t)$ has an alternative derivation as a compound distribution, since it admits of a random sum interpretation as follows. Consider the process $N(t)$ to be defined by

$$N(t) = \sum_{n=1}^{M(t)} Y_n,$$

where $M(t)$ is a regular Poisson process and $\{Y_n\}$ is a sequence of independent and identically distributed (IID) discrete random variables with probabilities

$$c_j = \Pr\{Y_n = j\} \qquad \text{(for all } n\text{)}$$
$$= \frac{\lambda_j}{\sum_i \lambda_i};$$

that is, occurrences happen according to a Poisson process $\{M(t)\}$ but are not necessarily singles in that their size is j with probability c_j. Then, by the laws of probability,

$$\Pr\{N(t) = m\} = \sum_{k=0}^{m} \left[\Pr\{M(t) = k\} \cdot \Pr\left\{ \sum_{n=1}^{k} Y_n = m \right\} \right]$$
$$= \sum_{k=0}^{m} e^{-\lambda t} \frac{(\lambda t)^k}{k!} c_m^{(k)}$$

In other words, the compound approach looks at the process as one Poisson stream with a randomly varying batch size, whereas the multiple approach looks at the process as the sum of Poisson streams, each with a constant batch size.

Poisson–exponential streams are special cases of a larger class of problems called *renewal processes*. An ordinary renewal process arises from any sequence of non-negative IID random variables. Many of the properties that we have derived for Poisson–exponential sequences can also be derived in a renewal context. Some other specific results will be needed later in the text, particularly when the input is an arbitrary G stream, but these will be derived as needed by direct probabilistic arguments. The reader particularly interested in renewal theory is referred to Chapter 7 of Ross (2007), Chapter 3 of Resnick (1992), Chapter 9 of Çinlar (1975), or Chapter 5 of Heyman and Sobel (1982).

In subsequent chapters of the book, the Poisson process and its associated characteristics will play a key role in the development of many queueing models. This is true not only because of the many mathematically agreeable properties of the Poisson–exponential, but also because many real-life situations in fact do obey the appropriate requirements. Though it may seem at first glance that the demands of exponential interoccurrence times are rather stringent, this is not the case.

A strong argument in favor of exponential inputs is the one that often occurs in the context of reliability. It is the result of the well-known fact that the limit of a binomial distribution is Poisson, which says that if a mechanism consists of many parts, each of which can fail with only a small probability, and if the failures for the different parts are mutually independent and identical, then the total flow of failures can be considered Poisson. Another view that favors the exponential comes from the theory of extreme values. Here the exponential appears quite frequently as the limiting distribution of the (normalized) *first-order statistic* of random samples drawn from continuous populations (see Problem 1.24 for one such example).

There is also an additional argument that comes out of information theory. It is that the exponential distribution is the one that provides the least information, where information content or negative entropy of the distribution $f(x)$ is defined as $\int f(x) \log f(x)\, dx$. It can easily be shown that the exponential distribution has least information or highest entropy, and is therefore the most random law that can be used, and thus certainly provides a reasonably conservative approach. We treat the topic of choosing the appropriate probability model in more detail in Chapters 6 and 8.

1.9 Stochastic Processes and Markov Chains

A stochastic process is the mathematical abstraction of an empirical process whose development is governed by probabilistic laws (the Poisson process is one example). From the point of view of the mathematical theory of probability, a stochastic process is best defined as a family of random variables, $\{X(t), t \in T\}$, defined over some index set or parameter space T. The set T is sometimes also called the time range, and $X(t)$ denotes the state of the process at time t. Depending on the nature of the time range, the process is classified as a discrete-parameter or continuous-parameter process as follows:

1. If T is a countable sequence, for example, $T = \{0, \pm 1, \pm 2, \ldots\}$ or $T = \{0, 1, 2, \ldots\}$, then the stochastic process $\{X(t), t \in T\}$ is said to be a discrete-parameter process defined on the index set T.

2. If T is an interval or an algebraic combination of intervals, for example, $T = \{t\colon -\infty < t < +\infty\}$ or $T = \{t\colon 0 < t < +\infty\}$, then the stochastic process $\{X(t), t \in T\}$ is called a continuous-parameter process defined on the index set T.

1.9.1 Markov Process

A discrete-parameter stochastic process $\{X(t), t = 0, 1, 2, \ldots)$ or a continuous-parameter stochastic process $\{X(t), t > 0)$ is said to be a Markov process if, for any set of n time points $t_1 < t_2 < \cdots < t_n$ in the index set or time range of the process, the conditional distribution of $X(t_n)$, given the values of $X(t_1), X(t_2), X(t_3), \ldots, X(t_{n-1})$, depends only on the immediately preceding value, $X(t_{n-1})$; more precisely, for any real numbers $x_1, x_2, \ldots, x_n$,

$$\begin{aligned}\Pr\{X(t_n) \le x_n | X(t_1) = x_1, \ldots, X(t_{n-1}) = x_{n-1}\}\\ = \Pr\{X(t_n) \le x_n | X(t_{n-1}) = x_{n-1}\}.\end{aligned}$$

In nonmathematical language one says that, given the "present" condition of the process, the "future" is independent of the "past," and the process is thus "memoryless."

Markov processes are classified according to:

1. the nature of the index set of the process (whether discrete or continuous parameter), and

2. the nature of state space of the process (whether discrete or continuous parameter).

Table 1.5 Classification of Markov Processes

State Space	Type of Parameter	
	Discrete	Continuous
Discrete	(Discrete-parameter) Markov chain	Continuous-parameter Markov chain
Continuous	Discrete-parameter Markov process	Continuous-parameter Markov process

A real number x is said to be a state of a stochastic process $\{X(t), t \in T\}$ if there exists a time point t in T such that the $\Pr\{x - h < X(t) < x + h\}$ is positive for every $h > 0$. The set of possible states constitutes the state space of the process. If the state space is discrete, the Markov process is generally called a Markov chain, although some authors reserve the term "chain" for only those Markov processes with both discrete state space and discrete-parameter space. In this text, we shall say that a discrete-parameter Markov process with discrete state space is a plain Markov chain, and that a continuous-parameter Markov process with discrete state space is a continuous-time Markov chain. (Multivariate extensions can be nicely formulated for vector states $\boldsymbol{x}$.)

A Markov chain is finite if the state space is finite; otherwise it is a denumerable or infinite Markov chain. Since a discrete-parameter process is observed at a countable number of time points, let the successive observations be denoted by $X_0, X_1, X_2, \ldots, X_n, \ldots$ where X_n is the random variable whose value represents the state of the system at the nth time point. An arbitrary sequence of random variables $\{X_n\}$ is thus a Markov chain if each random variable X_n is discrete and the following holds: For any integer $m > 2$ and any set of m points $n_1 < n_2 < \cdots < n_m$, the conditional distribution of X_{n_m}, given values of $X_{n_1}, X_{n_2}, \ldots, X_{n_{m-1}}$ depends only on $X_{n_{m-1}}$, the immediately preceding value; that is,

$$\begin{aligned}\Pr\{X_{n_m} &= x_{n_m} | X_1 = x_{n_1}, \ldots, X_{n_{m-1}} = x_{n_{m-1}}\} \\ &= \Pr\{X_{n_m} = x_{n_m} | X_{n_{m-1}} = x_{n_{m-1}}\}.\end{aligned}$$

When a Markov process has a continuous-state space and discrete-parameter space, we call it a discrete-parameter Markov process. If both the state space and parameter space are continuous, it is called a continuous-parameter Markov process. Table 1.5 summarizes our classification scheme for Markov processes.

An important generalization of the Markov chain, which is very useful in queueing, is the semi-Markov process (SMP) or Markov renewal process (MRP). The state transitions in an SMP form a discrete Markov chain, but the times between successive transitions are random variables. If these random variables are distributed exponentially for a continuous-parameter case or geometrically for a discrete-parameter case, with mean dependent on the current state only, the SMP reduces to a Markov process because of the memoryless property of these random variables.

1.9.2 Discrete-Parameter Markov Chains

Consider a sequence of random variables, $\{X_n, n = 0, 1, 2, \ldots | X_n = 0, 1, 2, \ldots\}$, which forms a Markov chain with discrete-parameter space; that is, for all n,

$$\Pr\{X_n = j | X_0 = i_0, X_1 = i_1, \ldots, X_{n-1} = i_{n-1}\} = \Pr\{X_n = j | X_{n-1} = i_{n-1}\}.$$

If the value of the random variable X_n is j, then the system is said to be in state j after n steps or transitions. The conditional probabilities $\Pr\{X_n = j | X_{n-1} = i\}$ are called the *single-step transition probabilities* or just *transition probabilities*. If these probabilities are independent of n, then the chain is said to be *homogeneous* and the probabilities $\Pr\{X_n = j | X_{n-1} = i\}$ can be written as p_{ij}. The matrix formed by placing p_{ij} in the (i, j) location is known as the *transition matrix* or *chain matrix* (call it $\boldsymbol{P}$). For homogeneous chains, the m-step transition probabilities

$$\Pr\{X_{n+m} = j | X_n = i\} = p_{ij}^{(m)}$$

are also independent of n. The unconditional probability of state j at the nth trial will be written as

$$\Pr\{X_n = j\} = \pi_j^{(n)},$$

so that the initial distribution is given by $\pi_j^{(0)}$.

From the basic laws of probability, one can easily show that the matrix formed by the elements $\{p_{ij}^{(m)}\}$, say, $\boldsymbol{P}^{(m)}$, can be found simply by multiplying $\boldsymbol{P}^{(m-k)}$ by $\boldsymbol{P}^{(k)}$ for any value of k, $0 < k < m$. This is the matrix equivalent of the well-known Chapman–Kolmogorov (CK) equations for this Markov process, namely,

$$p_{ij}^{(m)} = \sum_r p_{ir}^{(m-k)} p_{rj}^{(k)} \quad (0 < k < m)$$

or, in matrix notation,

$$\boldsymbol{P}^{(m)} = \boldsymbol{P}^{(m-k)} \boldsymbol{P}^{(k)}. \tag{1.20}$$

Letting $k = m - 1$ in (1.20) gives

$$\boldsymbol{P}^{(m)} = \boldsymbol{P} \cdot \boldsymbol{P}^{(m-1)}, \tag{1.21}$$

and, continuing this procedure recursively, we can easily show that

$$\boldsymbol{P}^{(m)} = \boldsymbol{P} \cdot \boldsymbol{P} \cdots \boldsymbol{P} = \boldsymbol{P}^m$$

Hence $\boldsymbol{P}^{(m)}$ can be obtained by multiplying the matrix $\boldsymbol{P}$ by itself m times.

Often one is interested in the probabilities of being in state j after m transitions regardless of the starting state. If the vector $\boldsymbol{\pi}^{(m)}$ is created from the probabilities $\{\pi_j^{(m)}\}$, then

$$\boldsymbol{\pi}^{(m)} = \boldsymbol{\pi}^{(m-1)} \boldsymbol{P}, \tag{1.22}$$

which, when used recursively, gives

$$\boldsymbol{\pi}^{(m)} = \boldsymbol{\pi}^{(0)}\boldsymbol{P}^m \tag{1.23}$$

for the initial state vector $\boldsymbol{\pi}^{(0)}$.

Defining the matrix $\boldsymbol{Q}$ as $\boldsymbol{P} - \boldsymbol{I}$, where $\boldsymbol{I}$ is the identity matrix, and subtracting $\boldsymbol{\pi}^{(m-1)}$ from both sides of (1.22) gives

$$\boldsymbol{\pi}^{(m)} - \boldsymbol{\pi}^{(m-1)} = \boldsymbol{\pi}^{(m-1)}\boldsymbol{Q}. \tag{1.24}$$

Note that $\boldsymbol{P}$ is always a *stochastic matrix* (i.e., its rows sum to one), while $\boldsymbol{Q}$ has rows that sum to zero.

1.9.3 Continuous-Parameter Markov Chains

We now consider a continuous-parameter Markov chain $\{X(t), t \in T\}$ for $T = \{t\colon 0 \le t < \infty\}$. Consider any times $s > t > u \ge 0$ and states i and j; then

$$p_{ij}(u, s) = \sum_r p_{ir}(u, t)p_{rj}(t, s), \tag{1.25}$$

where $p_{ij}(u, s)$ is the probability of moving from state i to j in the time beginning at u and ending at s, and the summation is over all states of the chain. This result should be fairly intuitive and says that the chain can reach state j at time s by starting from state i at time u and stopping off at time t at any other possible state r. This is the CK equation for the continuous process [analogous to (1.20) for the discrete process]. In matrix notation, (1.25) becomes

$$\boldsymbol{P}(u, s) = \boldsymbol{P}(u, t)\boldsymbol{P}(t, s).$$

Letting $u = 0$ and $s = t + \Delta t$ in (1.25) gives

$$p_{ij}(0, t + \Delta t) = \sum_r p_{ir}(0, t)p_{rj}(t, t + \Delta t).$$

Defining $p_i(0)$ as the probability that the chain starts in state i at time 0 and $p_j(t)$ as the unconditional probability that the chain is in state j at time t regardless of starting state, we multiply the above equation by $p_i(0)$ and sum over all states i to get

$$\sum_i p_i(0)p_{ij}(0, t + \Delta t) = \sum_r \sum_i p_{ir}(0, t)p_i(0)p_{rj}(t, t + \Delta t),$$

or

$$p_j(t + \Delta t) = \sum_r p_r(t)p_{rj}(t, t + \Delta t). \tag{1.26}$$

For the Poisson process treated earlier,

$$p_{rj}(t, t + \Delta t) = \begin{cases} \lambda \Delta t + o(\Delta t) & (r = j - 1, j \ge 1), \\ 1 - \lambda \Delta t + o(\Delta t) & (r = j), \\ o(\Delta t) & (\text{elsewhere}). \end{cases}$$

Substituting this into (1.26) gives

$$p_j(t+\Delta t) = [\lambda\,\Delta t + o(\Delta t)]p_{j-1}(t) \\ + [1 - \lambda\,\Delta t + o(\Delta t)]p_j(t) + o(\Delta t) \quad (j \geq 1),$$

which is (1.7). What we did in Section 1.7 was to derive the CK equation for the Poisson process from scratch, appealing to the same basic probability arguments that yield the general CK equations (1.25) and (1.26).

There is an additional theory that takes one from the CK equation to two differential equations, which are called the forward and backward equations. If the transition probability functions $p(u,s)$ of the chain have the additional properties that there exist continuous functions $q_i(t)$ and $q_{ij}(t)$ such that

$$\Pr\{\text{a change of state in } (t, t+\Delta t)\} = 1 - p_{ii}(t, t+\Delta t) = q_i(t)\Delta t + o(\Delta t), \tag{1.27a}$$

$$p_{ij}(t, t+\Delta t) = q_{ij}(t)\Delta t + o(\Delta t), \tag{1.27b}$$

then under some mild regularity conditions (e.g., see Ross, 1996), (1.25) leads to (see Problem 1.20)

$$\frac{\partial}{\partial t}p_{ij}(u,t) = -q_j(t)p_{ij}(u,t) + \sum_{r\neq j} p_{ir}(u,t)q_{rj}(t) \tag{1.28a}$$

and

$$\frac{\partial}{\partial u}p_{ij}(u,t) = q_i(u)p_{ij}(u,t) - \sum_{r\neq i} q_{ir}(u)p_{rj}(u,t). \tag{1.28b}$$

These two differential equations are known, respectively, as Kolmogorov's forward and backward equations. Consider (1.28a) further. Let $u = 0$, and assume a homogeneous process so that $q_i(t) = q_i$ and $q_{ij}(t) = q_{ij}$ for all t. We then obtain

$$\frac{dp_{ij}(0,t)}{dt} = -q_j p_{ij}(0,t) + \sum_{r\neq j} p_{ir}(0,t)q_{rj}.$$

Multiplying both sides of the above equation by $p_i(0)$ and summing over all i yields

$$\frac{dp_j(t)}{dt} = -q_j p_j(t) + \sum_{r\neq j} p_r(t)q_{rj}.$$

In matrix notation, this equation can be written as

$$\boldsymbol{p}'(t) = \boldsymbol{p}(t)\boldsymbol{Q}, \tag{1.29}$$

where $\boldsymbol{p}(t)$ is the vector $(p_0(t), p_1(t), p_2(t), \ldots)$, $\boldsymbol{p}'(t)$ is the vector of its derivatives, and

$$\boldsymbol{Q} = \begin{pmatrix} -q_0 & q_{01} & q_{02} & q_{03} & \cdots \\ q_{10} & -q_1 & q_{12} & q_{13} & \cdots \\ q_{20} & q_{21} & -q_2 & q_{23} & \cdots \\ \cdot & \cdot & \cdot & \cdot & \\ \cdot & \cdot & \cdot & \cdot & \\ \cdot & \cdot & \cdot & \cdot & \end{pmatrix}.$$

Note from (1.27b) that $q_i = \sum_{j \neq i} q_{ij}$, since

$$\sum_j p_{ij}(t, t + \Delta t) = 1,$$

which implies that

$$1 - q_i \Delta t + o(\Delta t) + \sum_{j \neq i} [q_{ij} \Delta t + o(\Delta t)] = 1,$$

or

$$-q_i \Delta t + o(\Delta t) = -\sum_{j \neq i} [q_{ij} \Delta t + o(\Delta t)],$$

so that $q_i = \sum_{j \neq i} q_{ij}$.

The matrix $\boldsymbol{Q}$ can also be looked at as

$$\boldsymbol{Q} = \lim_{\Delta t \to 0} \frac{\boldsymbol{P}(t, t + \Delta t) - \boldsymbol{I}}{\Delta t},$$

where

$$\boldsymbol{P}(t, t + \Delta t) = \{p_{ij}(t, t + \Delta t)\}.$$

Thus $\boldsymbol{Q}$ plays a similar role for continuous-parameter Markov chains to the one $\boldsymbol{Q} = \boldsymbol{P} - \boldsymbol{I}$ played for discrete-parameter Markov chains. The matrix $\boldsymbol{Q}$ is often called the *intensity matrix* (or *infinitesimal generator*) of the continuous-parameter Markov chain.

Again referring to the Poisson process, we can use (1.29) to get (1.12) directly by noting that $q_j = \lambda$, $q_{rj} = \lambda$ for $r = j - 1$ and $j \geq 1$, and $q_{ij} = 0$ elsewhere, so that

$$\frac{dp_j(t)}{dt} = -\lambda p_j(t) + \lambda p_{j-1}(t).$$

Equation (1.11) can be obtained similarly.

Since the state space of any queue is composed of nonnegative integers (representing the number of customers present), a large percentage of queueing problems can be categorized as continuous-parameter (time) Markov chains. Many such models have the additional *birth–death* property that the net change across an infinitesimal time interval can never be other than $-1, 0$, or $+1$, and that

$$\Pr\{\text{increase } n \to n + 1 \text{ in } (t, t + \Delta t)\} = \lambda_n \Delta t + o(\Delta t) \quad (n \geq 0),$$
$$\Pr\{\text{decrease } n \to n - 1 \text{ in } (t, t + \Delta t)\} = \mu_n \Delta t + o(\Delta t) \quad (n \geq 1).$$

Hence

$$\Pr\{\text{no change in } (t, t+\Delta t)\} = 1 - (\lambda_n + \mu_n)\Delta t + o(\Delta t)$$

and

$$\begin{aligned} q_{n,n+1} &= \lambda_n, & & \\ q_{n,n-1} &= \mu_n & & (\mu_n \neq 0), \\ q_{rj} &= 0 & & (\text{elsewhere}), \\ q_n &= \lambda_n + \mu_n & & (q_0 = \lambda_0), \end{aligned}$$

so that it is possible to get either a forward or a backward Kolmogorov equation.

Substituting for q_i and q_{ij}, the matrix $\boldsymbol{Q}$ is

$$\boldsymbol{Q} = \begin{pmatrix} -\lambda_0 & \lambda_0 & 0 & 0 & \cdots \\ \mu_1 & -(\lambda_1+\mu_1) & \lambda_1 & 0 & \cdots \\ 0 & \mu_2 & -(\lambda_2+\mu_2) & \lambda_2 & \cdots \\ \cdot & \cdot & \cdot & \cdot & \\ \cdot & \cdot & \cdot & \cdot & \\ \cdot & \cdot & \cdot & \cdot & \end{pmatrix},$$

and using the Kolmogorov forward equation (1.29), we obtain a set of differential–difference equations for the birth–death process, namely,

$$\begin{aligned} \frac{dp_j(t)}{dt} &= -(\lambda_j + \mu_j)p_j(t) + \lambda_{j-1}p_{j-1}(t) + \mu_{j+1}p_{j+1}(t) \quad (j \geq 1), \\ \frac{dp_0(t)}{dt} &= -\lambda_0 p_0(t) + \mu_1 p_1(t). \end{aligned} \tag{1.30}$$

Note that setting $\mu = 0$ and $\lambda_i = \lambda$ for all i also gives the Poisson equations (1.11) and (1.12) of the previous section. The Poisson process is thus often called the *pure birth* process (see Problem 1.21).

1.9.4 Imbedded Markov Chains

In many of the situations in this text requiring the use of a continuous-parameter queueing model, we can often get quite satisfactory results by instead looking at the state of the queue only at certain selected times, leading to an *imbedded* discrete-parameter Markov chain. To illustrate this, consider the birth–death process *only* at transition times; that is, we create an imbedded stochastic process that turns out to be a discrete-parameter Markov chain. The transition probability matrix $\boldsymbol{P} = \{p_{ij}\}$, where p_{ij} is the probability that, given the process is in state i and a transition occurs, it goes next to state j, can be shown to be (see Problem 1.23)

$$p_{ij} = \begin{cases} \dfrac{\lambda_i}{\lambda_i + \mu_i} & (j = i+1, i \geq 1), \\ \dfrac{\mu_i}{\lambda_i + \mu_i} & (j = i-1, i \geq 1), \\ 1 & (i = 0, j = 1), \\ 0 & \text{otherwise.} \end{cases} \tag{1.31}$$

Another way to view a continuous-time Markov chain is as a process that traverses from state to state in continuous time, the holding times being exponential (required to satisfy the Markov property) random variables with mean $1/q_i$ [in the birth–death example, $1/q_i = 1/(\lambda_i + \mu_i)$] for holding in state i. At times of transition, we have a discrete-parameter Markov chain with transition probabilities $p_{ij} = q_{ij}/q_i$ [in the birth–death example, p_{ij} is given by (1.31)].

1.9.5 Long-Run Behavior of Markov Processes

We are often interested in the behavior of a Markov process after a long period of time, particularly in whether its behavior "settles down" probabilistically. We discuss three related concepts having to do with long-run behavior, namely, *limiting distributions, stationary distributions*, and *ergodicity*.

Consider a discrete-parameter Markov chain, and suppose that

$$\lim_{m\to\infty} p_{ij}^{(m)} = \pi_j \quad \text{(for all } i\text{)};$$

that is, after a long time, the probability that the process is in state j given that it started in state i is independent of the starting state i. This means that $\boldsymbol{P}^m$ approaches a limit as m goes to infinity, namely, that all rows of $\boldsymbol{P}^m$ become identical. We call the $\{\pi_j\}$, the *limiting* or *steady-state* probabilities of the Markov chain.

Consider now the unconditional state probabilities after m steps, as given by $\boldsymbol{\pi}^{(m)} = \boldsymbol{\pi}^{(0)}\boldsymbol{P}^{(m)}$; that is,

$$\pi_j^{(m)} = \sum_i \pi_i^{(0)} p_{ij}^{(m)}.$$

Then

$$\begin{aligned}\lim_{m\to\infty} \pi_j^{(m)} &= \lim_{m\to\infty} \sum_i \pi_i^{(0)} p_{ij}^{(m)} = \sum_i \pi_i^{(0)} \lim_{m\to\infty} p_{ij}^{(m)} \\ &= \sum_i \pi_i^{(0)} \pi_j = \pi_j \sum_i \pi_i^{(0)} = \pi_j,\end{aligned}$$

and hence $\pi_j^{(m)}$ goes to the same limit as $p_{ij}^{(m)}$ and is independent of starting-state probabilities and the time parameter m. When these unconditional limiting probabilities exist, they can be found as follows. From (1.22),

$$\lim_{m\to\infty} \boldsymbol{\pi}^{(m)} = \lim_{m\to\infty} \boldsymbol{\pi}^{(m-1)}\boldsymbol{P}.$$

Letting $\boldsymbol{\pi} = (\pi_0, \pi_1, \ldots)$ represent the limiting probability vector, we have

$$\lim_{m\to\infty} \boldsymbol{\pi}^{(m)} = \lim_{m\to\infty} \boldsymbol{\pi}^{(m-1)} = \boldsymbol{\pi},$$

so that

$$\boxed{\boldsymbol{\pi} = \boldsymbol{\pi}\boldsymbol{P}, \quad \text{or} \quad \boldsymbol{0} = \boldsymbol{\pi}\boldsymbol{Q}.} \tag{1.32}$$

From this, together with the *boundary condition* that $\sum_j \pi_j = 1$, we can obtain the $\{\pi_j\}$. These well-known equations are called the *stationary equations* of the Markov chain, and their solution is called the *stationary distribution*. The equations (1.32) will play a major role in the solution to some of the more advanced queueing models treated in Chapter 5. Note that the boundary condition may be written in vector notation as $\boldsymbol{\pi e} = 1$, for $\boldsymbol{e}$ a column vector with all elements equal to one.

It is possible in some cases to get solutions to (1.32) even when no limiting distribution exists. Thus when a limiting distribution exists, this implies a solution to (1.32), and the resulting stationary distribution is the limiting distribution. But the converse is not true; that is, a solution to (1.32) does not imply the existence of a limiting distribution. We illustrate this with the following examples.

■ **EXAMPLE 1.1**

Consider a degenerate discrete-parameter stochastic process that sequentially alternates between two states, -1 and $+1$. If we call -1 state 0 and $+1$ state 1, the transition probability matrix is

$$\boldsymbol{P} = \begin{pmatrix} 0 & 1 \\ 1 & 0 \end{pmatrix}.$$

We can solve (1.32) and get the stationary distribution vector by

$$(\pi_0, \pi_1) = (\pi_0, \pi_1)\begin{pmatrix} 0 & 1 \\ 1 & 0 \end{pmatrix},$$

which gives $\pi_0 = \pi_1$. Using the fact that $\pi_0 + \pi_1 = 1$ (for a valid probability distribution), we have the stationary probability distribution $\pi_0 = \pi_1 = \frac{1}{2}$. But is this also the limiting distribution? Intuitively we would think not, since the process keeps alternating forever and at any time in the future, the probability that it will be found in a particular state (say, -1) is either one or zero, depending on the particular value of m chosen. This is easily verified by successive multiplication of the matrix $\boldsymbol{P}$, yielding

$$\boldsymbol{P}^{(m)} = \begin{cases} \begin{pmatrix} 1 & 0 \\ 0 & 1 \end{pmatrix}, & m \text{ even}, \\ \begin{pmatrix} 0 & 1 \\ 1 & 0 \end{pmatrix}, & m \text{ odd}. \end{cases}$$

Thus even though a solution to $\boldsymbol{\pi} = \boldsymbol{\pi P}$ exists, there is no $\lim_{m\to\infty} p_{ij}^{(m)}$. Furthermore, $\lim_{m\to\infty} p_{ij}^{(m)}$ does not exist unless $\boldsymbol{\pi}^{(0)} = (\frac{1}{2}, \frac{1}{2})$, that is, unless the initial probability distribution is set equal to the stationary distribution. For this $\boldsymbol{\pi}^{(0)}$ (and only this $\boldsymbol{\pi}^{(0)}$),

$$\boldsymbol{\pi}^{(1)} = (\tfrac{1}{2}, \tfrac{1}{2})\begin{pmatrix} 0 & 1 \\ 1 & 0 \end{pmatrix} = (\tfrac{1}{2}, \tfrac{1}{2}),$$

and hence

$$\pi^{(m)} = (\tfrac{1}{2}, \tfrac{1}{2}) \quad \text{for all } m.$$

In the previous example, what we have done by using $\pi^{(0)} = (\frac{1}{2}, \frac{1}{2})$, the stationary solution, is to make this stochastic process strictly stationary. A strictly stationary stochastic process is defined as follows: For all k and h, the joint probability distribution of $X(t_1), X(t_2), \ldots, X(t_k)$ (called a *finite-dimensional distribution* of order k) is equal to the joint probability distribution of $X(t_1+h), X(t_2+h), \ldots, X(t_k+h)$. Note that stationary processes therefore possess time-independent distribution functions. The solution to $\pi = \pi P$ does not imply strict stationarity, but strict stationarity does imply that $\pi^{(m)}$ is time-independent. The process in Example 1.1 is not in general strictly stationary, but can be made so by using an initial probability vector equal to the stationary vector $(\frac{1}{2}, \frac{1}{2})$. While strict stationarity guarantees the time independence of $\pi^{(m)}$, the process still does not possess a steady state, since $p_{ij}^{(m)}$ never becomes independent of the starting state i or the time parameter m.

■ **EXAMPLE 1.2**

Consider now a discrete-parameter Markov chain similar to the one of Example 1.1 but with two possible states, 0 and 1, and transition probability matrix

$$P = \begin{pmatrix} \frac{1}{2} & \frac{1}{2} \\ \frac{1}{2} & \frac{1}{2} \end{pmatrix}.$$

Note that at each transition, there is a probability $\frac{1}{2}$ of staying in the current state and a probability $\frac{1}{2}$ that the chain will change state. Again, it is easy to get the stationary distribution by $\pi = \pi P$ and $\pi e = 1$, which yields $\pi_0 = \pi_1 = \frac{1}{2}$.

Now let us see if a limiting distribution exists. Successive multiplication of P yields

$$P^{(m)} = \begin{pmatrix} \frac{1}{2} & \frac{1}{2} \\ \frac{1}{2} & \frac{1}{2} \end{pmatrix} \quad \text{for all } m.$$

Thus $\pi = (\frac{1}{2}, \frac{1}{2})$ is the limiting (steady-state) distribution vector, namely,

$$\lim_{m\to\infty} p_{i0}^{(m)} = \tfrac{1}{2} \qquad (i = 0, 1),$$

$$\lim_{m\to\infty} p_{i1}^{(m)} = \tfrac{1}{2} \qquad (i = 0, 1).$$

This process is also not, in general, stationary unless $\pi^{(0)} = (\frac{1}{2}, \frac{1}{2})$, but it does, as we have shown, possess a steady state, since $\lim_{m\to\infty} p_{ij}^{(m)} = \pi_j$, regardless of which $\pi^{(0)}$ is used.

Let us complicate the example slightly by changing P to

$$P = \begin{pmatrix} \frac{1}{3} & \frac{2}{3} \\ \frac{2}{3} & \frac{1}{3} \end{pmatrix}.$$

The stationary solution is still $\pi = \left(\frac{1}{2}, \frac{1}{2}\right)$, which is easy to verify by solving $\pi = \pi P$ and $\pi e = 1$. We must now see whether $P^{(m)}$ goes to a limit; that is, whether $\lim_{m\to\infty} p_{ij}^{(m)}$ exists. Successive multiplication of P here by itself yields

$$P \cdot P = P^2 = \begin{pmatrix} \frac{1}{3} & \frac{2}{3} \\ \frac{2}{3} & \frac{1}{3} \end{pmatrix} \begin{pmatrix} \frac{1}{3} & \frac{2}{3} \\ \frac{2}{3} & \frac{1}{3} \end{pmatrix} = \begin{pmatrix} \frac{5}{9} & \frac{4}{9} \\ \frac{4}{9} & \frac{5}{9} \end{pmatrix}.$$

Now

$$P \cdot P^2 = P^3 = \begin{pmatrix} \frac{1}{3} & \frac{2}{3} \\ \frac{2}{3} & \frac{1}{3} \end{pmatrix} \begin{pmatrix} \frac{5}{9} & \frac{4}{9} \\ \frac{4}{9} & \frac{5}{9} \end{pmatrix} = \begin{pmatrix} \frac{13}{27} & \frac{14}{27} \\ \frac{14}{27} & \frac{13}{27} \end{pmatrix},$$

and continuing on with this procedure, we see that $P^{(m)}$ converges to a matrix all of whose entries are $\frac{1}{2}$. Thus the stationary distribution is indeed the steady-state distribution. Once again, we can make the process completely stationary by letting $\pi^{(0)} = \left(\frac{1}{2}, \frac{1}{2}\right)$, since then $\pi^{(0)} P = \left(\frac{1}{2}, \frac{1}{2}\right)$ and $\pi^{(m)} = \left(\frac{1}{2}, \frac{1}{2}\right)$ for all m. If we were to use any other $\pi^{(0)}$, we would find that $\pi^{(m)}$ equals $\left(\frac{1}{2}, \frac{1}{2}\right)$ only in the limit.

For continuous-parameter processes, the stationary solution can be obtained from

$$\mathbf{0} = pQ, \tag{1.33}$$

where Q is the intensity matrix as defined in (1.29). The same concepts of stationarity and steady state apply for the continuous-parameter case, with t replacing m in the limiting process. Direct determination of the existence of a steady-state solution is more difficult here, since it would involve obtaining the solution to the differential equations of (1.29) and then taking the $\lim_{t\to\infty} p'(t)$. If it exists, it will, of course, equal the p obtained from (1.33). This is a considerably more difficult task than successive multiplication of the transition probability matrix P required for the discrete-parameter case.

At the end of this section, we shall present some theorems that will allow us to determine the existence of a limiting distribution. Once we know a limiting distribution exists, it can be obtained from $\pi = \pi P$ or $\mathbf{0} = pQ$, and the respective boundary conditions $\pi e = 1$ or $pe = 1$.

1.9.6 Ergodicity

Closely associated with the concepts of limiting and stationary distributions is the idea of *ergodicity*, which has to do with the information contained in one infinitely long sample path of a process (e.g., see Papoulis, 1991). Ergodicity is important in that it deals with the problem of determining measures of a stochastic process $X(t)$ from a single realization, as is often done in analyzing simulation output. That is $X(t)$ is ergodic in the most general sense if, with probability one, all its "measures" can be determined or well approximated from a single realization, $x_0(t)$, of the process. Since statistical measures of the process are usually expressed as time averages, this

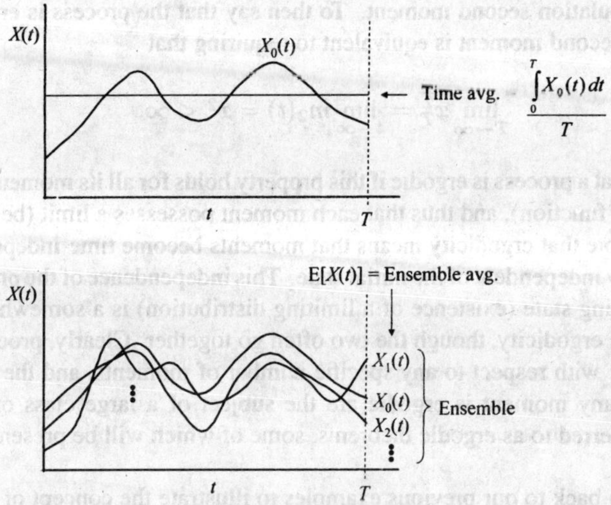

Figure 1.7 Ergodicity.

is often stated as follows: $X(t)$ is ergodic if time averages equal ensemble averages, that is, expected values (see Figure 1.7). One may not always be interested in all, but sometimes only in certain measures (or moments) of a process. We can then define ergodicity with respect to these moments, and a process might thus be ergodic for certain moments, but not for others. However, in queueing theory, we are typically interested in fully ergodic processes, that is, processes that are ergodic with respect to all moments.

When dealing with stationary processes, statistical averages are independent of time; hence concern with respect to ergodicity centers only on convergence of time averages (e.g., see Karlin and Taylor, 1975, pp. 474 ff., or Heyman and Sobel, 1982, pp. 366 ff.). In queueing theory, the processes of interest are in general not stationary, and thus our interest in ergodicity involves the convergence of both time and ensemble averages.

Mathematically now, the time average of the square of a realization of a process, for example, would be written as

$$\overline{x_T^2} = \frac{1}{T}\int_0^T [x_0(t)]^2 \, dt. \tag{1.34}$$

This is then the second moment of the sample function $x_0(t)$. The ensemble average at time t would be denoted by

$$\mathrm{E}[\{X(t)\}]^2 = \lim_{n\to\infty} \frac{\sum_{i=0}^{n}\{x_i(t)\}^2}{n} = m_2(t) \tag{1.35}$$

and is the population second moment. To then say that the process is ergodic with respect to its second moment is equivalent to requiring that

$$\lim_{T\to\infty} \overline{x_T^2} = \lim_{t\to\infty} m_2(t) = \overline{x^2} < \infty. \tag{1.36}$$

We thus say that a process is ergodic if this property holds for all its moments (ergodic in distribution function), and thus that each moment possesses a limit (becomes free of "time"). Note that ergodicity means that moments become time-independent, but not necessarily independent of the initial state. This independence of the process from its actual starting state (existence of a limiting distribution) is a somewhat stronger condition than ergodicity, though the two often go together. Clearly, processes need not be ergodic with respect to any specific number of moments, and the conditions under which any moment is ergodic are the subject of a large class of theorems commonly referred to as ergodic theorems, some of which will be presented later in this section.

We now go back to our previous examples to illustrate the concept of ergodicity. We shall show that the process of Example 1.1 is not ergodic by showing that the relationship (1.36) does not hold for the first moment (this is all that is required to prove nonergodicity, since for a process to be ergodic it must be ergodic in *all* its moments).

Consider the equivalent of (1.34) for the first moment of our alternating process. The average value of the process after $m-1$ transitions [assume it starts in state $+1$, i.e., $\pi^{(0)} = (1, 0)$] is

$$\overline{x_m} = \frac{1}{m}\sum_{i=0}^{m-1} x_i$$
$$= \begin{cases} 0 & (m \text{ even}), \\ 1/m & (m \text{ odd}). \end{cases}$$

The limit of $\overline{x_m}$ as m goes to infinity is clearly zero.

Since each path must be identical, the ensemble average after m transitions is

$$m_1(m) = \mathrm{E}[X_m] = \begin{cases} +1 & (m \text{ even}), \\ -1 & (m \text{ odd}). \end{cases}$$

The limit of the ensemble average as m goes to infinity does not equal zero; in fact, it does not even exist. Hence the process is not ergodic in its first moment and is thus not ergodic in any way.

If we use $\pi^{(0)} = (\frac{1}{2}, \frac{1}{2})$, however, then the limit of $m_1(m)$ as m goes to infinity is also zero and the process is indeed ergodic even though, as we have previously shown, no limiting distribution exists. We note that stationary processes [e.g., our problem when $\pi^{(0)} = (\frac{1}{2}, \frac{1}{2})$] are ergodic. But we shall see shortly that ergodic processes need not be stationary.

Now let us see if the process of Example 1.2 is ergodic. We first consider the time average of the kth power of the process. We need to calculate

$$\lim_{m\to\infty} \overline{x_m^k} = \lim_{m\to\infty} \frac{\sum_{i=0}^{m-1} x_i^k}{m}.$$

Now x_i^k, the value of the kth power of the process at the ith transition, is an observation from a random variable taking on values 0 or 1, each with probability $\frac{1}{2}$. Thus, ignoring the starting state (it washes out in the limit, since x_0^k is either zero or one and $\lim_{m\to\infty} x/m = 0$), we have that $\lim_{m\to\infty} \overline{x_m^k}$ is the limit of the average value of $m-1$ IID Bernoulli random variables with parameter $\frac{1}{2}$. By the strong law of large numbers, this quantity goes to the Bernoulli mean of $\frac{1}{2}$.

Let us now consider the ensemble average at time m, $\mathrm{E}[X_m^k] = m_k(m)$, which is the kth population moment of a Bernoulli random variable with parameter $\frac{1}{2}$ and can be calculated as

$$\mathrm{E}[X_m^k] = \sum_{i=0}^{1} x_i^k p_i = 0^k(\tfrac{1}{2}) + 1^k(\tfrac{1}{2}) = \tfrac{1}{2} \quad (k = 1, 2, 3, \ldots).$$

The limit of $\mathrm{E}[X_m^k]$ as m goes to infinity is also $\frac{1}{2}$; hence the process is ergodic. Note that ergodicity of this process holds regardless of $\boldsymbol{\pi}^{(0)}$, so that even though in general the process is not stationary, it is nevertheless ergodic.

It is generally not easy to show ergodicity or nonergodicity by direct methods, as in the foregoing examples, and the theorems to follow will aid in this task. Prior to stating the key theorems that will enable us to determine when a solution to the stationary equation exists, if a limiting distribution exists, and whether or not the process is ergodic, we first present some definitions required for characterizating discrete-parameter Markov chains.

Two states, i and j, are said to *communicate* ($i \leftrightarrow j$) if i is accessible from j ($j \to i$) and j is accessible from i ($i \to j$). A chain is said to be *irreducible* if all of its states communicate, that is, if there exists an n such that $p_{ij}^{(n)} > 0$ for all pairs (i, j).

The period of a return state k of a chain is defined as the greatest common divisor (GCD) of the set of integers $\{n\}$ for which $p_{kk}^{(n)} > 0$. A state is said to be aperiodic if this GCD is 1, that is, if it has period 1. A chain is said to be aperiodic if each of its states is aperiodic. For example, if a process can return to a certain state in 5 or 7 time units, this state is aperiodic, since possible return times are 5, 7, 10, 12, 14, 15, 17, 19, 20, 21, 22, 24, 25, 26, 27, ..., and this sequence has greatest divisor equal to one. On the other hand, if the return times are 3 and 6, then we get a sequence of possible returns of $3, 6, 9, 12, \ldots$, meaning that the state is periodic with period equal to three.

Define $f_{jj}^{(n)}$ as the probability that a chain starting at state j returns for the first time to j in n transitions. Hence the probability that the chain ever returns to j is

$$f_{jj} = \sum_{n=1}^{\infty} f_{jj}^{(n)}.$$

If $f_{jj} = 1$, then j is said to be a *recurrent* state; if $f_{jj} < 1$, then j is a *transient* state. When $f_{jj} = 1$,

$$m_{jj} = \sum_{n=1}^{\infty} n f_{jj}^{(n)}$$

is the *mean recurrence time*. If $m_{jj} < \infty$, then j is known as a *positive recurrent state*, while if $m_{jj} = \infty$, then j is a *null recurrent state*.

To illustrate, consider the following transition probability matrix, with six states numbered $0 \rightarrow 5$:

$$\begin{pmatrix} \frac{1}{6} & 0 & 0 & \frac{1}{2} & \frac{1}{3} & 0 \\ 0 & 1 & 0 & 0 & 0 & 0 \\ \frac{1}{5} & 0 & \frac{2}{5} & \frac{1}{5} & 0 & \frac{1}{5} \\ \frac{1}{4} & 0 & 0 & \frac{3}{4} & 0 & 0 \\ \frac{1}{2} & 0 & 0 & \frac{1}{2} & 0 & 0 \\ 0 & \frac{1}{4} & \frac{1}{4} & 0 & 0 & \frac{1}{2} \end{pmatrix}.$$

Now state 1 is recurrent, since if the process starts there, it stays there forever. State 5 is transient (can visit a few times and then never return), as is state 2. States 0, 3, 4 are recurrent—once one is entered the process moves among these forever. These states form a recurrent *class*.

As another example, consider the imbedded birth–death transition probability matrix given by (1.31). We see that it is irreducible, may be recurrent (e.g., if $\mu_i \geq \lambda_i$ for all $i \geq$ some value n), and has period two.

Define $f_{ij}^{(n)}$, $i \neq j$, as the probability that the first passage from state i to state j occurs in exactly n steps. Then the probability that state j is ever reached from i is

$$f_{ij} = \sum_{n=1}^{\infty} f_{ij}^{(n)}.$$

The expected value of the sequence $\{f_{ij}^{(n)}, n = 1, 2 \ldots\}$ of first-passage probabilities for a fixed pair (i, j), $i \neq j$, is denoted by m_{ij} and is called the *mean first-passage time*; that is,

$$m_{ij} = \sum_{n=1}^{\infty} n f_{ij}^{(n)} \quad (i \neq j).$$

There are an extensive number of theorems in the literature which permit one to determine the presence of recurrence in a Markov chain and to calculate the mean recurrence time whenever appropriate (e.g., see Çinlar, 1975). The following theorems, which we state without proofs, tie much of the theory together and relate the concepts of ergodicity, limiting probabilities, and stationary probabilities.

Theorem 1.1.

(a) *In an irreducible and positive recurrent discrete-parameter Markov chain, a nondegenerate solution to the stationary equations*

$$\boldsymbol{\pi} = \boldsymbol{\pi P}, \quad \boldsymbol{\pi e} = 1$$

always exists, where the solution vector $\boldsymbol{\pi} = \{\pi_j\}$ *is such that* $\pi_j = 1/m_{jj}$.

Furthermore, when all moments of this stationary distribution are finite, we have the following:

(b) *If the starting probability vector* $\boldsymbol{\pi}^{(0)}$ *is set equal to the stationary probability vector* $\boldsymbol{\pi}$, *the above chain becomes a stationary stochastic process and hence ergodic.*

(c) *If the above chain is aperiodic as well as irreducible and positive recurrent, the process is ergodic and has a limiting probability distribution equal to the stationary distribution.*

Note that existence of a limiting distribution is the strongest condition, ergodicity is somewhat weaker, and a nondegenerate solution to the stationary equations is the weakest of the three conditions.

Theorem 1.2. *An irreducible, aperiodic chain is positive recurrent if there exists a nonnegative solution of the system*

$$\sum_{j=0}^{\infty} p_{ij} x_j \le x_i - 1 \quad (i \ne 0)$$

such that

$$\sum_{j=0}^{\infty} p_{0j} x_j < \infty.$$

Theorem 1.3. *For a continuous-parameter Markov chain, the imbedded chain (at points of transition) need not be aperiodic as long as the mean holding times in all states are bounded, for Theorem 1.1 to be valid.*

The reason aperiodicity is not required for the imbedded Markov chain is that since the process is continuous in time, the time between transitions varies continuously, and as long as it is bounded, the transition times "wash out" any periodicity that may come from the imbedded process.

Returning to the birth–death process and considering its imbedded Markov chain, we see that the matrix $\boldsymbol{P}$ as given by (1.31) is irreducible and positive recurrent as long as certain conditions hold (e.g., $\mu_i > \lambda_i$ for all $i \ge n$). Furthermore, if the mean state holding times, $1/(\lambda_i + \mu_i)$, are bounded (λ_i and/or $\mu_i \ge \epsilon > 0$ for all i), then Theorem 1.3 is satisfied and the process is ergodic and possesses a limiting (steady-state) probability distribution. Such processes are sometimes also referred to as achieving *statistical equilibrium* (see Cooper, 1981).

When dealing with Markov chains, some authors (e.g., Feller, 1957, pp. 353ff.; Heyman and Sobel, 1982, pp. 230ff.) define a positive recurrent aperiodic state as ergodic, and thus the necessary conditions for Theorem 1.1 to hold, that is, irreducibility and positive recurrence, can be stated as requiring all states of a Markov chain to be ergodic. While it is certainly true that such chains are ergodic as defined

Table 1.6 Long-Run Behavior Concepts

Concept	Definition	Relationship/Interpretation
Stationary distribution (SD)	Solution to $\boldsymbol{\pi} = \boldsymbol{\pi P}, \boldsymbol{\pi e} = 1$ $(\mathbf{0} = \boldsymbol{pQ}, \boldsymbol{pe} = 1)$.	If all states communicate, π_j can be interpreted as the long-run percentage of time the process spends in state j.
Stationary stochastic process (SSP)	State distributions invariant over time.	If all states communicate, a solution to $\boldsymbol{\pi} = \boldsymbol{\pi P}$ exists and the process is ergodic.
Ergodic process (EP)	Time averages equal ensemble averages in limit.	A solution to $\boldsymbol{\pi} = \boldsymbol{\pi P}$ exists. EPs become SSPs in limit as time goes to infinity.
Limiting or steady-state or statistical equilibrium distribution	Limiting state probability distribution independent of time and initial state.	Process is also ergodic. Stationarity and independence of starting state achieved in limit. The limiting distribution is the SD.

by (1.36) and the text immediately following it, we believe it is somewhat confusing to define ergodicity as synonymous with the existence of limiting distributions, since it is a somewhat weaker condition, as we have previously mentioned.

To summarize the long-run behavior concepts of Markov chains studied in this section, we recap the preceding definitions in Table 1.6.

1.10 Introduction to the QtsPlus Software

Today, spreadsheets are an indispensable tool for engineers and operations research specialists. Several papers have discussed the application of spreadsheets in the various operations research disciplines, such as optimization and queueing theory (Bodily, 1986; Leon et al., 1996; Grossman, 1999). To facilitate learning queueing theory, a collection of spreadsheet queueing models, collectively known as *QtsPlus*, is available with this textbook. We now introduce QtsPlus and some of its capabilities. See Appendix E for instructions to install and run the software.

We illustrate how to use QtsPlus to solve Example 1.2. Follow the instructions in Appendix E to start the software. Once it is active, select the **Basic** model category from the drop-down list provided, then select the **Discrete-Time Markov Chain** model from the list of models available under the **Basic** category. Once the

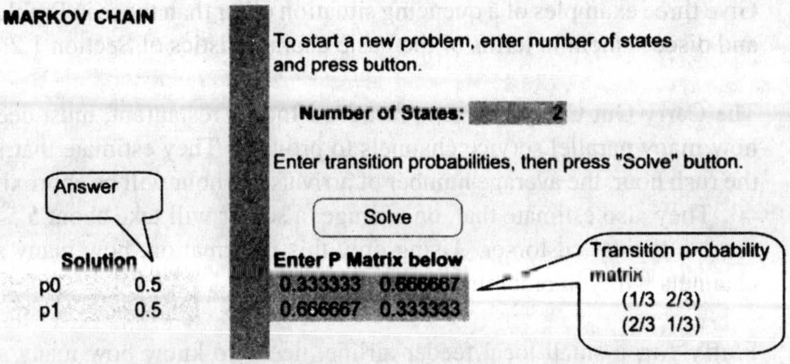

Figure 1.8 QtsPlus solution to Example 1.2.

workbook, called marchain.xls, is open, enter 2 into the input field: **Number of States**. If a pop-up message box appears asking,

> This will cause existing model parameters to be discarded. Do you wish to continue?

Press the **Yes** button to set up a new **P** matrix. Using Excel formulas, fill the respective cells of the ***P*** matrix on the worksheet with the initial parameters as shown below.

$$\begin{array}{cc} =1/3 & =2/3 \\ =2/3 & =1/3 \end{array}$$

Press the **Solve** button. The answer appears on the left side of the worksheet (Figure 1.8) and coincides with the stationary solution $\pi = (\frac{1}{2}, \frac{1}{2})$ previously calculated.

PROBLEMS

1.1. Discuss the following queueing situations in terms of the characteristics given in Section 1.2.

(a) Aircraft landing at an airport.
(b) Supermarket checkout procedures.
(c) Post-office or bank customer windows.
(d) Toll booths on a bridge or highway.
(e) Gasoline station with several pump islands.
(f) Automatic car wash facility.
(g) Telephone calls coming into a customer information system.
(h) Appointment patients coming into a doctor's office.
(i) Tourists wishing a guided tour of the White House.
(j) Electronic components on an assembly line consisting of three operations and one inspection at end of line.
(k) Processing of programs coming from a number of independent sources on a local area network into a central computer.

1.2. Give three examples of a queueing situation other than those in Problem 1.1, and discuss them in terms of the basic characteristics of Section 1.2.

1.3. The Carry Out Curry House, a fast-food Indian restaurant, must decide on how many parallel service channels to provide. They estimate that, during the rush hour, the average number of arrivals per hour will be approximately 40. They also estimate that, on average, a server will take about 5.5 min to serve a typical customer. Using only this information, how many service channels will you recommend they install?

1.4. Fluffy Air, a small local feeder airline, needs to know how many slots to provide for telephone callers to be placed on hold. They plan to have enough answerers so that the average waiting time on hold for a caller will be 75 seconds during the busiest period of the day. They estimate the average call-in rate to be 3 per minute. What would you advise?

1.5. The Outfront BBQ Rib Haven does carry out only. During peak periods, two servers are on duty. The owner notices that during these periods, the servers are almost never idle. She estimates the percent time idle of each server to be 1 percent. Ideally, the percent idle time would be 10 percent to allow time for important breaks.

(a) If the owner decides to add a third server during these times, how much idle time would each server have then?

(b) Suppose that by adding the third server, the pressure on the servers is reduced, so they can work more carefully, but their service output rate is reduced by 20 percent. What now is the percent time each would be idle?

(c) Suppose, instead, the owner decides to hire an aid (at a much lower salary) who servers as a gofer for the two servers, rather than hiring another full server. This allows the two servers to decrease their average service time by 20 percent (relative to the original service rate). What now is the percent idle time of each of the two servers?

1.6. The Happidaiz frozen-yogurt stand does a thriving business on warm summer evenings. Even so, there is only a single person on duty at all times. It is known that the service time (dishing out the yogurt and collecting the money) is normally distributed with mean 2.5 min and standard deviation 0.5 min. (Although the normal distribution allows for negative values, the standard deviation with respect to the mean is small so that negative values are more than 4 standard deviations below the mean and the probability of negative values is essentially zero.) You arrive on a particular evening to get your favorite crunchy chocolate yogurt cone and find 8 people ahead of you. Estimate the average time until you get the first lick. What is the probability that you will have to wait more than $\frac{1}{2}$ hour? [*Hint*: Remember that the sum of normal random variables is itself normally distributed.]

1.7. A certain football league consists of 32 teams. Each team has 67 active players. There is a draft each year for teams to acquire new players. Each team acquires 7 new players per year in the draft. The number of active players on each team must always be 67. Thus each team must cut some existing players each year to make room for the new players.

(a) Assuming that a football player can only join a team by being selected in the draft, estimate the average career length of a football player in the league.

(b) Now suppose that a player can join a team in one of two ways: (1) by being selected in the draft, as before, or (2) by signing directly with a team outside the draft. Suppose further that the average career length of a football player is known to be 3.5 years. Under the same assumptions as before, estimate the average number of players who enter the league each year *without being drafted.*

1.8. The following table gives enrollment statistics for undergraduates at a university. From this data, estimate the average length of time that an undergraduate is enrolled at the university (this average should include students who graduate as well as students who transfer or drop out).

Year	New Students		Total
	First-yr Students	Transfer Students	Enrollment
1	1,820	2,050	16,800
2	1,950	2,280	16,700
3	1,770	2,220	17,100
4	1,860	2,140	16,400
5	1,920	2,250	17,000

1.9. You are selling your home. You observe that at any given time there are typically about 50 homes for sale in your area. New homes go on the market at a rate of about 5 per week. About how long will it take to sell your home? What assumptions are made to arrive at your answer?

1.10. Table 1.7 gives observations regarding customers at a single-server FCFS queue.

(a) Compute the average time in the queue and the average time in the system.

(b) Calculate the average system waiting time of those customers who had to wait for service (i.e., exclude those who were immediately taken into service). Calculate the average length of the queue, the average number in the system, and the fraction of idle time of the server.

1.11. Items arrive at an initially unoccupied inspection station at a uniform rate of one every 5 min. With the time of the first arrival set equal to 5, the chronological times for inspection completion of the first 10 items were observed to be 7, 17, 23, 29, 35, 38, 39, 44, 46, and 60, respectively. By

Table 1.7 Data for Problem 1.10

Customer	Interarrival Time	Service Time
1	1	3
2	9	7
3	6	9
4	4	9
5	7	10
6	9	4
7	5	8
8	8	5
9	4	5
10	10	3
11	6	6
12	12	3
13	6	5
14	8	4
15	9	9
16	5	9
17	7	8
18	8	6
19	8	8
20	7	3

manual simulation of the operation for 60 min, using these data, develop sample results for the mean number in system and the percentage idle time experienced.

1.12. Table 1.8 lists the arrival times and service durations for customers in a FCFS single-server queue. From this data, compute L_q (the time-average number in queue) and $L_q^{(A)}$ (the average number in queue as seen by arriving customers). For L_q, use a time horizon of [0, 15.27], where 15.27 is the time that the last customer exits the system. Assume the system is empty at $t = 0$.

1.13. Derive (1.13) of Section 1.7 by the sequential use of (1.12); then employ mathematical induction to prove (1.14).

1.14. Given the probability function found for the Poisson process in (1.14), find its moment generating function, $M_{N(t)}(\theta)$, that is, the expected value of $e^{\theta N(t)}$. Then use this MGF to show that the mean and variance both equal λt.

Table 1.8 Data for Problem 1.12

Arrival Time (min)	Service Duration (min)
1	2.22
2	1.76
3	2.13
4	0.14
5	0.76
6	0.70
7	0.47
8	0.22
9	0.18
10	2.41
11	0.41
12	0.46
13	1.37
14	0.27
15	0.27

1.15. Derive the Poisson process by using the third assumption that the numbers of arrivals in nonoverlapping intervals are statistically independent and then applying the binomial distribution.

1.16. Prove that the Poisson process has stationary increments.

1.17. By the use of the arguments of Sections 1.7 and 1.8, find the distribution of the counting process associated with IID Erlang interoccurrence times.

1.18. Assume that arrivals can occur singly or in batches of two, with the batch size following the probability distribution

$$f(1) = p, \; f(2) = 1 - p \quad (0 < p < 1)$$

and with the time between successive batches following the exponential probability distribution

$$a(t) = \lambda e^{-\lambda t} \quad (t > 0).$$

Show that the probability distribution for the number of arrivals in time t is the compound Poisson distribution given by

$$p_n(t) = e^{-\lambda t} \sum_{k=0}^{\lfloor n/2 \rfloor} \frac{p^{n-2k}(1-p)^k(\lambda t)^{n-k}}{(n-2k)!k!},$$

where $\lfloor n/2 \rfloor$ is the greatest integer $\leq n/2$.

1.19. **(a)** You are given two Poisson processes with intensities λ_1 and λ_2. Find the probability that there is an occurrence of the first stream before the second, starting at time $t = 0$.

(b) A queueing system is being observed. We see that all m identical, exponential servers are busy, with n more customers waiting, and decide to shut off the arrival stream. On average, how long will it take for the system to empty completely?

1.20. Verify (1.28a) and (1.28b) by using (1.27b) in (1.25). [*Hint*: To obtain (1.28a), let $s = t + \Delta t$. What is required to obtain (1.28b)?]

1.21. Derive the Poisson equations (1.11) and (1.12) by considering the Poisson process as a pure birth process and by using the Kolmogorov forward equation (1.29) with the proper q_i and q_{ij}.

1.22. **(a)** Compute the stationary probability distribution for Examples 1.1 and 1.2.

(b) Compute the stationary probability distribution for a Markov chain with the following single-step transition probability matrix:

$$\begin{pmatrix} .25 & .20 & .12 & .43 \\ .25 & .20 & .12 & .43 \\ 0 & .25 & .20 & .55 \\ 0 & 0 & .25 & .75 \end{pmatrix}.$$

1.23. Derive the imbedded transition probabilities for the birth–death process as given by (1.31). [*Hint*: Use results of Problem 1.19(a) or the law of conditional probability: $\Pr\{A|B,C\} = \Pr\{A,B|C\}/\Pr\{B|C\}$.]

1.24. Consider the first-order statistic (call it $T_{(1)}$) of a random sample of size n drawn from a uniform (0, 1) population. Its CDF is found to be

$$F(t) = 1 - (1-t)^n.$$

Show that the random variable $nT_{(1)}$ converges in law to an exponential as $n \to \infty$.

1.25. Suppose you have learned that an $M/G/1/K$ has a blocking probability of $p_k = .1$, that $\lambda = \mu = 1$, and that $L = 5$. Find λ_{eff}, W, W_q, p_0, and ρ_{eff}.

1.26. In choosing the proper distributions to represent interarrival and service times, the *coefficient of variation* (CV) can often be useful. The CV is defined as the ratio of the standard deviation to the mean and provides a measure of the relative spread of a distribution. For example, service consisting of routine tasks should have a relatively small spread around the mean (CV ≤ 1), whereas service consisting of diverse tasks (some quick, some time consuming) should have a relatively large spread around the mean (CV ≥ 1). The exponential distribution, widely used in queue modeling, has CV = 1 (standard deviation = mean). Two other distributions often employed in queueing are the *Erlang* distribution and the *mixed-exponential* distribution (the *hyperexponential* distribution is a special case). The Erlang is a two-parameter distribution, having a type or shape parameter k (an integer ≥ 1) and a scale parameter, which we shall denote by β. The mean of the Erlang is the product $k\beta$, and its standard deviation is the product $\beta\sqrt{k}$. The CV for an Erlang is then $1/\sqrt{k} \leq 1$. When $k = 1$, the Erlang reduces to the exponential distribution. The mixed-exponential distribution function is a convex linear combination of exponential distributions, mixed according to some probability distribution (e.g., we select from one exponential population, mean μ_1, with probability p, and from a second exponential, mean μ_2, with probability $1 - p$). The CV for a mixed-exponential distribution can be shown to be always >1. Using the software, solve the following problems:

(a) Data taken on a server who provides espresso to customers at the Betterbean Boutique show that the mean time to serve a customer is 2.25 min with a standard deviation of 1.6 min. What is the probability that service takes more than 5 min? [*Hint*: Find the closest integer value for k and then solve for β.]

(b) Data collected at a small post office in the rural town of Arlingrock reveal that the clerk has two types of customers—those who desire to purchase stamps only and those who require other more complicated functions. The distributions of service times for each type of customer can be well approximated by the exponential distribution; the stamp-only customers take on average 1.06 min, while the nonstamp customers take on average 3.8 min. What is the probability that a stamp customer requires more than 5 min? What is the probability that a nonstamp customer takes more than 5 min? If 15% of all arrivals are stamp-only customers, what is the probability that the next customer in line requires more than 5 min?

CHAPTER 2

SIMPLE MARKOVIAN QUEUEING MODELS

The purpose of this chapter is to develop a broad class of simple queueing models using the theory of birth–death processes. A birth–death process is a specific type of continuous-time Markov chain, the structure of which leads to a straightforward solution for the steady-state probabilities $\{p_n\}$. Examples of queues that can be modeled as birth–death processes are $M/M/1$, $M/M/c$, $M/M/c/K$, $M/M/c/c$, $M/M/\infty$, and variations of these queues with state-dependent arrival and service rates. We first begin with the general theory of birth–death process. Then, we apply these results to obtain measures of effectiveness for the queueing systems given above.

2.1 Birth–Death Processes

A birth–death process is a specific type of continuous-time Markov chain. It consists of a set of states $\{0, 1, 2, \ldots\}$, typically denoting the "population" of some system. State transitions occur as unit jumps up or down from the current state. More specifically, when the system is in state $n \geq 0$, the time until the next arrival (or "birth") is an exponential random variable with rate λ_n. At an arrival, the system moves from state n to state $n + 1$. When the system is in state $n \geq 1$, the time until

Fundamentals of Queueing Theory, Fourth Edition.
By D. Gross, J. F. Shortle, J. M. Thompson, and C. M. Harris
Copyright © 2008 John Wiley & Sons, Inc.

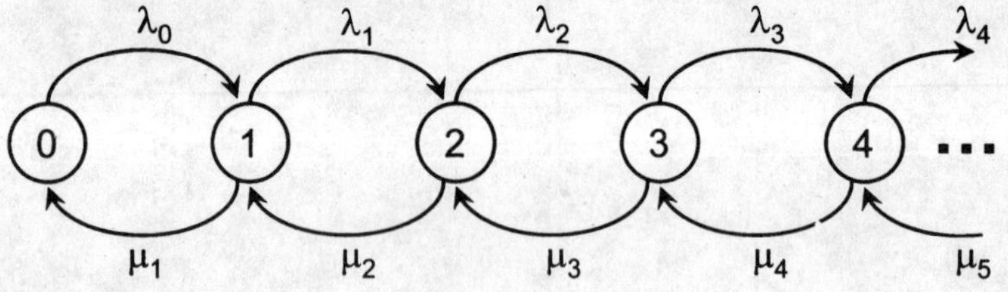

Figure 2.1 Rate transition diagram for a birth–death process.

the next departure (or "death") is an exponential random variable with rate μ_n. At a departure, the system moves from state n to state $n-1$. This is a continuous-time Markov chain, and its rate transition diagram is given in Figure 2.1.

In queueing theory, the states denote the number of customers in the system. "Births" correspond to customer arrivals and "deaths" correspond to customer departures. For example, the $M/M/1$ queue (as we will discuss in the next section) is a birth–death process with $\lambda_n = \lambda$ and $\mu_n = \mu$.

Now we apply the theory of continuous-time Markov chains to analyze the birth–death process. Let p_n denote the long-term fraction of time the system is in state n. Then, as discussed in Section 1.9, a solution for $\{p_n\}$ exists and can be determined from $\mathbf{0} = \boldsymbol{pQ}$ (1.33), subject to certain conditions on λ_n and μ_n. For the birth–death process, the vector-matrix equation $\mathbf{0} = \boldsymbol{pQ}$ can be written in component form as

$$\begin{aligned} 0 &= -(\lambda_n + \mu_n)p_n + \lambda_{n-1}p_{n-1} + \mu_{n+1}p_{n+1} \quad (n \geq 1), \\ 0 &= -\lambda_0 p_0 + \mu_1 p_1, \end{aligned}$$

or

$$\begin{aligned} (\lambda_n + \mu_n)p_n &= \lambda_{n-1}p_{n-1} + \mu_{n+1}p_{n+1} \quad (n \geq 1), \\ \lambda_0 p_0 &= \mu_1 p_1. \end{aligned} \tag{2.1}$$

These equations can also be obtained using the concept of *flow balance*. The basic idea is this: In steady state, the rate of transitions out of a given state must equal the rate of transitions into that state. As we illustrate in a moment, the left side of (2.1) is the rate of transitions out of state n, and the right side of (2.1) is the rate of transitions into state n. Thus (2.1) is simply balancing the rate of transitions into and out of state n.

We explain this more carefully as follows. When the system is in state n, the average arrival (or "birth") rate is λ_n arrivals per unit time. Since the system is in state n a fraction p_n of the time, then $\lambda_n p_n$ is the long-term rate of transitions from n to $n+1$. Likewise, when in state n, the average departure (or "death") rate is μ_n departures per unit time. Thus $\mu_n p_n$ is the long-term rate of transitions from n to $n-1$. Since a transition out of state n can be either upward or downward, $(\lambda_n + \mu_n)p_n$ is the long-term rate of transitions out of state n.

Similarly, since transitions *into* state n can occur either from the state below $(n-1)$ or the state above $(n+1)$, the long-term rate of transitions into state n is $\lambda_{n-1}p_{n-1} + \mu_{n+1}p_{n+1}$. Thus (2.1) balances the rate of transitions into and out of state $n \geq 1$.

The second equation $\lambda_0 p_0 = \mu_1 p_1$ represents a flow balance for the boundary state 0. This state is slightly different from the other states since no departures can occur when there are 0 in the system (i.e., there are no transitions from 0 to -1), and no arrivals can occur resulting in 0 in the system (i.e., there are no transitions from -1 to 0).

The Chapman–Kolmogorov differential equations (1.29) can also be obtained using a flow balance, since if we are not in steady state, the flow rates are not equal and the difference in the rates of flow is the rate of change of the system state with respect to time, namely, $p_n'(t)$.

Now, to find a solution for (2.1) we first rewrite the equations as

$$\begin{aligned} p_{n+1} &= \frac{\lambda_n + \mu_n}{\mu_{n+1}} p_n - \frac{\lambda_{n-1}}{\mu_{n+1}} p_{n-1} \quad (n \geq 1), \\ p_1 &= \frac{\lambda_0}{\mu_1} p_0. \end{aligned} \tag{2.2}$$

If follows that

$$\begin{aligned} p_2 &= \frac{\lambda_1 + \mu_1}{\mu_2} p_1 - \frac{\lambda_0}{\mu_2} p_0 = \frac{\lambda_1 + \mu_1}{\mu_2} \frac{\lambda_0}{\mu_1} p_0 - \frac{\lambda_0}{\mu_2} p_0 \\ &= \frac{\lambda_1 \lambda_0}{\mu_2 \mu_1} p_0. \end{aligned}$$

Similarly,

$$\begin{aligned} p_3 &= \frac{\lambda_2 + \mu_2}{\mu_3} p_2 - \frac{\lambda_1}{\mu_3} p_1 = \frac{\lambda_2 + \mu_2}{\mu_3} \frac{\lambda_1 \lambda_0}{\mu_2 \mu_1} p_0 - \frac{\lambda_1 \lambda_0}{\mu_3 \mu_1} p_0 \\ &= \frac{\lambda_2 \lambda_1 \lambda_0}{\mu_3 \mu_2 \mu_1} p_0. \end{aligned}$$

The pattern that appears to be emerging is that

$$\boxed{\begin{aligned} p_n &= \frac{\lambda_{n-1}\lambda_{n-2}\cdots\lambda_0}{\mu_n \mu_{n-1}\cdots\mu_1} p_0 \quad (n \geq 1) \\ &= p_0 \prod_{i=1}^{n} \frac{\lambda_{i-1}}{\mu_i}. \end{aligned}} \tag{2.3}$$

To verify that this is, in fact, the correct formula for all $n \geq 0$, we apply mathematical induction on (2.3). First, (2.3) is correct for $n = 0$, since $\prod_{i=1}^{n}(\cdot)$ is assumed by default to be 1 when $n = 0$. We have also shown that (2.3) is correct for $n = 1, 2,$

and 3. Now, we show that if (2.3) is correct for $n = k \geq 0$, then it is also correct for $n = k + 1$. Starting from (2.2)

$$\begin{aligned} p_{k+1} &= \frac{\lambda_k + \mu_k}{\mu_{k+1}} p_k - \frac{\lambda_{k-1}}{\mu_{k+1}} p_{k-1} \\ &= \frac{\lambda_k + \mu_k}{\mu_{k+1}} p_0 \prod_{i=1}^{k} \frac{\lambda_{i-1}}{\mu_i} - \frac{\lambda_{k-1}}{\mu_{k+1}} p_0 \prod_{i=1}^{k-1} \frac{\lambda_{i-1}}{\mu_i} \\ &= \frac{p_0 \lambda_k}{\mu_{k+1}} \prod_{i=1}^{k} \frac{\lambda_{i-1}}{\mu_i} + \frac{p_0 \mu_k}{\mu_{k+1}} \prod_{i=1}^{k} \frac{\lambda_{i-1}}{\mu_i} - \frac{p_0 \mu_k}{\mu_{k+1}} \prod_{i=1}^{k} \frac{\lambda_{i-1}}{\mu_i} \\ &= p_0 \prod_{i=1}^{k+1} \frac{\lambda_{i-1}}{\mu_i}. \end{aligned}$$

The second equality follows from the induction hypothesis. Thus the proof by induction is complete. Since probabilities must sum to 1, it follows that

$$\boxed{p_0 = \left(1 + \sum_{n=1}^{\infty} \prod_{i=1}^{n} \frac{\lambda_{i-1}}{\mu_i}\right)^{-1}}. \tag{2.4}$$

From (2.4), we see that a necessary and sufficient condition for the existence of a steady-state solution is the convergence of the infinite series

$$1 + \sum_{n=1}^{\infty} \prod_{i=1}^{n} \frac{\lambda_{i-1}}{\mu_i}.$$

As we will see, (2.3) and (2.4) are extremely useful in analyzing a variety of queueing models.

Before proceeding to these models, we give an alternate derivation of these results using a different set of balance equations. Previously, we derived (2.3) and (2.4) starting from (2.1). The equations in (2.1) are called *global* balance equations, since they equate the total mean flow into each state with the total mean flow out of that state.

We now give an alternate set of balance equations, which can be used to obtain the same results. To illustrate, place an artificial line between states $n - 1$ and n, as shown in Figure 2.2.

Just as mean flows into and out of a state must be equal in steady state, so also mean flows across the barrier must be equal in steady state. This can be seen as follows. If the system starts in state 0, then the first transition across the barrier is to the right. The next transition across the barrier must then be to the left, followed by a transition to the right, and so forth. Thus, except for possibly the last transition, every right-transition corresponds with exactly one left-transition. In the long term, the rate of transitions from $n - 1$ to n ($\lambda_{n-1} p_{n-1}$) must equal the rate of transitions from n to $n - 1$ ($\mu_n p_n$). This yields

$$\lambda_{n-1} p_{n-1} = \mu_n p_n \qquad (n \geq 1), \tag{2.5}$$

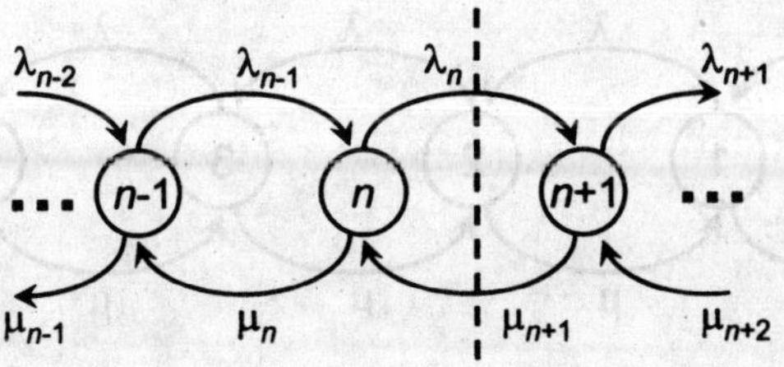

Figure 2.2 Flow balance between states.

or

$$p_n = \frac{\lambda_{n-1}}{\mu_n} p_{n-1} \qquad (n \geq 1).$$

By iteratively applying the above equation, we can derive (2.3):

$$\begin{aligned} p_n &= \frac{\lambda_{n-1}}{\mu_n} p_{n-1} \\ &= \frac{\lambda_{n-1}\lambda_{n-2}}{\mu_n \mu_{n-1}} p_{n-2} \\ &\vdots \\ &= \frac{\lambda_{n-1}\lambda_{n-2}\cdots\lambda_0}{\mu_n \mu_{n-1}\cdots\mu_1} p_0. \end{aligned}$$

The equations in (2.5) are called *detailed* balance equations, and they relate the mean flows between two states. In this case, it was somewhat easier to obtain (2.3) using the detailed balance equations (2.5) rather than the global balance equations (2.1).

It is not true for all Markov chains that the mean flows between two states are equal. The equating of these adjacent flows relates to something called *reversibility*, a concept that becomes particularly useful later in our work on queueing networks (see Section 4.1.1 and also Section 5.2.2). The flow balance between two states works here because of the birth–death characteristic that only adjacent states can directly communicate. For more general Markovian models, this is not necessarily true. However, for all Markovian models, equating the total flow out of a state with the total flow into the state always yields the global balance equations, from which the $\{p_n\}$ can be determined.

2.2 Single-Server Queues ($M/M/1$)

This section considers a single-server $M/M/1$ queue in steady state. Interarrival times and service times are assumed to be exponentially distributed with density functions given, respectively, as

$$\begin{aligned} a(t) &= \lambda e^{-\lambda t}, \\ b(t) &= \mu e^{-\mu t}. \end{aligned}$$

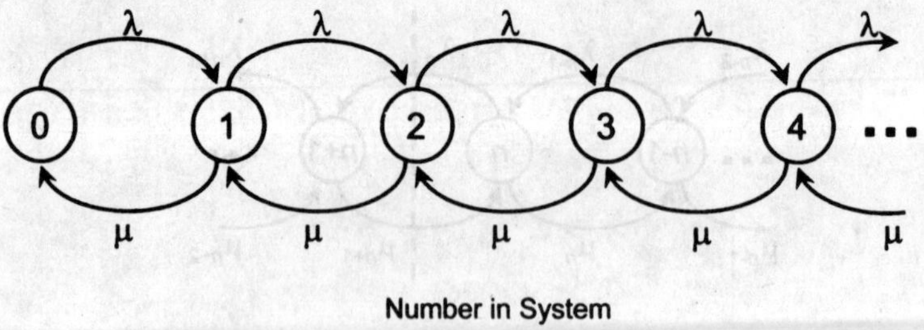

Figure 2.3 Rate transition diagram for the $M/M/1$ queue.

Let n denote the number of customers in the system. Arrivals can be considered as "births" to the system, and departures can be considered as "deaths." The rate of arrivals λ is fixed, regardless of the number in the system. The rate of the server μ is fixed, regardless of the number in the system (provided there is at least one customer in the system). Thus the $M/M/1$ queue is a birth–death process with $\lambda_n = \lambda$ $(n \geq 0)$ and $\mu_n = \mu$ $(n \geq 1)$; see Figure 2.3.

The flow-balance equations (2.1) for this system are

$$\begin{aligned}(\lambda + \mu)p_n &= \mu p_{n+1} + \lambda p_{n-1} \quad (n \geq 1),\\ \lambda p_0 &= \mu p_1.\end{aligned} \tag{2.6}$$

Alternatively, these can be written as

$$\begin{aligned}p_{n+1} &= \frac{\lambda + \mu}{\mu} p_n - \frac{\lambda}{\mu} p_{n-1} \quad (n \geq 1),\\ p_1 &= \frac{\lambda}{\mu} p_0.\end{aligned} \tag{2.7}$$

We now present three methods for solving (2.6) and (2.7). The first and probably the most straightforward is an iterative procedure. The second involves generating functions. The third involves the concept of linear operators and is analogous to methods used for differential equations. The reason for presenting three methods of solution is that one may be more successful than the others, depending on the particular model at hand. For the $M/M/1$ queue, all three methods work equally well, and we take this opportunity to illustrate their use.

2.2.1 Solving for $\{p_n\}$ Using an Iterative Method

In this section, we iteratively use the balance equations given by (2.6) and (2.7) to obtain a sequence of state probabilities, $p_1, p_2, p_3, \ldots$, each in terms of p_0. When we believe that we have enough information about the form of these state probabilities, we make a conjecture on their general form for all states n. Then, we attempt to verify that our conjecture is correct using mathematical induction.

In fact, this is precisely what we did for the general birth–death process in Chapter 1. Specifically, we verified that (2.3) is the appropriate formula for the steady-state probability p_n for any birth–death process with birth rates $\{\lambda_n, n = 0, 1, 2, \ldots\}$ and death rates $\{\mu_n, n = 1, 2, 3, \ldots\}$. Since the $M/M/1$ system is a birth–death process with constant birth and death rates, we can directly apply (2.3) with $\lambda_n = \lambda$ and $\mu_n = \mu$ for all n. It follows that

$$p_n = p_0 \prod_{i=1}^{n} \left(\frac{\lambda}{\mu}\right) = p_0 \left(\frac{\lambda}{\mu}\right)^n, \quad (n \geq 1).$$

To get p_0, we use the fact that the probabilities $\{p_n\}$ must sum to 1:

$$1 = \sum_{n=0}^{\infty} p_n = \sum_{n=0}^{\infty} p_0 \left(\frac{\lambda}{\mu}\right)^n = p_0 \sum_{n=0}^{\infty} \rho^n.$$

In the last step, we have used our earlier definition from Section 1.5 that $\rho = \lambda/\mu$ for single-server queues, where ρ is the traffic intensity or utilization. Then,

$$p_0 = \frac{1}{\sum_{n=0}^{\infty} \rho^n}.$$

Now, $\sum_{n=0}^{\infty} \rho^n$ is a geometric series that converges if and only if $\rho < 1$. Making use of the well-known expression for a geometric series, we have

$$\sum_{n=0}^{\infty} \rho^n = \frac{1}{1-\rho} \quad (\rho < 1),$$

which implies that

$$p_0 = 1 - \rho \quad (\rho = \lambda/\mu < 1). \tag{2.8}$$

This is consistent with the general result for p_0 that we derived previously in Section 1.5 for all $G/G/1$ queues. In summary, the full steady-state solution for the $M/M/1$ system is the *geometric* probability function

$$\boxed{p_n = (1-\rho)\rho^n \quad (\rho = \lambda/\mu < 1).} \tag{2.9}$$

We emphasize that the existence of a steady-state solution depends on the condition that $\rho < 1$, or equivalently, $\lambda < \mu$. This makes intuitive sense, for if $\lambda > \mu$, the mean arrival rate is greater than the mean service rate, so the server gets further and further behind. In other words, the system size increases without bound over time. It is not as intuitive, however, to explain why no steady-state solution exists when $\lambda = \mu$. One possible way to explain infinite buildup when $\lambda = \mu$ is that as the queue grows, it is more and more difficult for the server to decrease the queue because the average service rate is no higher than the average arrival rate.

2.2.2 Solving for $\{p_n\}$ Using Generating Functions

In this section, we use the probability generating function $P(z) = \sum_{n=0}^{\infty} p_n z^n$ (z complex with $|z| \leq 1$), to find the steady-state probabilities $\{p_n\}$. The basic procedure is to first find a closed expression for $P(z)$ using (2.7). Then, we expand $P(z)$ in a power series. The probabilities $\{p_n\}$ are then obtained as the coefficients in this power series.

To start, we rewrite (2.7) in terms of ρ:

$$p_{n+1} = (\rho + 1)p_n - \rho p_{n-1} \quad (n \geq 1), \tag{2.10}$$

$$p_1 = \rho p_0. \tag{2.11}$$

Now, we multiply both sides of the first line by z^n:

$$p_{n+1}z^n = (\rho + 1)p_n z^n - \rho p_{n-1} z^n,$$

or

$$z^{-1}p_{n+1}z^{n+1} = (\rho + 1)p_n z^n - \rho z p_{n-1} z^{n-1}.$$

The previous equation is valid for $n \geq 1$. Summing each of these equations from $n = 1$ to ∞ gives

$$z^{-1}\sum_{n=1}^{\infty} p_{n+1}z^{n+1} = (\rho + 1)\sum_{n=1}^{\infty} p_n z^n - \rho z \sum_{n=1}^{\infty} p_{n-1} z^{n-1}.$$

Since the generating function is defined as $P(z) = \sum_{n=0}^{\infty} p_n z^n$, the previous equation can be rewritten as

$$z^{-1}[P(z) - p_1 z - p_0] = (\rho + 1)[P(z) - p_0] - \rho z P(z). \tag{2.12}$$

We know from (2.11) that $p_1 = \rho p_0$, so

$$z^{-1}[P(z) - (\rho z + 1)p_0] = (\rho + 1)[P(z) - p_0] - \rho z P(z).$$

Solving for $P(z)$ gives

$$P(z) = \frac{p_0}{1 - z\rho}. \tag{2.13}$$

To find p_0, we use the boundary condition that the probabilities $\{p_n\}$ sum to 1. Consider $P(1)$, which can be seen to be

$$P(1) = \sum_{n=0}^{\infty} p_n 1^n = \sum_{n=0}^{\infty} p_n = 1.$$

Thus from (2.13), we have

$$P(1) = 1 = \frac{p_0}{1 - \rho}, \tag{2.14}$$

so that $p_0 = 1 - \rho$.

Because the $\{p_n\}$ are probabilities, $P(z) > 0$ when z is real and $z > 0$. Therefore $P(1) > 0$. This implies from (2.14) that $P(1) = p_0/(1-\rho) > 0$. Therefore ρ must be less than one, since p_0 is a probability and is greater than zero. In summary,

$$\boxed{P(z) = \frac{1-\rho}{1-\rho z} \quad (\rho < 1, \quad |z| \le 1).} \tag{2.15}$$

It is easy to expand (2.15) as a power series by simple long division or to recognize it as the sum of a geometric series, since $|\rho z| < 1$. That is,

$$\frac{1}{1-\rho z} = 1 + \rho z + (\rho z)^2 + (\rho z)^3 + \cdots,$$

and thus the probability generating function is

$$P(z) = \sum_{n=0}^{\infty} (1-\rho)\rho^n z^n. \tag{2.16}$$

Now, p_n is the coefficient in front of z^n, so

$$p_n = (1-\rho)\rho^n \quad (\rho = \lambda/\mu < 1),$$

which is what was previously obtained in (2.9).

We make a number of concluding observations about the algebraic form of the generating function given in (2.15). First, this expression is the quotient of two (simple) polynomials (i.e., it is a rational function). The numerator is a constant, and the denominator is a linear form $1 - \rho z$. It follows that the denominator has the single zero of $1/\rho$. This is the reciprocal of the traffic intensity, and its value is greater than one. We will be seeing more generating functions as we develop more queueing models in the text, and these comments will be important in comparison with other models.

Finally, we note that for some models, it is relatively easy to find a closed expression for $P(z)$, but quite difficult to find its series expansion to obtain the $\{p_n\}$. However, even if the series expansion cannot be found, $P(z)$ still provides useful information. For example, $dP(z)/dz$ evaluated at $z = 1$ gives the expected number in the system, $L = \sum_{n=0}^{\infty} np_n$.

2.2.3 Solving for $\{p_n\}$ Using Operators

In this section, we use the theory of linear difference equations to solve for $\{p_n\}$. Consider the first equation given in (2.7), which can be written as

$$p_{n+1} = (\rho + 1)p_n - \rho p_{n-1} \quad (n \ge 1).$$

This is a linear difference equation. It expresses p_{n+1} as a function of the two previous values, p_n and p_{n-1}. The equation is like a differential equation, but it involves differences of successive values of p, rather than derivatives. The index n is discrete, analogous to the continuous parameter, typically t representing time, in a differential equation. A general solution to (2.7) can be obtained using the theory of linear difference equations. We summarize this theory next. It is similar in many respects to the theory for solving linear differential equations.

We begin by defining a linear operator D on the sequence $\{a_0, a_1, a_2, \ldots\}$:

$$Da_n \equiv a_{n+1} \quad \text{(for all } n\text{)}.$$

The composition of operators yields

$$D^m a_n = a_{n+m} \quad \text{(for all } n \text{ and } m\text{)}.$$

Then a general linear difference equation with constant coefficients

$$C_n a_n + C_{n+1} a_{n+1} + \cdots + C_{n+k} a_{n+k} = 0 \tag{2.17}$$

can be written as

$$C_n a_n + C_{n+1} D a_n + \cdots + C_{n+k} D^k a_n = 0.$$

For example, a second-order difference equation of the form

$$C_2 a_{n+2} + C_1 a_{n+1} + C_0 a_n = 0 \tag{2.18}$$

can be written as

$$(C_2 D^2 + C_1 D + C_0) a_n = 0. \tag{2.19}$$

The quadratic in D gives the characteristic equation for this difference equation: $C_2 r^2 + C_1 r + C_0 = 0$. The roots of this equation determine the form of the solution for $\{a_n\}$. In particular, if the characteristic equation has two distinct real roots r_1 and r_2, then

$$a_n = d_1 r_1^n \quad \text{and} \quad a_n = d_2 r_2^n$$

are both solutions to (2.18), where d_1 and d_2 are arbitrary constants. This can be verified by the substitution of $d_1 r_1^n$ and $d_2 r_2^n$ into (2.18). For example, letting $a_n = d_1 r_1^n$ we have, upon substitution in (2.18),

$$C_2 d_1 r_1^{n+2} + C_1 d_1 r_1^{n+1} + C_0 d_1 r_1^n = 0,$$

or

$$d_1 r_1^n (C_2 r_1^2 + C_1 r_1 + C_0) = 0.$$

The right term is zero, since r_1 is a root of the characteristic equation, by assumption. Similarly, one can show that $a_n = d_2 r_2^n$ is a solution and hence that their sum, $d_1 r_1^n + d_2 r_2^n$, is also a solution. It can be shown in a manner similar to that used for ordinary linear differential equations that this sum is the most general solution.

This approach is directly applicable to the solution of the steady-state difference equations (2.7), since these are linear difference equations of a form like (2.18). If (2.7) is rewritten as

$$\mu p_{n+2} - (\lambda + \mu)p_{n+1} + \lambda p_n = 0 \quad (n \geq 0),$$

then the $\{p_n\}$ are the solution to

$$[\mu D^2 - (\lambda + \mu)D + \lambda]p_n = 0, \tag{2.20}$$

subject to the boundary conditions

$$p_1 = \frac{\lambda}{\mu} p_0 = \rho p_0,$$

$$\sum_{n=0}^{\infty} p_n = 1.$$

This quadratic in D factors easily to give

$$(D - 1)(\mu D - \lambda)p_n = 0,$$

and hence

$$p_n = d_1(1)^n + d_2(\lambda/\mu)^n = d_1 + d_2\rho^n, \tag{2.21}$$

where d_1 and d_2 are to be found with the use of the boundary conditions. We see from (2.21) that

$$p_1 = d_1 + d_2\rho \quad \text{and} \quad p_2 = d_1 + d_2\rho^2.$$

But d_1 must be zero to allow the probabilities to sum to one, so that $p_1 = d_2\rho$. Now we also know from (2.11) that $p_1 = \rho p_0$. Thus $d_2 = p_0$ and

$$p_n = \rho^n p_0.$$

The empty probability p_0 is now found as in the first method by summing the $\{p_n\}$ over all n and is

$$p_0 = 1 - \rho \quad (\rho < 1).$$

2.2.4 Measures of Effectiveness

The steady-state probability distribution for the system size allows us to calculate the system's measures of effectiveness. Two of immediate interest are the expected number in the system and the expected number in the queue at steady state. To derive these, let N represent the random variable "number of customers in the system in steady state" and L represent its expected value. We can then write

$$L = \mathrm{E}[N] = \sum_{n=0}^{\infty} np_n$$

$$= (1 - \rho)\sum_{n=0}^{\infty} n\rho^n. \tag{2.22}$$

Consider the summation

$$\sum_{n=0}^{\infty} n\rho^n = \rho + 2\rho^2 + 3\rho^3 + \cdots$$
$$= \rho(1 + 2\rho + 3\rho^2 + \cdots)$$
$$= \rho \sum_{n=1}^{\infty} n\rho^{n-1}.$$

We observe that $\sum_{n=1}^{\infty} n\rho^{n-1}$ is simply the derivative of $\sum_{n=0}^{\infty} \rho^n$ with respect to ρ, since the summation and differentiation operations may be interchanged, as the functions are sufficiently well behaved. Since $\rho < 1$,

$$\sum_{n=0}^{\infty} \rho^n = \frac{1}{1-\rho};$$

hence

$$\sum_{n=1}^{\infty} n\rho^{n-1} = \frac{d[1/(1-\rho)]}{d\rho}$$
$$= \frac{1}{(1-\rho)^2}.$$

So the expected number in the system at steady state is then

$$L = \frac{\rho(1-\rho)}{(1-\rho)^2},$$

or simply

$$\boxed{L = \frac{\rho}{1-\rho} = \frac{\lambda}{\mu - \lambda}.} \tag{2.23}$$

If the random variable "number in queue in steady state" is denoted by N_q and its expected value by L_q, then we have

$$L_q = \sum_{n=1}^{\infty} (n-1)p_n = \sum_{n=1}^{\infty} np_n - \sum_{n=1}^{\infty} p_n$$
$$= L - (1 - p_0) = \frac{\rho}{1-\rho} - \rho = \frac{\rho^2}{1-\rho}.$$

Note that $L_q = L - (1 - p_0)$ holds for all single-channel, one-at-a-time service queues, since no assumptions were made in the derivation as to the input and service distributions; this can also be seen from (1.3). Thus the mean queue length is

$$\boxed{L_q = \frac{\rho^2}{1-\rho} = \frac{\lambda^2}{\mu(\mu - \lambda)}.} \tag{2.24}$$

We might also be interested in the expected queue size of nonempty queues, which we denote by L'_q; that is, we wish to ignore the cases where the queue is empty. Another way of looking at this measure is to view it as the expected size of the queues that form from time to time. We can write

$$L'_q = \mathrm{E}[N_q | N_q \neq 0]$$
$$= \sum_{n=1}^{\infty} (n-1)p'_n = \sum_{n=2}^{\infty} (n-1)p'_n,$$

where p'_n is the conditional probability distribution of n in the system given the queue is not empty, or $p'_n = \Pr\{n \text{ in system } | n \geq 2\}$. From the laws of conditional probability,

$$p'_n = \frac{\Pr\{n \text{ in system and } n \geq 2\}}{\Pr\{n \geq 2\}}$$
$$= \frac{p_n}{\sum_{n=2}^{\infty} p_n} \quad (n \geq 2)$$
$$= \frac{p_n}{1-(1-\rho)-(1-\rho)\rho}$$
$$= \frac{p_n}{\rho^2}.$$

The probability distribution $\{p'_n\}$ is the distribution $\{p_n\}$ normalized by omitting the cases $n = 0$ and 1. Thus

$$L'_q = \sum_{n=2}^{\infty} (n-1)\frac{p_n}{\rho^2}$$
$$= \frac{L - p_1 - (1 - p_0 - p_1)}{\rho^2}.$$

Hence

$$\boxed{L'_q = \frac{1}{1-\rho} = \frac{\mu}{\mu - \lambda}.} \tag{2.25}$$

As a side observation, it is not by coincidence that

$$\Pr\{n \text{ in system} \geq 2\} = \rho^2,$$

because it can easily be established for all n that

$$\Pr\{N \geq n\} = \rho^n.$$

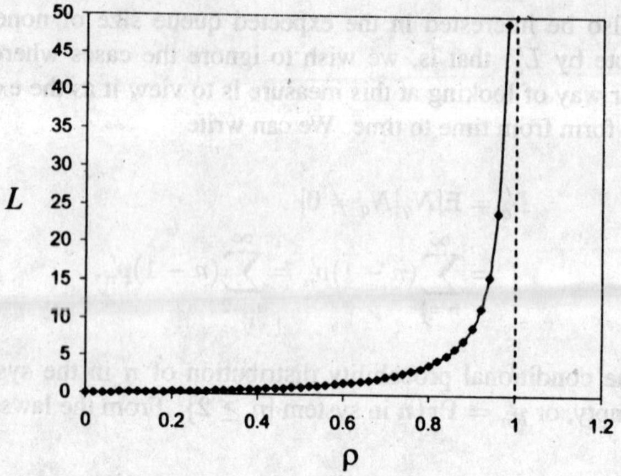

Figure 2.4 Mean number in system for $M/M/1$ queue.

The proof is as follows:

$$\Pr\{N \geq n\} = \sum_{k=n}^{\infty}(1-\rho)\rho^k$$

$$= (1-\rho)\rho^n \sum_{k=n}^{\infty} \rho^{k-n}$$

$$= \frac{(1-\rho)\rho^n}{1-\rho} = \rho^n.$$

To complete the basic part of this presentation, we recall from Chapter 1 that the expected steady-state system waiting time W and line delay W_q can be found easily from L and L_q by using Little's formulas, $L = \lambda W$ and $L_q = \lambda W_q$. In the case of the $M/M/1$ queue, it thus follows from (2.23) and (2.24) that

$$\boxed{W = \frac{L}{\lambda} = \frac{\rho}{\lambda(1-\rho)} = \frac{1}{\mu - \lambda}} \tag{2.26}$$

and

$$\boxed{W_q = \frac{L_q}{\lambda} = \frac{\rho}{\mu(1-\rho)} = \frac{\rho}{\mu - \lambda}.} \tag{2.27}$$

Figure 2.4 graphically shows the relationship between the average system size L and the queue utilization ρ. As seen, L is an increasing function of ρ. For low values of ρ, L increases somewhat slowly, and then as ρ gets closer to 1, L increases very rapidly, eventually growing to an infinite limit as $\rho \to 1$.

Practically, this implies that it is not desirable to have a queueing system with ρ close to 1. Although it may be desirable to maximize server productivity – that is, to keep the server nearly always busy – this comes at the cost of extreme delays. In addition, the *variability* of system size increases rapidly in ρ as well (see Problem 2.3), so system predictability degrades as $\rho \to 1$. A lower value of ρ is usually better to balance delays with idle server time.

While the formulas for L, L_q, W, and W_q are simple and useful, there are many practical limitations. The formulas were developed under specific assumptions: Poisson arrivals, exponential service times, steady-state conditions, and queue stability. In particular, we note the following:

1. Equation (2.23) is valid only when $\rho < 1$. When $\rho > 1$, (2.23) yields a negative number, which is meaningless with respect to the system size.

2. Equation (2.23) is valid only in *steady state*. In many systems, the arrival and/or service rates may change so that the system is not able to reach steady state. As an example, consider a brief thunderstorm over an airport. During the thunderstorm, the arrival rate of airplanes exceeds the reduced airport capacity (so $\rho > 1$ during this period). The queue rapidly increases during this period. Once the bad weather passes, the service rate returns to normal and the queue begins to drain off. These transient effects are not well captured by the $M/M/1$ model.

Similar remarks apply to the other measures of effectiveness, L_q, W_q, and W.

■ **EXAMPLE 2.1**

Ms. H. R. Cutt runs a one-person hair salon. She does not make appointments, but runs the salon on a first-come, first-served basis. She finds that she is extremely busy on Saturday mornings, so she is considering hiring a part-time assistant and even possibly moving to a larger building. Having obtained a master's degree in operations research (OR) prior to embarking upon her career, she elects to analyze the situation carefully before making a decision.

She thus keeps careful records for a succession of Saturday mornings and finds that customers seem to arrive according to a Poisson process with a mean arrival rate of 5 per hour. Because of her excellent reputation (what else would you expect from someone with an M.S. in OR?), customers were always willing to wait. The data further showed that customer processing time (aggregated female and male) was exponentially distributed with an average of 10 min.

Cutt first decided to calculate the average number of customers in the shop and the average number of customers waiting for a haircut. From the data, $\lambda = 5$/h and $\mu = \frac{1}{10}$/ min $= 6$/h. This gives $\rho = \frac{5}{6}$. From (2.23) and (2.24) she finds $L = 5$ and $L_q = 4\frac{1}{6}$. The average number waiting when there is at least one person waiting is found from (2.25) as $L'_q = 6$. She is also interested in the percentage of time an arrival can walk right in without having to wait at all, which happens when no one is in the shop. The probability of this is $p_0 = 1 - \rho = \frac{1}{6}$. Hence approximately 16.7% of the time Cutt is idle and a customer can get into the chair without waiting. Because of the Poisson process governing arrivals and its completely random property, as discussed in Section 1.7, the percentage of customers that can go directly into service is also 16.7%. Thus 83.3% of the customers must wait prior to getting into the chair.

Cutt's waiting room has only four seats at present. She is interested in the probability that a customer, upon arrival, will not be able to find a seat and have to stand. This can easily be calculated as

$$\Pr\{\text{finding no seat}\} = \Pr\{N \geq 5\} = \rho^5 \doteq 0.402.$$

This tells Cutt that a little over 40% of the time a customer cannot find a seat and also that 40% of the customers will have to stand upon arrival. Cutt is also interested in learning how much time customers spend waiting, and the average system waiting time and line delay are easily computed from (2.26) and (2.27) to be $W = 1/(\mu - \lambda) = 1$ h and $W_q = \rho/(\mu - \lambda) = \frac{5}{6}$ h, not very pleasing outcomes.

To get even more information on the nature of customer waiting, Cutt has decided that she would like to know the precise probability that the line delay is more than 45 min. But to do this, she needs to have the probability distribution function for the waiting time in queue, which is something she has forgotten from her M.S. days. So she has decided to go through a complete derivation of the result by herself. This follows.

2.2.5 Waiting-Time Distribution

Let T_q denote the random variable "time spent waiting in the queue" (in steady state) and $W_q(t)$ represent its cumulative probability distribution. Up to now the queue discipline has had no effect on our derivations. When considering individual waiting times, however, queue discipline must be specified, and we are here assuming that it is first come, first served (FCFS).

The queue-waiting-time random variable has an interesting property in that it is part discrete and part continuous. Waiting time in queue is, for the most part, a continuous random variable, except that there is a nonzero probability that the wait is zero. This occurs when the system is empty and an arriving customer begins service immediately upon arrival. Let q_n denote the probability (in steady state) that an arriving customer finds n in the system (just prior to arrival). Then,

$$\begin{aligned} W_q(0) &= \Pr\{T_q \leq 0\} = \Pr\{T_q = 0\} \\ &= \Pr\{\text{system empty at an arrival}\} = q_0. \end{aligned}$$

The probabilities $\{q_n\}$ are not always the same as the probabilities $\{p_n\}$ with which we have been working. Recall, p_n is the fraction of time that there are n in the system. This is not necessarily the same as the probability q_n that an arriving customer finds n in the system. For Poisson input, $q_n = p_n$, but that is not always the case (e.g., the $G/M/1$ queue in Section 5.3). Thus

$$W_q(0) = p_0 = 1 - \rho.$$

It then remains to find $W_q(t)$ for $t > 0$. Consider $W_q(t)$, the probability of a customer waiting a time less than or equal to t for service. If there are n units in the system upon arrival, then in order for the customer to go into service at a time between 0 and t, all n units must have been served by time t. Since the service distribution is memoryless, the distribution of the time required for n completions is independent of the time of the current arrival and is the convolution of n exponential random variables. This is an Erlang type-n distribution (see Section 3.3.1 for further discussion on the Erlang distribution). In addition, since the input is Poisson, $p_n = q_n$, as mentioned previously. Therefore

$$\begin{aligned} W_q(t) &= \Pr\{T_q \le t\} = W_q(0) \\ &\quad + \sum_{n=1}^{\infty} \Pr\{n \text{ completions in} \le t \mid \text{arrival found } n \text{ in system}\} \cdot p_n \\ &= 1 - \rho + (1-\rho) \sum_{n=1}^{\infty} \rho^n \int_0^t \frac{\mu(\mu x)^{n-1}}{(n-1)!} e^{-\mu x}\, dx \\ &= 1 - \rho + \rho \int_0^t \mu(1-\rho) e^{-\mu x} \sum_{n=1}^{\infty} \frac{(\mu x \rho)^{n-1}}{(n-1)!}\, dx \\ &= 1 - \rho + \rho \int_0^t \mu(1-\rho) e^{-\mu(1-\rho)x}. \end{aligned}$$

From the last line, we see that $W_q(t)$ represents the mixture of a discrete random variable and a continuous random variable. Specifically, the term on the right is the CDF of an exponential random variable, with parameter $\mu(1-\rho)$, weighted by the factor ρ. The term on the left $(1-\rho)$ is the CDF of a random variable that takes on the constant value of zero, weighted by the factor $(1-\rho)$. Thus $W_q(t)$ is the mixture of an exponential distribution and a discrete probability mass at zero.

$W_q(t)$ can be further simplified to yield a more convenient form:

$$W_q(t) = 1 - \rho e^{-(\mu-\lambda)t} \qquad (t \ge 0). \tag{2.28}$$

We already know from (2.27) that the mean of this distribution is $W_q = \rho/(\mu-\lambda)$. In verification of this result, we recompute the mean as

$$W_q = \int_0^\infty [1 - W_q(t)]\, dt = \int_0^\infty \rho e^{-\mu(1-\rho)t}\, dt = \frac{\rho}{\mu-\lambda}.$$

In a similar manner, the CDF of the total time in system can be derived. Let T denote the total time an arriving customer spends in the system (in steady state). Let $W(t)$ denote the CDF of T and let $w(t)$ denote the PDF. It can be shown (Problem 2.5) that T is an exponential random variable with mean $1/(\mu-\lambda)$. That is,

$$\begin{aligned} W(t) &= 1 - e^{-(\mu-\lambda)t} \qquad (t \ge 0), \\ w(t) &= (\mu-\lambda) e^{-(\mu-\lambda)t} \qquad (t > 0). \end{aligned} \tag{2.29}$$

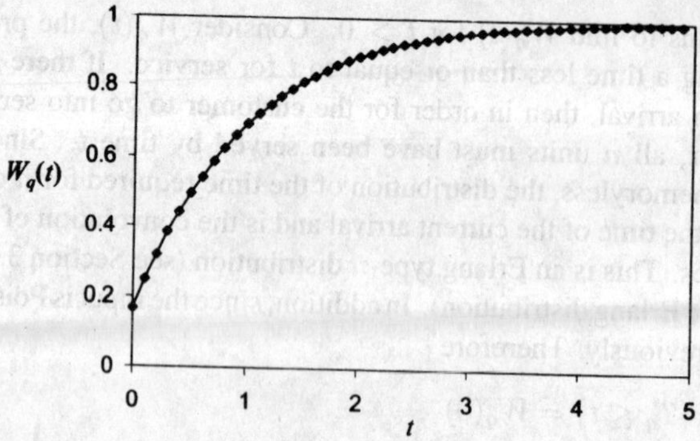

Figure 2.5 Sample CDF of line delay in $M/M/1$ queue.

The derivation of (2.29) closely follows that of $W_q(t)$ except that $n+1$ service completions are required in time $\leq t$.

The CDFs, $W(t)$ and $W_q(t)$, are discipline-dependent, as can be seen in their derivations by the requirement that a customer's waiting time is determined by the amount of time it takes to serve the customers found upon arrival. In contrast, the derivations of the *average* measures associated with the $M/M/1$ queue – L, L_q, W, W_q, and p_n – do not depend on the order in which customers are served. Thus all of these measures are valid for the general discipline $M/M/1/\infty/\text{GD}$ model.

Finally, we emphasize that the measures of effectiveness used here are all calculated for steady state. These results are not applicable to the initial few customers who arrive soon after opening. Also, the results do not apply when the arrival rate or service rate varies in time, as is the case, for example, when there is a rush of customers at a particular time of day.

■ **EXAMPLE 2.2**

Figure 2.5 shows a plot of the CDF $W_q(t)$ as experienced by Cutt's customers ($\mu = 6$, $\lambda = 5$). The jump discontinuity at $t = 0$ results from the fact that there is a nonzero probability that an arrival finds an empty system. Using (2.28), the probability that an arriving customer waits more than 45 min before receiving service is $\frac{5}{6}e^{-3/4} \doteq 0.3936$. In a similar manner, the distribution of the *total* time in the system (including service) can be calculated using (2.29). This distribution does not have a jump discontinuity at $t = 0$.

2.3 Multiserver Queues ($M/M/c$)

We now turn our attention to the multiserver $M/M/c$ model: Arrivals are Poisson with rate λ, there are c servers, and each server has an independently and identically distributed exponential service-time distribution with mean $1/\mu$.

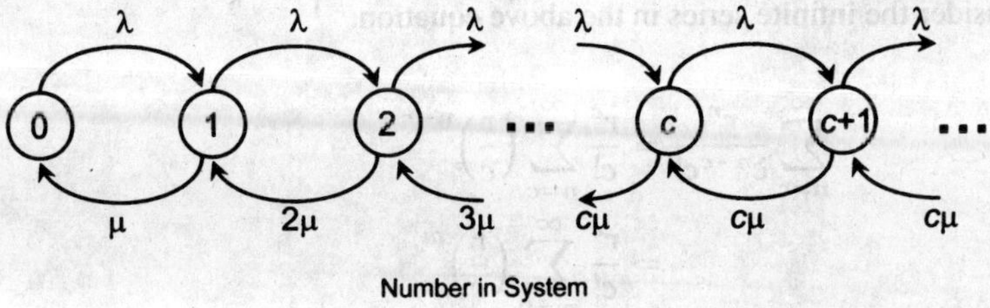

Figure 2.6 Rate transition diagram for the $M/M/c$ queue.

Like the $M/M/1$ queue, this queue can be modeled as a birth–death process (Figure 2.6). Since the arrival rate is constant, the "birth" rate is $\lambda_n = \lambda$ for all n, regardless of the number of customers in the system. In contrast, the rate of service completions (or "deaths") depends on the number in the system. If there are c or more customers in the system, then all c servers must be busy. Since each server processes customers with rate μ, the combined service-completion rate for the system is $c\mu$. When there are fewer than c customers in the system, $n < c$, only n of the c servers are busy and the combined service-completion rate for the system is $n\mu$. Hence μ_n may be written as

$$\mu_n = \begin{cases} n\mu & (1 \le n < c), \\ c\mu & (n \ge c). \end{cases} \tag{2.30}$$

Using the prior theory developed for birth–death processes, we can insert the values for λ_n and μ_n into (2.3) to obtain the steady-state probabilities p_n:

$$\boxed{p_n = \begin{cases} \dfrac{\lambda^n}{n!\mu^n} p_0 & (0 \le n < c), \\ \dfrac{\lambda^n}{c^{n-c}c!\mu^n} p_0 & (n \ge c). \end{cases}} \tag{2.31}$$

We see that p_n has the form of a Poisson random variable for $0 \le n < c$ and the form of a geometric random variable for $n \ge c$. In order to find p_0, we again use the condition that the probabilities must sum to 1, which gives

$$p_0 = \left(\sum_{n=0}^{c-1} \frac{\lambda^n}{n!\mu^n} + \sum_{n=c}^{\infty} \frac{\lambda^n}{c^{n-c}c!\mu^n} \right)^{-1}.$$

As in Section 1.5, we let $r = \lambda/\mu$ and $\rho = r/c = \lambda/c\mu$. Then we have

$$p_0 = \left(\sum_{n=0}^{c-1} \frac{r^n}{n!} + \sum_{n=c}^{\infty} \frac{r^n}{c^{n-c}c!} \right)^{-1}.$$

Now consider the infinite series in the above equation:

$$\begin{aligned}\sum_{n=c}^{\infty} \frac{r^n}{c^{n-c}c!} &= \frac{r^c}{c!}\sum_{n=c}^{\infty}\left(\frac{r}{c}\right)^{n-c} \\ &= \frac{r^c}{c!}\sum_{m=0}^{\infty}\left(\frac{r}{c}\right)^{m} \\ &= \frac{r^c}{c!}\frac{1}{1-r/c} \qquad (r/c = \rho < 1).\end{aligned}$$

Therefore we can write

$$p_0 = \left(\frac{r^c}{c!(1-\rho)} + \sum_{n=0}^{c-1}\frac{r^n}{n!}\right)^{-1} \qquad (r/c = \rho < 1). \tag{2.32}$$

Here, the condition for the existence of a steady-state solution is $\lambda/c\mu < 1$. That is, the mean arrival rate must be less than the mean maximum potential service rate of the system. This is intuitively what we would expect. Also, when $c = 1$, (2.32) reduces to (2.8), the analogous equation for the $M/M/1$ queue.

We can now derive measures of effectiveness for the $M/M/c$ model utilizing the steady-state probabilities given by (2.31) and (2.32) in a manner similar to that used for the $M/M/1$ model in Section 2.2.3. We first consider the expected queue size L_q, as it is computationally easier to determine than L, since we have only to deal with p_n for $n \geq c$. Thus

$$\begin{aligned}L_q &= \sum_{n=c+1}^{\infty}(n-c)p_n = \sum_{n=c+1}^{\infty}(n-c)\frac{r^n}{c^{n-c}c!}p_0 \\ &= \frac{r^c p_0}{c!}\sum_{n=c+1}^{\infty}(n-c)\rho^{n-c} = \frac{r^c p_0}{c!}\sum_{m=1}^{\infty} m\rho^m = \frac{r^c \rho p_0}{c!}\sum_{m=1}^{\infty} m\rho^{m-1} \\ &= \frac{r^c \rho p_0}{c!}\frac{d}{d\rho}\sum_{m=1}^{\infty}\rho^m = \frac{r^c \rho p_0}{c!}\frac{d}{d\rho}\left(\frac{1}{1-\rho} - 1\right) \\ &= \frac{r^c \rho p_0}{c!(1-\rho)^2}.\end{aligned}$$

Thus

$$L_q = \left(\frac{r^c \rho}{c!(1-\rho)^2}\right)p_0. \tag{2.33}$$

To find L, we employ Little's formula to get W_q, then use W_q to find $W = W_q + 1/\mu$, and finally employ Little's formula again to calculate $L = \lambda W$. Thus we get

$$W_q = \frac{L_q}{\lambda} = \left(\frac{r^c}{c!(c\mu)(1-\rho)^2} \right) p_0, \tag{2.34}$$

$$W = \frac{1}{\mu} + \left(\frac{r^c}{c!(c\mu)(1-\rho)^2} \right) p_0, \tag{2.35}$$

and

$$L = r + \left(\frac{r^c \rho}{c!(1-\rho)^2} \right) p_0. \tag{2.36}$$

The final result for L could have been obtained directly from L_q by using $L = L_q + r$, which we showed in Section 1.5 to be valid for any *G/G/c* system.

We now obtain $W_q(0)$, the probability that a customer has *zero* delay in queue before receiving service. Equivalently, $1 - W_q(0)$ is the probability that a customer has *nonzero* delay in queue before receiving service. This measure of congestion is often used in managing call centers. In such systems, an arriving caller waits "on hold" in a virtual queue when all servers are busy. In queue, the caller typically hears music or informational announcements. $1 - W_q(0)$ does not directly indicate the *length* of time a customer waits in queue, but rather the probability that a customer is not able to immediately access a server.

To find $W_q(0)$, let T_q represent the random variable "time spent waiting in queue" (in steady state) and $W_q(t)$ its CDF. Then,

$$\begin{aligned} W_q(0) &= \Pr\{T_q = 0\} = \Pr\{< c-1 \text{ in system}\} \\ &= \sum_{n=0}^{c-1} p_n = p_0 \sum_{n=0}^{c-1} \frac{r^n}{n!}. \end{aligned}$$

To evaluate $\sum r^n/n!$, recall that it appears in the expression for p_0 in (2.32), so that

$$\sum_{n=0}^{c-1} \frac{r^n}{n!} = \frac{1}{p_0} - \frac{r^c}{c!(1-\rho)}.$$

This gives

$$W_q(0) = p_0 \left(\frac{1}{p_0} - \frac{r^c}{c!(1-\rho)} \right) = 1 - \frac{r^c p_0}{c!(1-\rho)}. \tag{2.37}$$

Equivalently, the probability that an arriving customer has a *nonzero* wait in queue is

$$C(c,r) \equiv 1 - W_q(0) = \frac{r^c}{c!(1-\rho)} \Bigg/ \left(\frac{r^c}{c!(1-\rho)} + \sum_{n=0}^{c-1} \frac{r^n}{n!} \right). \tag{2.38}$$

This is sometimes called the *Erlang-C* formula, which we denote by $C(c, r)$. This formula gives the probability that an arriving customer is delayed in the queue (i.e., has positive, nonzero wait in the queue), as a function of the parameters c and r. The formula is based on the $M/M/c$ model and thus carries with it all of the corresponding assumptions. In particular, the model ignores complexities such as abandonments, retrials, and nonstationary arrivals – factors that may be important in modeling call centers. Also, the model assumes an infinite queue size. For call centers, this corresponds to an infinite number of available access lines into the center.

■ **EXAMPLE 2.3**

Calls to a technical support center arrive according to a Poisson process with rate 30 per hour. The time for a support person to serve one customer is exponentially distributed with a mean of 5 minutes. The support center has 3 technical staff to assist callers. What is the probability that a customer is able to immediately access a support staff, without being delayed on hold? (Assume that customers do not abandon their calls.)

For this problem, $\lambda = 30$, $\mu = 12$, and $c = 3$. Then $r = 2.5$ and $\rho = 5/6$. From (2.38),

$$C(c,r) = \frac{2.5^3}{3!(1-5/6)} \bigg/ \left(\frac{2.5^3}{3!(1-5/6)} + 1 + \frac{2.5}{1!} + \frac{2.5^2}{2!} \right) \doteq .702.$$

Since $C(c, r)$ represents the probability of positive delay, the probability of no delay is .298.

Now suppose that the call center wishes to increase the probability of non-delayed calls to 90%. How many servers are needed? To answer this, we incrementally increase c until $1 - C(c, r) \geq .90$. It is found that $c = 6$ servers is the minimum number of servers that satisfies this requirement.

We now obtain the complete probability distributions of the waiting times, $W(t)$ and $W_q(t)$, in a manner similar to that of Section 2.2.4. [Note that $W(t)$ and $W_q(t)$ were not needed to obtain the average values W and W_q.] For $T_q > 0$ and assuming FCFS,

$$W_q(t) = \Pr\{T_q \leq t\} = W_q(0) + \sum_{n=c}^{\infty} \Pr\{n - c + 1 \text{ completions in } \leq t \mid \text{arrival found } n \text{ in system}\} \cdot p_n.$$

Now when $n \geq c$, the system output is Poisson with mean rate $c\mu$, so that the time between successive completions is exponential with mean $1/c\mu$, and the distribution of the time for the $n-c+1$ completions is Erlang type $n-c+1$. Thus we can write

$$\begin{aligned}
W_q(t) &= W_q(0) + p_0 \sum_{n=c}^{\infty} \frac{r^n}{c^{n-c}c!} \int_0^t \frac{c\mu(c\mu x)^{n-c}}{(n-c)!} e^{-c\mu x}\, dx \\
&= W_q(0) + \frac{r^c p_0}{(c-1)!} \int_0^t \mu e^{-c\mu x} \sum_{n=c}^{\infty} \frac{(\mu r x)^{n-c}}{(n-c)!}\, dx \\
&= W_q(0) + \frac{r^c p_0}{(c-1)!} \int_0^t \mu e^{-\mu x(c-r)}\, dx \\
&= W_q(0) + \frac{r^c p_0}{(c-1)!(c-r)} \int_0^t \mu(c-r) e^{-\mu(c-r)x}\, dx \\
&= W_q(0) + \frac{r^c p_0}{c!(1-\rho)} \left(1 - e^{-(c\mu-\lambda)t}\right).
\end{aligned}$$

Putting this result together with (2.37), we find that

$$\boxed{W_q(t) = 1 - \frac{r^c p_0}{c!(1-\rho)} e^{-(c\mu-\lambda)t}.} \tag{2.39}$$

From (2.39), we note that

$$\Pr\{T_q > t\} = 1 - W_q(t) = \frac{r^c p_0}{c!(1-\rho)} e^{-(c\mu-\lambda)t},$$

so that the conditional probability $\Pr\{T_q > t | T_q > 0\} = e^{-(c\mu-\lambda)t}$. As with the $M/M/1$ queue, $W_q(t)$ is the mixture of a discrete probability mass at zero and an exponential distribution.

Letting $c = 1$ reduces (2.39) to the equation for $W_q(t)$ of the $M/M/1$ model given in (2.28). Similar statements would be true for the other $M/M/c$ measures of effectiveness. We leave as an exercise (Problem 2.14) to show that

$$W_q = \mathrm{E}[T_q] = \int_0^{\infty} [1 - W_q(t)]\, dt = \left(\frac{r^c}{c!(c\mu)(1-\rho)^2}\right) p_0,$$

as given by (2.34).

To find the formula for the CDF of the system waiting time, we first split the situation into two separate possibilities, namely, those customers having no wait in queue [occurring with probability $W_q(0)$] and those customers having a positive wait in queue [occurring with probability $1 - W_q(0)$].

The time in the system for the first class of customers is just the time in service, since there is no wait in the queue. For these customers, the CDF of the time in the system is identical to the CDF of the time in service, an exponential distribution with mean $1/\mu$.

For the second class of customers, the time in the system is the sum of the wait in queue plus the time in service. Thus the CDF of the time in the system is the convolution of (a) an exponential distribution with mean $1/(c\mu - \lambda)$ and (b) an exponential distribution with mean $1/\mu$. The first distribution is the conditional distribution of T_q given that $T_q > 0$ (see earlier in this section). This convolution can also be written as the difference of the two exponential functions (see Problem 2.19),

$$\Pr\{T \le t\} = \frac{c(1-\rho)}{c(1-\rho)-1}(1-e^{-\mu t}) - \frac{1}{c(1-\rho)-1}(1-e^{-(c\mu-\lambda)t}).$$

Thus the overall CDF of the $M/M/c$ system waits may be written as

$$\begin{aligned} W(t) &= W_q(0)[1-e^{-\mu t}] + [1-W_q(0)] \\ &\quad \times \left(\frac{c(1-\rho)}{c(1-\rho)-1}(1-e^{-\mu t}) - \frac{1}{c(1-\rho)-1}(1-e^{-(c\mu-\lambda)t}) \right) \\ &= \frac{c(1-\rho)-W_q(0)}{c(1-\rho)-1}(1-e^{-\mu t}) - \frac{1-W_q(0)}{c(1-\rho)-1}(1-e^{-(c\mu-\lambda)t}). \end{aligned}$$

We now illustrate these developments with an example.

■ EXAMPLE 2.4

City Hospital's eye clinic offers free vision tests every Wednesday evening. There are three ophthalmologists on duty. A test takes, on average, 20 min, and the actual time is found to be approximately exponentially distributed around this average. Clients arrive according to a Poisson process with a mean of 6/h, and patients are taken on a first-come, first-served basis. The hospital planners are interested in knowing (1) the average number of people waiting; (2) the average amount of time a patient spends at the clinic; and (3) the average percentage idle time of each of the doctors. Thus we wish to calculate L_q, W, and the percentage idle time of a server.

We begin by calculating p_0, since this factor appears in all the formulas derived for the measures of effectiveness. We have that $c = 3$, $\lambda = 6$/h, and $\mu = 1/(20 \text{ min}) = 3$/h. Thus $r = \lambda/\mu = 2$, $\rho = \frac{2}{3}$, and, from (2.32),

$$p_0 = \left(1 + 2 + \frac{2^2}{2!} + \frac{2^3}{3!(1-\frac{2}{3})}\right)^{-1} = \frac{1}{9}.$$

From (2.33), we find that

$$L_q = \left(\frac{(2^3)(\frac{2}{3})}{3!(1-\frac{2}{3})^2}\right)\left(\frac{1}{9}\right) = \frac{8}{9},$$

and from (2.33) and (2.35) that

$$W = \frac{1}{\mu} + \frac{L_q}{\lambda} = \frac{1}{3} + \frac{\frac{8}{9}}{6} = \frac{13}{27} \text{ h} \doteq 28.9 \text{ min}.$$

Next, we have already shown (see Table 1.2) that the long-term average fraction of idle time for any server in an $M/M/c$ is equal to $1-\rho$. For this problem, therefore, each physician is idle $\frac{1}{3}$ of the time, since the traffic intensity is $\rho = \frac{2}{3}$. Given the three servers on duty, two of them will be busy at any time (on average), since $r = 2$. Furthermore, the fraction of time that there is at least one idle doctor can be computed here as $p_0 + p_1 + p_2 = \Pr\{T_q = 0\} = \frac{5}{9}$.

2.4 Choosing the Number of Servers

In managing a queueing system, it is often desirable to determine an appropriate number of servers c for the system. A larger number of servers improves quality of service to the customers but incurs a higher cost to the queue owner. The problem is to find the number of servers that adequately balances the quality and cost of service. This section gives a simple approximation that can be helpful in choosing the number of servers for an $M/M/c$ queue. The approximation works for queues with a large number of servers.

Before discussing the approximation, we first observe that in steady state the number of servers must be greater than the offered load r. Otherwise, the queue is unstable. Thus we can write

$$c = r + \Delta,$$

where $\Delta > 0$ is the number of additional servers used in excess of the offered load (Δ may need to be a fraction in order to make c an integer). Thus the problem of choosing the number of servers c is equivalent to choosing the number of servers Δ in excess of the offered load.

To motivate ideas for choosing c (or Δ), consider an $M/M/c$ queue with offered load $r = 9$, $c = 12$ servers, and traffic intensity $\rho = 0.75$. Suppose that the owner of the queue has observed the system for a long time and is satisfied with its overall performance, considering both the congestion experienced by the customers and the cost of paying the servers.

Now suppose that the offered load quadruples to $r = 36$. How many new servers should the owner hire? There are several lines of reasoning that can be taken to answer this question.

1. Choose c to maintain (approximately) a constant traffic intensity ρ. In the baseline case, there are 4 servers available for every 3 customers in service, on average ($\rho = 0.75$). It may seem reasonable to keep this ratio constant. So, if the traffic level quadruples, the number of servers should also quadruple to $c = 48$.

2. Choose c to maintain (approximately) a constant measure of congestion. An example congestion measure is $1 - W_q(0)$, the probability that a customer is delayed in the queue, which can be obtained from the Erlang-C formula (2.38). In the baseline case, this turns out to be $1 - W_q(0) = 0.266$. If the traffic quadruples to $r = 36$, then the minimum number of servers needed to maintain (or improve) this level of service is $c = 42$. More generally, if α is the maximum desired fraction of callers delayed in the queue, c is chosen as

follows:

$$\text{Find the smallest } c \text{ such that } 1 - W_q(0) \le \alpha. \tag{2.40}$$

This is the inverse problem of evaluating (2.38). The number of servers can be found by increasing c until $1 - W_q(0) \le \alpha$ or by using a binary search on c.

3. Choose c to maintain (approximately) a constant "padding" of servers. In the baseline case, $c = r + 3$. In other words, there are 3 extra servers beyond the minimum number needed to handle the offered load $r = 9$. With the increased traffic intensity $r = 36$, let $c = r + 3 = 39$.

These three approaches have sometimes been called the "quality domain," the "quality and efficiency domain" (QED), and the "efficiency domain" (Gans et al., 2003). The three approaches hold (approximately) constant ρ, $1 - W_q(0)$, and Δ, respectively. Table 2.1 summarizes the system parameters for the three approaches in this example.

Table 2.1 Example Performance Measures for Various Choices of c

	c	ρ	$1 - W_q(0)$	Δ
Baseline	12	0.75	0.266	3
1. Quality domain	48	0.75	0.037	12
2. Quality and efficiency domain	42	0.86	0.246	6
3. Efficiency domain	39	0.92	0.523	3

In the quality domain, there is an emphasis on providing a high level of service at the expense of server cost. The number of servers is held approximately proportional to the offered load. Yet, as the offered load increases, the probability of delay in the queue decreases. Specifically, as $r \to \infty$, $1 - W_q(0) \to 0$ (holding ρ constant). In other words, large queueing systems that have been scaled using the quality-domain approach have very little queueing delay.

In the efficiency domain, there is an emphasis on minimizing cost at the expense of service quality. Here, the number of excess servers is held approximately fixed. As the offered load increases, congestion in the queue increases. Specifically, holding Δ constant, as $r \to \infty$, $\rho \to 1$ and $1 - W_q(0) \to 1$. In other words, large queueing systems that have been scaled using the efficiency-domain approach are nearly always congested.

The QED provides a balance between the quality and efficiency domains. In this domain, the objective is to maintain a fixed quality of service. That is, $1 - W_q(0)$ is held approximately constant as the system grows. In contrast, in the quality domain $1 - W_q(0)$ goes to 0; in the efficiency domain $1 - W_q(0)$ goes to 1.

One difficulty with this approach is that the choice for c is not a simple formula. In particular, c is chosen according to (2.40), which requires inverting (2.38). In contrast, the formulas for c in the other domains are simple and intuitive: In the quality domain, $c = r/\rho$, and in the efficiency domain, $c = r + \Delta$.

The main result of this section is that although there is no simple formula for choosing c in the QED, there is a simple *approximate* formula. The basic idea is that the number of excess servers should increase with the *square root* of the offered load. Specifically, the solution to (2.40) when the offered load is large is approximately

$$\boxed{c \approx r + \beta\sqrt{r} \quad \text{or} \quad \Delta \approx \beta\sqrt{r},} \tag{2.41}$$

where β is a constant. Intuitively, (2.41) implies that to achieve a fixed quality of service, the extra number of servers (beyond the offered load) should be approximately proportional to the square root of the offered load. In the preceding example, r increased by a factor of 4. Thus to maintain the same quality of service, Δ should increase by a factor of 2, which is exactly what was observed in the example (Δ increased from 3 to 6).

The square-root law (2.41) is justified theoretically by the following theorem (Halfin and Whitt, 1981).

Theorem. *Consider a sequence of $M/M/c$ queues indexed by the parameter $n = 1, 2, \ldots$. Suppose that queue n has $c_n = n$ servers and offered load r_n. Then*

$$\lim_{n\to\infty} C(c_n = n, r_n) = \alpha, \qquad 0 < \alpha < 1, \tag{2.42}$$

if and only if

$$\lim_{n\to\infty} \frac{n - r_n}{\sqrt{n}} = \beta, \quad \beta > 0, \tag{2.43}$$

where $C(c, r)$ is the Erlang-C formula, and α and β are constants related via

$$\alpha = \frac{\phi(\beta)}{\phi(\beta) + \beta\Phi(\beta)}, \tag{2.44}$$

where $\phi(\cdot)$ and $\Phi(\cdot)$ are the PDF and CDF of a standard normal random variable.

Intuitively, (2.42) states that the quality of service $C(c, r) = 1 - W_q(0)$ is approximately constant across the sequence of queues. Equation (2.43) is roughly the square-root law in (2.41). That is, $n - r_n \approx \beta\sqrt{n}$ or $n \approx r_n + \beta\sqrt{n}$; the square-root law replaces $\sqrt{n}$ with $\sqrt{r_n}$. In summary, the sequence of queues has approximately the same quality of service provided that the excess number of servers grows with the square root of the offered load. The parameters α and β can be interpreted as constants representing the quality of service: α is the probability of nonzero delay in the queue $\alpha = 1 - W_q(0)$, while β is a constant related to α via the relationship in (2.44).

The square-root law can be used in a relative sense without specifying the precise values of these constants. For example, if the provider is satisfied with the current level of service (perhaps the level of service is not even measured), and if the offered load doubles, then the number of servers in excess of the offered load should be increased by roughly a factor of $\sqrt{2}$.

Alternatively, the constants can be used to approximate absolute levels of service. Specifically, to choose the number of servers c that achieves a quality of service α, first find β that satisfies (2.44) and then use (2.41) to estimate c. A similar approximation is to let β equal the $(1-\alpha)$ quantile of the standard normal distribution (Kolesar and Green, 1998).

EXAMPLE 2.5

For an $M/M/c$ queue with $\lambda = 200$ and $\mu = 1$, find the minimum number of servers c so that the probability of nonzero delay is less than .01. We first compute the exact value of c so that the probability of nonzero delay in (2.38) is less than .01. By trial and error (say, using the QtsPlus software), this is found to be $c = 235$.

Alternatively, using the approximation, the value of β that satisfies (2.44) with $\alpha = .01$ is found using numerical root-finding to be about 2.375. Then from (2.41), the approximate number of servers required is $c \approx 200 + 2.375\sqrt{200} \doteq 233.6$ which is close to the exact value. Or, if we let β be the $(1 - .01)$ quantile of the standard normal distribution, then $\beta = 2.326$, which gives $c \approx 200 + 2.326\sqrt{200} \doteq 232.9$.

Figure 2.7 compares the square-root approximation (2.41) and the exact values from (2.40). The x-axis is the offered load r to an $M/M/c$ queue. The y-axis is the minimum number of servers c so that $1 - W_q(0) \leq \alpha$. The individual points are the exact values found from the inversion problem in (2.40). The solid lines are approximate values obtained from the square-root law in (2.41). As seen, the approximation works quite well for these examples. In Section 6.6.1, we talk more about designing queueing systems and how to choose the best value for c, where service costs for both provider and customer are considered.

2.5 Queues with Truncation ($M/M/c/K$)

We now take up the parallel-server birth–death model $M/M/c/K$, in which there is a limit K placed on the number allowed in the system at any time. The approach here is identical to that of the infinite-capacity $M/M/c$ except that the arrival rate λ_n must now be 0 whenever $n \geq K$. It then follows from (2.31) that the steady-state system-size probabilities are given by

$$p_n = \begin{cases} \dfrac{\lambda^n}{n!\mu^n} p_0 & (0 \leq n < c), \\[2ex] \dfrac{\lambda^n}{c^{n-c}c!\mu^n} p_0 & (c \leq n \leq K). \end{cases} \tag{2.45}$$

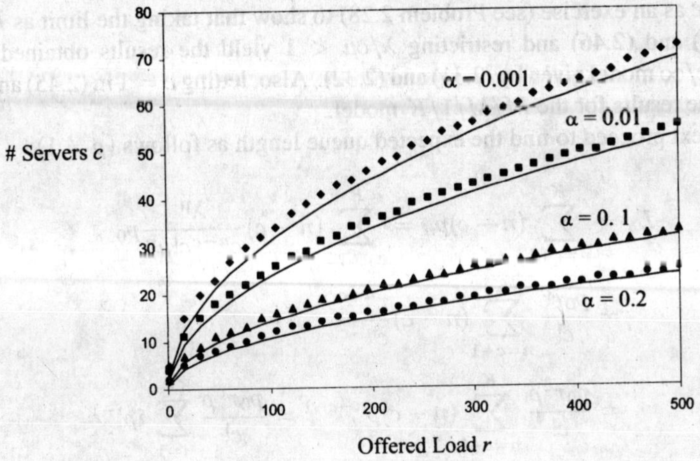

Figure 2.7 Comparison of square-root law (solid lines) with exact values (individual points).

Similar to the $M/M/c$ queue, p_n has a Poisson form for $0 \leq n < c$ and a geometric form for $c \leq n \leq K$. The usual boundary condition that the probabilities sum to 1 yields p_0. Again, the computation is nearly identical to that for the $M/M/c$, except that now both series in the computation are finite and thus there is no requirement that the traffic intensity ρ be less than 1. So

$$p_0 = \left(\sum_{n=0}^{c-1} \frac{\lambda^n}{n!\mu^n} + \sum_{n=c}^{K} \frac{\lambda^n}{c^{n-c}c!\mu^n} \right)^{-1}.$$

To simplify, consider the second summation above, with $r = \lambda/\mu$ and $\rho = r/c$:

$$\sum_{n=c}^{K} \frac{r^n}{c^{n-c}c!} = \frac{r^c}{c!} \sum_{n=c}^{K} \rho^{n-c}$$

$$= \begin{cases} \dfrac{r^c}{c!} \left(\dfrac{1-\rho^{K-c+1}}{1-\rho} \right) & (\rho \neq 1), \\ \dfrac{r^c}{c!}(K-c+1) & (\rho = 1). \end{cases}$$

Thus

$$p_0 = \begin{cases} \left[\dfrac{r^c}{c!} \left(\dfrac{1-\rho^{K-c+1}}{1-\rho} \right) + \displaystyle\sum_{n=0}^{c-1} \frac{r^n}{n!} \right]^{-1} & (\rho \neq 1), \\ \left[\dfrac{r^c}{c!}(K-c+1) + \displaystyle\sum_{n=0}^{c-1} \frac{r^n}{n!} \right]^{-1} & (\rho = 1). \end{cases} \tag{2.46}$$

We leave as an exercise (see Problem 2.28) to show that taking the limit as $K \to \infty$ in (2.45) and (2.46) and restricting $\lambda/c\mu < 1$ yield the results obtained for the $M/M/c/\infty$ model given by (2.31) and (2.32). Also, letting $c = 1$ in (2.45) and (2.46) yields the results for the $M/M/1/K$ model.

We next proceed to find the expected queue length as follows ($\rho \neq 1$):

$$\begin{aligned} L_q &= \sum_{n=c+1}^{K} (n-c)p_n = \sum_{n=c+1}^{K} (n-c)\frac{\lambda^n}{c^{n-c}c!\mu^n}p_0 \\ &= \frac{p_0 r^c}{c!}\sum_{n=c+1}^{K}(n-c)\frac{r^{n-c}}{c^{n-c}} \\ &= \frac{p_0 r^c \rho}{c!}\sum_{n=c+1}^{K}(n-c)\rho^{n-c-1} = \frac{p_0 r^c \rho}{c!}\sum_{i=1}^{K-c} i\rho^{i-1} \\ &= \frac{p_0 r^c \rho}{c!}\frac{d}{d\rho}\left(\sum_{i=0}^{K-c}\rho^i\right) = \frac{p_0 r^c \rho}{c!}\frac{d}{d\rho}\left(\frac{1-\rho^{K-c+1}}{1-\rho}\right), \end{aligned}$$

or

$$\boxed{L_q = \frac{p_0 r^c \rho}{c!(1-\rho)^2}[1-\rho^{K-c+1}-(1-\rho)(K-c+1)\rho^{K-c}].} \tag{2.47}$$

For $\rho = 1$, it is necessary to employ L'Hôpital's rule twice.

To obtain the expected system size, recall from our work with the unrestricted $M/M/c$ model that $L = L_q + r$. However, for the finite-waiting-space case, we need to adjust this result (and Little's formula as well), since a fraction p_K of the arrivals do not join the system, because they have come when there is no waiting space left. Thus the actual rate of arrivals to join the system must be adjusted accordingly. Since Poisson arrivals see time averages (the PASTA property), it follows that the effective arrival rate seen by the servers is $\lambda(1 - p_K)$. We henceforth denote any such adjusted input rate as λ_{eff}. The relationship between L and L_q must therefore be reframed for this model to be $L = L_q + \lambda_{\text{eff}}/\mu = L_q + \lambda(1-p_K)/\mu = L_q + r(1-p_K)$. We know that the quantity $r(1-p_K)$ must be less than c, since the average number of customers in service must be less than the total number of available servers. This suggests the definition of something called $\rho_{\text{eff}} = \lambda_{\text{eff}}/c\mu$, which would thus have to be less than 1 for any $M/M/c$ model even though no such restriction exists on the value of $\rho = \lambda/c\mu$.

Expected values for waiting times can readily be obtained by use of Little's formula as

$$\boxed{\begin{aligned} W &= \frac{L}{\lambda_{\text{eff}}} = \frac{L}{\lambda(1-p_K)}, \\ W_q &= W - \frac{1}{\mu} = \frac{L_q}{\lambda_{\text{eff}}}. \end{aligned}} \tag{2.48}$$

For $M/M/1/K$, all of the above measures of effectiveness reduce to considerably simpler expressions, with key results of

$$p_0 = \begin{cases} \dfrac{1-\rho}{1-\rho^{K+1}} & (\rho \neq 1), \\ \dfrac{1}{K+1} & (\rho = 1), \end{cases} \tag{2.49}$$

$$p_n = \begin{cases} \dfrac{(1-\rho)\rho^n}{1-\rho^{K+1}} & (\rho \neq 1), \\ \dfrac{1}{K+1} & (\rho = 1), \end{cases} \tag{2.50}$$

and

$$L_q = \begin{cases} \dfrac{\rho}{1-\rho} - \dfrac{\rho(K\rho^K + 1)}{1-\rho^{K+1}} & (\rho \neq 1), \\ \dfrac{K(K-1)}{2(K+1)} & (\rho = 1), \end{cases} \tag{2.51}$$

with $L = L_q + (1 - p_0)$. Note that this final relationship implies that $1 - p_0 = \lambda(1 - p_K)/\mu$, which when rewritten as $\mu(1 - p_0) = \lambda(1 - p_K)$ verifies that the system's effective output rate must equal its effective input rate.

The derivation of the waiting-time CDF is somewhat complicated, since the series are finite, although they can be expressed in terms of cumulative Poisson sums, as we shall show. Also, it is now necessary to derive the arrival-point probabilities $\{q_n\}$, since the input is no longer Poisson because of the size truncation at K, and $q_n \neq p_n$.

We use Bayes' theorem to determine the q_n, so that

$$\begin{aligned} q_n &\equiv \Pr\{n \text{ in system}|\text{arrival about to occur}\} \\ &= \frac{\Pr\{\text{arrival about to occur}|n \text{ in system}\} \cdot p_n}{\sum_{n=0}^{K} \Pr\{\text{arrival about to occur}|n \text{ in system}\} \cdot p_n} \\ &= \lim_{\Delta t \to 0} \left\{ \frac{[\lambda \Delta t + o(\Delta t)]p_n}{\sum_{n=0}^{K-1} [\lambda \Delta t + o(\Delta t)]p_n} \right\} \\ &= \lim_{\Delta t \to 0} \left\{ \frac{[\lambda + o(\Delta t)/\Delta t]p_n}{\sum_{n=0}^{K-1} [\lambda + o(\Delta t)/\Delta t]p_n} \right\} \\ &= \frac{\lambda p_n}{\lambda \sum_{n=0}^{K-1} p_n} \\ &= \frac{p_n}{1-p_K} \quad (n \leq K-1). \end{aligned}$$

We note in passing that had this same analysis been performed for $M/M/c/\infty$, then the final portion of the equation above would be equal to p_n, since p_K goes to 0 when the capacity constraint is removed (i.e., K goes to ∞). Thus it follows for $M/M/c/\infty$ that $q_n = p_n$.

Finally, to get the CDF $W_q(t)$ for the line delays, we note, in a fashion similar to the derivation leading to (2.39), that

$$W_q(t) = \Pr\{T_q \le t\} = W_q(0) + \sum_{n=c}^{K-1} \Pr\{n-c+1 \text{ completions in } \le t \mid \text{arrival found } n \text{ in system}\} \cdot q_n,$$

since there cannot be arrivals joining the system whenever they encounter K customers. It follows that

$$\begin{aligned} W_q(t) &= W_q(0) + \sum_{n=c}^{K-1} q_n \int_0^t \frac{c\mu(c\mu x)^{n-c}}{(n-c)!} e^{-c\mu x}\, dx \\ &= W_q(0) + \sum_{n=c}^{K-1} q_n \left(1 - \int_t^\infty \frac{c\mu(c\mu x)^{n-c}}{(n-c)!} e^{-c\mu x}\, dx\right). \end{aligned}$$

For the simplification of (1.15), Section 1.7, we have shown that

$$\int_t^\infty \frac{\lambda(\lambda x)^m}{m!} e^{-\lambda x}\, dx = \sum_{i=0}^{m} \frac{(\lambda t)^i e^{-\lambda t}}{i!}.$$

Letting $m = n - c$ and $\lambda = c\mu$ gives

$$\int_t^\infty \frac{c\mu(c\mu x)^{n-c}}{(n-c)!} e^{-c\mu x}\, dx = \sum_{i=0}^{n-c} \frac{(c\mu t)^i e^{-c\mu t}}{i!}$$

and hence

$$\begin{aligned} W_q(t) &= W_q(0) + \sum_{n=c}^{K-1} q_n - \sum_{n=c}^{K-1} q_n \sum_{i=0}^{n-c} \frac{(c\mu t)^i e^{-c\mu t}}{i!} \\ &= 1 - \sum_{n=c}^{K-1} q_n \sum_{i=0}^{n-c} \frac{(c\mu t)^i e^{-c\mu t}}{i!}. \end{aligned}$$

■ **EXAMPLE 2.6**

Consider an automobile emission inspection station with three inspection stalls, each with room for only one car. It is reasonable to assume that cars wait in such a way that when a stall becomes vacant, the car at the head of the line pulls up to it. The station can accommodate at most four cars waiting (seven in the station) at one time. The arrival pattern is Poisson with a mean of one car every minute during the peak periods. The service time is exponential with mean

6 min. I. M. Fussy, the chief inspector, wishes to know the average number in the system during peak periods, the average wait (including service), and the expected number per hour that cannot enter the station because of full capacity.

Using minutes as the basic time unit, $\lambda = 1$ and $\mu = \frac{1}{6}$. Thus we have $r = 6$ and $\rho = 2$ for this $M/M/3/7$ system. We first calculate p_0 from (2.46) and find that

$$p_0 = \left(\sum_{n=0}^{2} \frac{6^n}{n!} + \frac{6^3}{3!} \frac{1-2^5}{1-2} \right)^{-1}$$

$$= \frac{1}{1141} \doteq 0.00088.$$

From (2.47) we then get

$$L_q = \frac{p_0(6^3)(2)}{3!}[1 - 2^5 + 5(2^4)] = \frac{3528}{1141} \doteq 3.09 \text{ cars.}$$

Then $L = L_q + r(1 - p_K)$, so

$$L = \frac{3528}{1141} + 6\left(1 - \frac{6^7}{(3^4)(3!)(1141)}\right) = \frac{9606}{1141} \doteq 6.06 \text{ cars.}$$

To find the average wait during peak periods, (2.48) gives that

$$W = \frac{L}{\lambda_{\text{eff}}} = \frac{L}{\lambda(1 - p_7)} = \frac{L}{1 - p_0 6^7/(3^4 3!)} \doteq 12.3 \text{ min}.$$

The expected number of cars per hour that cannot enter the station is given by

$$60\lambda p_k = 60 p_7 = \frac{60 p_0 6^7}{3^4 3!} \doteq 30.4 \text{ cars/h}.$$

This might suggest an alternative setup for the inspection station.

2.6 Erlang's Loss Formula ($M/M/c/c$)

The special case of the truncated queue $M/M/c/K$ with $K = c$, that is, where no line is allowed to form, gives rise to a stationary distribution known as Erlang's first formula. This stationary distribution can be obtained from (2.45) and (2.46) with $K = c$ as

$$p_n = \frac{(\lambda/\mu)^n}{n!} \Bigg/ \left(\sum_{i=0}^{c} \frac{(\lambda/\mu)^i}{i!} \right), \qquad (0 \le n \le c). \tag{2.52}$$

When $n = c$, the resultant formula for p_c is called *Erlang's loss formula* or the *Erlang-B formula*. This is the probability of a full system at any time in steady state.

Since the input to the $M/M/c/c$ queue is Poisson, p_c is also the fraction of arriving customers who find the system full and are lost from the system.

$$B(c,r) \equiv p_c = \frac{r^c/c!}{\sum_{i=0}^{c} r^i/i!}, \qquad (r = \lambda/\mu). \tag{2.53}$$

Here, we use the notation $B(c,r)$ for the Erlang loss formula to emphasize the dependence on c and r.

The original physical situation that motivated Erlang (1917) to devise this model was the simple telephone network. Incoming calls arrive as a Poisson stream, service times are mutually independent exponential random variables, and all calls that arrive finding every trunk line busy (i.e., getting a busy signal) are turned away. The model has always been of great value in telecommunications design.

But the great importance of this formula lies in the very surprising fact that (2.52) is valid for *any M/G/c/c, independent* of the form of the service-time distribution. That is, the steady-state system probabilities are only a function of the mean service time, not of the underlying CDF. Erlang was also able to deduce the formula for the case when service times are constant. While this result was later shown to be correct, his proof was not quite valid. Later works by Vaulot (1927), Pollaczek (1932), Palm (1938), Kosten (1948), and others smoothed out Erlang's 1917 proof and supplied proofs for the general service-time distribution. A further addition to this sequence of papers was work by Takács (1969), which supplied some additional results for the problem. We shall prove the validity of Erlang's loss formula for general service in Chapter 5, Section 5.2.2.

2.6.1 Computational Issues

The Erlang-B formula can cause numerical problems on a computer if implemented directly from (2.53). In particular, when the number of servers c is large, terms like $c!$ can exceed computer limits. For example, 171! exceeds the largest double-precision number, which is about $1.8 \cdot 10^{308}$. Applications like call centers often require values of c greater than 171. This section gives alternate formulas that are more practical to implement on a computer. In addition, the alternate formulas for computing $B(c,r)$ can be used to compute measures of congestion for the $M/M/c$ queue, such as the Erlang-C formula (2.38), L_q, L, W_q, and W (2.33)–(2.36).

For the Erlang-B formula, it can be shown (Problem 2.37) that $B(c,r)$ satisfies the following iterative relationship:

$$B(c,r) = \frac{rB(c-1,r)}{c+rB(c-1,r)}, \qquad c \geq 1, \tag{2.54}$$

with initial condition $B(c=0,r)=1$. Thus, to calculate $B(c,r)$ for a given value of c, one starts with $B(0,r)=1$ and then iteratively applies (2.54) until the desired

c is reached. This method avoids numerical overflow that may be encountered with direct application of (2.53).

Although the Erlang-B formula applies to the $M/M/c/c$ (or $M/G/c/c$) queue, it can also be used in the computation of congestion measures for the $M/M/c$ queue. For example, let $C(c,r) = 1 - W_q(0)$ be the Erlang-C probability of delay in an $M/M/c$ queue (2.38). Then, $C(c,r)$ can be written as a function of $B(c,r)$ as follows (Problem 2.38):

$$C(c,r) = \frac{cB(c,r)}{c - r + rB(c,r)}. \tag{2.55}$$

Thus, to calculate $C(c,r)$, one can first calculate $B(c,r)$ iteratively using (2.54) and then apply (2.55) to get $C(c,r)$. Furthermore, $C(c,r)$ can be used in the computation of L_q, L, W_q, and W for the $M/M/c$ queue. For example, (2.33) can be rewritten:

$$L_q = C(c,r)\frac{\rho}{1-\rho} = C(c,r)\frac{r}{c-r}. \tag{2.56}$$

Similarly, W_q, W, and L (2.34)–(2.36) can all be expressed in terms of $C(c,r)$.

■ **EXAMPLE 2.7**

With $\lambda = 6$, $\mu = 3$, and $c = 4$, calculate the fraction of customers blocked for an $M/M/c/c$ queue; calculate $1 - W_q(0)$ and L_q for an $M/M/c$ queue. The offered load is $r = 6/3 = 2$ and the traffic intensity is $\rho = 1/2$. The fraction of customers blocked for an $M/M/c/c$ queue is $B(4,2)$. This can be calculated iteratively using (2.54):

$$\begin{aligned}
B(0,2) &= 1\\
B(1,2) &= 2 \cdot B(0,2)/(1 + 2 \cdot B(0,2)) = 2/3,\\
B(2,2) &= 2 \cdot (2/3)/(2 + 2 \cdot (2/3)) = 2/5,\\
B(3,2) &= 2 \cdot (2/5)/(3 + 2 \cdot (2/5)) = 4/19,\\
B(4,2) &= 2 \cdot (4/19)/(4 + 2 \cdot (4/19)) = 2/21.
\end{aligned}$$

Alternatively, $B(4,2)$ can be calculated from (2.53):

$$\begin{aligned}
B(4,2) &= \frac{2^4/4!}{1 + 2 + 2^2/2! + 2^3/3! + 2^4/4!}\\
&= \frac{16/24}{(24 + 48 + 48 + 32 + 16)/24} = \frac{2}{21}.
\end{aligned}$$

The probability of delay $1 - W_q(0)$ for an $M/M/c$ queue is $C(4,2)$, which can be calculated using (2.55):

$$C(4,2) = 4 \cdot (2/21)/(4 - 2 + 2 \cdot (2/21)) = 4/23.$$

The average number in queue L_q can be calculated using (2.56):

$$L_q = C(4,2) \cdot 2/(4-2) = 4/23.$$

Alternatively, $1 - W_q(0)$ and L_q can be calculated using (2.32), (2.33), and (2.38).

$$\frac{1}{p_0} = 1 + 2 + \frac{2^2}{2!} + \frac{2^3}{3!} + \frac{2^4}{4!(1-0.5)} = \frac{92}{12} = \frac{23}{3},$$

$$1 - W_q(0) = \frac{2^4}{4! \cdot 0.5} \cdot \frac{3}{23} = \frac{4}{23}, \quad L_q = \frac{2^4 \cdot 0.5}{4!(0.5)^2} \cdot \frac{3}{23} = \frac{16}{12} \cdot \frac{3}{23} = \frac{4}{23}.$$

2.7 Queues with Unlimited Service ($M/M/\infty$)

We now treat a queueing model for which there is unlimited service, that is, an infinite number of servers available. This model is often referred to as the ample-server problem. A self-service situation is a good example of the use of such a model.

We make use of the general birth–death results with $\lambda_n = \lambda$ and $\mu_n = n\mu$, for all n, which yields

$$p_n = \frac{r^n}{n!} p_0, \qquad p_0 = \left(\sum_{n=0}^{\infty} \frac{r^n}{n!} \right)^{-1}.$$

The infinite series in the expression for p_0 is equal to e^r. Therefore

$$\boxed{p_n = \frac{r^n e^{-r}}{n!} \quad (n \geq 0),} \tag{2.57}$$

which is a Poisson distribution with mean $r = \lambda/\mu$. The value of λ/μ is not restricted in any way for the existence of a steady-state solution. It also turns out (we show this in Section 5.2.3) that (2.57) is valid for *any* $M/G/\infty$ model. That is, p_n depends only on the mean service time and not on the form of the service-time distribution. It is not surprising that this is true here in light of a similar result we mentioned previously for $M/M/c/c$, since p_n of (2.57) could have been obtained from (2.52) by taking the limit as $c \to \infty$.

The expected system size is the mean of the Poisson distribution of (2.57) and is thus found as $L = r = \lambda/\mu$. Since we have as many servers as customers in the system, $L_q = 0 = W_q$. The average waiting time in the system is merely the average service time, so that $W = 1/\mu$, and the waiting-time distribution function $W(t)$ is identical to the service-time distribution, namely, exponential with mean $1/\mu$.

■ **EXAMPLE 2.8**

Television station KCAD in a large western metropolitan area wishes to know the average number of viewers it can expect on a Saturday evening prime-time program. It has found from past surveys that people turning on their television sets on Saturday evening during prime time can be described rather well by a Poisson distribution with a mean of 100,000/h. There are five major TV stations in the area, and it is believed that a given person chooses among these essentially at random. Surveys have also shown that the average person tunes in for 90 min and that viewing times are approximately exponentially distributed.

Since the mean arrival rate (people tuning in KCAD during prime time on Saturday evening) is $100,000/5 = 20{,}000$/h and the mean service time is 90 min or 1.5 h, it follows that the average number of viewers during prime time is $L = 20{,}000/\frac{2}{3} = 30{,}000$ people.

2.8 Finite-Source Queues

In previous models, we have assumed that the population from which arrivals come (the calling population) is infinite, since the number of arrivals in any time interval is a Poisson random variable with a denumerably infinite sample space. We now treat a problem where the calling population is finite of size M, and future event occurrence probabilities are functions of the system state. A typical application of this model is that of machine repair, where the calling population is the machines, an arrival corresponds to a machine breakdown, and the repair technicians are the servers. We assume c servers are available, that the service times are identical exponential random variables with mean $1/\mu$, and that the arrival process is described as follows. If a calling unit is not in the system at time t, the probability it will have entered by time $t + \Delta t$ is $\lambda \Delta t + o(\Delta t)$; that is, the time a calling unit spends outside the system is exponential with mean $1/\lambda$.

Because of these assumptions, we can use the birth–death theory developed previously. Figure 2.8 shows the rate transition diagram for this queue, where the state n denotes the number of customers in the system. (For the machine repair problem, "number in system" corresponds to the number of broken machines.) The birth and death rates are

$$\lambda_n = \begin{cases} (M-n)\lambda & (0 \le n < M), \\ 0 & (n \ge M) \end{cases}$$

and

$$\mu_n = \begin{cases} n\mu & (0 \le n < c), \\ c\mu & (n \ge c). \end{cases}$$

Using (2.3) yields (with, as usual, $r = \lambda/\mu$)

$$p_n = \begin{cases} \dfrac{M!/(M-n)!}{n!} r^n p_0 & (1 \le n < c), \\[2ex] \dfrac{M!/(M-n)!}{c^{n-c}\, c!} r^n p_0 & (c \le n \le M), \end{cases}$$

or equivalently,

$$p_n = \begin{cases} \dbinom{M}{n} r^n p_0 & (1 \le n < c), \\[2ex] \dbinom{M}{n} \dfrac{n!}{c^{n-c}\, c!} r^n p_0 & (c \le n \le M). \end{cases} \tag{2.58}$$

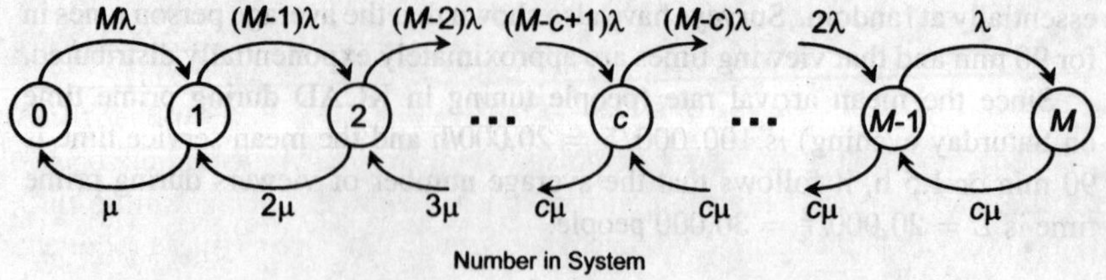

Figure 2.8 Rate transition diagram for a finite-source queue.

The algebraic form of the $\{p_n\}$ does not allow an easy closed-form calculation of p_0. Instead, we first calculate each of the coefficients multiplying p_0. That is, let a_n be the coefficient in front of p_0 (so $p_n = a_n p_0$). Then,

$$p_0 = \frac{1}{1 + a_1 + a_2 + a_3 + \cdots + a_M}.$$

To find the average number of customers in the system (if we are dealing with the machine breakdown problem, we are interested in machines "down" for repair), we use the definition of expected value and get

$$L = \sum_{n=1}^{M} np_n = p_0 \sum_{n=1}^{M} na_n.$$

To obtain L_q and the expected waiting-time measures, W and W_q, we first find the effective mean rate of arrivals into the system. As noted earlier, the mean arrival rate when the system is in state n is $(M - n)\lambda$. The *overall* arrival rate is the sum of the state-dependent rates weighted by p_n. That is,

$$\lambda_{\text{eff}} = \sum_{n=0}^{M} (M - n)\lambda p_n = \lambda(M - L). \tag{2.59}$$

Equation (2.59) is certainly intuitive, since, on average, L are in the system and hence, on average, $M - L$ are outside and each has a mean arrival rate of λ. For L_q, we know from Little's formula that

$$\boxed{L_q = L - \frac{\lambda_{\text{eff}}}{\mu} = L - r(M - L).} \tag{2.60}$$

Alternatively, $L_q = \sum_{n=c+1}^{M} (n - c)p_n$ gives the same result. It follows from Little's formula that

$$\boxed{W = \frac{L}{\lambda(M - L)} \quad \text{and} \quad W_q = \frac{L_q}{\lambda(M - L)}.} \tag{2.61}$$

For the single-server version of this problem, the expression for the system-state probabilities found in (2.58) reduces to

$$p_n = \binom{M}{n} n!\, r^n p_0 \qquad (0 \le n \le M),$$

and the rest of the analysis is identical.

There is an important *invariance* result for finite-source queues, similar in importance to the fact that $M/G/c/c$ steady-state probabilities are independent of the form of G. It is that (2.58) is valid for any finite-source system with exponential service, independent of the nature of the distribution of time to breakdown, as long as the lifetimes are independent with mean $1/\lambda$. The interested reader is referred to Bunday and Scraton (1980) for details of the proof. Furthermore, the $M/G/c/c$-type result also holds in that if the number of repair technicians equal the number of machines, the repair distribution can be G, as long as the failure times are exponential.

EXAMPLE 2.9

The Train SemiConductor Company uses five robots in the manufacture of its circuit boards. The robots break down periodically, and the company has two repair people to do service when robots fail. When one is fixed, the time until the next breakdown is thought to be exponentially distributed with a mean of 30 h. The shop always has enough of a work backlog to ensure that all robots in operating condition will be working. The repair time for each service is thought to be exponentially distributed with a mean of 3 h. The shop manager wishes to know the average number of robots operational at any given time, the expected downtime of a robot that requires repair, and the expected percentage of idle time of each repairer.

To answer any of these questions, we must first calculate p_0. In this example, $M = 5, c = 2, \lambda = \frac{1}{30}$, and $\mu = \frac{1}{3}$, and thus $r = \lambda/\mu = \frac{1}{10}$. We use (2.58) and obtain the five $\{a_n\}$ multipliers as $a_1 = 5/10 = \frac{1}{2}$; $a_2 = \frac{1}{10}$; $a_3 = 15/1000 = \frac{3}{200}$; $a_4 = 15/10,000 = \frac{3}{2000}$; and $a_5 = 15/200,000 = \frac{3}{40000}$. It thus follows that

$$p_0 = \left(1 + \tfrac{1}{2} + \tfrac{1}{10} + \tfrac{3}{200} + \tfrac{3}{2000} + \tfrac{3}{40000}\right)^{-1} = \tfrac{40000}{64663} \doteq 0.619.$$

The average number of operational robots is $M - L$, where

$$L = p_0\left(1 \cdot \tfrac{1}{2} + 2 \cdot \tfrac{1}{10} + 3 \cdot \tfrac{3}{200} + 4 \cdot \tfrac{3}{2000} + 5 \cdot \tfrac{3}{40000}\right) = \tfrac{30055}{64663} \doteq 0.465.$$

Thus 5 − 0.465, or 4.535, robots are in operating condition on average. The expected downtime can be found from (2.61) and is

$$W = \frac{\frac{30055}{64663}}{\frac{1}{30}\left(5 - \frac{30055}{64663}\right)} = \frac{901650}{29326} \doteq 3.075 \text{ h}.$$

The average fraction of idle time of each server is

$$p_0 + \tfrac{1}{2}p_1 = p_0\left(1 + \tfrac{1}{2}a_1\right) = p_0\left(1 + \tfrac{1}{4}\right) = \tfrac{50000}{64663} \doteq 0.773,$$

so each repair person is idle approximately 77% of the time.

The manager, because of the long idle time, is interested in knowing the answer to the same questions if the repair force is reduced to one person. The results are

$$p_0 \doteq 0.564, \quad L \doteq 0.640,$$
$$M - L \doteq 4.360, \quad W \doteq 4.400 \text{ h}.$$

Since over four robots are expected operational at any time under both situations and the increase in downtime is about one hour with only one repair person, the manager might well decide to move one of them to work elsewhere.

2.8.0.1 Models with Spares. The finite-source model can be generalized to include the use of spares. We assume now that there are M machines in operation plus an additional Y spares. When a machine in operation fails, a spare is immediately substituted for it (if available). If no spare is available when the failure occurs, then the system becomes short. Once a machine is repaired, it becomes a spare, unless the system is short, in which case the repaired machine goes immediately into service. At any given time, there are at most M machines in operation, so the rate of failures is at most $M\lambda$ (i.e., spares that are not in operation do not contribute to the failure rate). For this model, λ_n is different from before and is given by

$$\lambda_n = \begin{cases} M\lambda & (0 \le n < Y), \\ (M - n + Y)\lambda & (Y \le n < Y + M), \\ 0 & (n \ge Y + M), \end{cases}$$

where n represents the number of failed machines. Again considering c technicians, we have

$$\mu_n = \begin{cases} n\mu & (0 \le n < c), \\ c\mu & (n \ge c). \end{cases}$$

We first assume $c \le Y$ and use (2.3) to find that (with $r = \lambda/\mu$)

$$p_n = \begin{cases} \dfrac{M^n}{n!} r^n p_0 & (0 \le n < c), \\[2ex] \dfrac{M^n}{c^{n-c}c!} r^n p_0 & (c \le n < Y), \\[2ex] \dfrac{M^Y M!}{(M - n + Y)!c^{n-c}c!} r^n p_0 & (Y \le n \le Y + M). \end{cases} \tag{2.62}$$

If Y is very large, we essentially have an infinite calling population with mean arrival rate $M\lambda$. Letting Y go to infinity in (2.62) yields the $M/M/c/\infty$ results of (2.31) with $M\lambda$ for λ.

When $c > Y$, we have

$$p_n = \begin{cases} \dfrac{M^n}{n!} r^n p_0 & (0 \le n \le Y), \\ \dfrac{M^Y M!}{(M-n+Y)!n!} r^n p_0 & (Y+1 \le n < c), \\ \dfrac{M^Y M!}{(M-n+Y)!c^{n-c}c!} r^n p_0 & (c \le n \le Y+M). \end{cases} \tag{2.63}$$

When $Y = 0$ (i.e., no spares), then (2.63) reduces to (2.58).

Observe that the direct use of (2.63) to get the coefficients $\{a_n\}$ might get particularly messy when M and Y are large. Fortunately, we can avoid such difficulty by using the recursion relating p_{n+1} to p_n, which comes quite naturally out of the birth–death formulation and is a direct consequence of local balance. The balance equation may be written in *recursive* form as

$$p_{n+1} = \left(\frac{\lambda_n}{\mu_{n+1}}\right) p_n,$$

so that it follows for the no-spares problem that

$$p_{n+1} = \begin{cases} \dfrac{M-n}{n+1} r p_n & (0 \le n \le c-1), \\ \dfrac{M-n}{c} r p_n & (c \le n \le M-1). \end{cases}$$

Similar recursions can be developed when $Y > 0$, and, indeed, all of the more complicated birth–death modules in the text's software use this sort of recursion to do their computations.

The empty-system probability, p_0, can be found as previously for the no-spares model by once more using the fact that the probabilities must sum to 1, so that the computation of p_0 is made up of finite sums. The same is true for L and L_q. To obtain results for W and W_q comparable to (2.61), we must again obtain the *effective* mean arrival rate λ_{eff}. To get λ_{eff}, one can use (2.60) directly or obtain it using logic similar to that used for (2.59), namely,

$$\begin{aligned} \lambda_{\text{eff}} &= \sum_{n=0}^{Y-1} M\lambda p_n + \sum_{n=Y}^{Y+M} (M-n+Y)\lambda p_n \\ &= \lambda\left(M - \sum_{n=Y}^{Y+M} (n-Y)p_n\right). \end{aligned} \tag{2.64}$$

The calculations for the spares model conclude in a similar fashion to those of the no-spares case.

To close this section, we derive the complete distribution for the waiting time. The standard procedure in the past has been to find the waiting time of an arriving customer conditioned on the existence of n in the system at the point of arrival, and then unconditioning with respect to the stationary distribution $\{q_n\}$, where $\{q_n\}$ are the state probabilities given an arrival occurs. Here, we have $q_n \neq p_n$, and therefore, before we can obtain the CDF of the system or line wait, we have to relate the general-time probability p_n to the probability q_n that an *arrival* finds n in the system. For the general finite-source queue (machine-repair problem without spares), the two probabilities are related as

$$q_n = \frac{(M-n)p_n}{k},$$

where k is an appropriate normalizing constant determined from summing the $\{q_n\}$ to 1. To prove this, we again use Bayes' theorem as in the $M/M/c/K$ situation of Section 2.5.

$$\begin{aligned}
q_n &= \Pr\{n \text{ in system} \mid \text{arrival is about to occur}\} \\
&= \frac{\Pr\{n \text{ in system}\}\Pr\{\text{arrival is about to occur} \mid n \text{ in system}\}}{\Pr\{\text{arrival is about to occur}\}} \\
&= \frac{\Pr\{n \text{ in system}\}\Pr\{\text{arrival is about to occur} \mid n \text{ in system}\}}{\sum_n(\Pr\{n \text{ in system}\}\Pr\{\text{arrival is about to occur} \mid n \text{ in system}\})} \\
&= \lim_{\Delta t \to 0} \frac{p_n[(M-n)\lambda\Delta t + o(\Delta t)]}{\sum_n p_n[(M-n)\lambda\Delta t + o(\Delta t)]} \\
&= \frac{(M-n)p_n}{\sum_n (M-n)p_n} = \frac{(M-n)p_n}{M-L}.
\end{aligned}$$

As an interesting sidelight it can be shown (see Problem 2.44) that $q_n(M)$, the arrival-point probability for the no-spares case with M machines, equals $p_n(M-1)$, the general-time probability for the no-spares machine-repair situation with $M-1$ machines. This is not necessarily the case when there are spares. In fact, when spares are present, $q_n(M)$ can be shown to be (see Problem 2.45)

$$q_n = \begin{cases} \dfrac{Mp_n}{M - \sum_{i=Y}^{Y+M}(i-Y)p_i} & (0 \le n \le Y-1), \\[2ex] \dfrac{(M-n+Y)p_n}{M - \sum_{i=Y}^{Y+M}(i-Y)p_i} & (Y \le n \le Y+M-1). \end{cases} \tag{2.65}$$

This is not equal to $p_n(M-1)$, but rather is equal to $p_n(Y-1)$; that is, if we reduce the population size by one by reducing spares, not operating machines, then the general-time probabilities of the reduced population equal the arrival-point probabilities of the original population.

The waiting-time distributions again turn out to be in terms of cumulative Poisson sums, as in the $M/M/c/K$ model. The analysis proceeds as follows:

$$\begin{aligned}
W_q(t) &= \Pr\{T_q \le t\} = W_q(0) \\
&\quad + \sum_{n=c}^{Y+M-1} [\Pr\{n-c+1 \text{ completions in } \le t \mid \text{arrival found } n \text{ in system}\} \cdot q_n] \\
&= W_q(0) + \sum_{n=c}^{Y+M-1} q_n \int_0^t \frac{c\mu(c\mu x)^{n-c}}{(n-c)!} e^{-c\mu x}\,dx \\
&= W_q(0) + \sum_{n=c}^{Y+M-1} q_n \left[1 - \int_t^\infty \frac{c\mu(c\mu x)^{n-c}}{(n-c)!} e^{-c\mu x}\,dx\right] \\
&= 1 - \sum_{n=c}^{Y+M-1} q_n \sum_{i=0}^{n-c} \frac{(c\mu t)^i}{i!} e^{-c\mu t}.
\end{aligned}$$

2.9 State-Dependent Service

In this section we treat Markovian queues with state-dependent service; that is, the mean service rate depends on the state of the system (number in the system). In many real situations, the server (or servers) may speed up when seeing a long line forming. On the other hand, it may happen if the server is inexperienced that he/she/it becomes flustered and the mean service rate actually decreases as the system becomes more congested. It is these types of situations that are now considered.

The first model we consider is one in which a single server has two mean rates, say, *slow* and *fast*. Work is performed at the slow rate until there are k in the system, at which point there is a switch to the fast rate (e.g., the service mechanism might be a machine with two speeds). We still assume the service times are Markovian, but the mean rate μ_n now explicitly depends on the system state n. Furthermore, no limit on the number in the system is imposed. Thus μ_n is given as

$$\mu_n = \begin{cases} \mu_1 & (1 \le n < k), \\ \mu & (n \ge k). \end{cases} \tag{2.66}$$

Assuming the arrival process is Poisson with parameter λ and utilizing (2.3), we have

$$p_n = \begin{cases} \rho_1^n p_0 & (0 \le n < k), \\ \rho_1^{k-1} \rho^{n-k+1} p_0 & (n \ge k), \end{cases} \tag{2.67}$$

where $\rho_1 = \lambda/\mu_1$ and $\rho = \lambda/\mu < 1$. Because the probabilities must sum to 1, it follows that

$$p_0 = \left(\sum_{n=0}^{k-1} \rho_1^n + \sum_{n=k}^{\infty} \rho_1^{k-1} \rho^{n-k+1}\right)^{-1},$$

so that

$$p_0 = \begin{cases} \left(\dfrac{1-\rho_1^k}{1-\rho_1} + \dfrac{\rho\rho_1^{k-1}}{1-\rho}\right)^{-1} & (\rho_1 \neq 1, \rho < 1), \\ \left(k + \dfrac{\rho}{1-\rho}\right)^{-1} & (\rho_1 = 1, \rho < 1). \end{cases} \tag{2.68}$$

If $\mu_1 = \mu$, then (2.67) and (2.68) reduce to the equations of the $M/M/1$ queue. To find the expected system size, we proceed as in Section 2.2.4 (assuming $\rho_1 \neq 1$):

$$\begin{aligned} L = \sum_{n=0}^{\infty} np_n &= p_0 \left(\sum_{n=0}^{k-1} n\rho_1^n + \sum_{n=k}^{\infty} n\rho_1^{k-1}\rho^{n-k+1}\right) \\ &= p_0 \left[\rho_1 \sum_{n=0}^{k-1} n\rho_1^{n-1} + \rho_1 \left(\frac{\rho_1}{\rho}\right)^{k-2} \sum_{n=k}^{\infty} n\rho^{n-1}\right] \\ &= p_0 \left[\rho_1 \frac{d}{d\rho_1} \sum_{n=0}^{k-1} \rho_1^n + \rho_1 \left(\frac{\rho_1}{\rho}\right)^{k-2} \frac{d}{d\rho} \sum_{n=k}^{\infty} \rho^n\right] \\ &= p_0 \left[\rho_1 \frac{d}{d\rho_1} \left(\frac{1-\rho_1^k}{1-\rho_1}\right) + \rho_1 \left(\frac{\rho_1}{\rho}\right)^{k-2} \frac{d}{d\rho} \left(\frac{1}{1-\rho} - \frac{1-\rho^k}{1-\rho}\right)\right]. \end{aligned}$$

So, finally,

$$L = p_0 \left(\frac{\rho_1[1 + (k-1)\rho_1^k - k\rho_1^{k-1}]}{(1-\rho_1)^2} + \frac{\rho\rho_1^{k-1}[k - (k-1)\rho]}{(1-\rho)^2}\right). \tag{2.69}$$

We can find L_q using the last two relationships of Table 1.2 as

$$L_q = L - (1 - p_0),$$

and W and W_q from Little's formulas as

$$W = L/\lambda \quad \text{and} \quad W_q = L_q/\lambda.$$

Note that the relation $W = W_q + 1/\mu$ cannot be used here, since μ is not constant but depends on the system-state switch point k. However, by combining the above equations, we see that

$$W = W_q + \frac{1 - p_0}{\lambda},$$

which implies that the expected service time is $(1 - p_0)/\lambda$.

■ **EXAMPLE 2.10**

Sonia Schine and John B. Goode have invented and applied for a patent on a machine that polishes automobiles. They have formed a partnership called the

Goode–Schine Garage and have rented an old building in which they have set up their machine. Since this is a part-time job for both partners, the garage is open on Saturdays only. Customers are taken on a first-come, first served basis, and since their garage is in a low-density population and traffic area, there is virtually no limit on the number of customers who can wait. The car-polishing machine can run at two speeds. At the low speed, it takes 40 min, on average, to polish a car. On the high speed, it takes only 20 min on average. Once a switch is made, the actual times can be assumed to follow an exponential distribution.

It is estimated that customers will arrive according to a Poisson process with a mean interarrival time of 30 min. Ms. Schine has had a course in queueing theory and decides to calculate the effect of two policies: switching to high speed if there are any customers waiting (i.e., two or more in the system) versus switching to high speed only when more than one customer is waiting (three or more in the system). The machine speeds can be switched at any time, even while the machine is in operation. It is desired to know the average waiting time under the two policies.

It is therefore necessary to calculate W for the case when $k = 2$ and then for $k = 3$. We must first calculate p_0 from (2.68), then L from (2.69), and finally W from Little's formula. Before doing these computations, we first calculate ρ_1 and ρ to be

$$\rho_1 = \frac{\lambda}{\mu_1} = \frac{\frac{1}{30}}{\frac{1}{40}} = \frac{4}{3} \quad \text{and} \quad \rho = \frac{\lambda}{\mu} = \frac{\frac{1}{30}}{\frac{1}{20}} = \frac{2}{3}.$$

For case 1, $k = 2$ and

$$p_0 = \tfrac{1}{5} = 0.2, \quad L = \tfrac{12}{5} = 2.4 \text{ cars},$$
$$W = L/\lambda = 2.4/\tfrac{1}{30} = 72 \text{ min} = 1 \text{ h } 12 \text{ min}.$$

For case 2, $k = 3$ and

$$p_0 = \tfrac{3}{23} \doteq 0.13, \quad L = 204/69 \doteq 2.96 \text{ cars},$$
$$W = L/\tfrac{1}{30} \doteq 89 \text{ min} = 1 \text{ h } 29 \text{ min}.$$

Schine feels that the average wait of 17 more minutes for switching speeds at three rather than two might not have an adverse effect on their clientele. However, it costs more to run the machine at the higher speed. In fact, it is estimated that it costs \$15 per operating hour to operate the machine at low speed and \$24 per operating hour to operate at high speed. Thus the expected cost of operation when switching at k is given by

$$\begin{aligned} C(k) &= 15 \sum_{n=1}^{k-1} p_n + 24 \sum_{n=k}^{\infty} p_n \\ &= 15 \sum_{n=1}^{k-1} \rho_1^n p_0 + 24 \left(1 - \sum_{n=0}^{k-1} \rho_1^n p_0\right) \end{aligned}$$

For case 1 we have

$$C(2) = 15(\tfrac{4}{3})(\tfrac{1}{5}) + 24[1 - \tfrac{1}{5} - (\tfrac{4}{3})(\tfrac{1}{5})] \doteq \$16.80/\text{h},$$

while for case 2 the average operating cost per hour is

$$C(3) = 15(\tfrac{3}{23})[\tfrac{4}{3} + \tfrac{16}{9}] + 24[1 - \tfrac{3}{23} - (\tfrac{4}{3})(\tfrac{3}{23}) - (\tfrac{16}{9})(\tfrac{3}{23})] \doteq \$17.22/\text{h}.$$

Thus it is cheaper to switch at $k = 2$ even though the hourly cost per operating hour is higher, since switching at $k = 2$ yields a higher idle-time probability p_0, which more than makes up for the higher high-speed operating cost. In addition, this provides better customer service in that W is reduced by 17 min. If, however, the high-speed operating cost were even higher, it might turn out that switching at $k = 3$ could be more economical (see Problem 2.50). It should be noted that the costs of \$16.80/h and \$17.22/h, respectively, are not true costs per operating hour, but rather average costs per hour including hours or portions of hours when the machine is idle. It should further be noted that to obtain a true optimal operating policy, the cost of customer wait should also be included. This topic will be taken up in greater detail in Chapter 6, Section 6.5.1.

The model above can be generalized to a c-server system with a rate switch at $k > c$ (see Problem 2.52). One can generalize the model by having two rate switches (or even more if desired) at, say, k_1 and k_2 (see Problem 2.53). It is also possible to derive results for a multiserver, multirate, state-dependent model of this type.

Yet another possible type of model with state-dependent service is one where the mean service rate changes whenever the system size changes. We again assume a single-server, Markovian state-dependent-service model, possibly with μ_n given by

$$\mu_n = n^\alpha \mu.$$

From the general steady-state birth–death solution of (2.3), we have

$$p_n = \frac{\lambda^n}{n^\alpha (n-1)^\alpha (n-2)^\alpha \cdots (1)^\alpha \mu^n} p_0$$
$$= \frac{r^n}{(n!)^\alpha} p_0 \qquad (r = \lambda/\mu),$$

where

$$p_0 = \left(\sum_{n=0}^{\infty} \frac{r^n}{(n!)^\alpha} \right)^{-1}.$$

The infinite series for p_0 converges for any r as long as $\alpha > 0$, but it is not obtainable in closed form unless $\alpha = 1$. (In fact, when $\alpha = 1, \sum r^n/n! = e^r$, which reduces the model to the ample-server case presented in Section 2.7.) Thus to evaluate p_0 (and thus all the other mean measures of effectiveness) in the general case, numerical methods must be used. See Gross and Harris (1985) for a discussion of possible procedures for numerically calculating p_0.

2.10 Queues with Impatience

The intent of this section is to discuss the effects of customer impatience upon the development of waiting lines of the $M/M/c$ type. These concepts may easily be extended to other Markovian models in a reasonably straightforward fashion and will not be explicitly pursued. However, some examples of impatience are discussed later for the $M/G/1$ queue.

Customers are said to be impatient if they tend to join the queue only when a short wait is expected and tend to remain in line if the wait has been sufficiently small. The impatience that results from an excessive wait is just as important in the total queueing process as the arrivals and departures. When this impatience becomes sufficiently strong and customers leave before being served, the manager of the enterprise involved must take action to reduce the congestion to levels that customers can tolerate. The models subsequently developed find practical application in this attempt of management to provide adequate service for its customers with tolerable waiting.

Impatience generally takes three forms. The first is balking, the reluctance of a customer to join a queue upon arrival; the second reneging, the reluctance to remain in line after joining and waiting; and the third jockeying between lines when each of a number of parallel lines has its own queue.

2.10.1 $M/M/1$ Balking

In real practice, it often happens that arrivals become discouraged when the queue is long and do not wish to wait. One such model is the $M/M/c/K$; that is, if people see K ahead of them in the system, they do not join. Generally, unless K is the result of a physical restriction such as no more places to park or room to wait, people will not act quite like that voluntarily. Rarely do all customers have exactly the same discouragement limit all the time.

Another approach to balking is to employ a series of monotonically decreasing functions of the system size multiplying the average rate λ. Let b_n be this function, so that $\lambda_n = b_n\lambda$ and

$$0 \le b_{n+1} \le b_n \le 1 \quad (n > 0, \quad b_0 \equiv 1).$$

For example, using (2.3) when $c = 1$ gives

$$\begin{aligned} p_n &= p_0 \prod_{i=1}^{n} \frac{\lambda_{i-1}}{\mu_i} \\ &= p_0 \left(\frac{\lambda}{\mu}\right)^n \prod_{i=1}^{n} b_{i-1}. \end{aligned}$$

Possible examples that may be useful for the discouragement function b_n are $1/(n+1)$, $1/(n^2+1)$, and $e^{-\alpha n}$. People are not always discouraged because of queue size, but may attempt to estimate how long they would have to wait. If the queue is moving quickly, then the person may join a long one. On the other hand, if the queue is slow-moving, a customer may become discouraged even if the line is short. Now if n people are in the system, an estimate for the average waiting time might be n/μ, if the customer had an idea of μ. We usually do, so a plausible balking function might thus be $b_n = e^{-\alpha n/\mu}$. The $M/M/1/K$ model is a special case of balking where $b_i = 1$ for $0 \le i \le K-1$ and 0 otherwise.

2.10.2 $M/M/1$ Reneging

Customers who tend to be impatient may not always be discouraged by excessive queue size, but may instead join the queue to see how long their wait may become, all the time retaining the prerogative to renege if their estimate of their total wait is intolerable. We now consider a single-channel birth–death model where both reneging and the simple balking of the previous section exist, which gives rise to a reneging function $r(n)$ defined by

$$r(n) = \lim_{\Delta t \to 0} \frac{\Pr\{\text{unit reneges during}\Delta t \text{ when there are } n \text{ customers present}\}}{\Delta t},$$
$$r(0) = r(1) \equiv 0.$$

This new process is still birth–death, but the death rate must now be adjusted to $\mu_n = \mu + r(n)$. Thus it follows from (2.3) that

$$p_n = p_0 \prod_{i=1}^{n} \frac{\lambda_{i-1}}{\mu_i}$$
$$= p_0 \lambda^n \prod_{i=1}^{n} \frac{b_{i-1}}{\mu + r(i)} \qquad (n \ge 1),$$

where

$$p_0 = \left(1 + \sum_{n=1}^{\infty} \lambda^n \prod_{i=1}^{n} \frac{b_{i-1}}{\mu + r(i)}\right)^{-1}.$$

A good possibility for the reneging function $r(n)$ is $e^{\alpha n/\mu}, n \ge 2$. A waiting customer would probably estimate the average system waiting time as n/μ if $n-1$ customers were in front of him, assuming an estimate for μ were available. Again, the probability of a renege would be estimated by a function of the form $e^{\alpha n/\mu}$.

As was mentioned in the introduction to this section, there is yet an additional form of impatience called jockeying, that is, moving back and forth among the several subqueues before each of several multiple channels. Although this phenomenon is quite interesting and clearly applicable to numerous real-life situations, jockeying is analytically difficult to pursue very far, especially when we have more than two channels, since the probability functions become too complicated and the general concepts become hazy. Even though partial results can be obtained for the two-channel case, no particular insight into the multichannel cases is gained from it. If, however, the reader is specifically interested in this subject, we refer to Koenigsberg (1966).

2.11 Transient Behavior

In this section we consider the transient behavior of three specific queueing systems, namely, $M/M/1/1$ (no one allowed to wait), $M/M/1/\infty$, and $M/M/\infty$ (ample service). This discussion is restricted to these three models, since the mathematics becomes extremely complicated with the slightest relaxation of Poisson-exponential assumptions, and it is our feeling that the exhibition of some fairly simple results is sufficient for our purposes. Even these three transient derivations vary greatly in difficulty. The $M/M/1/1$ solution can be found fairly easily, but the problem becomes much more complicated when the restriction on waiting room is relaxed, or multiple servers are considered.

2.11.1 Transient Behavior of $M/M/1/1$

The derivation of the transient probabilities $\{p_n(t)\}$ that at an arbitrary time t there are n customers in a single-channel system with Poisson input, exponential service, and no waiting room is a straightforward procedure, since $p_n(t) = 0$ for all $n > 1$. It begins in the usual fashion from the birth-death differential equations as given by (1.30), with $\lambda_0 = \lambda, \lambda_n = 0, n > 0$, and $\mu_1 = \mu$:

$$\begin{aligned} \frac{dp_1(t)}{dt} &= -\mu p_1(t) + \lambda p_0(t), \\ \frac{dp_0(t)}{dt} &= -\lambda p_0(t) + \mu p_1(t). \end{aligned} \tag{2.70}$$

These differential-difference equations can be solved easily in view of the fact that it is always true that

$$p_0(t) + p_1(t) = 1.$$

Hence (2.70) is equivalent to

$$\frac{dp_1(t)}{dt} \equiv p_1'(t) = -\mu p_1(t) + \lambda[1 - p_1(t)].$$

So

$$p_1'(t) + (\lambda + \mu)p_1(t) = \lambda.$$

This is an ordinary first-order linear differential equation with constant coefficients. Its solution can be obtained from the discussion in Section 1.7 as

$$p_1(t) = Ce^{-(\lambda+\mu)t} + \frac{\lambda}{\lambda + \mu}.$$

To determine C, we use the boundary value of $p_1(t)$ at $t = 0$, which is $p_1(0)$. Thus

$$C = p_1(0) - \frac{\lambda}{\lambda + \mu},$$

and consequently

$$\begin{aligned} p_1(t) &= \frac{\lambda}{\lambda+\mu}(1 - e^{-(\lambda+\mu)t}) + p_1(0)e^{-(\lambda+\mu)t}, \\ p_0(t) &= \frac{\mu}{\lambda+\mu}(1 - e^{-(\lambda+\mu)t}) + p_0(0)e^{-(\lambda+\mu)t}, \end{aligned} \tag{2.71}$$

since $p_0(t) = 1 - p_1(t)$ for all t.

The stationary solution can be found directly from (2.70) in the usual way by letting the derivatives equal zero and then, using the fact that $p_0 + p_1 = 1$, solving for p_0 and p_1 ($M/M/1/K$ with $K = 1$). Also, the limiting (steady-state, equilibrium) solution can be found as the limit of the transient solution of (2.71) as t goes to ∞, and we find that

$$p_1 = \frac{\rho}{\rho+1} \quad \text{and} \quad p_0 = \frac{1}{\rho+1}.$$

Existence of the limiting distribution is always assured, independent of the value of $\rho = \lambda/\mu$, and thus it is identical to the stationary distribution (to see this, put $K = 1$ in the p_n expression for the $M/M/1/K$ of Section 2.5).

To get a better feel for the behavior of this queueing system for small values of time, let us graph $p_1(t)$ from (2.71). First rewrite (2.71) in the form

$$p_1(t) = p_1 + be^{-ct},$$

where,

$$p_1 = \frac{\lambda}{\lambda+\mu} = \frac{\rho}{\rho+1}, \quad b = p_1(0) - p_1, \quad \text{and} \quad c = \lambda + \mu.$$

Figure 2.9 shows a sample graph of $p_1(t)$ for a case where $b > 0$ ($\lambda = 0.2$, $\mu = 0.4$, $p_1(0) = 0.7$). We see that $p_1(t)$ is asymptotic to p_1. In addition, if the initial probability $p_1(0)$ equals the stationary probability p_1, then $b = 0$ and $p_1(t)$ equals the constant p_1 for all t. In other words, the queueing process can be translated into steady state at any time by starting the process in equilibrium. This property is, in fact, true for any ergodic queueing system, independent of any assumptions about its parameters.

2.11.2 Transient Behavior of $M/M/1/\infty$

The transient derivation for $M/M/1/\infty$ is quite a complicated procedure, so presentation of it is in outline form only. A more complete picture of the details can be found in Gross and Harris (1985) and Saaty (1961). The solution of this problem postdated that of the basic Erlang work by nearly half a century, with the first published solution due to Ledermann and Reuter (1954), in which they used spectral analysis for the general birth–death process. In the same year, an additional paper appeared on the solution of this problem by Bailey (1954), and later one by Champernowne (1956). Bailey's approach to the time-dependent problem was via generating functions for

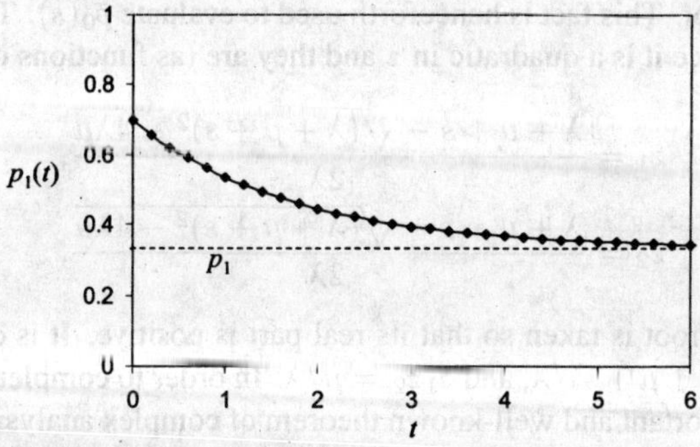

Figure 2.9 Illustration of transient solution $p_1(t)$.

the partial differential equation, and Champernowne's was via complex combinatorial methods. It is Bailey's approach that has been the most popular over the years, and this is basically the one we take. Remember that the key thing that makes this problem more difficult than may seem at first is that we are dealing with an *infinite* system of linear differential equations.

To begin, let it be assumed that the initial system size at time 0 is i. That is, if $N(t)$ denotes the number in the system at time t, then $N(0) = i$. The differential–difference equations governing the system size are given in (1.30) as

$$\begin{aligned} p_n'(t) &= -(\lambda+\mu)p_n(t) + \lambda p_{n-1}(t) + \mu p_{n+1}(t) \quad (n > 0), \\ p_0'(t) &= -\lambda p_0(t) + \mu p_1(t). \end{aligned} \tag{2.72}$$

It turns out that we solve these time-dependent equations using a combination of probability generating functions, partial differential equations, and Laplace transforms. Define

$$P(z,t) = \sum_{n=0}^{\infty} p_n(t) z^n \quad (z \text{ complex}),$$

such that the summation is convergent in and on the unit circle (i.e., for $|z| \le 1$), with its Laplace transform defined as

$$\bar{P}(z,s) = \int_0^\infty e^{-st} P(z,t)\, dt \quad (\text{Re } s > 0).$$

After the generating function is formed from (2.72) — yielding a partial differential equation — it is found when the Laplace transform is taken that

$$\bar{P}(z,s) = \frac{z^{i+1} - \mu(1-z)\bar{p}_0(s)}{(\lambda+\mu+s)z - \mu - \lambda z^2}, \tag{2.73}$$

where $\bar{p}_0(s)$ is the Laplace transform of $p_0(t)$.

Since the Laplace transform $\bar{P}(z,s)$ converges in the region $|z| \le 1, \text{Re } s > 0$, wherever the denominator of the right-hand side of (2.73) has zeros in that region, so must the numerator. This fact is henceforth used to evaluate $\bar{p}_0(s)$. The denominator has two zeros, since it is a quadratic in z and they are (as functions of s)

$$\begin{aligned} z_1 &= \frac{\lambda+\mu+s-\sqrt{(\lambda+\mu+s)^2-4\lambda\mu}}{2\lambda}, \\ z_2 &= \frac{\lambda+\mu+s+\sqrt{(\lambda+\mu+s)^2-4\lambda\mu}}{2\lambda}, \end{aligned} \tag{2.74}$$

where the square root is taken so that its real part is positive. It is clear that $|z_1| < |z_2|$, $z_1 + z_2 = (\lambda+\mu+s)/\lambda$, and $z_1 z_2 = \mu/\lambda$. In order to complete the derivation, the following important and well-known theorem of complex analysis due to Rouché is employed.

Rouché's Theorem. *If f(z) and g(z) are functions analytic inside and on a closed contour C and if $|g(z)| < |f(z)|$ on C, then f(z) and $f(z)+g(z)$ have the same number of zeros inside C.*

A proof of this theorem may be found in any book on complex variables. For $|z| = 1$ and Re $s > 0$ we see that

$$|f(z)| \equiv |(\lambda+\mu+s)z| = |\lambda+\mu+s| > \lambda+\mu \geq |\mu+\lambda z^2| \equiv |g(z)|.$$

Hence from Rouché's theorem, $(\lambda+\mu+s)z-\mu-\lambda z^2$ has only one zero in the unit circle. This zero is obviously z_1, since $|z_1| < |z_2|$. Thus equating the numerator of the right-hand side of (2.73) to zero for $z = z_1$ gives

$$\bar{p}_0(s) = \frac{z_1^{i+1}}{\mu(1-z_1)}.$$

When this transform for $p_0(t)$ is inserted into (2.73) and the result written in infinite series form, we find [remembering that $i = N(0)$]

$$\bar{P}(z,s) = \frac{1}{\lambda z_2}\sum_{j=0}^{i} z_1^j z^{i-j}\sum_{k=0}^{\infty}\left(\frac{z}{z_2}\right)^k + \frac{z_1^{i+1}}{\lambda z_2(1-z_1)}\sum_{k=0}^{\infty}\left(\frac{z}{z_2}\right)^k \qquad (|z/z_2| < 1).$$

Now, the transform of $p_n(t)$, $\bar{p}_n(s)$, is the coefficient of z^n in the Laplace transform of the generating function $P(z,t)$, $\bar{P}(z,s)$. So the next step in the process is to find $\bar{p}_n(s)$, and then this is, in turn, inverted to get $p_n(t)$, utilizing key properties of the transforms of Bessel functions in this last step. The final result is, in fact, in terms of modified Bessel functions of the first kind, $I_n(y)$, and is

$$\begin{aligned} p_n(t) = e^{-(\lambda+\mu)t}\Bigg[&\rho^{(n-i)/2} I_{n-i}(2t\sqrt{\lambda\mu}) + \rho^{(n-i-1)/2} I_{n+i+1}(2t\sqrt{\lambda\mu}) \\ &+ (1-\rho)\,\rho^n \sum_{j=n+i+2}^{\infty} \rho^{-j/2} I_j(2t\sqrt{\lambda\mu})\Bigg] \end{aligned} \tag{2.75}$$

for all $n \geq 0$, where

$$I_n(y) = \sum_{k=0}^{\infty} \frac{(y/2)^{n+2k}}{k!(n+k)!} \quad (n > -1)$$

is the infinite series for the modified Bessel function of the first kind. Abate and Whitt (1989) point out that computing values with (2.75) can be challenging, since the expression involves an infinite sum of modified Bessel functions. For a discussion of various numerical approaches to analyzing the transient $M/M/1$ queue, see Abate and Whitt (1989).

We can show, using properties of the Bessel functions, that (2.75) tends to the stationary result $p_n = (1-\rho)\rho^n$ as $t \to \infty$ when $\rho = \lambda/\mu < 1$. When $\lambda/\mu \geq 1, p_n(t) \to 0$ for all n, so that only when $\lambda/\mu < 1$ do we get a valid steady-state probability distribution. This agrees with our previous steady-state result of (2.9), as it should, since this is an ergodic system.

2.11.3 Transient Behavior of $M/M/\infty$

It turns out for this model that the development of the transient solution is not too difficult a task. To begin, let it be assumed that the initial system size at time 0 is 0, so that $N(0) = 0$. The differential–difference equations governing the system size are derived from (1.30) with $\lambda_n = \lambda$ and $\mu_n = n\mu$ as

$$\begin{aligned} p_n'(t) &= -(\lambda + n\mu)p_n(t) + \lambda p_{n-1}(t) + (n+1)\mu p_{n+1}(t) \quad (n > 0), \\ p_0'(t) &= -\lambda p_0(t) + \mu p_1(t). \end{aligned} \tag{2.76}$$

It turns out that we can solve these time-dependent equations using a combination of probability generating functions and partial differential equations, without requiring Laplace transforms. It is found that the generating function of the probabilities $\{p_n(t)\}$ is (see, e.g., Gross and Harris, 1985)

$$P(z,t) = \sum_{n=0}^{\infty} p_n(t) z^n = \exp\left((z-1)(1-e^{-\mu t})\frac{\lambda}{\mu}\right). \tag{2.77}$$

To obtain the state probabilities it is necessary to expand (2.77) in a power series, $a_0 + a_1 z + a_2 z^2 + \cdots$, where the coefficients a_n then are the transient probabilities $p_n(t)$ we desire, since we are expanding a probability generating function. To do this we can use a Taylor series expansion about zero (Maclaurin series), and we find that

$$p_n(t) = \frac{1}{n!}\left((1-e^{-\mu t})\frac{\lambda}{\mu}\right)^n \exp\left(-(1-e^{-\mu t})\frac{\lambda}{\mu}\right) \quad (n \geq 0).$$

It is easily seen that letting $t \to \infty$ yields the steady-state solution, which is equivalent to the Poisson distribution of (2.57),

$$p_n = \frac{(\lambda/\mu)^n e^{-\lambda/\mu}}{n!}.$$

We remind the reader that, in general, analytical solutions for transient queueing situations are extremely difficult to obtain. We do treat a few special cases briefly in later chapters, cases such as $M/G/1$, $G/M/1$, and $M/G/\infty$. However, since transient solutions require solving sets of differential equations, numerical methods can often be successfully employed. We treat this topic in some detail in Chapter 8, Section 8.1.2.

2.12 Busy-Period Analysis

This section includes analysis of the busy period for $M/M/1$ and $M/M/c$ queues. A *busy period* begins when a customer arrives at an idle channel and ends when the channel next becomes idle. A *busy cycle* is the sum of a busy period and an adjacent idle period, or equivalently, the time between two successive departures leaving an empty system, or two successive arrivals to an empty system. Since the arrivals are assumed to follow a Poisson process, the distribution of the idle period is exponential with mean $1/\lambda$; hence the CDF of the busy cycle for the $M/M/1$ is the convolution of this negative exponential with the CDF of the busy period itself. Therefore the CDF of the busy period is sufficient to describe the busy cycle also and is found as follows.

The CDF of the busy period is determined by considering the original $M/M/1$ differential equations given in (2.72) with an absorbing barrier imposed at zero system size (i.e., $\lambda_0 = 0$ in the birth–death equations) and an initial size of 1 [i.e., $p_1(0) = 1$]. Then it should be clear that $p_0(t)$ will, in fact, be the required busy period CDF and $p_0'(t)$ the density. The necessary equations are

$$\begin{aligned} p_0'(t) &= \mu p_1(t) \quad \text{[because of absorbing barrier]}, \\ p_1'(t) &= -(\lambda+\mu)p_1(t) + \mu p_2(t) \quad \text{[because of absorbing barrier]}, \\ p_n'(t) &= -(\lambda+\mu)p_n(t) + \lambda p_{n-1}(t) + \mu p_{n+1}(t) \quad \text{[same as (2.72)]}. \end{aligned}$$

In a fashion identical to the $M/M/1$ transient, it can be shown (see Gross and Harris, 1985) that the Laplace transform of the generating function is

$$\bar{P}(z,s) = \frac{z^2 - (\mu - \lambda z)(1-z)(z_1/s)}{\lambda(z-z_1)(z_2-z)}, \tag{2.78}$$

where z_1 and z_2 have the same values as in (2.74). Now the Laplace transform of $p_0(t)$, $\bar{p}_0(s)$, is the first coefficient of the power series $\bar{P}(z,s)$ and is thus found as $\bar{P}(0,s)$. This gives

$$\bar{p}_0(s) = \frac{2\mu}{s[\lambda+\mu+s+\sqrt{(\lambda+\mu+s)^2-4\lambda\mu}]}.$$

From the properties of Laplace transforms and Bessel functions, it can then be shown that the busy period's density function is

$$p_0'(t) = \frac{\sqrt{\mu/\lambda}\,e^{-(\lambda+\mu)t}I_1(2\sqrt{\lambda\mu}t)}{t}.$$

To get the average length of the busy period, $E[T_{bp}]$, we simply find the value of the negative of the derivative of the transform of $p_0'(t)$, $s\bar{p}_0(s)$, evaluated at $s = 0$. But an attractive alternative way to find the mean length of the busy period is to use the simple steady-state ratio argument that

$$\frac{1-p_0}{p_0} = \frac{E[T_{bp}]}{E[T_{idle}]} = \frac{E[T_{bp}]}{1/\lambda}.$$

Since $p_0 = 1 - \lambda/\mu$, it follows that the expected lengths of the busy period and busy cycle, respectively, are

$$E[T_{bp}] = \frac{1}{\mu - \lambda} \quad \text{and} \quad E[T_{bc}] = \frac{1}{\lambda} + \frac{1}{\mu - \lambda}. \tag{2.79}$$

Equation (2.79) holds for all $M/G/1$-type queues, since the exponential service property played no role in the derivation.

It is not too difficult to extend the notion of the busy period conceptually to the multichannel case. Recall that for one channel a busy period is defined to begin with the arrival of a customer at an idle channel and to end when the channel next becomes idle. In an analogous fashion, let us define an i-channel busy period for $M/M/c$ $(0 \le i \le c)$ to begin with an arrival at the system at an instant when there are $i-1$ in the system and to end at the very next point in time when the system size dips to $i-1$. Let us say that the case where $i = 1$ (an arrival to an empty system) defines the system busy period. In fashion similar to that for $M/M/1$, use $T_{b,i}$ to denote the random variable "length of the i-channel busy period." Then the CDF of $T_{b,i}$ is determined by considering the original $M/M/c$ differential–difference equations of (2.72) with an absorbing barrier imposed at a system size of $i-1$ and an initial size of i. Then it should be clear that $p_{i-1}(t)$ will, in fact, be the required CDF, and its derivative the density. The necessary equations are

$$\begin{aligned}
p_{i-1}'(t) &= i\mu p_i(t) \quad \text{[because of absorbing barrier]},\\
p_i'(t) &= -(\lambda + i\mu)p_i(t) + (i+1)\mu p_{i+1}(t) \quad \text{[because of absorbing barrier]},\\
p_n'(t) &= -(\lambda + n\mu)p_n(t) + \lambda p_{n-1}(t) + (n+1)\mu p_{n+1}(t) \quad (i < n < c),\\
p_n'(t) &= -(\lambda + c\mu)p_n(t) + \lambda p_{n-1}(t) + c\mu p_{n+1}(t) \quad (n \ge c).
\end{aligned}$$

Proceeding further gets us bogged down in great algebraic detail. Any resultant CDF will be in terms of modified Bessel functions, but with enough time and patience, $p_{i-1}'(t), \bar{p}_{i-1}(s)$, and $E[T_{b,i}]$ can be obtained.

PROBLEMS

2.1. You are told that a small single-server, birth–death-type queue with finite capacity cannot hold more than three customers. The three arrival or birth rates are $(\lambda_0, \lambda_1, \lambda_2) = (3, 2, 1)$, while the service or death rates are $(\mu_1, \mu_2, \mu_3) = (1, 2, 2)$. Find the steady-state probabilities $\{p_i, i =$

$0, 1, 2, 3\}$ and L. Then determine the average or effective arrival rate $\lambda_{\text{eff}} = \sum \lambda_i p_i$, and the expected system waiting time W.

2.2. The finite-capacity constraint of Problem 2.1 has been pushed up to 10 now, with the arrival rates known to be (4, 3, 2, 2, 3, 1, 2, 1, 2, 1) and service rates to be (1, 1, 1, 2, 2, 2, 3, 3, 3, 4). Do as before and find $p_i, i = 0, 1, 2, \ldots, 10$, the mean system size L, the effective arrival rate $\lambda_{\text{eff}} = \sum \lambda_i p_i$, and the expected system waiting time W.

2.3. For an $M/M/1$ queue, derive the variance of the number of customers in the system in steady state.

2.4. For the $M/M/1$ and $M/M/c$ queues, find $\text{E}[T_q | T_q > 0]$, that is, the expected time one must wait in the queue, given that one must wait at all.

2.5. Derive $W(t)$ and $w(t)$ (the total-waiting-time CDF and its density) as given by the equations (2.29).

2.6. Equation (1.4) is valid for any queueing system and can be used to prove Little's formula for the $G/G/1$ queue in a somewhat different manner than that used in Section 1.5.1. The approach is to plot the cumulative count of arrivals on the same graph as the cumulative count of departures. Then it can be seen that the area between these two step functions from the beginning of a busy period to the beginning of the next (a busy cycle) is the accumulated total of the system waiting times of all the customers who have entered into the system during this busy cycle. Use this argument to derive an empirical version of Little's formula over a busy cycle.

2.7. What effect does doubling λ and μ have on L, L_q, W, and W_q in an $M/M/1$ model?

2.8. A graduate research assistant "moonlights" in the food court in the student union in the evenings. He is the only one on duty at the counter during the hours he works. Arrivals to the counter seem to follow the Poisson distribution with mean of 10/h. Each customer is served one at a time and the service time is thought to follow an exponential distribution with a mean of 4 min. Answer the following questions.

(a) What is the probability of having a queue?

(b) What is the average queue length?

(c) What is the average time a customer spends in the system?

(d) What is the probability of a customer spending more than 5 min in the queue before being waited on?

(e) The graduate assistant would like to spend his idle time grading papers. If he can grade 22 papers an hour on average when working continuously, how many papers per hour can he average while working his shift?

2.9. A rent-a-car maintenance facility has capabilities for routine maintenance (oil change, lubrication, minor tune-up, wash, etc.) for only one car at a time. Cars arrive there according to a Poisson process at a mean rate of three per day, and service time to perform this maintenance seems to have an exponential distribution with a mean of $\frac{7}{24}$ day. It costs the company a fixed \$375 a day to operate the facility. The company estimates loss in profit on a car of \$25/day for every day the car is tied up in the shop. The company, by changing certain procedures and hiring faster mechanics, can decrease the mean service time to $\frac{1}{4}$ day. This also increases their operating costs. Up to what value can the operating cost increase before it is no longer economically attractive to make the change?

2.10. Parts arrive at a painting machine according to a Poisson process with rate λ/h. The machine can paint one part at a time. The painting process appears to be exponential with an average service time of $1/\mu$ h. It costs the company about $\$C_1$ per part per hour spent in the system (i.e., being painted or waiting to be painted). The cost of owning and operating the painting machine is strictly a function of its speed. In particular, a machine that works at an average rate of μ costs $\$\mu C_2$/h, whether or not it is always in operation. Determine the value of μ that minimizes the cost of the painting operation.

2.11. Find the probability that k or more are in an $M/M/c/\infty$ system for any $k \geq c$.

2.12. For the $M/M/c/\infty$ model, give an expression for p_n in terms of p_c instead of p_0, and then derive L_q in terms of ρ and p_c.

2.13. **(a)** Our local fast-food emporium, Burger Bliss, has yet to learn a lot about queueing theory. So it does not require that all waiting customers form a single line, and instead they make every arrival randomly choose one of *three* lines formed before each server during the weekday lunch period. But they are so traditional about managing their lines that barriers have been placed between the lines to prevent jockeying. Suppose that the overall stream of incoming customers has settled in at a constant rate of 60/h (Poisson-distributed) and that the time to complete a customer's order is well described by an exponential distribution of mean 150 seconds, independent and identically from one customer to the next. Assuming steady state, what is the average total system size?

(b) The BB has now agreed that it is preferable to have one line feeding the three servers, so the barriers have been removed. What is the expected steady-state system size now?

2.14. Verify the formula for the mean line delay W_q of the $M/M/c/\infty$ queue.

2.15. The Outfront BBQ Rib Haven does carry out only. During peak periods, two servers are on duty. The owner notices that during these periods, the servers are almost never idle. She estimates the percent time idle of each server to be 1%. Ideally, the percent idle time would be 10% to allow time for important breaks.

(a) If the owner decides to add a third server during these times, how much idle time would each server have then?

(b) Suppose that by adding the third server, the pressure on the servers is reduced, so they can work more carefully, but their service output rate is reduced by 20%. What now is the percent time each would be idle?

(c) Suppose, instead, the owner decides to hire an aid (at a much lower salary) who servers as a gofer for the two servers, rather than hiring another full server. This allows the two servers to decrease their average service time by 20% (relative to the original service rate). What now is the percent idle time of each of the two servers?

2.16. In the spring of 2006, George Mason University's basketball team advanced to the Final Four of the NCAA tournament – only the second double-digit seed ever to do so. To celebrate the event, the campus book store ordered tee shirts. On the day of the sale, demand for shirts was steady throughout the day and fairly well described by a Poisson process with a rate of 66 per hour. There were four cash registers in operation and the average time of a transaction was 3.5 minutes. Service times were approximately exponentially distributed.

(a) What was the average length of the line for shirts?

(b) How long, on average, did customers wait in line to get a shirt?

(c) What fraction of customers spent more than 30 minutes in the store to get a shirt?

2.17. You are the owner of a small book store. You have two cash registers. Customers wait in a single line to purchase books at one of the two registers. Customers arrive according to a Poisson process with rate $\lambda = 30$ per hour. The time to complete the purchase transactions for one customer follows an exponential distribution with mean 3 minutes.

(a) Determine W, W_q, L, and L_q for this system.

(b) Suppose that you pay the register clerks \$10 per hour and that each customer on average purchases books that give you a net \$2 profit. What is the hourly rate that you make money?

(c) Now suppose that you offer a \$2 rebate to every customer who joins the queue and finds 4 or more people *in the queue*. Now, what is the hourly rate that you make money?

2.18. For an $M/M/2$ queue with $\lambda = 60/\text{h}$ and $\mu = 0.75/\text{ min}$, calculate L, L_q, W, W_q, $\Pr\{N \geq k\}$, and $\Pr\{T_q > t\}$ for $k = 2, 4$ and $t = 0.01, 0.03$ h.

2.19. For $M/M/c/\infty$, derive the distribution function of the system waiting time for those customers for whom $T_q > 0$, remembering that their wait is the sum of a line delay and a service time.

2.20. For the $M/M/c/K$ queue, calculate L, L_q, W, W_q, p_K, $\Pr\{N \geq k\}$, and $\Pr\{T_q \geq t\}$ for $\lambda = 2/\text{min}$, $\mu = 45/\text{h}$, $c = 2$, $K = 6$, $k = 2$ and 4, and $t = 0.01$ and 0.02 h.

2.21. The office of the Deputy Inspector General for Inspection and Safety administers the Air Force Accident and Incident Investigation and Reporting Program. It has established 25 investigation teams to analyze and evaluate each accident or incident to make sure it is properly reported to accident investigation boards. Each of these teams is dispatched to the locale of the accident or incident as each requirement for such support occurs. Support is only rendered those commands that have neither the facilities nor qualified personnel to conduct such services. Each accident or incident will require a team being dispatched for a random amount of time, apparently exponential with mean of 3 weeks. Requirements for such support are received by the Deputy Inspector General's office as a Poisson process with mean rate of 347/yr. At any given time, two teams are not available due to personnel leaves, sickness, and so on. Find the expected time spent by an accident or incident in and waiting for evaluation.

2.22. An organization is presently involved in the establishment of a telecommunication center so that it may provide a more rapid outgoing message capability. Overall, the center is responsible for the transmission of outgoing messages and receives and distributes incoming messages. The center manager at this time is primarily concerned with determining the number of transmitting personnel required at the new center. Outgoing message transmitters are responsible for making minor corrections to messages, assigning numbers when absent from original message forms, maintaining an index of codes and a 30-day file of outgoing messages, and actually transmitting the messages. It has been predetermined that this process is exponential and requires a mean time of 28 min/message. Transmission personnel will operate at the center 7 h/day, 5 days/week. All outgoing messages will be processed in the order they are received and follow a Poisson process with a mean rate of 21 per 7-h day. Processing on messages requiring transmission must be started within an average of 2 h from the time they arrive at the center. Determine the minimum number of transmitting personnel to accomplish this service criterion. If the service criterion were to require the probability of any message waiting for the start of processing for more than 3 h to be less than .05, how many transmitting personnel would be required?

2.23. A small branch bank has two tellers, one for receipts and one for withdrawals. Customers arrive to each teller's cage according to a Poisson distribution with a mean of 20/h. (The total mean arrival rate at the bank is 40/h.) The service time of each teller is exponential with a mean of 2 min. The bank manager is considering changing the setup to allow each teller to handle both withdrawals and deposits to avoid the situations that arise from time to time when the queue is sizable in front of one teller while the other is idle. However, since the tellers would have to handle both receipts and withdrawals, their efficiency would decrease to a mean service time of 2.4 min. Compare the present system with the proposed system with respect to the total expected number of people in the bank, the expected time a customer would have to spend in the bank, the probability of a customer having to wait more than 5 min, and the average idle time of the tellers.

2.24. The Hott Too Trott Heating and Air Conditioning Company must choose between operating two types of service shops for maintaining its trucks. It estimates that trucks will arrive at the maintenance facility according to a Poisson distribution with mean rate of one every 40 min and believes that this rate is independent of which facility is chosen. In the first type of shop, there are dual facilities operating in parallel; each facility can service a truck in 30 min on average (the service time follows an exponential distribution). In the second type there is a single facility, but it can service a truck in 15 min on average (service times are also exponential in this case). To help management decide, they ask their operations research analyst to answer the following questions:

(a) How many trucks, on average, will be in each of the two types of facilities?

(b) How long, on average, will a truck spend in each of the two types of facilities?

(c) Management calculates that each minute a truck must spend in the shop reduces contribution to profit by two dollars. They also know from previous experience in running dual-facility shops that the cost of operating such a facility is one dollar per minute (including labor, overhead, etc.). What would the operating cost per minute have to be for operating the single-facility shop in order for there to be no difference between the two types of shops?

2.25. The ComPewter Company, which leases out high-end computer workstations, considers it necessary to overhaul its equipment once a year. Alternative 1 is to provide two separate maintenance stations where all work is done by hand (one machine at a time) for a total annual cost of $750,000. The maintenance time for a machine has an exponential distribution with a mean of 6 h. Alternative 2 is to provide one maintenance station with mostly automatic equipment involving an annual cost of $1 million. In this case, the maintenance time for a machine has an exponential distribution with a mean of 3 h. For both alternatives, the machines arrive according to a Poisson input with a mean arrival rate of one every 8 h (since the company leases such a large number of machines, we can consider the machine population as infinite). The cost of downtime per machine is $150/h. Which alternative should the company choose? Assume that the maintenance facilities are always open and that they work $(24)(365) = 8760$ h/yr.

2.26. Show the following:

(a) An $M/M/1$ is always better with respect to L than an $M/M/2$ with the same ρ.

(b) An $M/M/2$ is always better than two independent $M/M/1$ queues with the same service rate but each getting half of the arrivals.

2.27. For Problem 2.26(a), show that the opposite is true when considering L_q. In other words, faced with a choice between two M/M systems with identical arrival rates, one with two servers and one with a single server who can work twice as fast as each of the two servers, which is the preferable system?

2.28. Show for the $M/M/c/K$ model that taking the limit for p_n and p_0 as $K \to \infty$ and restricting $\lambda/c\mu < 1$ in (2.45), (2.46), and (2.47) yield the results obtained for the $M/M/c/\infty$ model.

2.29. Show that the $M/M/c/K$ equations (2.45)–(2.47) reduce to those for $M/M/1/K$ when $c = 1$.

2.30. For the $M/M/3/K$ model, compute L_q as K goes from 3 to "∞" for each of the following ρ values: 1.5, 1, 0.8, 0.5. Comment.

2.31. Find the probability that a customer's wait in queue exceeds 20 min for an $M/M/1/3$ model with $\lambda = 4$/h and $1/\mu = 15$ min.

2.32. A small drive-it-through-yourself car wash, in which the next car cannot go through the washing procedure until the car in front is completely finished, has a capacity to hold on its grounds a maximum of 10 cars (including the one in wash). The company has found its arrivals to be Poisson with mean rate of 20 cars/h, and its service times to be exponential with a mean of 12 min. What is the average number of cars lost to the firm every 10-h day as a result of its capacity limitations?

2.33. Under the assumption that customers will not wait if no seats are available, Example 2.1's hair salon proprietor Cutt can rent, on Saturday, the conference room of a small computer software firm adjacent to her shop for \$30.00 (cost of cleanup on a Saturday). Her shop is open on Saturdays from 8:00 A.M. to 2:00 P.M., and her marginal profit is \$6.75 per customer. This office can seat an additional four people. Should Cutt rent?

2.34. The Fowler-Heir Oil Company operates a crude-oil unloading port at its major refinery. The port has six unloading berths and four unloading crews. When all berths are full, arriving ships are diverted to an overflow facility 20 miles down river. Tankers arrive according to a Poisson process with a mean of one every 2 h. It takes an unloading crew, on average, 10 h to unload a tanker, the unloading time following an exponential distribution. Tankers waiting for unloading crews are served on a first-come, first-served basis. Company management wishes to know the following:

(a) On average, how many tankers are at the port?

(b) On average, how long does a tanker spend at the port?

(c) What is the average arrival rate at the overflow facility?

(d) The company is considering building another berth at the main port. Assume that construction and maintenance costs would amount to X dollars per year. The company estimates that to divert a tanker to the overflow port when the main port is full costs Y dollars. What is the relation between X and Y for which it would pay for the company to build an extra berth at the main port?

2.35. Fly-Bynite Airlines has a telephone exchange with three lines, each manned by a clerk during its busy periods. During their peak three hours per 24-h period, many callers are unable to get into the exchange (there is no provision for callers to hold if all servers are busy). The company estimates, because of severe competition, that 60% of the callers not getting through use another airline. If the number of calls during these peak periods is roughly Poisson with a mean of 20 calls/h and each clerk spends on average 6 min with a caller, his service time being approximately exponentially distributed, and the average customer spends $210/trip, what is the average daily loss due to the limited service facilities? (We may assume that the number of people not getting through during off-peak hours is negligible.) If a clerk's pay and fringe benefits cost the company $24/h and a clerk must work an 8-h shift, what is the optimum number of clerks to employ? The three peak hours occur during the 8-h day shift. At all other times, one clerk can handle all the traffic, and since the company never closes the exchange, exactly one clerk is used on the off shifts. Assume that the cost of adding lines to the exchange is negligible.

2.36. A call center has 24 phone lines and 3 customer service representatives. Suppose that calls arrive to the center according to a Poisson process with rate $\lambda = 15$ per hour. The time to process each call is exponential with a mean of 10 minutes. If all of the service representatives are busy, an arriving customer is placed on hold, but ties up one of the phone lines. If all of the phone lines are tied up, the customer receives a busy signal and the call is lost.

(a) What is the average time that a customer spends on hold?

(b) What is the average number of lines busy at one time?

(c) Suppose that you pay $0.03 per minute for each call to your center (including the time on hold). Also, the cost for each lost call is estimated at $20. Fixing the number of service representatives, what is the optimal number of phone lines you should have?

2.37. Prove the iterative relationship in (2.54) for the Erlang-B formula.

2.38. Prove the relationship in (2.55) between the Erlang-B and Erlang-C formulas.

2.39. Show that the steady-state probabilities obtained for the ample-server model ($M/M/\infty$) can also be developed by taking the limit as $c \to \infty$ in the results for the $M/M/c$ model.

2.40. The Good Writers Correspondence Academy offers a go-at-your-own-pace correspondence course in good writing. New applications are accepted at any time, and the applicant can enroll immediately. Past records indicate applications follow a Poisson distribution with a mean of 8/month. An applicant's mean completion time is found to be 10 weeks, with the distribution of completion times being exponential. On average, how many pupils are enrolled in the school at any given time?

2.41. An application of an $M/M/\infty$ model to the field of *inventory control* is as follows. A manufacturer of a very expensive, rather infrequently demanded item uses the following inventory control procedure. She keeps a safety stock of S units on hand. The customer demand for units can be described by a Poisson process with mean λ. Every time a request for a unit is made (a customer demand), an order is placed at the factory to manufacture another (this is called a one-for-one ordering policy). The amount of time required to manufacture a unit is exponential with mean $1/\mu$. There is a carrying cost for inventory on shelf of $\$h$ per unit per unit time held on shelf (representing capital tied up in inventory which could be invested and earning interest, insurance costs, spoilage, etc.) and a shortage cost of $\$p$ per unit (a shortage occurs when a customer requests a unit and there is none on shelf, i.e., safety stock is depleted to zero). It is assumed that customers who request an item but find that there is none immediately available will wait until stock is replenished by orders due in (this is called backordering or backlogging); thus one can look at the charge $\$p$ as a discount given to the customer because he must wait for his request to be satisfied. The problem, then, becomes one of finding the optimal value of S that minimizes total expected costs per unit time; that is, find the S that minimizes

$$\mathrm{E}[C] = h\sum_{z=1}^{S} zp(z) + p\lambda \sum_{z=-\infty}^{0} p(z) \quad (\$/\text{unit time}),$$

where z is the steady-state on-hand inventory level (+ means items on shelf, − means items in backorder) and $p(z)$ is the probability frequency function. Note that $\sum_{z=1}^{S} zp(z)$ is the average value of the safety stock and $\lambda \sum_{z=-\infty}^{0} p(z)$ is the expected number of backorders per unit time, since the second summation is the fraction of time there is no on-shelf safety stock and λ is the average request rate. If $p(z)$ could be determined, one could optimize E[C] with respect to S.

(a) Show the relationship between Z and N, where N denotes the number of orders outstanding, that is, the number of orders currently being processed at the factory. Hence relate $p(z)$ to p_n.

(b) Show that the $\{p_n\}$ are the steady-state probabilities of an $M/M/\infty$ queue if one considers the order-processing procedure as the queueing system. State explicitly what the input and service mechanisms are.

(c) Find the optimum S for $\lambda = 8$/month, $1/\mu = 3$ days, $h = \$50$/unit per month held, and $p = \$500$ per unit backordered.

2.42. Farecard machines that dispense tickets for riding on the subway have a mean operating time to breakdown of 45 h. It takes a technician on average 4 h to repair a machine, and there is one technician at each station. Assume that the time to breakdown and the time to repair are exponentially distributed. What is the number of installed machines necessary to assure that the probability of having at least five operational is greater than .95?

2.43. For the machine repair model with spares, calculate all the usual measures of effectiveness for a problem with $M = 10, Y = 2, \lambda = 1, \mu = 3.5, c = 3$. Find the $\Pr\{N \geq k\}$ for $k = 2,4$.

2.44. Show for the basic machine repair model (no spares) that $q_n(M)$, the failure (arrival) point probabilities for a population of size M, equal $p_n(M-1)$, the general-time probabilities for a population of size $M-1$. The $\{q_n\}$ are sometimes referred to as *inside observer probabilities*, while the $\{p_n\}$ are referred to as *outside observer probabilities*.

2.45. Derive $q_n(M)$ given by (2.65) for a machine repair problem with spares, and show that this is *not* equal to $p_n(M-1)$, but is equal to $p_n(Y-1)$ (i.e., the steady-state probabilities for a case where M is the same and the number of spares is reduced by one). The algebra is quite messy, so show it only for a numerical example ($M = 2, Y = 1, c = 1, \lambda/\mu = 1$). While that is no proof, the statement can be shown to hold in general (see Sevick and Mitrani, 1979, or Lavenberg and Reiser, 1979).

2.46. A coin-operated dry-cleaning store has five machines. The operating characteristics of the machines are such that any machine breaks down according to a Poisson process with mean breakdown rate of one per day. A repairman can fix a machine according to an exponential distribution with a mean repair time of one-half day. Currently, three repairmen are on duty. The manager, Lew Cendirt, has the option of replacing these three repairmen with a super-repairman whose salary is equal to the total of the three regulars, but who can fix a machine in one-third the time, that is, in one-sixth day. Should he be hired?

2.47. Suppose that each of five machines in a given shop breaks down according to a Poisson law at an average rate of one every 10 h, and the failures are repaired one at a time by two maintenance people operating as two channels, such that each machine has an exponentially distributed servicing requirement of mean 5 h.

(a) What is the probability that exactly one machine will be up at any one time?

(b) If performance of the workforce is measured by the ratio of average waiting time to average service time, what is this measure for the current situation?

(c) What is the answer to (a) if an identical spare machine is put on reserve?

2.48. Find the steady-state probabilities for a machine-repair problem with M machines, Y spares, and c technicians ($c \leq Y$) but with the following discipline: If no spares are on hand and a machine fails ($n = Y + 1$), the remaining $M - 1$ machines running are stopped until a machine is repaired; that is, if the machines are to run, there must be M running simultaneously.

2.49. Very often in real-life modeling, even when the calling population is finite, an infinite-source model is used as an approximation. To compare the two models, calculate L for Example 2.9 assuming the calling population (number of machines) is infinite. Also calculate L for an exact model when the number of machines equals 10 and 5, respectively, for $M\lambda = \frac{1}{3}$ in both cases, and compare to the calculations from an approximate infinite-source model. How do you think ρ affects the approximation? [*Hint*: When using an infinite-source model as an approximation to a finite-source model, λ must be set equal to $M\lambda$.]

2.50. Find the average operating costs per hour of Example 2.10 when the following conditions prevail:

(a) C_1 = low-speed cost = \$25/(operating hour); C_2 = high-speed cost = \$50/(operating hour).

(b) C_1 = \$25/(operating hour); C_2 = \$60/(operating hour).

(c) Evaluate (b) for $k = 4$. What now is the best policy?

2.51. Assume we have a two-state, state-dependent service model as described in Section 2.9 with $\rho_1 = \frac{4}{3}$ and $\rho = \frac{2}{3}$. Suppose that the customers are lawn-treating machines owned by the Green Thumb Lawn Service Company and these machines require, at random times, greasing on the company's two-speed greasing machine. Furthermore, suppose that the cost per operating hour of the greaser at the lower speed, C_1, is \$25, and at the high speed, C_2, is \$110. Also, the company estimates the cost of downtime of a lawn treater to be \$5/h. What is the optimal switch point k? [*Hint*: Try several values of k starting at $k = 1$, and compute the total expected cost.]

2.52. Derive the steady-state system-size probabilities for a c-server model with Poisson input and exponential state-dependent service where the mean service rate switches from μ_1 to μ when $k > c$ are in the system.

2.53. Derive the steady-state system-size probabilities for a single-server model with Poisson input and exponential state-dependent service with mean rates $\mu_1(1 \leq n < k_1)$, $\mu_2(k_1 \leq n < k_2)$, and $\mu(n \geq k_2)$.

2.54. For the problem treated at the end of Section 2.9 where $\mu_n = n^\alpha \mu$, show for $\alpha \geq 1$ that the tail of the infinite series for calculating p_0 discarded by truncation at some value N is bounded by the tail of the series for e^r. So, given any $\epsilon > 0$, N can be found such that the discarded tail is less than ϵ. Furthermore, show that if $p_0(N)$ is the estimate of p_0 based on N terms (where N is such that the discarded tail is bounded by ϵ), then the error bounds on p_0 become

$$p_0 < p_0(N) < \frac{p_0}{1-\epsilon} \approx p_0(1+\epsilon).$$

2.55. It is known for an $M/M/1$ balking situation that the stationary distribution is given by the negative binomial

$$p_n = \binom{N+n-1}{N-1} x^n (1+x)^{-N-n} \quad (n \geq 0, \quad x > 0, \quad N > 1).$$

Find L, L_q, W, W_q, and b_n.

2.56. For an $M/M/1$ balking model, it is known that $b_n = e^{-\alpha n/\mu}$. Find p_n (for all n).

2.57. Consider a single-server queue where the arrivals are Poisson with rate $\lambda = 10$ per hour. The service distribution is exponential with rate $\mu = 5$ per hour. Suppose that customers balk at joining the queue when it is too long. Specifically, when there are n in the system, an arriving customer joins the queue with probability $1/(1+n)$. Determine the steady-state probability that there are n in the system.

2.58. Suppose that the $M/M/1$ reneging model of Section 2.10.2 has the balking function $b_n = 1/n$ for $0 \leq n \leq k$ and 0 for $n > k$, and a reneging function $r(n) = n/\mu$. Find the stationary system-size distribution.

2.59. Derive the steady-state $M/M/\infty$ solution directly from the transient.

2.60. Find the mean number in an $M/M/\infty$ system at time t, assuming the system is empty at $t = 0$.

2.61. For $\rho = 0.5, 0.9$, and 1 in an $M/M/1/\infty$ model with $\lambda = 1$, plot $p_0(t)$ versus t as t goes to infinity. Comment.

2.62. For $\rho = 0.5$ in an $M/M/1/\infty$ model with $\lambda = 1$, find L at $t = 3$.

2.63. (a) Prove that if $\bar{f}(s)$ is the Laplace transform of $f(t)$, then $\bar{f}(s+a)$ is the Laplace transform of $e^{-at} f(t)$.

(b) Show that the Laplace transform of a linear combination of functions is the same linear combination of the transforms of the functions: symbolically, $\mathfrak{L}[\sum_i a_i f_i(t)] = \sum_i a_i \mathfrak{L}[f_i(t)]$.

2.64. Use the properties of Laplace transforms to find the functions whose Laplace transforms are the following:

(a) $(s+1)/(s^2+2s+2)$.

(b) $1/(s^2-3s+2)$.

(c) $1/[s^2(s^2+1)]$.

(d) $e^{-s}/(s+1)$.

2.65. For the following generating functions (not necessarily *probability* generating functions), write the sequence they generate:

(a) $G(z) = 1/(1-z)$.

(b) $G(z) = z/(1-z)$.

(c) $G(z) = e^z$.

2.66. Show that the moment generating function of the sum of independent random variables is equal to the product of their moment generating functions.

2.67. Use the result of Problem 2.66 to show the following:

(a) The sum of two independent Poisson random variables is a Poisson random variable.

(b) The sum of two independent and identical exponential random variables has a gamma or Erlang distribution.

(c) The sum of two independent but nonidentical exponential random variables has a density that is a linear combination of the two original exponential densities.

CHAPTER 3

ADVANCED MARKOVIAN QUEUEING MODELS

This chapter continues the development of models that are amenable to analytic methods and is concerned especially with Markovian problems of the non-birth-death type. That is, we allow changes of more than one over infinitesimal time intervals but insist on retaining the memoryless Markovian property. The Chapman-Kolmogorov and backward and forward equations, plus the resultant balance equations, are all still valid, and together are the essence of the approach to solution for these non-birth-death Markovian problems.

3.1 Bulk Input ($M^{[X]}/M/1$)

In continuation of our relaxation of the simple assumptions underlying $M/M/1$, let it now be assumed, in addition to the assumption that the arrival stream forms a Poisson process, that the actual number of customers in any arriving module is a random variable X, which takes on the value n with probability c_n, where n is a positive integer, $0 < n < \infty$. It should be clear that this new queueing problem—let it be called $M^{[X]}/M/1$—is still Markovian in the sense that future behavior is a function only of the present and not the past.

Fundamentals of Queueing Theory, Fourth Edition.
By D. Gross, J. F. Shortle, J. M. Thompson, and C. M. Harris

Copyright © 2008 John Wiley & Sons, Inc.

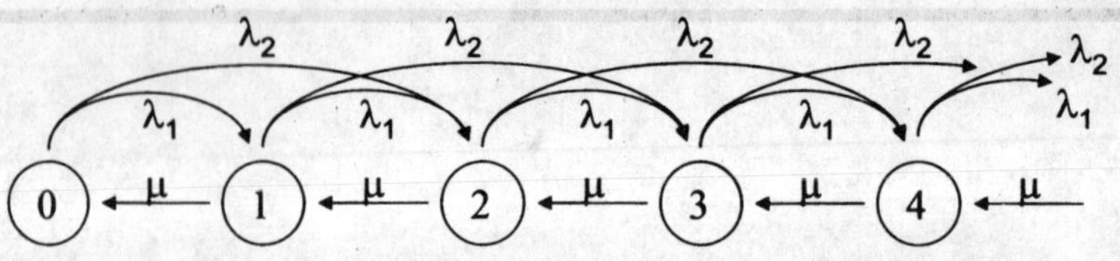

Figure 3.1 State transition rates for sample $M^{[X]}/M/1$ queue.

We now recall the discussions of Section 1.8, particularly those of the common generalizations of the Poisson process. If λ_n is the arrival rate of a Poisson process of batches of size n, then clearly $c_n = \lambda_n/\lambda$, where λ is the composite arrival rate of all batches and is equal to $\sum_{n=1}^{\infty} \lambda_n$. This total process, which arises from the overlap of the set of Poisson processes with rates $\{\lambda_n, n = 1, 2, \ldots\}$, is a *multiple* or *compound Poisson process.*

Figure 3.1 shows the state transition rates for a sample $M^{[X]}/M/1$ queue, where in this example the number X of customers per batch is either 1 or 2. For an arbitrary batch-size distribution X, a general set of rate balance equations can be derived (Problem 3.1):*

$$0 = -(\lambda + \mu)p_n + \mu p_{n+1} + \lambda \sum_{k=1}^{n} p_{n-k}c_k \qquad (n \geq 1),$$
$$0 = -\lambda p_0 + \mu p_1. \tag{3.1}$$

The last term in the general equation of (3.1) comes from the fact that a total of n in the system can arise from the presence of $n - k$ in the system followed by the arrival of a batch of size k. To solve the system of equations given by (3.1), we use a generating-function approach. Difference-equation methods are often used instead to solve the problem when the maximum batch is small. It is also not a very difficult matter to extend the results of this section to the $M^{[X]}/M/c$ model. This would be done in the same manner as *M/M/*1 is extended to *M/M/c*.

To complete the solution, we first define

$$C(z) = \sum_{n=1}^{\infty} c_n z^n \qquad (|z| \leq 1)$$

and

$$P(z) = \sum_{n=0}^{\infty} p_n z^n \qquad (|z| \leq 1)$$

as the generating functions of the batch-size probabilities $\{c_n\}$ and the steady-state probabilities $\{p_n\}$, respectively. The batch-size probabilities are usually known, so

*Assuming steady state. Under the proper parameter settings, the processes treated in this chapter meet the conditions of Theorem 1.3.

$C(z)$ can be thought of as an input. The objective then is to determine $P(z)$ from $C(z)$ and from this to find the unknown steady state probabilities $\{p_n\}$. Multiplying each equation of (3.1) by z^n and summing the results together gives

$$0 = -\lambda \sum_{n=0}^{\infty} p_n z^n - \mu \sum_{n=1}^{\infty} p_n z^n + \frac{\mu}{z} \sum_{n=1}^{\infty} p_n z^n + \lambda \sum_{n=1}^{\infty} \sum_{k=1}^{n} p_{n-k} c_k z^n. \qquad (3.2)$$

We observe that $\sum_{k=1}^{n} p_{n-k} c_k$ is the probability function for the sum of the steady-state system size and batch size, since this is merely a convolution formula for discrete random variables. It can easily be shown that the generating function of this sum is the product of the respective generating functions (a basic property of all generating functions), namely,

$$\sum_{n=1}^{\infty} \sum_{k=1}^{n} p_{n-k} c_k z^n = \sum_{k=1}^{\infty} c_k z^k \sum_{n=k}^{\infty} p_{n-k} z^{n-k} = C(z)P(z).$$

Hence (3.2) may be rewritten as

$$0 = -\lambda P(z) - \mu[P(z) - p_0] + \frac{\mu}{z}[P(z) - p_0] + \lambda C(z)P(z),$$

and thus

$$P(z) = \frac{\mu p_0(1 - z)}{\mu(1 - z) - \lambda z[1 - C(z)]} \qquad (|z| \le 1). \qquad (3.3)$$

To find p_0 from $P(1)$ and L from $P'(1)$, rewrite the generating function (3.3), with $r = \lambda/\mu$, as

$$P(z) = \frac{p_0}{1 - rz\bar{C}(z)}, \qquad \bar{C}(z) \equiv \frac{1 - C(z)}{1 - z}.$$

Then,

$$1 = P(1) = \frac{p_0}{1 - r\bar{C}(1)}$$

and

$$L = P'(1) = p_0 \, r \cdot \frac{\bar{C}(1) + \bar{C}'(1)}{(1 - r\bar{C}(1))^2}.$$

Now, $\bar{C}(1)$ and $\bar{C}'(1)$ can be found by applying L'Hôpital's rule to $\bar{C}(z)$: $\bar{C}(1) = \mathrm{E}[X]$ (after one application) and $\bar{C}'(1) = \mathrm{E}[X(X-1)]/2$ (after two applications). Therefore

$$\boxed{p_0 = 1 - r\mathrm{E}[X] = 1 - \rho}$$

and

$$\boxed{L = \frac{r\left(\mathrm{E}[X] + \mathrm{E}[X^2]\right)}{2(1-\rho)} = \frac{\rho + r\mathrm{E}[X^2]}{2(1-\rho)} \qquad (\rho = \lambda \mathrm{E}[X]/\mu).} \qquad (3.4)$$

Note that $\bar{C}(z)$ is the generating function of the complementary batch-size probabilities $\Pr\{X > n\} = 1 - C_n = \bar{C}_n$. This is because $1/(1-z)$ is the generating function of 1; also, $C(z)/(1-z)$ is the generating function of the cumulative probabilities C_n, which can be seen by noting that

$$\sum_{n=1}^{\infty} C_n z^n = \sum_{n=1}^{\infty}\left(\sum_{i=1}^{n} c_i\right) z^n = \sum_{i=1}^{\infty}\left(c_i z^i \sum_{n=i}^{\infty} z^{n-i}\right) = \left(\sum_{i=1}^{\infty} c_i z^i\right)\frac{1}{1-z}.$$

As expected, $\rho < 1$ is the necessary and sufficient condition for stationarity. The remaining usual measures of effectiveness may be found by using the single-server results (Table 1.2) that $L_q = L - (1 - p_0) = L - \rho$ and then Little's formulas. The individual state probabilities $\{p_n\}$ can often also be obtained by the direct inversion of the generating function of (3.3). Two useful examples are to let the batch sizes be constant (each of size K) or geometrically distributed.

EXAMPLE 3.1

X is a constant K. In this case, the formula for the mean system size simplifies to

$$L = \frac{\rho + K\rho}{2(1-\rho)} = \frac{K+1}{2}\frac{\rho}{1-\rho} \quad (\rho = \lambda K/\mu), \tag{3.5}$$

which is precisely equal to the $M/M/1$ mean system size multiplied by $(K+1)/2$. Since this is a single-server system, it follows that

$$L_q = L - \rho = \frac{2\rho^2 + (K-1)\rho}{2(1-\rho)}. \tag{3.6}$$

The inversion of $P(z)$ to get the individual $\{p_n\}$ is a reasonable task when K is small (similar to Example 3.3).

EXAMPLE 3.2

X is geometrically distributed. Suppose that

$$c_n = (1-\alpha)\alpha^{n-1} \quad (0 < \alpha < 1).$$

It follows that $\rho = \lambda \mathrm{E}[X]/\mu = r/(1-\alpha)$, so

$$C(z) = (1-\alpha)\sum_{n=1}^{\infty} \alpha^{n-1} z^n = \frac{z(1-\alpha)}{1-\alpha z}.$$

From (3.3),

$$
\begin{aligned}
P(z) &= \frac{(1-\rho)(1-z)}{1-z-rz[1-C(z)]} \\
&= \frac{(1-\rho)(1-z)}{1-z-rz[1-z(1-\alpha)/(1-\alpha z)]} \\
&= \frac{(1-\rho)(1-\alpha z)}{1-z[\alpha+(1-\alpha)\rho]} \\
&= (1-\rho)\left(\frac{1}{1-z[\alpha+(1-\alpha)\rho]} - \frac{\alpha z}{1-z[\alpha+(1-\alpha)\rho]}\right).
\end{aligned}
$$

Therefore utilizing the formula for the sum of a geometric series,

$$
P(z) = (1-\rho)\left(\sum_{n=0}^{\infty}\{[\alpha+(1-\alpha)\rho]z\}^n - \sum_{n=0}^{\infty}\alpha[\alpha+(1-\alpha)\rho]^n z^{n+1}\right),
$$

from which we get

$$
\begin{aligned}
P_n &= (1-\rho)\{[\alpha+(1-\alpha)\rho]^n - \alpha[\alpha+(1-\alpha)\rho]^{n-1}\} \\
&= (1-\rho)[\alpha+(1-\alpha)\rho]^{n-1}[(1-\alpha)\rho] \quad (n>0).
\end{aligned}
$$

■ EXAMPLE 3.3

Let us consider a multistage machine-line process that produces an assembly in quantity. After the first stage many items are found to have one or more defects, which must be repaired before they enter the second stage. It is the job of one worker to make the necessary adjustments to put the assembly back into the stream of the process. The number of defects per item is registered automatically, and it exceeds two an extremely small number of times. The interarrival times for both units with one and two defective parts are found to be governed closely by exponential distributions, with parameters $\lambda_1 = 1/\text{h}$ and $\lambda_2 = 2/\text{h}$, respectively. There are so many different types of parts that have been found defective that an exponential distribution does actually provide a good fit for the worker's service-time distribution, with mean $1/\mu = 10$ min.

But it is subsequently noted that the rates of defects have increased, although not continuously. It is therefore decided to put another person on the job, who will concentrate on repairing those units with two defects, while the original worker works only on singles. When to add the additional person will be decided on the basis of a cost analysis.

Now there are a number of alternative cost structures available, and it is decided by management that the expected cost of the system to the company will be based on the average delay time of assemblies in for repair, which is directly proportional to the average number of units in the system, L. To find L under the assumption that there are only two possible batch sizes, we first note that

$\lambda = \lambda_1 + \lambda_2 = 3$, $r = \frac{1}{2}$, $c_1 = \lambda_1/\lambda = \frac{1}{3}$, and $c_2 = \lambda_2/\lambda = \frac{2}{3}$. Thus the batch-size mean and second moment are equal, respectively, to $\mathrm{E}[X] = \frac{1}{3} + 2 \times \frac{2}{3} = \frac{5}{3}$ and to $\mathrm{E}[X^2] = \frac{1}{3} + 2^2 \times \frac{2}{3} = 3$, and it follows that the system utilization rate is $\rho = \lambda \mathrm{E}[X]/\mu = \frac{5}{6}$ and

$$L = \frac{\rho + r\mathrm{E}[X^2]}{2(1-\rho)} = \frac{\frac{5}{6} + \frac{3}{2}}{2\left(1 - \frac{5}{6}\right)} = 7.$$

Although not necessary for solving the above problem, the individual state probabilities may be found by writing out the generating function from (3.3), using the fact that here $C(z) = c_1 z + c_2 z^2 = \frac{1}{3}z + \frac{2}{3}z^2$:

$$\begin{aligned} P(z) &= \frac{\mu(1-\rho)(1-z)}{\mu(1-z) - \lambda z(1 - z/3 - 2z^2/3)} \\ &= \frac{1}{6 - 3z - 2z^2}. \end{aligned}$$

The roots of the denominator of this generating function are $(-3 \pm \sqrt{57})/4$, or 1.137 and -2.637 to three-decimal-place accuracy. Because the two roots are each greater than 1 in absolute value, it follows that there is a two-term linear partial-fraction expansion of $P(z)$, which turns out to be

$$\begin{aligned} P(z) &= \frac{1}{7.550}\left(\frac{1}{1.137 - z} + \frac{1}{2.637 + z}\right) \\ &= \frac{1}{7.550}\left(\frac{1/1.137}{1 - z/1.137} + \frac{1/2.637}{1 + z/2.637}\right) \\ &= \frac{1}{7.550}\left[\frac{1}{1.137}\sum_{n=0}^{\infty}\left(\frac{z}{1.137}\right)^n + \frac{1}{2.637}\sum_{n=0}^{\infty}\left(-\frac{z}{2.637}\right)^n\right]. \end{aligned}$$

Therefore

$$P_n = 0.116(0.880)^n + 0.050(-0.379)^n \quad (n \geq 0).$$

Now, if C_1 is the cost per unit time per waiting repair and C_2 the cost of a worker per unit time, then the expected cost per unit time, C, of the single-server system is $C = C_1 L + C_2$. If a second repairer now sets up a separate service channel, the additional cost of his or her time is incurred, over and above the cost of the items in the queue. In this case, we have two queues. The singlet line would be a standard $M/M/1$, but the doublets would not. However, a single Poisson stream of doublets is merely a special case of the multiple Poisson bulk-input model with $\lambda_1 = 0$. The expected number of required repairs in the system is then the sum of the expected values of the two streams. Since the first stream is a standard $M/M/1$, its expected length is

$$L_1 = \frac{\lambda_1/\mu}{1 - \lambda_1/\mu}.$$

To get the expected length of the second line, we use (3.5) with $K = 2$ and $\rho = 2\lambda_2/\mu$, so that

$$L_2 = \frac{3\lambda_2/\mu}{1 - 2\lambda_2/\mu},$$

and thus

$$L = L_1 + L_2 = \frac{\lambda_1/\mu}{1 - \lambda_1/\mu} + \frac{3\lambda_2/\mu}{1 - 2\lambda_2/\mu}.$$

Therefore the new expected cost is

$$C^* = C_1\left(\frac{\lambda_1/\mu}{1 - \lambda_1/\mu} + \frac{3\lambda_2/\mu}{1 - 2\lambda_2/\mu}\right) + 2C_2.$$

Hence any decision is based on the comparative magnitude of C and C^*, and an additional channel is invoked whenever $C^* < C$, or

$$\begin{aligned} C_1\left(\frac{\lambda_1/\mu}{1 - \lambda_1/\mu} + \frac{3\lambda_2/\mu}{1 - 2\lambda_2/\mu}\right) + C_2 < C_1 L &= C_1\left(\frac{\rho + r\mathrm{E}[X^2]}{2(1-\rho)}\right) \\ &= C_1\left(\frac{\lambda_1/\mu + 3\lambda_2/\mu}{1 - \lambda_1/\mu - 2\lambda_2/\mu}\right), \end{aligned}$$

that is,

$$C_2 < C_1\left(\frac{\lambda_1/\mu + 3\lambda_2/\mu}{1 - \lambda_1/\mu - 2\lambda_2/\mu} - \frac{\lambda_1/\mu}{1 - \lambda_1/\mu} - \frac{3\lambda_2/\mu}{1 - 2\lambda_2/\mu}\right),$$

and removed when the inequality is reversed. Using our values for the parameters gives a decision criterion of $C_2 < 19C_1/5$.

3.2 Bulk Service ($M/M^{[Y]}/1$)

This section considers a single-server Markovian queue with *bulk service*. We specifically assume that customers arrive according to an ordinary Poisson process, service times are exponential, there is a single server, customers are served FCFS, there is no waiting-capacity constraint, and customers are served K at a time. We consider two variations of this bulk-service model. The variations arise from how the system operates when there are less than K in the system. We call these two models the *full-batch* model and the *partial-batch* model.

In the full-batch model, the server processes exactly K customers at a time. If less than K customers are in the system, then the server remains idle until there are K customers, at which point the server processes the K customers simultaneously. The service time is the same for all customers in a batch, and this time is exponentially distributed with mean $1/\mu$. This model could represent, for example, a ferry that waits until there are exactly K cars on board before it departs.

In the partial-batch model, the server can process partial batches up to a maximum size of K. As before, customers are served K at a time, but now if there are less than

K in the system, the server begins service on these customers. Furthermore, when there are less than K in service, new arrivals immediately enter service up to the limit K and finish with the others, regardless of the entry time into service. The amount of time required for the service of any batch is an exponentially distributed random variable with mean $1/\mu$, whether or not the batch is of full size K. This model could represent, for example, a tour in which late arriving customers join the tour (up to a maximum of K in the tour) and finish as a group.

We refer to these models using the notation $M/M^{[K]}/1$, with an additional identifier to specify "full batch" or "partial batch." We start with the partial-batch model.

3.2.0.1 Partial-Batch Model. The basic model is a non-birth–death Markovian problem. The stochastic balance equations are given as (see Problem 3.1)

$$\begin{aligned} 0 &= -(\lambda+\mu)p_n + \mu p_{n+K} + \lambda p_{n-1} \quad (n \geq 1), \\ 0 &= -\lambda p_0 + \mu p_1 + \mu p_2 + \cdots + \mu p_{K-1} + \mu p_K. \end{aligned} \tag{3.7}$$

The first equation of (3.7) may be rewritten in operator notation as

$$[\mu D^{K+1} - (\lambda+\mu)D + \lambda]p_n = 0 \quad (n \geq 0); \tag{3.8}$$

hence if $(r_1, \ldots, r_{K+1})$ are the roots of the operator or characteristic equation, then

$$p_n = \sum_{i=1}^{K+1} C_i r_i^n \quad (n \geq 0).$$

Since $\sum_{n=0}^{\infty} p_n = 1$, each r_i must be less than one or $C_i = 0$ for all r_i not less than one. Let us now determine the number of roots less than one. For this, an appeal is made to Rouché's theorem (see Section 2.11.2). It can be found that there is exactly one root (say, r_0) in (0, 1) (see Problem 3.2). So

$$p_n = C r_0^n \quad (n \geq 0, \quad 0 < r_0 < 1).$$

Using the boundary condition that $\sum p_n$ must total one, we find that $C = p_0 = 1 - r_0$; hence

$$\boxed{p_n = (1-r_0)r_0^n \quad (n \geq 0, \quad 0 < r_0 < 1).} \tag{3.9}$$

Measures of effectiveness for this model can be obtained in the usual manner. The stationary solution has the same geometric form (2.9) as that of the $M/M/1$ queue, but with r_0 replacing ρ. Thus

$$\boxed{L = \frac{r_0}{1-r_0}, \quad W = \frac{L}{\lambda} = \frac{r_0}{\lambda(1-r_0)}.}$$

Each customer has an average service time of $1/\mu$, so W_q can be derived from W. Then, L_q can be derived from W_q using Little's law.

$$\boxed{W_q = W - \frac{1}{\mu}, \quad L_q = L - \frac{\lambda}{\mu}.}$$

Alternatively, L_q can be computed directly from the stationary probabilities to yield $L_q = r_0^K L$ (Problem 3.5). It can also be shown that this is equal to $L_q = L - \lambda/\mu$.

■ EXAMPLE 3.4

The Drive-It-Through-Yourself Car Wash decides to change its operating procedure. It installs new machinery that permits the washing of two cars at once (and one if no other cars wait). A car that arrives while a single car is being washed joins the wash and finishes with the first car. There is no waiting-capacity limitation. Arrivals are Poisson with mean 20/hour. The time to wash a car is exponentially distributed with a mean of 5 min. What is the average line length?

The given parameters are $\lambda = 20$/h, $\mu = 1/(5 \text{ min}) = 12$/h, and $K = 2$. The characteristic equation from (3.8) is

$$12r^3 - 32r + 20 = 4(3r^3 - 8r + 5) = 0.$$

One root is $r = 1$, and division by the factor $r - 1$ leaves

$$3r^2 + 3r - 5 = 0,$$

which has roots $r = (-3 \pm \sqrt{69})/6$. We select the positive root with absolute value less than 1, namely, $r_0 = (-3 + \sqrt{69})/6 \doteq 0.884$. Therefore

$$L \doteq \frac{0.884}{0.116} \doteq 7.6 \text{ cars} \quad \text{and} \quad L_q \doteq 7.6 - \frac{20}{12} \doteq 5.9 \text{ cars}.$$

3.2.0.2 Full-Batch Model. Let us now assume that the batch size must be exactly K, and if not, the server waits until such time to start. Then the equations in (3.7) must be slightly rewritten to read

$$\begin{aligned} 0 &= -(\lambda + \mu)p_n + \mu p_{n+K} + \lambda p_{n-1} \quad (n \geq K), \\ 0 &= -\lambda p_n + \mu p_{n+K} + \lambda p_{n-1} \quad (1 \leq n < K), \\ 0 &= -\lambda p_0 + \mu p_K. \end{aligned} \tag{3.10}$$

The first equation of this second approach to bulk service is identical to that of the first approach; hence

$$p_n = Cr_0^n \quad (n \geq K - 1, \quad 0 < r_0 < 1).$$

However, obtaining C (and p_0) is more complicated here, since the formula for p_n is valid only for $n \geq K - 1$. From the last equation of (3.10),

$$p_K = \frac{\lambda}{\mu} p_0 = Cr_0^K,$$

so

$$C = \frac{\lambda p_0}{\mu r_0^K} \quad \text{and} \quad p_n = \frac{p_0 \lambda r_0^{n-K}}{\mu} \quad (n \geq K - 1).$$

To get p_0, we use the $K-1$ stationary equations given in (3.10) as

$$\mu p_{n+K} = \lambda p_n - \lambda p_{n-1} \qquad (1 \le n < K).$$

The geometric form for p_n (when $n \ge K-1$) can be substituted for p_{n+K} in the previous equation giving

$$p_0 r_0^n = p_n - p_{n-1} \quad (1 \le n < K). \tag{3.11}$$

These equations can be solved by iteration starting with $n = 1$ or we can note that these are nonhomogeneous linear difference equations whose solutions are

$$p_n = C_1 + C_2 r_0^n.$$

Direct substitution into (3.11) implies that $C_2 = -p_0 r_0/(1-r_0)$. The boundary condition at $n = 0$ implies that $C_1 = p_0 - C_2$. This gives

$$p_n = \begin{cases} \dfrac{p_0(1-r_0^{n+1})}{1-r_0} & (1 \le n \le K-1), \\[2ex] \dfrac{p_0 \lambda r_0^{n-K}}{\mu} & (n \ge K-1). \end{cases} \tag{3.12}$$

[Either form for p_n is valid when $n = K-1$. This can be shown using (3.8).] To get p_0 we use the usual boundary condition that $\sum_{n=0}^{\infty} p_n = 1$. Hence from (3.12),

$$\begin{aligned} p_0 &= \left(1 + \sum_{n=1}^{K-1} \frac{1-r_0^{n+1}}{1-r_0} + \frac{\lambda}{\mu} \sum_{n=K}^{\infty} r_0^{n-K}\right)^{-1} \\ &= \left(1 + \frac{K-1}{1-r_0} - \frac{r_0^2(1-r_0^{K-1})}{(1-r_0)^2} + \frac{\lambda}{\mu(1-r_0)}\right)^{-1} \\ &= \left(\frac{\mu r_0^{K+1} - (\lambda+\mu)r_0 + \lambda + \mu K(1-r_0)}{\mu(1-r_0)^2}\right)^{-1}. \end{aligned}$$

But, from the characteristic equation in (3.8), we know that

$$\mu r_0^{K+1} - (\lambda+\mu)r_0 + \lambda = 0.$$

Thus

$$p_0 = \frac{\mu(1-r_0)^2}{\mu K(1-r_0)} = \frac{1-r_0}{K}. \tag{3.13}$$

The formulas for the $\{p_n\}$ can also be obtained via the probability generating function, which can be shown to be (see Problem 3.4)

$$P(z) = \frac{(1-z^K)\sum_{n=0}^{K-1} p_n z^n}{r z^{K+1} - (r+1)z^K + 1} \qquad (r = \lambda/\mu). \tag{3.14}$$

To make use of (3.14), it is necessary to eliminate the $p_n, n = 0, 1, \ldots, K-1$, from the numerator of the right-hand side. To do this we again appeal to Rouché's theorem (see Section 2.11.2). The generating function $P(z)$ has the property that it must converge inside the unit circle. We note that the denominator of $P(z)$ has $K+1$ zeros. Applying Rouché's theorem to the denominator (see Problem 3.6) tells us that K of these lie on or within the unit circle. One zero of the denominator is $z = 1$; thus $K-1$ lie within and must coincide with those of $\sum_{n=0}^{K-1} p_n z^n$ for $P(z)$ to converge, so that when a zero appears in the denominator it is canceled by one in the numerator. Hence this leaves one zero of the denominator (since there are a total of $K+1$) that lies outside the unit circle. We denote this by z_0.

A major observation is that the roots of the denominator are precisely the reciprocals of those of the characteristic equation (3.8). Thus $z_0 = 1/r_0$. This inverse relationship can be seen for the $M/M/1$ queue by comparing (2.13) and (2.21). In fact, it can be shown that the zeros of the characteristic equation of any system of linear difference equations with constant coefficients are the reciprocals of the poles of the system's generating function. This follows because of the way that the poles show up in a partial-fraction expansion in terms like $1/(a-z)$, which then can be rewritten as the infinite series $(1/a)\sum_{n=0}^{\infty}(z/a)^n$. We have already seen this sort of thing in Example 3.3.

Returning now to the generating function, we see that dividing the denominator by the product $(z-1)(z-z_0)$ results in a polynomial with $K-1$ roots inside the unit circle. These must therefore match the roots of $\sum_{n=0}^{K-1} p_n z^n$, so that the two polynomials differ by at most a multiplicative constant, and we may therefore write that

$$\sum_{n=0}^{K-1} p_n z^n = A\frac{\rho z^{K+1} - (\rho+1)z^K + 1}{(z-1)(z-z_0)}.$$

Substituting the right-hand side above into (3.14) yields

$$P(z) = \frac{A(1-z^K)}{(z-1)(z-z_0)} = \frac{A}{z_0 - z}\sum_{n=0}^{K-1} z_n.$$

Since $P(1) = 1$, it follows that

$$A = \frac{z_0 - 1}{K}$$

and thus

$$P(z) = \frac{(z_0 - 1)\sum_{n=0}^{K-1} z^n}{K(z_0 - z)}. \tag{3.15}$$

Prior to showing an example, we make a final comment on batch service queues. Chaudhry and Templeton (1983) have treated the generalized batch service policy such that if the number in the queue when the server becomes free (N) is less than k, the server waits until $N = k$ to start serving a batch. If $N > K$, the server takes the first K in the queue. Once service begins, future arrivals must wait until the server is free (unlike our first model). When $k = K$, this policy reduces to our second model. To treat this policy in general requires methods more advanced than those used in this chapter.

3.3 Erlangian Models

Up to now, all probabilistic queueing models studied have assumed Poisson input (exponential interarrival times) and exponential service times, or simple variations thereof. In many practical situations, however, the exponential assumptions may be rather limiting, especially the assumption concerning service times being distributed exponentially. In this section, we allow for a more general probability distribution for describing the input process and/or the service mechanism.

3.3.1 The Erlang Distribution

To begin, consider a random variable T that has the gamma probability density

$$f(t) = \frac{1}{\Gamma(\alpha)\beta^\alpha} t^{\alpha-1} e^{-t/\beta} \quad (\alpha, \beta > 0, \quad 0 < t < \infty),$$

where $\Gamma(\alpha) = \int_0^\infty t^{\alpha-1} e^{-t} dt$ is the gamma function and α and β are the parameters of the distribution. The mean and variance of this distribution are

$$E[T] = \alpha\beta,$$
$$\text{Var}[T] = \alpha\beta^2.$$

Now, we consider a special class of these distributions where α is restricted to be a positive integer. Specifically, α and β are related by

$$\alpha = k \text{ and } \beta = \frac{1}{k\mu},$$

where k is any arbitrary positive integer and μ is any arbitrary positive constant. This gives the Erlang family of probability distributions with PDF and CDF

$$f(t) = \frac{(\mu k)^k}{(k-1)!} t^{k-1} e^{-k\mu t} \quad (0 < t < \infty), \tag{3.16}$$

$$F(t) = 1 - \sum_{n=0}^{k-1} e^{-k\mu t} \frac{(k\mu t)^n}{n!} \quad (0 \le t < \infty). \tag{3.17}$$

The parameters of the Erlang are k and μ, and the mean and variance are

$$E[T] = \frac{1}{\mu}, \tag{3.18}$$

$$\text{Var}[T] = \frac{1}{k\mu^2}. \tag{3.19}$$

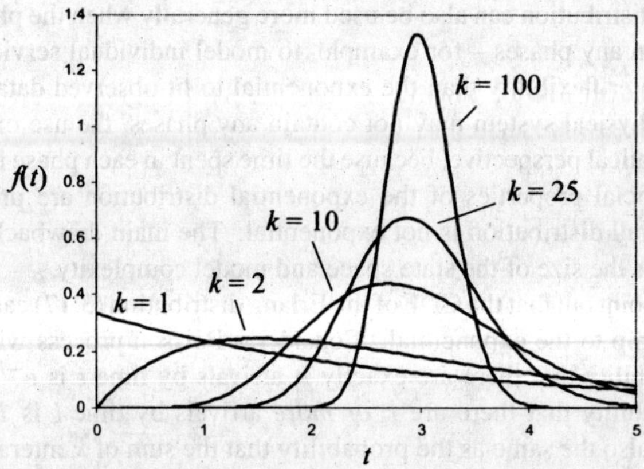

Figure 3.2 A family of Erlang distributions with mean 3.

For a particular value of k, the distribution is referred to as an Erlang type-k or E_k distribution. The Erlang family provides more flexibility in modeling than the exponential family, which only has one parameter. Figure 3.2 shows how the Erlang distribution varies with k, often called the shape parameter. When $k = 1$, the Erlang reduces to an exponential distribution with mean $1/\mu$. As k increases, the Erlang becomes more symmetrical and more closely centered around its mean. As $k \to \infty$, the Erlang becomes deterministic with value $1/\mu$. In practical situations, the Erlang family provides more flexibility in fitting a distribution to real data than the exponential family provides.

The Erlang distribution is also useful in queueing analysis because of its relationship to the exponential distribution. Specifically, we have the following property: The sum of k IID exponential random variables with mean $1/k\mu$ is an Erlang type-k distribution.

The proof of this property is left as an exercise; see Problem 2.67. This property makes it possible to take advantage of the Markovian property of the exponential distribution, even though the Erlang distribution itself is not Markovian (Section 1.8 showed that the exponential distribution was unique among continuous distributions in its Markovian property).

To demonstrate the applicability of the Erlang distribution, consider a technician who must perform four steps in a laboratory test, where each step is exponentially distributed with mean $1/4\mu$. Figure 3.3 gives a pictorial representation. The overall service function (time to complete the test) is Erlang type-4 with mean $1/\mu$. If the input process is Poisson, we have an $M/E_4/1$ queueing model. There are several implications of this model. First, all steps (or phases) of the service are independent and *identical*. Second, only one customer (or test) at a time is allowed in the service mechanism. That is, a customer enters phase 1 of the service, then progresses through the remaining phases, and must complete the last phase before the next customer enters the first phase. This rules out assembly-line-type models where, as soon as a customer finishes one phase of service, another can enter it.

The Erlang distribution can also be used more generally when the physical system does not contain any phases – for example, to model individual service times. The Erlang has greater flexibility than the exponential to fit observed data (Figure 3.2). Although the physical system may not contain any phases, the use of phases helps from a mathematical perspective, because the time spent in each phase is exponential. Thus the beneficial properties of the exponential distribution are preserved, even though the overall distribution is not exponential. The main drawback is that using phases increases the size of the state space and model complexity.

Finally, we point out that the CDF of the Erlang distribution (3.17) can be explained by its relationship to the exponential. Consider a Poisson process with arrival rate $k\mu$. The probability that there are exactly n arrivals by time t is $e^{-k\mu t}(k\mu t)^n/n!$. Thus the probability that there are k *or more* arrivals by time t is $F(t)$, given in (3.17). This is also the same as the probability that the sum of k interarrival times is less than or equal to t. Since the interarrival times are exponential with rate $k\mu$, this is the CDF of an Erlang type-k distribution.

3.3.2 Generalization to Phase-Type Distributions

The concept of phases can be generalized to include a much wider class of distributions than just the Erlang. Other distributions that use the concept of phases are the hyperexponential, hypoexponential or generalized Erlang, and Coxian distributions. For further information on phase-type distributions and their application in queueing theory, see Neuts (1981).

To show how this can be done, we demonstrate another method for generating an Erlang distribution. Consider the following three-state continuous-time Markov chain

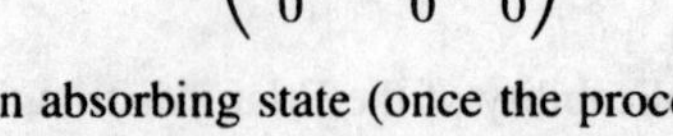

with $\boldsymbol{Q}$ matrix

$$\boldsymbol{Q} = \begin{pmatrix} -\mu & \mu & 0 \\ 0 & -\mu & \mu \\ 0 & 0 & 0 \end{pmatrix}.$$

In this chain, state 3 is an absorbing state (once the process enters state 3, it never leaves). Suppose the process starts in state 1. Let T be the time to absorption. Since the time in state 1 is exponential with mean $1/\mu$ and the time in state 2 is exponential

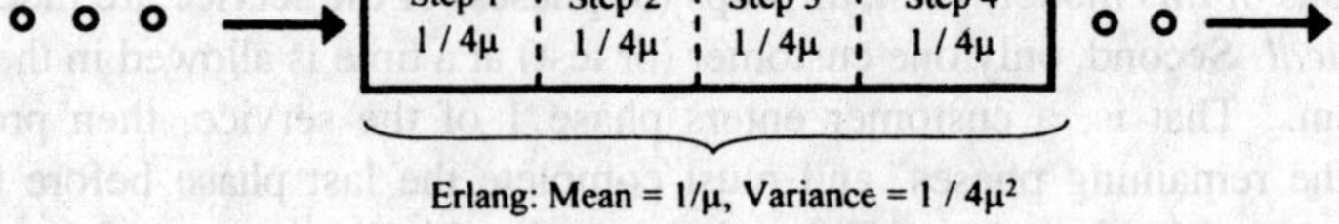

Figure 3.3 Use of the Erlang for phased service.

with mean $1/\mu$, the time to absorption is the sum of these two exponential random variables. Thus T follows an E_2 distribution (here, each phase has rate μ rather than 2μ as discussed in the previous section).

This demonstrates the basic idea. The phase-type distribution in question is defined as the time to absorption in a specified continuous-time Markov chain. By choosing different Markov chains and by choosing different initial probability vectors, we can construct different phase-type distributions. These distributions are not (in general) exponential, but they can be analyzed using the theory of continuous-time Markov chains, taking advantage of the underlying exponential distributions.

The hyperexponential distribution can also be constructed in a similar manner. Consider the Markov chain

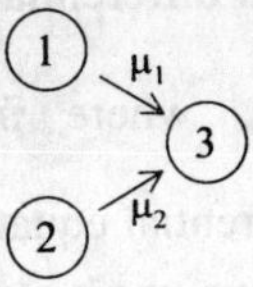

with $\boldsymbol{Q}$ matrix

$$\boldsymbol{Q} = \begin{pmatrix} -\mu_1 & 0 & \mu_1 \\ 0 & -\mu_2 & \mu_2 \\ 0 & 0 & 0 \end{pmatrix}.$$

Suppose that the system starts in state 1 with probability q and in state 2 with probability $1-q$. That is, the initial state vector is $\boldsymbol{p}(0) = (q, 1-q)$. Let T be the time to absorption. Then, T follows a hyperexponential (or H_2) distribution. In other words, with probability q, the time to absorption is exponential with mean $1/\mu_1$, and with probability $1-q$, the time to absorption is exponential with mean $1/\mu_2$.

For these examples, it was relatively easy to determine the distribution of T, because the Markov chains were simple. We now illustrate a more formal quantitative approach that can be applied to more complex Markov chains. We illustrate first with the hyperexponential model.

The Chapman–Kolmogorov equations for the hyperexponential Markov chain are

$$\begin{aligned} p_1(t+\Delta t) &= (1-\mu_1\Delta t)p_1(t) + o(\Delta t), \\ p_2(t+\Delta t) &= (1-\mu_2\Delta t)p_2(t) + o(\Delta t), \\ p_3(t+\Delta t) &= \mu_1\Delta t\, p_1(t) + \mu_2\Delta t\, p_2(t) + p_3(t) + o(\Delta t). \end{aligned}$$

The resulting differential equations are

$$\begin{aligned} p_1'(t) &= -\mu_1 p_1(t), \\ p_2'(t) &= -\mu_2 p_2(t), \\ p_3'(t) &= \mu_1 p_1(t) + \mu_2 p_2(t). \end{aligned}$$

These equations can also be obtained from $\boldsymbol{p}'(t) = \boldsymbol{p}(t)\boldsymbol{Q}$; see (1.29). The solution to the first differential equation is $p_1(t) = qe^{-\mu_1 t}$, where the constant q comes from the initial condition $p_1(0) = q$. Likewise, $p_2(t) = (1-q)e^{-\mu_2 t}$. Thus

$$p_3'(t) = q\mu_1 e^{-\mu_1 t} + (1-q)\mu_2 e^{-\mu_2 t}.$$

By definition, $p_3(t) = \Pr\{T \le t\}$. Thus $p_3'(t)$ is the PDF of T. From the equation above, we see that $p_3'(t)$ is the PDF of a two-term hyperexponential distribution, as expected.

Alternatively, the solutions for $p_1(t)$ and $p_2(t)$ can be derived directly from the matrix-vector formulation. First, let $\tilde{\boldsymbol{p}}(t) = (p_1(t), p_2(t))$ be the vector of state probabilities without the absorbing state. Let $\tilde{\boldsymbol{Q}}$ be the matrix consisting of the first two rows and columns of $\boldsymbol{Q}$. That is,

$$\tilde{\boldsymbol{Q}} = \begin{pmatrix} -\mu_1 & 0 \\ 0 & -\mu_2 \end{pmatrix}, \quad \text{and} \quad \boldsymbol{p}'(t) = \tilde{\boldsymbol{p}}(t)\tilde{\boldsymbol{Q}}.$$

The solution to this matrix system of differential equations is

$$\tilde{\boldsymbol{p}}(t) = \tilde{\boldsymbol{p}}(0)e^{\tilde{\boldsymbol{Q}}t}, \quad \text{where} \quad \tilde{\boldsymbol{p}}(0) = (q, 1-q).$$

This is analogous to the single differential equation $y'(t) = ay(t)$ whose solution is $y(t) = y(0)e^{at}$. To evaluate $e^{\tilde{\boldsymbol{Q}}t}$, we use a series expansion (similar to the scalar series expansion of e^x):

$$\begin{aligned} e^{\tilde{\boldsymbol{Q}}t} &= \boldsymbol{I} + \tilde{\boldsymbol{Q}}t + \frac{(\tilde{\boldsymbol{Q}}t)^2}{2!} + \cdots \\ &= \begin{pmatrix} 1 & 0 \\ 0 & 1 \end{pmatrix} + \begin{pmatrix} -\mu_1 t & 0 \\ 0 & -\mu_2 t \end{pmatrix} + \begin{pmatrix} \mu_1^2 t^2/2 & 0 \\ 0 & \mu_2^2 t^2/2 \end{pmatrix} + \cdots \\ &= \begin{pmatrix} e^{-\mu_1 t} & 0 \\ 0 & e^{-\mu_2 t} \end{pmatrix}. \end{aligned}$$

Thus

$$\tilde{\boldsymbol{p}}(t) = \tilde{\boldsymbol{p}}(0)e^{\tilde{\boldsymbol{Q}}t} = \left(qe^{-\mu_1 t}, (1-q)e^{-\mu_2 t}\right).$$

In other words, $p_1(t) = qe^{-\mu_1 t}$ and $p_2(t) = (1-q)e^{-\mu_2 t}$, as we found earlier. $\Pr\{T \le t\} = p_3'(t)$ is obtained as before using the last differential equation $p_3'(t) = \mu_1 p_1(t) + \mu_2 p_2(t)$.

■ EXAMPLE 3.5

We now apply these methods to the Markov chain corresponding to the E_2 distribution. The differential equations associated with this system are

$$\begin{aligned} p_1'(t) &= -\mu p_1(t), \\ p_2'(t) &= \mu p_1 - \mu p_2(t), \\ p_3'(t) &= \mu p_2(t). \end{aligned}$$

The solution to the first differential equation is $p_1(t) = e^{-\mu t}$ [where the constant in front comes from the initial condition $p_1(0) = 1$]. Then, the second differential equation becomes

$$p_2'(t) + \mu p_2(t) = \mu e^{-\mu t}.$$

The solution to this first-order, linear differential equation has the form $p_2(t) = Ae^{-\mu t} + Bte^{-\mu t}$. Substituting $p_2(t)$ into the differential equation gives $B = \mu$. Then, the boundary condition $p_2(0) = 0$ gives $A = 0$. Thus $p_2(t) = \mu te^{-\mu t}$. Finally, $p_3'(t)$ gives the E_2 density:

$$p_3'(t) = \mu p_2(t) = \mu^2 te^{-\mu t}.$$

This is the same as the PDF from (3.16), but with $k\mu$ replaced by μ. Alternatively, the same result can be obtained using the matrix formulation (Problem 3.25).

■ **EXAMPLE 3.6**

Hypoexponential distribution. Consider the following continuous-time Markov chain, where $\mu_1 \neq \mu_2$ and the system starts in state 1:

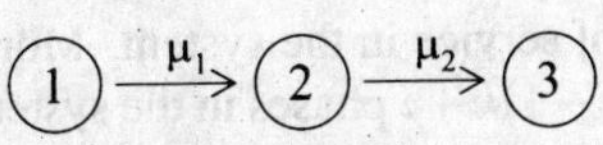

The associated $\boldsymbol{Q}$ matrix and initial conditions are

$$\boldsymbol{Q} = \begin{pmatrix} -\mu_1 & \mu_1 & 0 \\ 0 & -\mu_2 & \mu_2 \\ 0 & 0 & 0 \end{pmatrix}, \quad \boldsymbol{p}(0) = (1, 0, 0).$$

The time to absorption is the convolution of two *nonidentical* exponential random variables. (This is not an Erlang, since Erlangs require IID exponentials.) The derivation of $\boldsymbol{p}(t)$ is left as an exercise (Problem 3.11). It is found that

$$\begin{aligned} p_1(t) &= e^{-\mu_1 t}, \\ p_2(t) &= \frac{\mu_1}{\mu_2 - \mu_1} e^{-\mu_1 t} - \frac{\mu_1}{\mu_2 - \mu_1} e^{-\mu_2 t}, \\ p_3'(t) &= \mu_2 p_2(t) = \frac{\mu_1 \mu_2}{\mu_2 - \mu_1} e^{-\mu_1 t} - \frac{\mu_1 \mu_2}{\mu_2 - \mu_1} e^{-\mu_2 t}, \end{aligned} \tag{3.20}$$

where $p_3'(t)$ is the density function of the time to absorption, which is a two-term hypoexponential distribution.

Thus we see that this device provides a method for obtaining considerably more general distributions than merely Erlangs. We have more material on these matters in later sections.

3.3.3 Erlang Service Model ($M/E_k/1$)

We now consider a model in which the service time has an Erlang type-k distribution. More specifically, the overall service rate is assumed to be μ, so the rate of each service phase is $k\mu$. Even though the service may not actually consist of k phases, it is convenient in analyzing this model to consider the Erlang in this way. Let $p_{n,i}(t)$ be the probability that in steady state there are n customers in the system and the customer in service is in phase i $(i = 1, 2, \ldots, k)$. Here, we number the phases backward, so k is the first phase of service and 1 is the last (a customer leaving phase 1 actually leaves the system). We can write the following steady-state balance equations (Problem 3.9):

$$\begin{aligned}
0 &= -(\lambda + k\mu)p_{n,i} + k\mu p_{n,i+1} + \lambda p_{n-1,i} && (n \geq 2, 1 \leq i \leq k-1),\\
0 &= -(\lambda + k\mu)p_{n,k} + k\mu p_{n+1,1} + \lambda p_{n-1,k} && (n \geq 2),\\
0 &= -(\lambda + k\mu)p_{1,i} + k\mu p_{1,i+1} && (1 \leq i \leq k-1),\\
0 &= -(\lambda + k\mu)p_{1,k} + k\mu p_{2,1} + \lambda p_0,\\
0 &= -\lambda p_0 + k\mu p_{1,1}.
\end{aligned} \tag{3.21}$$

These equations are not particularly easy to handle because of their bivariate nature. However, it is not too difficult to obtain the expected measures of effectiveness (L, L_q, W, W_q) and the state probabilities from (3.21) by working directly on the process that counts phases of service in the system. More specifically, if the system is in state (n, i), there are $(n-1)k + i$ phases in the system. In other words, there are $n - 1$ customers in the queue (each requiring k phases) and the customer in service requires i more phases in service.

This is essentially equivalent to modeling the Erlang service queue as a constant bulk-input model (the $M^{[K]}/M/1$ of Section 3.1), where each input unit brings in $K = k$ phases and the (phase) service rate μ is replaced by $k\mu$ (see Problem 3.10). The two queues are not identical in all respects. In the Erlang queue, completed intermediate phases of service do not correspond to customer departures, whereas in the bulk queue, each individual customer departs the queue upon completion of service.

This connection can be used to determine the average wait in queue W_q for the $M/E_k/1$ queue. To obtain W_q, first observe that the average number of phases in the $M/E_k/1$ queue is the same as the average number of customers in the analogous $M^{[X]}/M/1$ queue, given in (3.5):

$$\frac{k+1}{2}\frac{\rho}{1-\rho}.$$

[Here, the service rate $k\mu$ replaces μ in (3.5), so $\rho = k\lambda/(k\mu) = \lambda/\mu$.] Since the average time to process each phase is $1/k\mu$, the average wait in queue is the average number of phases in the system multiplied by $1/k\mu$. This yields

$$\boxed{W_q = \frac{1 + 1/k}{2}\frac{\rho}{\mu(1-\rho)} \quad (\rho = \lambda/\mu).} \tag{3.22}$$

Problem 3.13 explores a variation of this argument applied to other queues. It follows that

$$\boxed{L_q = \lambda W_q = \frac{1 + 1/k}{2}\frac{\rho^2}{1-\rho},} \tag{3.23}$$

and $L = L_q + \rho, W = L/\lambda = W_q + 1/\mu$.

Now, to compute the steady-state probabilities themselves, we can immediately get the empty probability p_0, since we know for all single-channel, one-at a time-service queues that $p_0 = 1 - \rho$ (see Table 1.2). To get p_n, we convert (3.21) to a single-variable system using the transformation $(n, i) \leftrightarrow (n-1)k + i$. In other words, each state (n, i) is replaced with the single-variable state $(n-1)k + i$. This yields the following system of equations:

$$
\begin{aligned}
0 &= -(\lambda + k\mu)p_{(n-1)k+i} + k\mu p_{(n-1)k+i+1} + \lambda p_{(n-2)k+i} \qquad (3.24)\\
&\qquad\qquad (n \geq 1,\ 1 \leq i \leq k),\\
0 &= -\lambda p_0 + k\mu p_1,
\end{aligned}
$$

where any p turning out to have a negative subscript is assumed to be zero. Writing out the top equation in (3.24) sequentially starting at $n = 1, i = 1$ gives a simplified set of equations:

$$
\begin{aligned}
0 &= -(\lambda + k\mu)p_n + k\mu p_{n+1} + \lambda p_{n-k} \quad (n \geq 1),\\
0 &= -\lambda p_0 + k\mu p_1.
\end{aligned}
\qquad (3.25)
$$

This is precisely what (3.1) gives for a constant batch size k and service rate $k\mu$. Now if we let $p_n^{(p)}$ represent the probability of n in the bulk-input system defined by (3.25), then it follows that the probability of n in the Erlang service system, p_n, is given by

$$
p_n = \sum_{j=(n-1)k+1}^{nk} p_j^{(p)} \quad (n \geq 1). \qquad (3.26)
$$

The waiting-time CDF can also be obtained for this problem, but the method of solution is quite different from those presented so far. The results also follow nicely from the general theory of $G/G/1$ queues as presented later in Chapter 6, Section 6.1.

While we have here utilized the relation between the $M^{[k]}/M/1$ and the $M/E_k/1$, it is important to note that a similar partnership holds between the $M/M^{[k]}/1$ queue and the $E_k/M/1$, so that the previous bulk results of Section 3.2 can be useful in deriving results about the Erlang arrival model to be treated in the following section. Prior to considering an Erlang arrival model, we illustrate Erlang service models by the following three examples.

■ EXAMPLE 3.7

The Grabeur-Money Savings and Loan has a drive-up window. During the busy periods for drive-up service, customers arrive according to a Poisson distribution with a mean of 16/h. From observations on the teller's performance, the mean service time is estimated to be 2.5 min, with a standard deviation of $\frac{5}{4}$ min. It is thought that the Erlang would be a reasonable assumption for the distribution of the teller's service time. Also, since the building (and drive-up window) is located in a large shopping center, there is virtually no limit on the number of vehicles that can wait. The company officials wish to know, on average, how long a customer must wait until reaching the window for service, and how many vehicles are waiting for service.

The appropriate model, of course, is an $M/E_k/1$ model. To determine k, we first note that $1/\mu = 2.5$ min and that $\sigma^2 = 1/k\mu^2 = \frac{25}{16}$, which yields $k = 4$. Thus we have an $M/E_4/1$ model with $\rho = \frac{2}{3}$, so from (3.23) we have

$$L_q = \frac{5}{8}\frac{\frac{4}{9}}{1-\frac{2}{3}} = \frac{5}{6}, \qquad W_q = \frac{60}{16}\frac{5}{6} = \frac{25}{8} \text{ min}.$$

■ EXAMPLE 3.8

A small heating-oil distributor, the Phil R. Upp Company, has only one truck. The capacity of the truck is such that after delivering to a customer, it must return to be refilled. Customers call in for deliveries on an average of once every 50 min during the height of the heating season. The distribution of time between calls has been found to be approximately exponential. It has also been observed that it takes on average 20 min for the truck to get to a customer and 20 min for the truck to return. Both the time for driving out and the time for returning have approximately exponential distributions. The time it takes to unload and load the truck has been included in the driving times. W. A. R. Mup, company general manager, is considering the possibility of purchasing an additional truck and would like to know, under the current situation, how long on average a customer has to wait from the time he places a call until the truck arrives. All customers are served on a first-come, first-served basis.

The service time in this problem corresponds to the time a truck is tied up and is made up of two identical exponential stages, one consisting of loading and traveling out to the customer, and one consisting of unloading and traveling back to the terminal. Hence we have an $M/E_2/1$ model, where $\lambda = \frac{6}{5}$/h and $\mu = \frac{3}{2}$/h, so that $\rho = \frac{4}{5}$. The average time that a customer must wait from the time the call is placed until service starts (the truck begins loading for his delivery) is W_q and is given by (3.23) as

$$W_q = \frac{3}{4}\frac{\frac{16}{25}}{\frac{6}{5}(1-\frac{4}{5})} = 2 \text{ h}.$$

The average time it takes the truck to arrive once it is dispatched (loading for a given customer commences) is one-half an average service time, that is, 20 min. Thus the average wait from the time a customer calls in until the truck arrives and begins unloading is 2 h 20 min.

■ EXAMPLE 3.9

A manufacturer of a special electronic guidance-system component has a quality control checkpoint at the end of the production line to assure that the component is properly calibrated. If it fails the test, the component is sent to a special repair center, where it is readjusted. There are two specialists at the center, and each can adjust a component in an average of 5 min, their repair time being exponentially distributed. The average number of rejects from the quality control point per hour is 18, and the sequence of rejections appears to be well described by a Poisson process. The company can lease one machine that can adjust the component to the same degree of accuracy as the repair staff in exactly $2\frac{2}{3}$ min, that is, with no variation in repair time. The machine leasing costs are roughly equivalent to the salary and fringe-benefit costs for the staff. (If the repairers are replaced, they can be used elsewhere in the company and there will be no morale or labor problems.) Should the company lease the machine?

We wish to compare the expected waiting time W and system size L under each alternative. Alternative 1, keeping the staff of two, is an *M/M/*2 model, while alternative 2, leasing the machine, is an *M/D/*1 model. The calculations for the alternatives are as follows.

*M/M/*2:

$$\lambda = 18/\text{h}, \mu = 0.2/\text{ min} = 12/\text{h}, \qquad W_q = \tfrac{3}{28}\text{h} \doteq 6.4 \text{ min},$$
$$W \doteq 6.4 + 5 = 11.4 \text{ min} \quad \text{and} \quad L = \lambda W \doteq 3.42.$$

*M/D/*1: To obtain the results for this model, we use $\lim_{k\to\infty} M/E_k/1$;

$$\lambda = 18/\text{h}, \quad \mu = \tfrac{3}{8}/\text{ min} = 22.5/\text{h},$$
$$W_q = \lim_{k\to\infty}\left(\frac{1+1/k}{2}\frac{\rho}{\mu(1-\rho)}\right) = \frac{\rho}{2\mu(1-\rho)} = \frac{4}{45}\text{h} = \frac{16}{3} \text{ min},$$
$$W = \tfrac{16}{3} + \tfrac{8}{3} = 8 \text{ min} \quad \text{and} \quad L = \lambda W = 2.4.$$

Thus alternative 2 (the machine) is shown to be preferable.

3.3.4 Erlang Arrival Model ($E_k/M/1$)

As mentioned in the previous section, we can utilize the results of the second bulk-service model of Section 3.2 to develop results for the Erlang input model. We assume that the interarrival times are Erlang type-k distributed, with a mean of $1/\lambda$. We can look, therefore, at an arrival having passed through k phases, each with a mean time of $1/k\lambda$, prior to actually entering the system. Here we number the phases frontwards from 0 to $k-1$. Again, this is a device convenient for analysis that does not necessarily correspond to the actual arrival mechanism.

Let $p_n^{(P)}$ denote the number of *arrival phases* in the system in steady state. By this we mean the following: The number of arrival phases in the system includes k phases for each customer who has already arrived (but not yet departed) as well as the completed phases corresponding to the next arrival (even though this customer has not officially "arrived"). Then, the probabilities p_n, denoting the number of *customers* in the system in steady state, are given by a relation similar to (3.26):

$$p_n = \sum_{j=nk}^{nk+k-1} p_j^{(P)}. \tag{3.27}$$

This system is identical in structure to the full-batch bulk-service model given in Section 3.2. More specifically, the rate balance equations for $p_n^{(P)}$ in this model are identical to the rate balance equations for p_n in the bulk-service model, except with λ replaced by $k\lambda$ (and K replaced by k). Then, (3.12) and (3.13) imply

$$p_j^{(P)} = \rho(1-r_0)r_0^{j-k} \quad (j \geq k-1,\ \rho = \lambda/\mu), \tag{3.28}$$

where r_0 is the single root in $(0, 1)$ of the characteristic equation

$$\mu r^{k+1} - (k\lambda + \mu)r + k\lambda = 0, \tag{3.29}$$

which is the analog of (3.8). So, for $n \geq 1$,

$$\begin{aligned} p_n &= \sum_{j=nk}^{nk+k-1} p_j^{(P)} \\ &= \rho(1-r_0)(r_0^{nk-k} + r_0^{nk-k+1} + \cdots + r_0^{nk-1}) \\ &= \rho(1-r_0)r_0^{nk-k}(1 + r_0 + \cdots + r_0^{k-1}) \\ &= \rho(1-r_0^k)(r_0^k)^{n-1}. \end{aligned} \tag{3.30}$$

We see that p_n has a geometric form (as with the $M/M/1$ queue), but with r_0^k as the geometric parameter instead of ρ. In Chapter 5, Section 5.3, and Chapter 6, Section 6.4, we will show that the general-time steady-state system-size probabilities for all $G/M/1$ queues have a geometric form.

It follows from (3.30) that

$$L = \rho(1-r_0^k)\sum_{n=1}^{\infty} n(r_0^k)^{n-1} = \rho(1-r_0^k)\frac{1}{(1-r_0^k)^2}.$$

Thus

$$\boxed{L = \frac{\rho}{1-r_0^k}.}$$

From this, we can get $L_q = L - \rho$, $W = L/\lambda$, and $W_q = W - 1/\mu$.

We can also derive the waiting-time distribution for this model as follows (recall that this is a composite distribution, because there is a nonzero probability of no wait for service to commence). Let q_n denote the probability that an arriving customer finds n in the system. Because the arrival process is not Poisson, arrivals do not see time averages, so $q_n \neq p_n$. Instead, we find that

$$\begin{aligned} q_n &= \frac{\text{rate of customers arriving to find } n \text{ in the system}}{\text{total rate of arrivals}} \\ &= \frac{k\lambda \cdot \Pr\{n \text{ in system and arrival in phase } k-1\}}{\lambda} \\ &= kp^{(P)}_{nk+k-1}. \end{aligned}$$

From (3.28), we have, for $n \geq 0$,

$$\begin{aligned} q_n &= kp^{(P)}_{nk+k-1} \\ &= \frac{k\rho(1-r_0)}{r_0} r_0^{kn} \\ &= (1-r_0^k) r_0^{kn}, \end{aligned} \tag{3.31}$$

where the final step follows since r_0 is a root of (3.29). Now if there are n customers in the system upon arrival, the conditional waiting time is the time it takes to serve these n people, which is the convolution of n exponentials, each with mean $1/\mu$. This yields an Erlang type-n distribution, and the unconditional line-delay distribution function can be written as

$$\begin{aligned} W_q(t) &= q_0 + \sum_{n=1}^{\infty} q_n \int_0^t \frac{\mu(\mu x)^{n-1}}{(n-1)!} e^{-\mu x}\, dx \\ &= q_0 + \sum_{n=1}^{\infty} (1-r_0^k) r_0^{kn} \int_0^t \frac{\mu(\mu x)^{n-1}}{(n-1)!} e^{-\mu x}\, dx \\ &= q_0 + r_0^k \int_0^t (1-r_0^k)\mu e^{-\mu x} \sum_{n=1}^{\infty} \frac{(\mu x r_0^k)^{n-1}}{(n-1)!}\, dx \\ &= q_0 + r_0^k \int_0^t (1-r_0^k)\mu e^{-\mu(1-r_0^k)x}\, dx \\ &= q_0 + r_0^k [1 - e^{-\mu(1-r_0^k)t}]. \end{aligned}$$

The probability of no wait for service upon arrival is given by (3.31) as

$$q_0 = 1 - r_0^k,$$

and thus

$$\boxed{W_q(t) = 1 - r_0^k e^{-\mu(1-r_0^k)t} \quad (t \geq 0).} \tag{3.32}$$

■ EXAMPLE 3.10

Arrivals coming to a single-server queueing system are found to have an Erlang type-2 distribution with mean interarrival time of 30 min. The mean service time is 25 min, and service times are exponentially distributed. Find the steady-state system-size probabilities and the expected-value measures of effectiveness for this system.

The parameters for this system are $\lambda = 1/(30 \text{ min}) = 2/\text{h}$, $\mu = 1/(25 \text{ min}) = \frac{12}{5}/\text{h}$, and $k = 2$. The characteristic equation (3.29) is

$$\tfrac{12}{5}r_0^3 - \tfrac{32}{5}r_0 + 4 = \tfrac{1}{5}(12r_0^3 - 32r_0 + 20) = 0.$$

Upon simplifying we have

$$3r_0^3 - 8r_0 + 5 = 0.$$

This is the same operator equation as found in Example 3.4, and it has positive root in $(0, 1)$ approximately equal to 0.884. Thus from (3.30),

$$\begin{aligned} p_n &= \rho(1 - r_0^k)(r_0^k)^{n-1} \\ &\doteq 0.233(0.781)^n \quad (n \geq 1) \end{aligned}$$

and

$$p_0 = 1 - \rho = 1 - \tfrac{10}{12} = \tfrac{1}{6}.$$

The mean system size is

$$L = \frac{\rho}{1 - r_0^k} \doteq \frac{\frac{5}{6}}{1 - 0.781} \doteq 3.81,$$

and $W = L/\lambda \doteq 3.81/2 \doteq 1.91$ h, while $W_q = 3.81/2 - 5/12 \doteq 1.49$ h and $L_q \doteq 3.81 - 5/6 \doteq 2.98$.

3.3.5 $E_j/E_k/1$ Models

Consider an $E_j/E_k/1$ queue with arrival rate λ and service rate μ (i.e., each arrival phase has rate $j\lambda$ and each service phase has rate $k\mu$). The complete analysis of the $E_j/E_k/1$ queue is more complicated than either the $M/E_k/1$ or $E_k/M/1$, and we leave its details to Section 6.1. However, there are some things that can be said at this point that nicely connect the results of prior sections with those for the $E_j/E_k/1$ queue. The most important of these is that the solution to the $E_j/E_k/1$ queue is found in terms of roots to the following characteristic equation:

$$k\mu z^{j+k} - (j\lambda + k\mu)z^k + j\lambda = 0.$$

When $j = 1$, we have the $M/E_k/1$ queue, and the characteristic equation reduces to

$$k\mu z^{k+1} - (\lambda + k\mu)z^k + \lambda = 0,$$

which also can be obtained from (3.25). When $k = 1$, we have the $E_j/M/1$ queue, and the characteristic equation reduces to

$$\mu z^{j+1} - (j\lambda + \mu)z + j\lambda = 0,$$

which is the same as (3.29).

As in our earlier discussions, we can apply Rouché's theorem to help determine the location of the roots of this polynomial equation. It turns out that whenever $\lambda/\mu < 1$, there are k roots inside the complex unit circle plus the root of 1 on the boundary of the unit circle. Furthermore, by showing that the derivative of the polynomial cannot vanish inside the unit circle, it follows that all of the roots are distinct, with one always real and positive, while a second real and negative root exists whenever k is an even number. As a first illustration of this, consider the $M/E_2/1$ model of Example 3.8. Here, $\lambda = \frac{6}{5}, \mu = \frac{3}{2}, j = 1$, and $k = 2$. Thus the characteristic equation becomes

$$3z^3 - \tfrac{21}{5}z^2 + \tfrac{6}{5} = 0,$$

with roots 1, $(1 \pm \sqrt{11})/5$. The latter two roots have the requisite property that their absolute values are less than 1.

For a second illustration, suppose that the interarrival distribution for the previous example is now assumed to be a type-3 Erlang distribution. Then the characteristic equation becomes

$$3z^5 - \tfrac{33}{5}z^2 + \tfrac{18}{5} = \tfrac{3}{5}(5z^5 - 11z^2 + 6) = 0,$$

whose roots are found (using a mathematics package) to be 1 and (approximately) $0.914, -0.689, -0.612 \pm 1.237i$, so that there are two real roots with absolute values less than 1 and two other roots with absolute values more than 1.

Now, suppose that the service-time shape parameter for this problem is $k = 3$. Then the characteristic equation is

$$\tfrac{9}{2}z^6 - \tfrac{81}{10}z^3 + \tfrac{18}{5} = \tfrac{9}{10}(5z^6 - 9z^3 + 4) = 0,$$

whose roots are found to be 1 and (approximately) $0.928, -0.464 \pm 0.804i, -0.5 \pm 0.866i$, so that now there are two complex roots outside the unit circle, one real root less than 1, and two complex conjugate roots inside the unit circle.

The complete details on how these roots are manipulated to get state probabilities and a line-delay distribution will come later in the text. In the meantime, suffice it to say that the roots found in this process play a key role in obtaining the complete measures of effectiveness for the $E_j/E_k/1$ queue. The multiserver problems $M/E_k/c$, $E_k/M/c$, and $E_j/E_k/c$ will also be discussed later in Sections 6.1.1 and 7.2.2). This latter material also involves more complicated matrix-analytic methods utilizing phase-type distribution functions.

3.4 Priority Queue Disciplines

Up to this point, all the models considered have had the property that units proceed to service on a first-come, first-served basis. This is obviously not the only manner of service, and there are many alternatives, such as last-come, first-served, selection in random order, and selection by priority. In priority schemes customers with

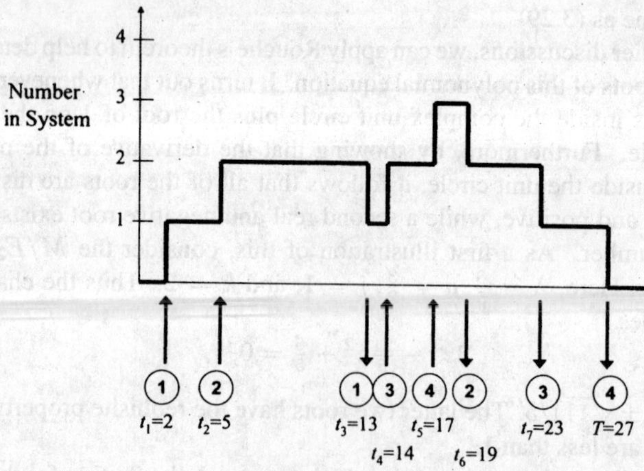

Figure 3.4 Number in a queueing system over time.

the highest priorities are selected for service ahead of those with lower priorities, independent of their time of arrival into the system. There are two further refinements possible in priority situations, namely, preemption and nonpreemption. In preemptive cases, the customer with the highest priority is allowed to enter service immediately even if another with lower priority is already present in service when the higher customer arrives. In addition, a decision has to be made whether to continue the preempted customer's service from the point of preemption when resumed or to start anew. On the other hand, a priority discipline is said to be nonpreemptive if there is no interruption and the highest-priority customer just goes to the head of the queue to wait its turn.

A considerable portion of real-life queueing situations contains priority considerations. Priority queues are generally more difficult to model than nonpriority ones, but the priority models should not be oversimplified merely to permit solution. Full consideration of priorities is absolutely essential when considering costs of queueing systems and optimal design.

Recall some of the arguments of Chapters 1 and 2. There is nothing in the $M/M/1$ (and related models) state-probability derivation that depends on queue discipline. Indeed, it can be shown that as long as selection for service is in no way a function of the relative size of the service time, then the $\{p_n\}$ are independent of queue discipline. The proof of Little's formula also remains unchanged, and since the average system size is unaltered, the average waiting time is likewise. But there will be changes in the waiting-time distribution.

It turns out that the waiting times are stochastically smallest under FCFS (all other things kept constant). That is, the introduction of any scheme of priorities not depending on service times is going to make all higher moments worse than under FCFS. With the background available to us at this point, it is difficult to prove this assertion in general. But we can give a basic idea of what is happening by reference to Figure 1.4, presented here with specific customer arrival and departure times as shown in Figure 3.4.

To illustrate the point, let us compare the sample paths under FCFS and under LCFS. Under FCFS (shown in the figure), the waiting times in system are, respectively, 11, 14, 9, and 10. The average waiting time is $44/4 = 11$. The sample unbiased variance is $14/3$.

Now, in the LCFS case, everything is the same, except that customer 4 is now served before customer 3 (that is, customer 4 departs at time 23, and customer 3 departs at time 27). In this case, the waiting times in system are, respectively, 11, 14, 13, and 6. The average waiting time is the same as before: $44/4 = 11$. The sample unbiased variance is now $38/3$, which is a substantial increase over the FCFS case.

In further expansion of some of these ideas, we note that the remaining total service or work required for a single server at any point during an arbitrary busy period is independent of the order of service as long as the system is conservative. That is, this is true whenever no service needs are created or destroyed within the system. This means, for example, that there is no renege in the midst of service, no preemption repeat when service is not exponential, no forced idleness of the server, and so on. Nonconservative systems will in fact violate Little's formula. In the multichannel case, work conservation results when every channel has the same service-time distribution. Otherwise, preemption may lead to a change in a customer's server and thus affect the waiting times of others more than normally.

We may also tie the concept of conservation to Little's formula. First we note that the remaining total work in a conserving system must be stable across the steady state. Thus for example, the expected gain in work present over a fixed time interval must be equal to the expected loss due to departing customers during the same period. We see then that Little's formula says that the work found by an arriving customer in the steady state under FCFS is always proportional to the general-time average queue size (i.e., $W = L/\lambda$). Furthermore, in work-conserving priority situations, the formula holds for the individual priority classes as well.

3.4.1 Nonpreemptive Systems with Two Classes

Suppose that customers arrive as a Poisson process to a single exponential channel and that each customer, upon arrival to the system, is assigned to one of two priority classes. (For further details on the queues in this section, see also Morse, 1958). The usual convention is to number the priority classes so that smaller numbers correspond to higher priorities. Suppose that the (Poisson) arrivals of the first or higher priority class have mean arrival rate λ_1 and that those of the second or lower priority class have mean arrival rate λ_2. The total arrival rate is $\lambda \equiv \lambda_1 + \lambda_2$. Also suppose that the first-priority customers are served ahead of the second-priority customers, but that there is no preemption.

From these assumptions, a system of balance equations may be established for the steady-state probabilities, defined as follows (where m and n are not both 0):

$$\begin{aligned} p_{mnr} \equiv \Pr\{&m \text{ priority-1 customers are in the system,} \\ &n \text{ priority-2 customers are in the system, and} \\ &\text{the customer in service is of priority } r = 1 \text{ or } 2\}. \end{aligned}$$

Let p_0 be the steady-state probability that the system is empty. Let $L^{(i)}$ be the average number of class-i customers in the system. Similarly, let $L_q^{(i)}$, $W_q^{(i)}$, and $W^{(i)}$ denote the corresponding measures of effectiveness for class-i customers.

3.4.1.1 Equal Service Rates. We first assume that the service rates of both classes are equal to μ. Define

$$\rho_1 = \frac{\lambda_1}{\mu}, \quad \rho_2 = \frac{\lambda_2}{\mu}, \quad \rho = \rho_1 + \rho_2 = \frac{\lambda}{\mu}.$$

We assume that $\rho < 1$. Then we have the following rate-balance equations (Problem 3.27):

$$\begin{aligned}
(\lambda+\mu)p_{mn2} &= \lambda_1 p_{m-1,n,2} + \lambda_2 p_{m,n-1,2} && (m \ge 1, n \ge 2),\\
(\lambda+\mu)p_{mn1} &= \lambda_1 p_{m-1,n,1} + \lambda_2 p_{m,n-1,1}\\
&\quad + \mu(p_{m+1,n,1} + p_{m,n+1,2}) && (m \ge 2, n \ge 1),\\
(\lambda+\mu)p_{m12} &= \lambda_1 p_{m-1,1,2} && (m \ge 1),\\
(\lambda+\mu)p_{1n1} &= \lambda_2 p_{1,n-1,1} + \mu(p_{2n1} + p_{1,n+1,2}) && (n \ge 1),\\
(\lambda+\mu)p_{0n2} &= \lambda_2 p_{0,n-1,2} + \mu(p_{1n1} + p_{0,n+1,2}) && (n \ge 2),\\
(\lambda+\mu)p_{m01} &= \lambda_1 p_{m-1,0,1} + \mu(p_{m+1,0,1} + p_{m12}) && (m \ge 2),\\
(\lambda+\mu)p_{012} &= \lambda_2 p_0 + \mu(p_{111} + p_{022}),\\
(\lambda+\mu)p_{101} &= \lambda_1 p_0 + \mu(p_{201} + p_{112}),\\
\lambda p_0 &= \mu(p_{101} + p_{012}).
\end{aligned} \tag{3.33}$$

Applying Little's formula to the server implies that ρ is the fraction of time the server is busy, or equivalently $p_0 = 1 - \rho$. The ordering of service does not affect the fraction of idle time. Similarly, the fraction of time the server is busy with a priority-r customer is ρ_r. Thus

$$\sum_{m=1}^{\infty}\sum_{n=0}^{\infty} p_{mn1} = \rho_1 \quad \text{and} \quad \sum_{m=0}^{\infty}\sum_{n=1}^{\infty} p_{mn2} = \rho_2.$$

Further, because the service distributions of the priority classes are both exponential with the same rate μ, the total number of customers in service has the same steady-state distribution as an $M/M/1$ queue. Thus

$$p_n = \sum_{m=0}^{n-1}(p_{n-m,m,1} + p_{m,n-m,2}) = (1-\rho)\rho^n \quad (n > 0).$$

It turns out, however, that obtaining a reasonable solution to these stationary equations is a very difficult, as we might have expected in view of the triple subscripts. The most we can do comfortably is obtain expected values via two-dimensional generating functions. Define

$$P_{m1}(z) = \sum_{n=0}^{\infty} z^n p_{mn1} \quad (m \geq 1),$$

$$P_{m2}(z) = \sum_{n=1}^{\infty} z^n p_{mn2} \quad (m \geq 0),$$

$$H_1(y, z) = \sum_{m=1}^{\infty} y^m P_{m1}(z) \quad (\text{with } H_1(1,1) = \rho_1),$$

$$H_2(y, z) = \sum_{m=0}^{\infty} y^m P_{m2}(z) \quad (\text{with } H_2(1,1) = \rho_2),$$

and

$$\begin{aligned} H(y,z) &= H_1(y,z) + H_2(y,z) + p_0 \\ &= \sum_{m=1}^{\infty}\sum_{n=0}^{\infty} y^m z^n p_{mn1} + \sum_{m=0}^{\infty}\sum_{n=1}^{\infty} y^m z^n p_{mn2} + p_0 \\ &= \sum_{m=1}^{\infty}\sum_{n=1}^{\infty} y^m z^n (p_{mn1} + p_{mn2}) + \sum_{m=1}^{\infty} y^m p_{m01} + \sum_{n=1}^{\infty} z^n p_{0n2} + p_0, \end{aligned}$$

where $H(y, z)$ is the joint generating function for the two classes, regardless of which type is in service. Note that $H(y, y) = p_0/(1 - \rho y)$ [with $H(1, 1) = 1$], since $H(y, z)$ collapses to the generating function of an $M/M/1$ queue when z is set equal to y and thus no priority distinction is made; see (2.13). Also, we have

$$\left.\frac{\partial H(y,z)}{\partial y}\right|_{y=z=1} = L^{(1)} = L_q^{(1)} + \rho_1 = \lambda_1 W^{(1)},$$

and

$$\left.\frac{\partial H(y,z)}{\partial z}\right|_{y=z=1} = L^{(2)} = L_q^{(2)} + \rho_2 = \lambda_2 W^{(2)}.$$

If we multiply equations in (3.33) by the appropriate powers of y and z and sum accordingly, it is found that

$$(1+\rho-\rho_1 y-\rho_2 z-1/y)\, H_1(y,z) = \frac{H_2(y,z)}{z} + \rho_1 y\, p_0 - P_{11}(z) - \frac{P_{02}(z)}{z}, \quad (3.34)$$

$$(1+\rho-\rho_1 y-\rho_2 z)\, H_2(y,z) = P_{11}(z) + \frac{P_{02}(z)}{z} - (\rho - \rho_2 z)p_0. \quad (3.35)$$

To determine the generating functions H_1 and H_2 fully, we need to know the values of $P_{11}(z)$, $P_{02}(z)$, and p_0. One equation relating $P_{11}(z)$, $P_{02}(z)$, and p_0 may be found by summing $z^n (n = 2, 3, \ldots)$ times the equation of (3.33) which involves p_{0n2}, and then using the final three equations of (3.33), thus obtaining

$$P_{11}(z) = (1 + \rho - \rho_2 z - 1/z)P_{02}(z) + (\rho - \rho_2 z)p_0.$$

Substitution of this equation into (3.34) and (3.35) gives H_1 and H_2 as functions of p_0 and $P_{02}(z)$, and thus also $H(y, z)$ as

$$\begin{aligned} H(y,z) &= H_1(y,z) + H_2(y,z) + p_0 \\ &= \frac{(1-y)p_0}{1-y-\rho y+\rho_1 y^2+\rho_2 yz} \\ &\quad + \frac{(1+\rho-\rho z+\rho_1 z)(z-y)P_{02}(z)}{z(1+\rho-\rho_1 y-\rho_2 z)(1-y-\rho y+\rho_1 y^2+\rho_2 yz)}. \end{aligned}$$

By employing the condition $H(1, 1) = 1$, it is found that $P_{02}(1) = \rho_2/(1 + \rho_1)$.

We next take the partial derivative of H with respect to y and evaluate at $(1, 1)$ to find $L^{(1)}$. [$\partial H(y, z)/\partial y$ cannot be evaluated directly at $(1, 1)$, so a limit must be taken instead.] In these steps, the exact functional relationship for $P_{02}(z)$ turns out not to be needed, and only $P_{02}(1)$ is required.

To obtain $L^{(2)}$, we could similarly take the partial derivative of H with respect to z and evaluate at $(1, 1)$. However, now the exact functional relationship for $P_{02}(z)$ is required [or more specifically $P'_{02}(1)$ is required]. An easier approach is to use the relation $L^{(1)} + L^{(2)} = \rho/(1 - \rho)$. That is, the total number of customers in the system is the same as in an $M/M/1$ queue (2.23), since the service distributions are the same for both customer classes. The other measures of effectiveness can then be obtained via standard relations $L_q^{(i)} = L^{(i)} - \rho_i$, $W_q^{(i)} = \lambda_i L_q^{(i)}$, and $W^{(i)} = \lambda_i L^{(i)}$. Without suffering through the details of the derivation, the final results for $L_q^{(i)}$ are

$$\boxed{\begin{aligned} L_q^{(1)} &= \frac{\lambda_1 \rho}{\mu - \lambda_1}, \\ L_q^{(2)} &= \frac{\lambda_2 \rho}{(\mu-\lambda_1)(1-\rho)}, \\ L_q &= \frac{\rho^2}{1-\rho}, \end{aligned}} \tag{3.36}$$

where $\rho = \lambda_1/\mu + \lambda_2/\mu$. Using the theory of multidimensional birth–death processes, Miller (1981) has shown that the actual probabilities for priority-1 customers are

$$p_{n_1} = (1-\rho)\left(\frac{\lambda_1}{\mu}\right)^{n_1} + \frac{\lambda_2}{\lambda_1}\left(\frac{\lambda_1}{\mu}\right)^{n_1}\left[1-\left(\frac{\mu}{\lambda_1+\mu}\right)^{n_1+1}\right] \qquad (n_1 \geq 0).$$

Some important observations may be made about the mean-value results.

1. Second-priority customers wait in queue longer than first-priority customers. This can be seen as follows:

$$W_q^{(2)} = \frac{\rho}{(\mu-\lambda_1)(1-\rho)} = \frac{\rho/(\mu-\lambda_1)}{1-\rho} = \frac{W_q^{(1)}}{1-\rho} > W_q^{(1)} \qquad (\text{when } \rho < 1).$$

2. As $\rho \to 1$, $L_q^{(2)} \to \infty$ (similarly, $W_q^{(2)}$, $W^{(2)}$, and $L^{(2)} \to \infty$). However, the corresponding means for the first-priority customers approach finite limits, assuming $\lambda_1/\mu < 1$ is held constant. The first-priority means go to ∞ only when $\lambda_1/\mu \to 1$. Thus it is possible that first-priority customers do not accumulate, even when an overall steady state does not exist.

3. Even though class-1 customers have priority, the presence of class-2 customers still creates delays for the class-1 customers. In particular,

$$\{L_q^{(1)} \text{ when } \lambda_2 = 0\} < \{L_q^{(1)} \text{ when } \lambda_2 > 0\}.$$

   This is because the class-1 customers cannot preempt class-2 customers who are already in service. However, if the class-1 customers have the power of preemption, then the class-2 customers do not effect the class-1 customers.

4. The average number in queue is the same as an $M/M/1$ queue. Similarly, the unconditional average wait, $W_q = (\lambda_1/\lambda)W_q^{(1)} + (\lambda_2/\lambda)W_q^{(2)}$ is the same an $M/M/1$ queue.

3.4.1.2 Unequal Service Rates. Now we assume that the service rates of the two classes are not necessarily equal. Specifically, priority-1 customers are served at a rate μ_1 and priority-2 customers are served at a rate μ_2. For this queue, define

$$\rho_1 \equiv \frac{\lambda_1}{\mu_1}, \quad \rho_2 \equiv \frac{\lambda_2}{\mu_2}, \quad \rho \equiv \rho_1 + \rho_2.$$

Balance equations similar to (3.33) can be obtained. The final results are (details may be found in Morse, 1958)

$$\boxed{\begin{aligned} L_q^{(1)} &= \frac{\lambda_1(\rho_1/\mu_1 + \rho_2/\mu_2)}{1-\rho_1}, \\ L_q^{(2)} &= \frac{\lambda_2(\rho_1/\mu_1 + \rho_2/\mu_2)}{(1-\rho_1)(1-\rho)}, \\ L_q &= L_q^{(1)} + L_q^{(2)}. \end{aligned}} \tag{3.37}$$

Miller (1981) gave the actual probabilities for priority-1 customers as

$$p_{n_1} = (1-\rho)\left(\frac{\lambda_1}{\mu_1}\right)^{n_1} + \frac{\lambda_2}{\lambda_1 + \mu_2 - \mu_1}\left[\left(\frac{\lambda_1}{\mu_1}\right)^{n_1} - \frac{\mu_1 \lambda_1^{n_1}}{(\lambda_1 + \mu_2)^{n_1+1}}\right] \quad (n_1 \geq 0).$$

3.4.1.3 Two-Class FCFS. We now consider a two-class model in which customers are served FCFS. In other words, there are two customer classes with respective arrival rates λ_1 and λ_2 and respective service rates μ_1 and μ_2. Service times are exponential and customers are served on a first-come first-served basis. This is not a priority model, but it is instructive to compare this two-class model with the priority models given in this section. Furthermore, this two-class FCFS model can be viewed as a single-class $M/H_2/1$ queue, where customers are grouped into a single arrival stream and the service distribution is a mixture of two exponential distributions (a hyperexponential distribution).

The analysis of this FCFS model is not terribly difficult (Problem 3.28), and the expected line lengths are

$$\boxed{\begin{aligned} L_q^{(1)} &= \frac{\lambda_1(\rho_1/\mu_1 + \rho_2/\mu_2)}{1-\rho}, \\ L_q^{(2)} &= \frac{\lambda_2(\rho_1/\mu_1 + \rho_2/\mu_2)}{1-\rho}, \\ L_q &= \frac{\lambda(\rho_1/\mu_1 + \rho_2/\mu_2)}{1-\rho}. \end{aligned}} \tag{3.38}$$

The L_q of (3.38) is always greater than that of the standard $M/M/1$ with mean service time equal to the weighted average of the respective means, namely, $1/\mu = (\lambda_1/\lambda)/\mu_1 + (\lambda_2/\lambda)/\mu_2$ (Problem 3.29). This is not surprising, because the service times now have a higher relative variability than exponential service times.

3.4.1.4 Comparison of Models. We first compare the priority queue (unequal service rates) of (3.37) with the nonpriority queue of (3.38). As expected, the imposition of priorities decreases the mean number of priority-1 customers in the queue and increases the mean number of priority-2 customers. A proof of this result is left as an exercise (Problem 3.31).

This result is quite intuitive – that the imposition of priorities helps one class and hurts the other. We now address how the imposition of priorities affects the *overall* performance of the system. Specifically, we compare the average overall number in the queue L_q between the two models. These can be rewritten as

$$\begin{aligned} L_q \text{ from } (3.37) &= \left(\frac{\rho_1/\mu_1 + \rho_2/\mu_2}{1-\rho}\right)\frac{\lambda - \lambda_1\rho}{1-\rho_1}, \\ L_q \text{ from } (3.38) &= \left(\frac{\rho_1/\mu_1 + \rho_2/\mu_2}{1-\rho}\right)\lambda. \end{aligned}$$

The L_q's (and also the W_q's) differ by the factor $(\lambda - \lambda_1\rho)/(\lambda - \lambda\rho_1)$. Thus there are fewer customers waiting in the priority queue (3.37) when

$$\frac{\lambda - \lambda_1\rho}{\lambda - \lambda\rho_1} < 1 \Leftrightarrow \lambda_1\rho > \lambda\rho_1 \Leftrightarrow \lambda_1\left(\frac{\lambda_1}{\mu_1} + \frac{\lambda_2}{\mu_2}\right) > (\lambda_1 + \lambda_2)\frac{\lambda_1}{\mu_1} \Leftrightarrow \mu_2 < \mu_1.$$

In summary, the priority queue results in less overall waiting (compared with the analogous FCFS queue with no priorities) when the first-priority customers have a faster service rate (or shorter service times). Conversely, the priority queue results in more overall waiting when the priority customers have longer service times. These

Table 3.1 Comparison of Nonpreemptive Priority Models

Versus	(a)	(b)	(c)
(b)	L_q : (b) = (a)		
(c)	L_q : (c) < (a) iff $\mu_1 > \mu_2$	Depends on parameters	
(d)	L_q : (d) ≥ (a)	N/A	$L_q^{(1)}$: (c) < (d) $L_q^{(2)}$: (d) < (c) L_q : (c) < (d) iff $\mu_1 > \mu_2$

Model	Parameters	Results
(a) $M/M/1$	λ, μ	(2.25) [a]
(b) Two priorities, one service rate	$\lambda_1, \lambda_2, \mu$	(3.36) [b]
(c) Two priorities, two service rates	$\lambda_1, \lambda_2, \mu_1, \mu_2$	(3.37)
(d) No priority, two service rates	$\lambda_1, \lambda_2, \mu_1, \mu_2$	(3.38)

[a] Use $\lambda = \lambda_1 + \lambda_2$ and $1/\mu = (\lambda_1/\lambda)/\mu_1 + (\lambda_2/\lambda)/\mu_2$.
[b] Use μ chosen to lie between μ_1 and μ_2.

comparative results have very important implications for the design of queueing systems and give rise to an optimal design rule called "the shortest processing time (SPT) rule" (see Schrage and Miller, 1966). That is, if the design criterion of a queue is the reduction of the total number waiting, or equivalently the overall mean delay, then priority should be given to the group of customers that has the faster service rate.

Finally, we compare the priority *two*-rate model of (3.37) with the priority *one*-rate model of (3.36). To make this comparison, we must make some choice for the value of the single rate μ. A natural choice is to equate the average service times. That is, μ is chosen so that

$$\frac{1}{\mu} = \left(\frac{\lambda_1}{\lambda}\right)\frac{1}{\mu_1} + \left(\frac{\lambda_2}{\lambda}\right)\frac{1}{\mu_2}.$$

Alternatively, we can choose μ to lie somewhere between μ_1 and μ_2. If the value of μ is chosen to equal the larger of μ_1 and μ_2, then it can be shown that all three measures, $L_q^{(1)}$, $L_q^{(2)}$, and L_q, are less in (3.36) (the one-rate model) than they are in (3.37) (the two-rate model). This comparison is completely reversed when $\mu = \min(\mu_1, \mu_2)$ (Problem 3.30). Any comparison when μ lies strictly between μ_1 and μ_2 depends on the values of the parameters involved, namely μ_1, μ_2, μ, λ_1, and λ_2.

Table 3.1 summarizes the comparison of models given in this section.

EXAMPLE 3.11

Our friend the hair-salon proprietor, Ms. H. R. Cutt, has decided to explore the possibility of giving priority to customers who wish only a trim cut. Ms. Cutt estimates that the time required to trim a customer (needed one-third of the time) is still exponential, but with a mean of 5 min, and that the removal of these customers from the total population increases the mean of the nonpriority class to 12.5 min, leaving the distribution exponential. Arrivals are still Poisson with rate $\lambda = 5$/h. If Cutt measures her performance by the value of the mean waiting time, will this change reduce average line waits?

We have $\lambda_1 = \lambda/3 = \frac{5}{3}$/h, $\lambda_2 = 2\lambda/3 = \frac{10}{3}$/h, $\mu_1 = 12$/h, and $\mu_2 = \frac{24}{5}$/h. This gives $\rho_1 = 5/36$, $\rho_2 = 25/36$, and $\rho = 30/36 = 5/6$. Substitution of the appropriate values into (3.37) gives

$$L_q^{(1)} = \frac{75}{248} \doteq 0.30, \qquad L_q^{(2)} = \frac{225}{62} \doteq 3.63, \qquad L_q \doteq 3.93.$$

The mean waiting time in queue (averaged over all customers) is $W_q = L_q/\lambda \doteq$ 47 min.

The correct model to compare this with is the no-priority two-service-rate model of (3.38) rather than the original $M/M/1$ model of Example 2.1. The results using this model are:

$$L_q^{(1)} = \frac{25}{16} \doteq 1.56, \qquad L_q^{(2)} = \frac{25}{8} \doteq 3.13, \qquad L_q \doteq 4.69.$$

The mean waiting time in queue (averaged over all customers) is $W_q = L_q/\lambda \doteq$ 56 min. In summary, implementing priorities reduces the number of priority-1 customers in the queue (so reduces the wait for these customers), but increases the number of priority-2 customers in the queue (so increases the wait for these customers). The *overall* wait in queue (averaged over all customers) is reduced by using priorities, illustrating the effect of the SPT rule.

Finally, if we compare this model to the $M/M/1$ model of Example 2.1 (where $\lambda = 5$/h and $\mu = 6$/h, matching the overall rates of this example), we find that $L_q = \frac{25}{6} \doteq 4.17$. Thus the two-service-rate model of (3.38) gives a larger L_q than what is obtained ignoring the fact that customers are of two types. This agrees with the previous discussion that the increased variability in the service times increases the system congestion.

3.4.2 Nonpreemptive Systems with Many Classes

As observed in the previous section, the determination of stationary probabilities in a nonpreemptive Markovian system is an exceedingly difficult matter, well near impossible when the number of priorities exceeds two. In light of this and the difficulty of handling multi-indexed generating functions when there are more than two priority classes, an alternative approach to obtaining the mean-value measures L and W is used, namely, a direct expected-value procedure.

Suppose that customers of the kth priority (the smaller the number, the higher the priority) arrive at a single channel queue according to a Poisson process with rate

λ_k ($k = 1, 2, \ldots, r$) and that these customers wait on a first-come, first-served basis within their respective priorities. Let the service distribution for the kth priority be exponential with mean $1/\mu_k$. A unit that begins service completes its service before another item is admitted, regardless of priorities.

We begin by defining.

$$\rho_k = \frac{\lambda_k}{\mu_k} \quad (1 \le k \le r), \qquad \sigma_k = \sum_{i=1}^{k} \rho_i \quad (\sigma_0 \equiv 0,\ \sigma_r \equiv \rho). \tag{3.39}$$

The system is stationary for $\sigma_r = \rho < 1$.

Consider a customer of priority i who arrives at the system. Upon arrival, assume that there are n_1 customers of priority 1 in the queue ahead of this new arrival, n_2 of priority 2, n_3 of priority 3, and so on. Let S_0 be the time required to finish the customer already in service (where this value could be zero in the event that the system is empty upon arrival). Let S_k be the time required to serve the n_k customers of priority k already in the queue ahead of the arriving customer (where $1 \le k \le i$). During the new customer's waiting time T_q, items of priority $k < i$ may arrive and go to service ahead of this current arrival. Let n'_k be the number of customers of priority k who arrive later and go ahead of the arriving customer (where $k < i$). Let S'_k be the time to serve these n'_k customers. Then,

$$T_q = \sum_{k=1}^{i-1} S'_k + \sum_{k=1}^{i} S_k + S_0.$$

Taking expected values of both sides gives

$$W_q^{(i)} \equiv \mathrm{E}[T_q] = \sum_{k=1}^{i-1} \mathrm{E}[S'_k] + \sum_{k=1}^{i} \mathrm{E}[S_k] + \mathrm{E}[S_0].$$

From the uniform property of the Poisson process, $\mathrm{E}[n_k]$ equals the time-average number of priority-k customers in the queue, or $L_q^{(k)}$. Little's formula then gives

$$\mathrm{E}[n_k] = L_q^{(k)} = \lambda_k W_q^{(k)}.$$

Because the service times are independent of n_k,

$$\mathrm{E}[S_k] = \frac{\mathrm{E}[n_k]}{\mu_k} = \frac{\lambda_k W_q^{(k)}}{\mu_k} = \rho_k W_q^{(k)}.$$

For a Poisson arrival process, the average number of priority-k arrivals during the current arrival's wait in queue is

$$\mathrm{E}[n'_k] = \lambda_k W_q^{(i)}.$$

Therefore

$$\mathrm{E}[S'_k] = \frac{\mathrm{E}[n'_k]}{\mu_k} = \frac{\lambda_k W_q^{(i)}}{\mu_k} = \rho_k W_q^{(i)}.$$

Combining these equations together, we have

$$W_q^{(i)} = W_q^{(i)} \sum_{k=1}^{i-1} \rho_k + \sum_{k=1}^{i} \rho_k W_q^{(k)} + \mathrm{E}[S_0],$$

or

$$W_q^{(i)} = \frac{\sum_{k=1}^{i} \rho_k W_q^{(k)} + \mathrm{E}[S_0]}{1 - \sigma_{i-1}}. \tag{3.40}$$

These equations are linear in $W_q^{(i)}$. The solution to (3.40) was found by Cobham (1954), after whom much of this analysis follows, by induction on i (see Problem 3.32). That solution is

$$W_q^{(i)} = \frac{\mathrm{E}[S_0]}{(1 - \sigma_{i-1})(1 - \sigma_i)}. \tag{3.41}$$

Now, S_0 (the remaining service time of the customer in service at the time of the arrival) has the value 0 if the system is idle; hence

$$\mathrm{E}[S_0] = \Pr\{\text{system is busy}\} \cdot \mathrm{E}[S_0 \mid \text{system is busy}].$$

The probability that the system is busy is

$$\lambda \cdot (\text{expected service time}) = \lambda \sum_{k=1}^{r} \frac{\lambda_k}{\lambda} \frac{1}{\mu_k} = \rho.$$

Also,

$$\begin{aligned} \mathrm{E}[S_0 \mid \text{system is busy}] &= \sum_{k=1}^{r} (\mathrm{E}[S_0 \mid \text{system busy with priority-}k\text{ customer}] \\ &\quad \times \Pr\{\text{system busy with priority-}k\text{ customer} \mid \text{system is busy}\}) \\ &= \sum_{k=1}^{r} \frac{1}{\mu_k} \frac{\rho_k}{\rho}. \end{aligned}$$

Therefore

$$\mathrm{E}[S_0] = \rho \sum_{k=1}^{r} \frac{1}{\mu_k} \frac{\rho_k}{\rho} = \sum_{k=1}^{r} \frac{\rho_k}{\mu_k}. \tag{3.42}$$

Plugging this into (3.41) finally gives

$$\boxed{W_q^{(i)} = \frac{\sum_{k=1}^{r} \rho_k / \mu_k}{(1 - \sigma_{i-1})(1 - \sigma_i)}.} \tag{3.43}$$

Note that (3.43) holds as long as $\sigma_r = \sum_{k=1}^{r} \rho_k < 1$. For the case $r = 2$, this formula reduces to the earlier result given in (3.37). The total expected queue size can also be obtained from this result using Little's formula as

$$L_q = \sum_{i=1}^{r} L_q^{(i)} = \sum_{i=1}^{r} \frac{\lambda_i \sum_{k=1}^{r} \rho_k / \mu_k}{(1-\sigma_{i-1})(1-\sigma_i)}.$$

For results on higher moments, see Kesten and Runnenburg (1957).

We now compare the priority queue and the ordinary $M/M/1$ queue. For the priority queue, the wait in queue averaged over *all* customers is

$$W_q \equiv \sum_{i=1}^{r} \frac{\lambda_i W_q^{(i)}}{\lambda}.$$

For the $M/M/1$ queue, it is natural to set the average service time equal to the average over all priority classes, namely,

$$\frac{1}{\mu} \equiv \sum_{i=1}^{r} \frac{\lambda_i}{\lambda} \frac{1}{\mu_i}. \tag{3.44}$$

Then, the expected-value measures for both queues are the same if and only if the μ_i's are identical. That is, for the priority queue,

$$W_q = \frac{\lambda}{\mu(\mu - \lambda)}$$

if and only if $\mu_i \equiv \mu$ for all i (see Problem 3.33). In fact, if the higher-priority units have faster service rates, then the average wait over all units is less than for the analogous nonpriority system [i.e., with service rate defined by (3.44)]. Again, this illustrates the SPT rule mentioned earlier. If the opposite is true, that is, lower priorities have faster service, then the analogous nonpriority model gives the lower average wait and system size (see Problem 3.34). These differences increase as saturation is approached. If priorities and service-rate rankings are mixed, then the result depends on the pairings of priorities and service rates and on the actual values of the average service times. Thus, if the overriding requirement in the design of a queueing system is the reduction of the delay for one specific set of items, then this class should be given priority. If, however, the criterion for design is simply to reduce the average wait in queue of all units, then it helps to give priority to that class of units which tends to have the fastest service rate. For a further discussion of the effect of priorities on delay, see Morse (1958) and Jaiswal (1968).

Finally, we give some comments on when there are no differences between the state probabilities and measures of effectiveness for the ordinary $M/M/1$ queue and the priority $M/M/1$ queue (as well as the $M/M/1$ queue with general queue discipline): The same state probabilities and measures hold for arbitrary queue disciplines provided (1) all arrivals stay in the queue until served, (2) the mean service time of all units is the same, (3) the server completes service before it starts on the next item, and (4) the service channel always admits a waiting customer immediately upon the completion of another.

3.4.2.1 General Service Distributions. In the derivation of (3.43), we only once used the assumption of exponential service. (The assumption of Poisson arrivals, on the other hand, was used multiple times). Specifically, for exponential service, we used

$$\mathrm{E}[S_0 \mid \text{system busy with priority-}k\text{ customer}] = \frac{1}{\mu_k}.$$

It is not difficult to generalize the preceding results to a general service distribution. Let X_k be the random service time of a priority-k customer. Suppose X_k follows a general distribution with first moment $\mathrm{E}[X_k] = 1/\mu_k$ and second moment $\mathrm{E}[X_k^2]$. Then,

$$\mathrm{E}[S_0 \mid \text{system busy with priority-}k\text{ customer}] = \frac{\mathrm{E}[X_k^2]\,\mu_k}{2}.$$

The derivation of this result is left as an exercise (Problem 5.7). The result is also the average residual time of a renewal process with interrenewal distribution following X_k (e.g., Ross, 2007). Then, (3.42) becomes

$$\mathrm{E}[S_0] = \rho \sum_{k=1}^{r} \frac{\mathrm{E}[X_k^2]\,\mu_k}{2}\frac{\lambda_k/\mu_k}{\rho} = \sum_{k=1}^{r}\frac{\mathrm{E}[X_k^2]\,\lambda_k}{2} = \frac{\lambda}{2}\sum_{k=1}^{r}\frac{\lambda_k}{\lambda}\mathrm{E}[X_k^2] = \frac{\lambda \mathrm{E}[S^2]}{2}.$$

Thus

$$\boxed{W_q^{(i)} = \frac{\lambda \mathrm{E}[S^2]/2}{(1-\sigma_{i-1})(1-\sigma_i)}.} \qquad (3.45)$$

3.4.2.2 Continuous Priority Classes. Phipps (1956) extended Cobham's models by allowing the number of priorities to be continuous, rather than discrete. He applied this idea to prioritize units based on their actual service time. That is, an arriving unit whose service time is x is assigned to the priority class x, where x is a continuous parameter (replacing the discrete parameter i). Thus units with shorter service times are processed first. This is similar to the shortest-processing-time (SPT) rule given earlier. However, in the earlier models, units were prioritized based on their *average* service times (i.e., the average service time of the class to which a customer belongs). Here, they are prioritized based on their *actual* service times. This assumes that the exact service time of a unit can be known before it enters service. Again, we assume there is no preemption.

Extending Cobham's results to the continuous case essentially involves replacing the discrete parameter i with the continuous parameter x and making the appropriate changes of summation to integration. As in the previous models, we assume that the arrival process is Poisson with rate λ. Let S be the random service time of an arbitrary customer, with CDF $B(t)$. Then, the arrival rate of customers with service times between x and $x + dx$ (i.e., the analog of λ_i) is

$$\lambda_x\, dx = \lambda \frac{dB(x)}{dx} dx = \lambda\, dB(x).$$

The service rate of units in class x is $\mu_x = 1/x$, since the service time of a priority-x unit is exactly x, by assumption. Then, the continuous analog of σ_i in (3.39) is

$$\sigma_x = \int_0^x \frac{\lambda_y}{\mu_y}\,dy = \int_0^x y\lambda\,dB(y).$$

The continuous analog of (3.45) is

$$W_q^{(x)} = \frac{\lambda \mathrm{E}[S^2]/2}{\left(1 - \lambda \int_0^x y\,B(y)\right)^2},$$

and the expected number of customers in the queue is

$$L_q = \int_0^\infty L_q^{(x)}\,dx = \int_0^\infty \lambda_x W_q^{(x)}\,dx.$$

As an example, if the service distribution is exponential with mean $1/\mu$, then these results simplify to

$$W_q^{(x)} = \frac{\lambda/\mu^2}{\left(1 - \lambda \int_0^x y\mu e^{-\mu y}\,dy\right)^2} = \frac{\lambda/\mu^2}{\left(1 - \frac{\lambda}{\mu}[1 - e^{-\mu x}(1+\mu x)]\right)^2}$$

and

$$L_q = \frac{\lambda^2}{\mu}\int_0^\infty e^{-\mu x}\left(1 - \frac{\lambda}{\mu}[1 - e^{-\mu x}(1+\mu x)]\right)^{-2} dx.$$

3.4.2.3 Multiple Servers. The analysis for the multiple-server case is very similar to that of the preceding model except that it must now be assumed that service is governed by identical exponential distributions for each priority at each of c channels. Unfortunately, for multichannels we must assume no service-time distinction between priorities, or else the mathematics becomes quite intractable.

Let us define

$$\rho_k = \frac{\lambda_k}{c\mu} \quad (1 \le k \le r), \qquad \sigma_k = \sum_{i=1}^{k} \rho_i \quad (\sigma_r \equiv \rho = \lambda/c\mu).$$

Again the system is completely stationary for $\rho < 1$, and

$$W_q^{(i)} = \sum_{k=1}^{i-1} \mathrm{E}[S_k'] + \sum_{k=1}^{i} \mathrm{E}[S_k] + \mathrm{E}[S_0],$$

where, as before, S_k is the time required to serve n_k items of the kth priority in the line ahead of the item, S_k' is the service time of the n_k' items of priority k which arrive during $W_q^{(i)}$, and S_0 is the amount of time remaining until the next server becomes available. The first two terms on the right-hand side of the $W_q^{(i)}$ equation are exactly the same as in the single-channel case, except that the system service rate $c\mu$ is used in place of the single-service rate μ_k throughout the argument.

To derive $E[S_0]$, consider

$$E[S_0] = \Pr\{\text{all channels busy}\} \cdot E[S_0 \mid \text{all channels busy}].$$

The probability that all channels are busy is [from (2.31)]

$$\sum_{n=c}^{\infty} p_n = p_0 \sum_{n=c}^{\infty} \frac{(c\rho)^n}{c^{n-c}c!} = p_0 \frac{(c\rho)^c}{c!(1-\rho)},$$

and

$$E[S_0 \mid \text{all channels busy}] = 1/c\mu$$

from the memorylessness of the exponential. Thus from (2.32),

$$E[S_0] = \frac{(c\rho)^c}{c!(1-\rho)(c\mu)} \left(\sum_{n=0}^{c-1} \frac{(c\rho)^n}{n!} + \frac{(c\rho)^c}{c!(1-\rho)} \right)^{-1}.$$

Therefore from (3.43),

$$W_q^{(i)} = \frac{E[S_0]}{(1-\sigma_{i-1})(1-\sigma_i)} = \frac{\left[c!(1-\rho)(c\mu)\sum_{n=0}^{c-1}(c\rho)^{(n-c)}/n! + c\mu\right]^{-1}}{(1-\sigma_{i-1})(1-\sigma_i)},$$

and the expected line wait taken over all priorities is thus

$$W_q = \sum_{i=1}^{r} \frac{\lambda_i}{\lambda} W_q^{(i)}.$$

3.4.3 Preemptive Priorities

This section modifies the Markovian model of Section 3.4.1 so that units of higher priority *preempt* units of lower priority in service. Lower priority units that are ejected from service cannot reenter service until the system is free of all higher priority units. Generally, for such queues, we must specify how the system handles ejected units that receive only partial service. Two common assumptions are (1) ejected units must start over thereby losing the partial work already completed, or (2) ejected units resume service from the point of interruption. Since we assume here that service times are exponential, this issue is irrelevant in view of the memoryless property.

For a Markovian preemptive queue, the system state is completely determined by the number of customers of each class in the system. For non-preemptive queues, we also had to specify the class of the customer in service. Here, the class of the customer in service is always the highest priority class of customers in the system, so the extra parameter is not needed in the state space.

We first consider a two-class system. Let p_{mn} be the steady-state probability that there are m units of priority 1 in the system with arrival rate λ_1 and service rate μ_1, and n units of priority 2 in the system with arrival rate λ_2 and service rate μ_2. Under these assumptions, a system of difference equations may be derived for the stationary probabilities ($\lambda = \lambda_1 + \lambda_2$ and $\rho = \lambda_1/\mu_1 + \lambda_2/\mu_2 < 1$), namely,

$$\begin{aligned}
\lambda p_{00} &= \mu_1 p_{10} + \mu_2 p_{01} \\
(\lambda + \mu_1) p_{m0} &= \lambda_1 p_{m-1,0} + \mu_1 p_{m+1,0}, \\
(\lambda + \mu_2) p_{0n} &= \mu_1 p_{1,n} + \lambda_2 p_{0,n-1} + \mu_2 p_{0,n+1}, \\
(\lambda + \mu_1) p_{mn} &= \lambda_1 p_{m-1,n} + \lambda_2 p_{m,n-1} + \mu_1 p_{m+1,n}.
\end{aligned} \tag{3.46}$$

There are $2^2 = 4$ classes of equations in (3.46) corresponding to states of the form (00), $(m0)$, $(0n)$, and (mn). In general, when the number of preemptive priorities is k, there are 2^k classes of equations (e.g., see Problem 3.37 for the three-class case).

One approach to obtain performance measures for this system is to derive various steady-state partial generating functions from the balance equations. From this, the moments of the number of units in the system can be obtained (e.g., White and Christie, 1958). The results are

$$\boxed{\begin{aligned}
L^{(1)} &= \frac{\rho_1}{1-\rho_1}, \\
L^{(2)} &= \frac{\rho_2 - \rho_1\rho_2 + \rho_1\rho_2(\mu_2/\mu_1)}{(1-\rho_1)(1-\rho_1-\rho_2)},
\end{aligned}}$$

where $L^{(i)}$ is the average number of class-i customers in the system in steady state. The class-1 customers are not affected in any way by the presence of the class-2 customers. Thus the class-1 customers are effectively operating as if they were in an $M/M/1$. This can be shown formally from the steady-state equations in (3.46) (Problem 3.38). In particular, $L^{(1)}$ matches the $M/M/1$ formula in (2.23).

In the case of r customer classes and general service distributions, where preempted customers resume service from the point of interruption, the result generalizes to (e.g., Jaiswal, 1968, Avi-Itzhak and Naor, 1963)

$$\boxed{L^{(i)} = \frac{\rho_i}{1-\sigma_{i-1}} + \frac{\lambda_i \sum_{j=1}^{i} \lambda_j \mathrm{E}[S_j^2]}{2(1-\sigma_{i-1})(1-\sigma_i)},}$$

where S_j is a random service time of a class-j customer, $\rho_i = \lambda_i \mathrm{E}[S_i]$, and $\sigma_i = \sum_{j=1}^{i} \rho_i$.

3.5 Retrial Queues

This section describes a class of models called *retrial queues*, named for the repeated attempts that customers make to access service. As an example, suppose that a customer is trying to call a local store. If the customer gets a busy signal, the

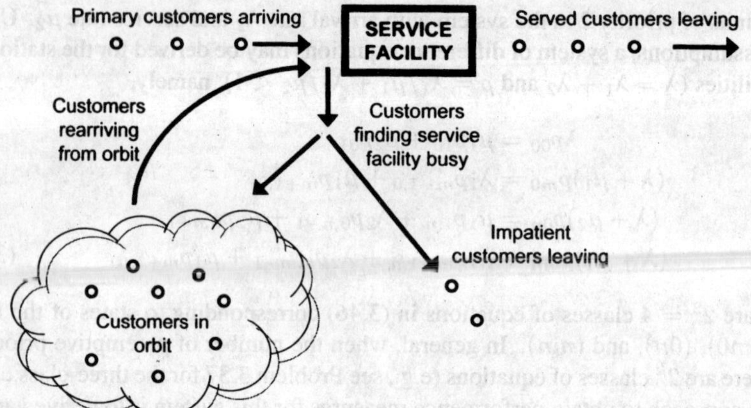

Figure 3.5 Basic model for a retrial queue.

customer may put the phone down and try again later, say, in 15 minutes. Hopefully, the server becomes free before the next call. However, other customers may also call in during this time, so the first customer may still get a busy signal on the second try.

This example motivates the basic conceptual model for a retrial queue, as shown in Figure 3.5. The main characteristics are:

1. An arriving customer, upon finding an available server, enters service.

2. An arriving customer, upon finding all servers busy, temporarily leaves the service facility. The customer may either:

   (a) Leave the system altogether (*impatience*).

   (b) Return later to the service facility. While away, the customer is said to be in *orbit*.

3. Customers in orbit cannot "see" the status of the service facility. They can only check the server status by "rearriving" at the service facility. Such an event is called a *retrial*.

4. Customers go back and forth from the orbit to the service facility until either service is received or they abandon the system.

In some regards, the orbit is like a queue, in that customers spend time waiting to be served. However, unlike a queue, a customer in orbit cannot monitor the status of the servers. In particular, once the server becomes free, there is a delay in time until a customer in orbit realizes that the server is free and begins service. Also, there is no concept of queueing order within the orbit. The order of service depends on the random order in which customers return to check the system status and the random chance that a server is available at the moment the customer returns to the system. In particular, customers do not generally receive service on a first-come first-served basis.

One limiting case is when the time spent in orbit for each customer is *instantaneous*. In other words, an arriving customer, upon finding all servers busy, goes to the orbit, then instantly returns to check the server status. In this case, the orbit behaves like a queue with random service discipline (i.e., each customer in the orbit is equally likely to be the next customer served).

Except for a few simple models, retrial queues are generally difficult to analyze analytically. In this section, we consider Markovian retrial queues, mostly with a single server. The material in this section is based largely on material in the book by Falin and Templeton (1997). For more advanced retrial models, including those involving general distributions, see this reference. For a bibliography on retrial queues, see Artalejo (1999).

3.5.1 $M/M/1$ Retrial Queue

The first model we consider is the $M/M/1$ retrial queue. In this model, customers arrive at a single server queue according to a Poisson process with rate λ. Service times are exponential with mean $1/\mu$. Any arriving customer, upon finding the server busy, enters the orbit and remains there for an exponentially distributed period of time with mean $1/\gamma$. All interarrival times (between primary arrivals), service times, and orbit times are independent. Customers repeat service attempts until the server is available. In this model, we assume that *no customers leave the system due to impatience*.

Let $N_s(t)$ be the number of customers in service at time t [since there is one server, $N_s(t) \in \{0, 1\}$]. Let $N_o(t)$ be the number of customers in orbit at time t. Then, $\{N_s(t), N_o(t)\}$ is a CTMC, with state space $\{i, n\}$, where $i \in \{0, 1\}$ and $n \in \{0, 1, 2, \ldots\}$. The total number of customers in the system is $N(t) = N_s(t) + N_o(t)$. Figure 3.6 shows the rate transitions between states. Let $p_{i,n}$ be the fraction of time that the system is in state $\{i, n\}$. Then, the rate balance equations are

$$(\lambda + n\gamma)p_{0,n} = \mu p_{1,n}, \qquad n \geq 0 \qquad (3.47)$$

$$(\lambda + \mu)p_{1,n} = \lambda p_{0,n} + (n+1)\gamma p_{0,n+1} + \lambda p_{1,n-1}, \qquad n \geq 1 \qquad (3.48)$$

$$(\lambda + \mu)p_{1,0} = \lambda p_{0,0} + \gamma p_{0,1}. \qquad (3.49)$$

We obtain steady-state solutions for this system using generating functions. The derivation follows that given in Falin and Templeton (1997). Portions of the derivation will be left as exercises. Define the following partial generating functions:

$$P_0(z) \equiv \sum_{n=0}^{\infty} z^n p_{0,n}, \quad P_1(z) \equiv \sum_{n=0}^{\infty} z^n p_{1,n}.$$

Multiply (3.47) by z^n and sum over $n \geq 0$:

$$\lambda \sum_{n=0}^{\infty} z^n p_{0,n} + \gamma \sum_{n=0}^{\infty} n z^n p_{0,n} = \mu \sum_{n=0}^{\infty} z^n p_{1,n}.$$

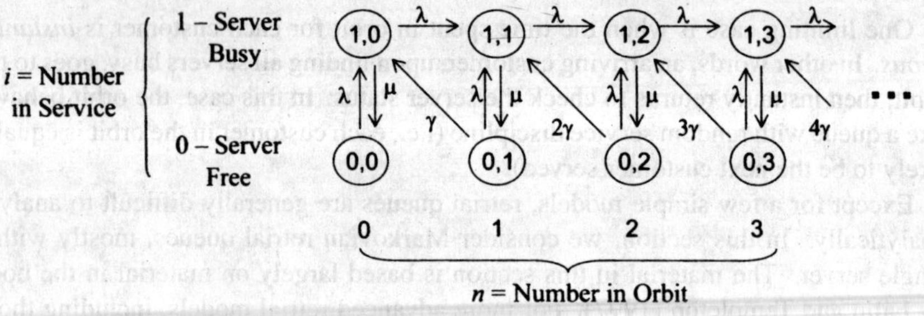

Figure 3.6 State transition rates for retrial queue.

This can be rewritten as

$$\lambda P_0(z) + z\gamma P_0'(z) = \mu P_1(z). \tag{3.50}$$

Similarly, multiply (3.48) by z^n, sum over $n \geq 1$, and add (3.49):

$$(\lambda + \mu)P_1(z) = \lambda P_0(z) + \gamma P_0'(z) + \lambda z P_1(z). \tag{3.51}$$

Solving for $P_1(z)$ in (3.50) and substituting into (3.51) gives (Problem 3.39)

$$P_0'(z) = \frac{\lambda\rho}{\gamma(1-\rho z)} P_0(z), \tag{3.52}$$

where $\rho = \lambda/\mu$. This is a separable differential equation, which we can write as follows:

$$\frac{P_0'(z)}{P_0(z)} = \frac{\lambda\rho}{\gamma(1-\rho z)}.$$

Integrating with respect to z gives

$$\ln P_0(z) = -\frac{\lambda}{\gamma}\ln(1-\rho z) + C'.$$

So,

$$P_0(z) = C(1-\rho z)^{-\lambda/\gamma}, \tag{3.53}$$

where $C = e^{C'}$. Now, $P_1(z)$ can be found by plugging $P_0(z)$ into (3.50):

$$\begin{aligned} P_1(z) &= \rho P_0(z) + \frac{\gamma}{\mu} z P_0'(z) \\ &= C\rho(1-\rho z)^{-\lambda/\gamma} + C\rho^2 z(1-\rho z)^{-(\lambda/\gamma)-1} \\ &= C\rho(1-\rho z)^{-(\lambda/\gamma)-1}. \end{aligned} \tag{3.54}$$

The constant C is found from the normalizing condition $P_0(1) + P_1(1) = 1$:

$$C = (1-\rho)^{(\lambda/\gamma)+1}.$$

Substituting this into (3.53) and (3.54) gives

$$\boxed{\begin{aligned} P_0(z) &= (1-\rho z)\left(\frac{1-\rho}{1-\rho z}\right)^{(\lambda/\gamma)+1} \\ P_1(z) &= \rho\left(\frac{1-\rho}{1-\rho z}\right)^{(\lambda/\gamma)+1} \end{aligned}} \tag{3.55}$$

To obtain the steady-state probabilities, we expand $P_0(z)$ and $P_1(z)$ in a power series using the binomial formula

$$(1+z)^m = \sum_{n=0}^{\infty}\binom{m}{n} z^n = \sum_{n=0}^{\infty}\frac{z^n}{n!}\prod_{i=0}^{n-1}(m-i). \tag{3.56}$$

The product is assumed to be 1 when $n = 0$. Expanding $P_0(z)$, from (3.53), gives

$$\begin{aligned} P_0(z) = C(1-\rho z)^{-\lambda/\gamma} &= C\sum_{n=0}^{\infty}\frac{(-\rho z)^n}{n!}\prod_{i=0}^{n-1}(-(\lambda/\gamma)-i) \\ &= \sum_{n=0}^{\infty}\left[C\frac{\rho^n}{n!\,\gamma^n}\prod_{i=0}^{n-1}(\lambda+i\gamma)\right] z^n. \end{aligned}$$

The coefficient in front of z^n is $p_{0,n}$. Similarly, $p_{1,n}$ can be found by expanding $P_1(z)$ (see Problem 3.40). In summary,

$$\boxed{\begin{aligned} p_{0,n} &= (1-\rho)^{(\lambda/\gamma)+1}\cdot\frac{\rho^n}{n!\,\gamma^n}\prod_{i=0}^{n-1}(\lambda+i\gamma), \\ p_{1,n} &= (1-\rho)^{(\lambda/\gamma)+1}\cdot\frac{\rho^{n+1}}{n!\,\gamma^n}\prod_{i=1}^{n}(\lambda+i\gamma). \end{aligned}} \tag{3.57}$$

The fraction of time the server is busy is $\sum_{n=0}^{\infty} p_{1,n}$. This can be found from the generating function, since $\sum_{n=0}^{\infty} p_{1,n} = P_1(1) = \rho$. The result could also have been derived without the generating function, by applying Little's law to the server: In steady state, the rate of customers completing service is λ. Thus the rate of customers beginning service is also λ. The average time in service is $1/\mu$. By Little's law, the average number in service is $\lambda/\mu = \rho$.

The mean number of customers in orbit can be derived from the partial generating functions in (3.55). First, we construct the generating function for the number of customers in orbit:

$$P(z) = \sum_{n=0}^{\infty} z^n(p_{0,n}+p_{1,n}) = P_0(z) + P_1(z).$$

Let L_o be the mean number of customers in orbit. Then, $L_o = P'(1)$. From this, it can be shown (Problem 3.41) that

$$\boxed{L_o = \frac{\rho^2}{1-\rho} \cdot \frac{\mu+\gamma}{\gamma}.} \tag{3.58}$$

L_o is the product of two terms: the average number in queue for an $M/M/1$ queue, $\rho^2/(1-\rho)$, and a term, $(\mu+\gamma)/\gamma$, that depends on the retrial rate γ. If γ is large, customers spend little time in orbit before making a retrial attempt. In the limit as $\gamma \to \infty$, customers spend no time in orbit before making a retrial attempt and thus are able to continuously monitor the status of the server. In this case, the system effectively behaves like an $M/M/1$ queue with random service order.

The mean time spent in orbit W_o (i.e., the mean time in orbit until finding the server idle and beginning service) can be derived from (3.58) using Little's law:

$$\boxed{W_o = \frac{L_o}{\lambda} = \frac{\rho^2}{\lambda(1-\rho)} \cdot \frac{\mu+\gamma}{\gamma} = \frac{\rho}{\mu-\lambda} \cdot \frac{\mu+\gamma}{\gamma}.} \tag{3.59}$$

The average time in system W and the average number of customers in the system L can be similarly derived:

$$W = W_o + \frac{1}{\mu} = \frac{\rho}{\mu-\lambda} \frac{\mu+\gamma}{\gamma} + \frac{1}{\mu} = \frac{\rho\mu(\mu+\gamma)+\gamma(\mu-\lambda)}{\mu\gamma(\mu-\lambda)} = \frac{\lambda\mu+\gamma\mu}{\mu\gamma(\mu-\lambda)}.$$

This simplifies to

$$\boxed{\begin{aligned} W &= \frac{1}{\mu-\lambda} \cdot \frac{\lambda+\gamma}{\gamma}, \\ L = \lambda W &= \frac{\rho}{1-\rho} \cdot \frac{\lambda+\gamma}{\gamma}. \end{aligned}}$$

L could also have been derived from $L = L_o + \rho$, where ρ corresponds to the average number of customers in service, as discussed earlier.

In all cases, the service measures are the product of the analogous measure for the $M/M/1$ queue and a term that goes to 1 as $\gamma \to \infty$. Conversely, as $\gamma \to 0$, the expected service measures go to ∞ because blocked customers spend an extremely long period of time in orbit before retrying.

3.5.2 $M/M/1$ Retrial Queue with Impatience

In the previous section, we assumed that each customer remains in the system until receiving service. In this section, we consider *impatient* customers who may abandon the system before receiving service. Specifically, we assume that each time a customer is denied service, the customer enters the orbit with probability q and abandons the system altogether with probability $1-q$. This choice is independent of all else.

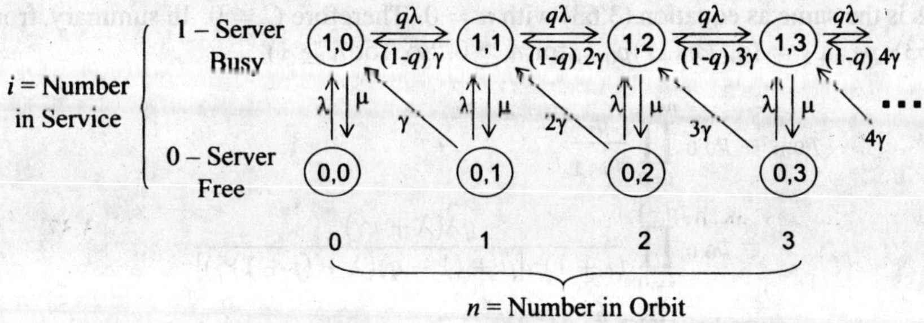

Figure 3.7 State transition rates for a retrial queue with impatience.

As before, we define the state space $\{i, n\}$, where $i \in \{0, 1\}$ is the number of customers in service and $n \in \{0, 1, 2, \ldots\}$ is the number of customers in orbit. Figure 3.7 shows the transition rates between states. The rate balance equations for this system are

$$(\lambda + n\gamma)p_{0,n} = \mu p_{1,n} \tag{3.60}$$

$$\begin{aligned}[q\lambda + \mu + (1-q)n\gamma]p_{1,n} &= \lambda p_{0,n} + q\lambda p_{1,n-1} + (n+1)\gamma p_{0,n+1} \\ &\quad + (1-q)(n+1)\gamma p_{1,n+1}\end{aligned} \tag{3.61}$$

$$(q\lambda + \mu)p_{1,0} = \lambda p_{0,0} + \gamma p_{0,1} + (1-q)\gamma p_{1,1}, \tag{3.62}$$

where (3.60) is valid for $n \geq 0$, and (3.61) is valid for $n \geq 1$. To solve these equations, we follow the approach given in Falin and Templeton (1997). Solving for $p_{1,n}$ in (3.60) and substituting the result into (3.61) gives (after a little algebra)

$$\begin{aligned}&(n+1)\gamma[\mu + (1-q)(\lambda + (n+1)\gamma)]p_{0,n+1} - q\lambda(\lambda + n\gamma)p_{0,n} \\ &\quad = n\gamma[\mu + (1-q)(\lambda + n\gamma)]p_{0,n} - q\lambda(\lambda + (n-1)\gamma)p_{0,n-1}, \qquad n \geq 1.\end{aligned}$$

This can be rewritten

$$x_{n+1}p_{0,n+1} - y_n p_{0,n} = x_n p_{0,n} - y_{n-1}p_{0,n-1}, \qquad n \geq 1,$$

where $x_n \equiv n\gamma[\mu + (1-q)(\lambda + n\gamma)]$ and $y_n \equiv q\lambda(\lambda + n\gamma)$. This implies that

$$x_{n+1}p_{0,n+1} - y_n p_{0,n} = C, \tag{3.63}$$

for some constant C (for all $n \geq 0$). The constant C can be determined from (3.62) as follows: Solve for $p_{1,0}$ and $p_{1,1}$ in (3.60) and substitute into (3.62). This gives

$$\gamma[\mu + (1-q)(\lambda + \gamma)]p_{0,1} - q\lambda^2 p_{0,0} = 0,$$

which can be rewritten

$$x_1 p_{0,1} - y_0 p_{0,0} = 0.$$

This is the same as equation (3.63) with $n = 0$. Therefore $C = 0$. In summary, from (3.63), $p_{0,n+1} = (y_n/x_{n+1})p_{0,n}$, for $n \geq 0$. So (for $n \geq 1$),

$$
\begin{aligned}
p_{0,n} &= p_{0,0} \prod_{i=0}^{n-1} \frac{y_i}{x_{i+1}} \\
&= p_{0,0} \prod_{i=0}^{n-1} \frac{q\lambda(\lambda + i\gamma)}{(i+1)\gamma[\mu + (1-q)(\lambda + (i+1)\gamma)]} \\
&= p_{0,0} \frac{q^n \lambda^n}{n!\gamma^n} \prod_{i=0}^{n-1} \frac{\lambda + i\gamma}{\mu + (1-q)(\lambda + \gamma + i\gamma)} \\
&= p_{0,0} \frac{q^n \lambda^n}{n!\gamma^n} \prod_{i=0}^{n-1} \frac{\gamma(\lambda/\gamma + i)}{(1-q)\gamma \left[\dfrac{\mu + (1-q)(\lambda+\gamma)}{(1-q)\gamma} + i \right]} \\
&= p_{0,0} \frac{q^n \lambda^n}{n!(1-q)^n \gamma^n} \prod_{i=0}^{n-1} \frac{(\lambda/\gamma + i)}{\left[\dfrac{\mu + (1-q)(\lambda+\gamma)}{(1-q)\gamma} + i \right]} \\
&= \frac{c^n}{n!} \cdot \frac{(a)_n}{(b)_n} p_{0,0},
\end{aligned}
$$

where

$$
a \equiv \frac{\lambda}{\gamma}, \; b \equiv \frac{\mu + (1-q)(\lambda+\gamma)}{(1-q)\gamma}, \; c \equiv \frac{q\lambda}{(1-q)\gamma}, \tag{3.64}
$$

and

$$
(a)_n \equiv a(a+1)(a+2)\cdots(a+n-1) \tag{3.65}
$$

is the *rising factorial function*. To obtain $p_{1,n}$, we use (3.60):

$$
\begin{aligned}
p_{1,n} &= \frac{\lambda + n\gamma}{\mu} p_{0,n} = \frac{\gamma(a+n)}{\mu} p_{0,n} \\
&= \frac{\gamma(a+n)}{\mu} \cdot \frac{c^n}{n!} \cdot \frac{a(a+1)\cdots(a+n-1)}{(b)_n} p_{0,0} \\
&= \frac{\gamma a}{\mu} \cdot \frac{c^n}{n!} \cdot \frac{(a+1)\cdots(a+n)}{(b)_n} p_{0,0} \\
&= \rho \frac{c^n}{n!} \cdot \frac{(a+1)_n}{(b)_n} p_{0,0}.
\end{aligned}
$$

Then, the partial generating functions are

$$
\begin{aligned}
P_0(z) &= \sum_{n=0}^{\infty} p_{0,n} z^n = p_{0,0} \sum_{n=0}^{\infty} \frac{(cz)^n}{n!} \cdot \frac{(a)_n}{(b)_n}, \\
P_1(z) &= \sum_{n=0}^{\infty} p_{1,n} z^n = p_{0,0} \sum_{n=0}^{\infty} \rho \frac{(cz)^n}{n!} \cdot \frac{(a+1)_n}{(b)_n}.
\end{aligned}
$$

This can be rewritten as

$$P_0(z) = \Phi(a, b; cz)\, p_{0,0},$$
$$P_1(z) = \rho\Phi(a+1, b; cz)\, p_{0,0}, \tag{3.66}$$

where Φ is the Kummer confluent function

$$\Phi(a, b; z) \equiv \sum_{n=0}^{\infty} \frac{(a)_n}{(b)_n} \cdot \frac{z^n}{n!}. \tag{3.67}$$

We can find $p_{0,0}$ with the normalizing condition $P_0(1) + P_1(1) = 1$, implying

$$p_{0,0} = \frac{1}{\Phi(a, b; c) + \rho\Phi(a+1, b; c)}. \tag{3.68}$$

The steady-state probabilities $p_{i,n}$ are obtained from the coefficients in front of z^n in the partial generating functions $P_0(z)$ and $P_1(z)$. In summary,

$$p_{0,n} = p_{0,0} \frac{c^n}{n!} \cdot \frac{(a)_n}{(b)_n}$$
$$p_{1,n} = p_{0,0}\, \rho \frac{c^n}{n!} \cdot \frac{(a+1)_n}{(b)_n}, \tag{3.69}$$

where a, b, c, $\Phi(\cdot)$, $(\cdot)_n$, and $p_{0,0}$ are given by (3.64), (3.65), (3.67), and (3.68). When $q = 1$, it can be shown that the equations for $p_{i,n}$ reduce to the same equations (3.57) for the $M/M/1$ retrial queue without impatience (see Problem 3.44). [However, one cannot directly plug in $q = 1$ since b and c in (3.64) both have $(1 - q)$ in the denominator.]

Various performance metrics can be derived from the partial generating functions. The fraction of time the server is busy is

$$p_1 \equiv \sum_{n=0}^{\infty} p_{1,n} = P_1(1) = \frac{\rho\Phi(a+1, b; c)}{\Phi(a, b; c) + \rho\Phi(a+1, b; c)} = \frac{\rho^*}{1+\rho^*}, \tag{3.70}$$

where

$$\rho^* = \rho \frac{\Phi(a+1, b; c)}{\Phi(a, b; c)}. \tag{3.71}$$

Note that (3.70) is similar in form to the server utilization of an $M/M/1/1$ queue, $p_1 = \rho/(1+\rho)$, but with ρ^* replacing ρ; see Section 2.6, equation (2.52), with $c = 1$. In general, ρ^* is greater than ρ, because blocked customers may reattempt service, which is not the case for an $M/M/1/1$ queue.

We now sketch a derivation for the mean number of customers in orbit L_o, using partial generating functions:

$$L_o = P_0'(1) + P_1'(1).$$

Portions of the derivation will be left as exercises. We will make use of the following properties of the Kummer function (e.g., Gradshteyn and Ryzhik, 2000, p. 1013, Equations 9.213 and 9.212.2 – 9.212.4):

$$\frac{d}{dz}\Phi(a,b;z) = \frac{a}{b}\Phi(a+1,b+1;z), \tag{3.72}$$

$$z\Phi(a+1,b+1;z) = b\Phi(a+1,b;z) - b\Phi(a,b;z), \tag{3.73}$$

$$a\Phi(a+1,b+1;z) = (a-b)\Phi(a,b+1;z) + b\Phi(a,b;z), \tag{3.74}$$

$$\begin{aligned}(a+1)\Phi(a+2,b+1;z) = (z+2a-b+1)\Phi(a+1,b+1;z)\\ +(b-a)\Phi(a,b+1;z).\end{aligned} \tag{3.75}$$

To evaluate $P_0'(1)$, first apply (3.72) to $P_0(z) = \Phi(a,b;cz)\,p_{0,0}$:

$$P_0'(z) = \frac{ca}{b}\Phi(a+1,b+1;cz)\,p_{0,0}.$$

Now evaluate at $z = 1$ and apply (3.73):

$$P_0'(1) = \frac{ca}{b}\Phi(a+1,b+1;c)\,p_{0,0} = a[\Phi(a+1,b;c) - \Phi(a,b;c)]\,p_{0,0}.$$

Similarly, to evaluate $P_1'(1)$, apply (3.72) and (3.75) to $P_1(z) = \rho\Phi(a+1,b;cz)\,p_{0,0}$:

$$\begin{aligned}P_1'(1) &= \rho\frac{c(a+1)}{b}\Phi(a+2,b+1;c)\,p_{0,0} \qquad (3.76)\\ &= \rho\frac{c}{b}\left[(c+2a-b+1)\Phi(a+1,b+1;c) + (b-a)\Phi(a,b+1;c)\right]\,p_{0,0}.\end{aligned}$$

Application of (3.73) and (3.74) gives (Problem 3.45)

$$P_1'(1) = \rho[(c+a-b+1)\Phi(a+1,b;c) - (a-b+1)\Phi(a,b;c)]\,p_{0,0}. \tag{3.77}$$

Combining $P_0'(1) + P_1'(1)$ gives

$$\begin{aligned}L_o = [\rho(c+a-b+1)+a]\Phi(a+1,b;c)\,p_{0,0}\\ -[\rho(a-b+1)+a]\Phi(a,b;c)\,p_{0,0}.\end{aligned} \tag{3.78}$$

Substituting the values for a, b, and c in (3.64) and $p_{0,0}$ in (3.68) gives (Problem 3.45)

$$\boxed{L_o = \frac{q}{1-q}\cdot\frac{\lambda+(\lambda-\mu)\rho^*}{\gamma(1+\rho^*)},} \tag{3.79}$$

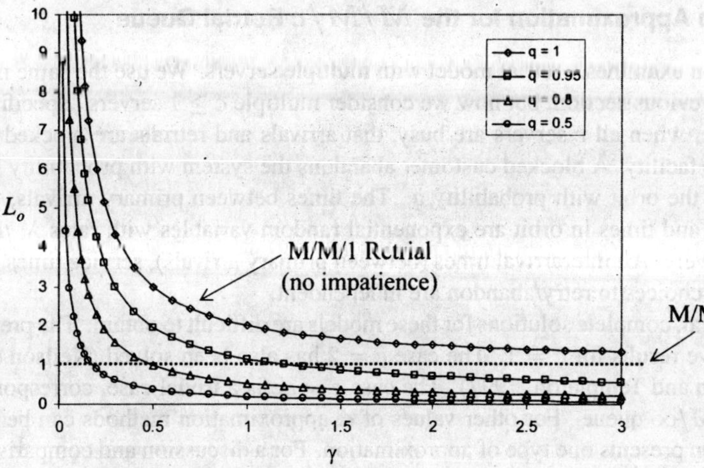

Figure 3.8 $M/M/1$ retrial queue with impatience.

where ρ^* is given in (3.71). Further performance metrics can also be determined. For example, the average number in the system is

$$L = L_o + p_1 = \frac{q}{1-q} \cdot \frac{\lambda + (\lambda - \mu)\rho^*}{\gamma(1+\rho^*)} + \frac{\rho^*}{1+\rho^*}.$$

The fraction of customers who eventually receive service (Problem 3.43) is

$$\text{Fraction of customers receiving service} = \frac{\mu p_1}{\lambda} = \frac{\rho^*}{\rho(1+\rho^*)}.$$

Metrics like W_o and W can be determined from L_o and L using Little's law (Problem 3.46).

Figure 3.8 shows an example of L_o as a function of γ, for various values of q. Here, $\lambda = (-1+\sqrt{5})/2 \doteq 0.618$ and $\mu = 1$. Several trends are evident from the figure. A higher retrial probability q yields more customers in orbit. The case $q = 1$ corresponds to the $M/M/1$ retrial queue without impatience (3.58). Lower values of γ yield more customers in orbit, since each customer waits longer before making a retrial attempt. As $\gamma \to 0$, $L_o \to \infty$. For a fixed $q < 1$, as $\gamma \to 0$, $L_o \to 0$. This is because a blocked customer reattempts service very quickly. The server is likely to still be busy on the reattempt. Because the customer has a nonzero probability of abandoning the system *at each reattempt*, a customer that is initially blocked is likely to never receive service. On the other hand, when $q = 1$, L_o approaches $L_q = \rho^2/(1-\rho)$ as $\gamma \to 0$, the queue wait for an $M/M/1$ queue, as discussed in the previous section.

3.5.3 An Approximation for the $M/M/c$ Retrial Queue

This section examines a retrial model with multiple servers. We use the same model as in the previous section, but now we consider multiple $c \geq 1$ servers. Specifically, we assume, when all c servers are busy, that arrivals and retrials are blocked from the service facility. A blocked customer abandons the system with probability $1 - q$ and enters the orbit with probability q. The times between primary arrivals, times in service, and times in orbit are exponential random variables with rates λ, μ, and γ, respectively. All interarrival times (between primary arrivals), service times, orbit times, and choices to retry/abandon are independent.

In general, complete solutions for these models are difficult to obtain. The previous section gave results for $c = 1$. The case $c = 2$ has also been solved (Keilson et al., 1968, Falin and Templeton, 1997). The case $c = \infty$ is a trivial case, corresponding to an $M/M/\infty$ queue. For other values of c, approximation methods can be used. This section presents one type of approximation. For a discussion and comparison of other approximations, see Wolff (1989). The approximation given here assumes that the retrial rate γ is low, so that the servers approximately reach steady state before the next retrial. The approximation does not work well when the retrial rate γ is high.

Similar to the previous models, we define the state space to be $\{i, n\}$, where $i \in \{0, 1, 2, \ldots, c\}$ is the number of customers in service and $n \in \{0, 1, 2, \ldots\}$ is the number of customers in orbit. Let $p_{i,n}$ be the fraction of time the system is in state $\{i, n\}$. Let p_i be the fraction of time that i servers are busy ($p_i = \sum_n p_{i,n}$). Define

$$\lambda_{r,i} \equiv \frac{1}{p_i} \sum_{n=0}^{\infty} (n\gamma) p_{i,n},$$

$$\lambda_r \equiv \sum_{i=0}^{c} \sum_{n=0}^{\infty} (n\gamma) p_{i,n} = \sum_{i=0}^{c} p_i \lambda_{r,i},$$

$$r_i \equiv \frac{\sum_{n=0}^{\infty} (n\gamma) p_{i,n}}{\sum_{i=0}^{c} \sum_{n=0}^{\infty} (n\gamma) p_{i,n}} = \frac{p_i \lambda_{r,i}}{\lambda_r}.$$

That is, $\lambda_{r,i}$ is the rate of retrial attempts during periods when exactly i servers are busy; λ_r is the overall rate of retrial attempts. r_i is the fraction of retrials that see i servers busy.

If the retrial rate γ is small, we expect the servers to approximately reach steady state before the next retrial. This motivates the following:

Assumption: Retrials see time averages.

In other words, the fraction of retrials that see i servers busy is the same as the fraction of time i servers are busy, so $r_i = p_i$. (If γ were high, this assumption would not be valid. In this case, a blocked customer would retry very quickly and thus would likely observe all c servers busy again. This is different than observing the servers at a *random* point in time, which would yield a lower probability of all servers busy.) From the previous definition of r_i, the assumption that $r_i = p_i$ is equivalent to the assumption that $\lambda_{r,i} = \lambda_r$. This says that the rate of retrials does not depend on the number of customers in service.

We now apply this assumption to some of the rate balance equations. First, equating rates between the system states where exactly i servers are busy and the states where exactly $i + 1$ servers are busy gives

$$(\lambda + \lambda_{r,i})p_i = (i+1)\mu p_{i+1}, \qquad i = 0, \ldots, c-1.$$

Under the assumption $\lambda_{r,i} = \lambda_r$,

$$p_i = \frac{(\lambda + \lambda_r)^i}{\mu^i \cdot i!} p_0 \qquad i = 0, \ldots, c.$$

In particular,

$$p_c = \frac{(\lambda + \lambda_r)^c}{\mu^c \cdot c!} p_0 = \frac{(\lambda + \lambda_r)^c}{\mu^c \cdot c!} \Bigg/ \sum_{i=0}^{c} \frac{(\lambda + \lambda_r)^i}{\mu^i \cdot i!}, \tag{3.80}$$

where the last equality follows from normalization.

Now, (3.80) is the Erlang-B formula (2.53) with $(\lambda + \lambda_r)$ replacing λ. (Recall that the Erlang-B formula is the blocking probability of an $M/M/c/c$ queue; see Section 2.6.) Thus we can derive (3.80) under a different assumption: Suppose that the stream of retrial attempts to the service facility is a Poisson process with rate λ_r and is independent of the primary arrivals. In this case, the combined stream of primary arrivals and retrial attempts to the service facility is a Poisson process with rate $\lambda + \lambda_r$. Thus the system behaves like an $M/M/c/c$ queue. So, the fraction of time p_c that all servers are busy is the Erlang-B formula with $(\lambda + \lambda_r)$ replacing λ.

In summary, we can derive (3.80) under any of the following assumptions: (1) Retrials see time averages, (2) the retrial rate is independent of the number in service, and (3) retrials follow a Poisson process, independent of the primary arrivals.

Although (3.80) specifies the form of p_c, we still need to find λ_r. To do this, equate the rates that customers enter and leave orbit:

$$q(\lambda + \lambda_{r,c})p_c = \lambda_r. \tag{3.81}$$

Assuming $\lambda_{r,i} = \lambda_r$ and using (3.80) gives

$$\frac{q(\lambda + \lambda_r)^{c+1}}{\mu^c \cdot c!} = \lambda_r \sum_{i=0}^{c} \frac{(\lambda + \lambda_r)^i}{\mu^i \cdot i!}. \tag{3.82}$$

This gives a polynomial of order $c + 1$. The equation can be solved numerically for λ_r.

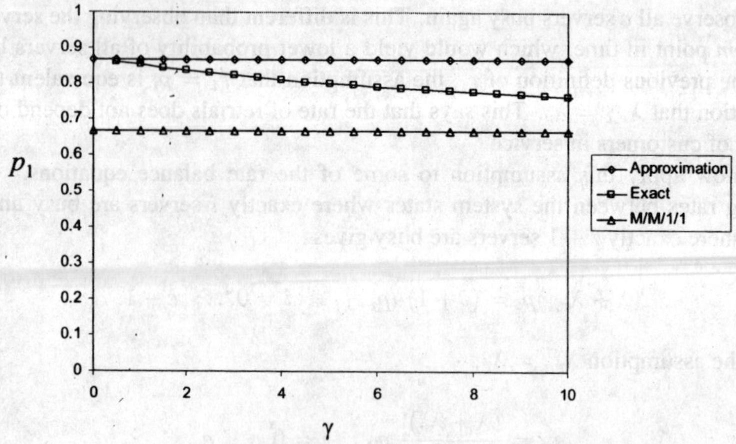

Figure 3.9 Example of retrial approximation.

■ **EXAMPLE 3.12**

Using the approximation, estimate p_1 when $c = 1$. First, define $\lambda_T \equiv \lambda + \lambda_r$. Then, (3.80) becomes

$$p_1 = \frac{\lambda_T}{\mu}\bigg/\left(1 + \frac{\lambda_T}{\mu}\right) = \frac{\lambda_T}{\mu + \lambda_T}.$$

Then, (3.81) becomes $q\lambda_T p_1 = \lambda_T - \lambda$ or $\lambda_T = \lambda/(1 - qp_1)$. Substituting this into the previous equation gives

$$p_1 = \frac{\lambda}{(1 - qp_1)\mu + \lambda}.$$

This is a quadratic equation, $q\mu p_1^2 - (\lambda + \mu)p_1 + \lambda = 0$, with solution

$$p_1 = \frac{\lambda + \mu - \sqrt{(\lambda + \mu)^2 - 4q\lambda\mu}}{2q\mu}. \tag{3.83}$$

We can compare this approximate result to the exact result from Section 3.5.2. Figure 3.9 shows an example where $\lambda = 2$, $\mu = 1$, and $q = 0.8$. The single-server utilization p_1 is calculated three ways: using the approximation (3.83), using the exact value (3.70), and using an approximating $M/M/1/1$ model with arrival rate λ, service rate μ, and no retrials ($q = 0$). For small values of γ, the exact value is close to the upper line, approximation (3.83). The accuracy degrades as γ gets larger. As $\gamma \to \infty$, the exact value approaches the lower line, the approximating $M/M/1/1$ model. The reason for this is that a blocked customer reattempts service very quickly and thus is likely to reattempt service multiple times while the server is still busy. Because the customer has a non-zero probability of abandoning the system *at each reattempt* (assuming

$q < 1$), a customer that is initially blocked is likely to never receive service. Thus the system effectively behaves like a queue where all blocked customers leave the system.

PROBLEMS

3.1. Use stochastic balance to obtain (3.1), (3.7), and (3.10).

3.2. Show that the operator equation in (3.8), $\mu r^{K+1} - (\lambda + \mu)r + \lambda = 0$, has exactly one root in the interval (0, 1), using Rouché's theorem. [*Hint*: Refer to Section 2.11.2, and set $g = r^{K+1}$ and $f = -(\lambda/\mu + 1)r + \lambda/\mu$.]

3.3. The moonlighting graduate assistant of Problem 2.8 decides that a more correct model for his short-order counter is an $M^{[X]}/M/1$, where the batch sizes are 1, 2, or 3 with equal probability, such that 5 batches arrive per hour on average (this maintains the previous total arrival rate of 10 customers/h). The mean service time remains at 4 min. Compare the average queue length and system size of this model with that of the previous $M/M/1$ model.

3.4. Consider an $M^{[2]}/M/1$ with service rate $\mu = 3$/h, in which all arrivals come in batches of two with frequency 1 batch/(2 h).

(a) Find the stationary system-size probability distribution, L, L_q, W, and W_q.

(b) Using the difference equations of (3.1), derive a closed-form expression for p_n.

3.5. For the partial-batch $M/M^{[Y]}/1$ model, show that $L_q = r_0^K L$ [compute directly using the steady-state probabilities in (3.9)]. Then, show that $L_q = r_0^K L$ is equivalent to the expression $L_q = L - \lambda/\mu$ derived in the text. [*Hint*: Use the characteristic equation derived from (3.8).]

3.6. Derive the probability generating function for the bulk-service model as given in (3.14).

3.7. Apply Rouché's theorem to the denominator of (3.14) to show that K zeros lie on or within the unit circle. [*Hint*: Show that K zeros lie on or within $|z| = 1 + \delta$ by defining $f(z)$ and $g(z)$ such that $f(z) + g(z)$ equals the denominator.]

3.8. A ferry loads cars for delivery across a river and must have a full ferry load of 10 cars. Loading is instantaneous, and the round-trip time is an exponential random variable with mean of 15 min. The cars arrive at the shore as a Poisson process with a mean of 30/h. On average, how many autos are waiting on the shore at any instant for a ferry?

3.9. For the $M/E_k/1$ queue, derive (3.21) by stochastic balance.

3.10. For the following queues, draw rate-transition diagrams for the corresponding Markov chains. Compare the two diagrams.

(a) The $M/E_k/1$ queue with arrival rate λ and service rate μ (i.e., the service rate of each phase is $k\mu$).

(b) The $M^{[X]}/M/1$ queue where the arrival rate of batches is λ, the number of customers per batch is a constant k, and the service rate of each customer is μ.

3.11. Complete the details of the derivation of (3.20), the two-term hypoexponential distribution, in Example 3.6. [*Hint*: Use direct evaluation of the Chapman–Kolmogorov differential equations.]

3.12. To show the effect of service-time variation on performance measures for the $M/E_k/1$ model, calculate and plot W_q versus $k(= 1, 2, 3, 4, 5, 10)$ for $\lambda = 1, \rho(= 1/\mu) = 0.5, 0.7, 0.9$.

3.13. The following is a variation of the logic used to derive (3.22). "The average number of customers in a queue is L. The average service time for one customer is $1/\mu$. An arriving customer sees on average L customers in the system. Thus the average wait in queue is $W_q = L/\mu$." For which of the following queues is this a valid argument: $M/M/1$, $M/M/c$, $M/G/1$, or $G/M/1$? State why or why not. If the argument is valid, verify that $W_q = L/\mu$.

3.14. To show the effect of arrival-time variation on performance measures for the $E_k/M/1$ model, calculate and plot W_q versus $k(= 1, 2, 3, 4, 5, 10)$ for $\lambda = 1, \rho(= 1/\mu) = 0.5, 0.7, 0.9$.

3.15. Derive the steady-state probability that a customer is in phase n for an $M/E_k/1/1$ model, that is, an Erlang service model where no queue is allowed to form.

3.16. A drive-through car-wash service has one station for washing cars. Cars arrive according to a Poisson process with rate $\lambda = 10$ per hour.

(a) If the time to wash a car is approximately a fixed value of 4 minutes, what is the expected wait in queue?

(b) You are thinking of buying a new car wash that senses the amount of dirt on the car and washes the car only as long as necessary. The new car wash is slightly faster on average, but there is variability in the time to wash each car. The time to wash a car has a mean of 3.5 minutes and a standard deviation of 2 minutes. Using an Erlang approximation for the service distribution, what is the expected wait in queue for the new car wash? Is it longer or shorter than with the existing car wash?

3.17. Consider the $M/E_k/c/c$ model.

(a) Derive the steady-state difference equation for this model. [*Hint*: Let $p_{n;s_1,s_2,\ldots,s_k}$ represent the probability of n in the system with s_1 channels in phase 1, s_2 in phase 2, etc.]

(b) Show that $p_n = p_0\rho^n/n!$, $\rho = \lambda/\mu$, is a solution to the problem. [*Hint*: First show that it is a solution to the equation of (a). Then show that

$$p_n = \sum_{s1+s2+\cdots+sk=n} p_{n;s_1,s_2,\ldots,s_k} = \frac{A\rho^n}{n!}$$

by utilizing the multinomial expansion $(x_1 + x_2 + \cdots + x_k)^n$, then setting $x_1 = x_2 = \cdots = x_k = 1$.]

(c) Compare this result with the $M/M/c/c$ results of Section 2.6, Equation (2.52), and comment.

3.18. Give a complete explanation of why the $M^{[k]}/M/1$ bulk-arrival model can be used to represent the $M/E_k/1$ model when customers are considered to be phases of service to be completed.

3.19. Consider a single-server model with Poisson input and an average customer arrival rate of 30/h. Currently, service is provided by a mechanism that takes *exactly* 1.5 min. Suppose the system can be served instead by a mechanism that has an exponential service-time distribution. What must be the mean service time for this mechanism to ensure (a) the same average time a customer spends in the system, or (b) the same average number of customers in the system?

3.20. For an $M/E_3/1$ model with $\lambda = 6$ and $\mu = 8$, find the probability of more than two in the system.

3.21. Consider the following model of a single bank teller. Customers arrive according to a Poisson process with rate λ. Customers are served on a FCFS basis. The time to serve each customer is exponential with rate μ. When the teller becomes idle (i.e., when there are no customers in the system), the teller begins a separate off-line task of counting money. The time to complete the money-counting task is exponential with rate γ. Customers who arrive while the teller is counting money must wait until the teller completes the task before receiving any service. If the teller completes the money-counting task before any customers arrive, the teller becomes idle until the next customer arrives. The teller then serves customers until a new idle moment (i.e., when no customers are in the system), at which time the teller again starts a new off-line task of counting money.

(a) Specify a continuous-time Markov chain for this system. That is, define a set of states and specify rate transitions between the states using a diagram. Then, give the rate balance equations for this system.

(b) Suppose that you have found the steady-state probabilities to your Markov chain. Give an expression for the average number of customers in the system (L) and the average number of customers in the queue (L_q) as a function of these probabilities.

3.22. A large producer of television sets has a policy of checking all sets prior to shipping them from the factory to a warehouse. Each line (large-screen, portable, etc.) has its own expert inspector. At times, the highest-volume line (color portables) has experienced a bottleneck condition (at least in the management's opinion), and a detailed study was made of the inspection performance. Sets were found to arrive at the inspector's station according to a Poisson distribution with a mean of 5/h. In carrying out the inspection, the inspector performs 10 separate tests, each taking, on average, 1 min. The times for each test were found to be approximately exponentially distributed. Find the average waiting time a set experiences, the average number of sets in the system, and the average idle time of the inspector.

3.23. The Rhodehawgg Trucking Company has a choice of hiring one of two individuals to operate its single-channel truck-washing facility. In studying their performances it was found that one man's times for completely washing a truck were approximately exponentially distributed with a mean rate of 6/day, while the other man's times were distributed according to an Erlang type 2 with a mean rate of 5/day. Which man should be hired when the arrival rate is 4/day?

3.24. Isle-Air Airlines offers air shuttle service between San Juan and Charlotte Amalie every 2 h. The procedure calls for no advance reservations but for passengers to purchase their tickets at the gate from which the shuttle leaves. It is found that passengers arrive according to a Poisson distribution with a mean of 18/h. There is one agent at the gate check-in counter, and a time study provided the following 50 observations on the processing time in minutes:

4.00, 1.44, 4.44, 1.74, 1.16, 4.20, 3.59, 2.14, 3.54, 2.56, 5.53, 2.02, 3.06, 1.66, 3.23, 4.84, 7.99, 3.07, 1.24, 3.40, 5.01, 2.78, 1.62, 5.19, 5.09, 3.78, 1.52, 3.94, 1.96, 6.20, 3.67, 3.37, 1.84, 1.60, 1.31, 5.64, 0.99, 3.06, 1.24, 3.11, 4.57, 0.90, 2.78, 1.64, 2.43, 5.26, 2.11, 4.27, 3.36, 4.76.

On average, how many are in the queue waiting for tickets, and what is the average wait in the queue? [*Hint*: Find the sample mean and variance of the observed service times, and see what distribution might fit.]

3.25. Complete the details of the matrix-vector derivation for the Erlang type-2 distribution, Example 3.5. [*Hint*: For this problem, it is easy to expand $e^{\tilde{Q}t}$ as an infinite series of matrices.]

3.26. A generalization of the inventory-control procedure of Problem 2.41 is as follows. Again using S as the safety-stock value, the policy is to place an order for an amount Q when the on-hand plus on-order inventory reaches a level $s(Q = S - s)$. Note that the one-for-one policy of Problem 2.41 is a special case for which $Q = 1$ or equivalently $s = S - 1$. This policy is called a trigger-point-order quantity policy and is sometimes also referred to as a continuous review (s, S) policy. Generally, a manufacturing setup cost of K dollars per order placed is also included, so that an additional cost term of $K\lambda/Q$ (dollars per unit time) is included in $\mathrm{E}[C]$. For this situation, again assuming Poisson demand and exponential lead times, describe the queueing model appropriate to the order-processing procedure. Then relate the steady-state probabilities resulting from the order processing queueing model to $p(z)$, the probability distribution of on-hand inventory. Finally, discuss the optimization procedure for E[C], now a function of two variables (either s and Q, S and Q, or S and s) and the practicality of using this type of analysis.

3.27. Derive the stationary equations (3.33) for the single exponential channel with two priorities.

3.28. Derive (3.38), the expected measures of performance for a single-server queue with two classes of customers and no priorities. (Class 1 customers arrive with rate λ_1 and have exponential service with rate μ_1. Class 2 customers arrive with rate λ_2 and have exponential service with rate μ_2.) [*Hint*: Consider the queue as an $M/G/1$ queue, where G is a mixed-exponential distribution. Then apply results from Chapter 5.]

3.29. Show that L_q of (3.38) is always greater than or equal to L_q of $M/M/1$, where $\rho = \lambda_1/\mu_1 + \lambda_2/\mu_2$ is used for the $M/M/1$ queue.

3.30. Compare the L_q's of the two-priority, two-rate case with those of the two-priority, one-rate case when the μ of the latter equals $\min(\mu_1, \mu_2)$.

3.31. For a two-class queue, show that the imposition of priorities decreases the mean number of priority-1 customers in the queue and increases the mean number of priority-2 customers in the queue. In other words, verify the results of cell [(c), (d)] in Table 3.1 for $L_q^{(1)}$ and $L_q^{(2)}$.

3.32. Carry out the induction that leads from (3.40) to (3.41).

3.33. For the nonpreemptive priority queue, suppose that $\mu_i \equiv \mu$ for all i. Under this condition, show that the average wait in queue across *all* customers is the same as the average wait in queue for the $M/M/1$ queue.

3.34. For the nonpreemptive priority queue with two classes, suppose that $\lambda_1 = \lambda_2 = 1$, $\mu_1 = 3$, and $\mu_2 = 2$ (i.e., the class with faster service has priority). Show that the expected queue wait W_q is lower than that of the ordinary $M/M/1$ queue with $1/\mu = 1/(2\mu_1) + 1/(2\mu_2)$. What happens when the class with slower service has priority?

3.35. Patients arrive at the emergency room of a hospital according to a Poisson process. A patient is classified into one of three types: high priority (i.e., a life is at stake), medium priority (i.e., the patient might be in pain but his or her life is not threatened), and low priority. On an average day, there are approximately 10 arrivals per hour. One-fifth of the arrivals are high priority, three-tenths are medium priority, and one-half are low priority. All registration matters are handled by a single clerk. The time to register a patient is exponentially distributed with a mean of 5.5 min (regardless of the classification of the patient). What is the average time to complete the registration of an arbitrary patient? (Assume that there is no preemption.)

3.36. What happens in Problem 3.35 when there is preemption? Compare the results here with the results in Problem 3.35.

3.37. Give the rate balance equations for a three-class preemptive Markovian queue. That is, generalize (3.46) to three priority classes.

3.38. For the two-class preemptive priority queue (Section 3.4.3), show from the rate balance equations (3.46) that the system behaves like an $M/M/1$ queue for the priority-1 customers.

3.39. For the $M/M/1$ retrial queue, derive (3.52) from (3.50) and (3.51).

3.40. Derive the steady-state probabilities $p_{1,n}$ (3.57) for the $M/M/1$ retrial queue from the partial generating function $P_1(z)$ (3.55) and the binomial formula (3.56).

3.41. Derive the average number of customers in orbit (3.58) for the $M/M/1$ retrial queue using the partial generating functions in (3.55).

3.42. A small store has a single phone line. Calls to the store are well modeled by an $M/M/1$ retrial queue with arrival rate $\lambda = 10$ per hour, service rate $\mu = 15$ per hour, and retrial rate $\gamma = 6$ per hour.

(a) What is the average length of time for a customer to reach a clerk at the store?

(b) What is the average rate that call *attempts* are made to the store? (Call attempts include calls that are answered and calls that receive a busy signal.)

(c) What is the fraction of call attempts that receive a busy signal?

(d) Suppose that you are measuring call attempts to the store and you are not aware that some customers are making redial attempts. That is, you assume each call attempt is from a distinct customer. You decide to model the system as an $M/M/1/1$ queue, where the arrival rate is the rate of call attempts found in (b). Based on these assumptions, what is the fraction of calls that receive a busy signal? Compare this to the actual result found in (c).

3.43. For the $M/M/1$ retrial queue with impatience, determine the fraction of customers who eventually receive service.

3.44. For the $M/M/1$ retrial queue with impatience, show that when $q = 1$, the steady-state probabilities in (3.69) reduce to the steady-state probabilities in (3.57) for the $M/M/1$ retrial queue (without impatience).

3.45. For the $M/M/1$ retrial queue with impatience, complete the following details for the derivation of L_o.

(a) Derive (3.77) from (3.76). [*Hint*: Write $(c + 2a - b + 1)$ as $(c + a - b + 1) + a$. Apply (3.73) to the term with $(c + a - b + 1)$ and (3.74) to the term with a.]
(b) Derive L_o (3.79) from (3.78).

3.46. For the $M/M/1$ retrial queue with impatience, determine the average time a customer spends in orbit, where the average is taken over:
(a) Customers who enter the system,
(b) Customers who enter the orbit (i.e., customers who are initially denied service and subsequently choose to enter the orbit).

3.47. Example 3.12 gives an approximate blocking probability p_1 for the $M/M/1$ retrial queue.
(a) In (3.83), p_1 is obtained as one of two solutions to a quadratic equation. Show that the correct solution was chosen.
(b) Show that p_1 goes to the appropriate blocking formula for an $M/M/1/1$ queue, as q goes to 0.

CHAPTER 4

NETWORKS, SERIES, AND CYCLIC QUEUES

In this chapter we present an introduction to the very important subject of queueing networks. This is an area of great research and application interest with many extremely difficult problems, far beyond the level of this text. As an introduction, we present some basic concepts and results that are quite useful in their own right, especially in queueing network design. Such problems have a special importance in modeling manufacturing facilities and computer/communication networks. The reader interested in delving into this topic further is referred to Bolch et al. (2006), Gelenbe and Pujolle (1998), Disney (1981), Kelly (1979), Lemoine (1977), van Dijk (1993), and Walrand (1988), for example.

Networks of queues can be described as a group of nodes (say, k of them), where each node represents a service facility of some kind with, let us say, c_i servers at node $i, i = 1, 2, \ldots, k$. In the most general case, customers may arrive from outside the system at any node and may depart from the system from any node. Thus customers may enter the system at some node, traverse from node to node in the system, and depart from some node, not all customers necessarily entering and leaving at the same nodes, or taking the same path once having entered the system. Customers may return to nodes previously visited, skip some nodes entirely, and even choose to

Fundamentals of Queueing Theory, Fourth Edition.
By D. Gross, J. F. Shortle, J. M. Thompson, and C. M. Harris
Copyright © 2008 John Wiley & Sons, Inc.

remain in the system forever. We will mainly be concerned with queueing networks with the following characteristics:

1. Arrivals from the "outside" to node i follow a Poisson process with mean rate γ_i.

2. Service (holding) times at each channel at node i are independent and exponentially distributed with parameter μ_i (a node's service rate may be allowed to depend on its queue length).

3. The probability that a customer who has completed service at node i will go next to node j (routing probability) is r_{ij} (independent of the state of the system), where $i = 1, 2, \ldots, k$, $j = 0, 1, 2, \ldots, k$, and r_{i0} indicates the probability that a customer will leave the system from node i.

Networks that have these properties are called *Jackson networks* (see Jackson, 1957, 1963). As we shall see later, their steady-state probability distributions have a most interesting and useful product-form solution. (In Section 7.3, we will relax the assumption of exponential service and Poisson arrivals. For such networks with general service and general arrival patterns, we will need approximation methods to analyze the networks.)

Cases for which $\gamma_i = 0$ for all i (no customer may enter the system from the outside) and $r_{i0} = 0$ for all i (no customer may leave the system) are referred to as *closed* Jackson networks (the general case described above is referred to as *open* Jackson networks). We have already studied a special closed system in Chapter 2, Section 2.8, namely, the machine-repair problem (finite-source queue). For that system, $i = 1, 2$, $j = 0, 1, 2$, $r_{12} = r_{21} = 1$, and all other $r_{ij} = 0$. Node 1 represents the operating (plus spare) machines, while node 2 represents the repair facility. Note that customers flow in a "circle," always from node 1 to node 2 and then back to node 1, and so on. Such closed network systems are also referred to as *cyclic* queues.

In this chapter, we first treat open networks where

$$\gamma_i = \begin{cases} \lambda & (i = 1), \\ 0 & \text{(elsewhere)} \end{cases}$$

and

$$r_{ij} = \begin{cases} 1 & (j = i + 1,\ 1 \le i \le k - 1), \\ 1 & (i = k,\ j = 0), \\ 0 & \text{(elsewhere)}. \end{cases}$$

These networks are called *series* or *tandem* queues, since the nodes can be viewed as forming a series system with flow always in a single direction from node to node. Customers may enter from the outside only at node 1 and depart only from node k. We will then generalize the series to a true open network and finally come back to the case of closed queueing networks, including the special cyclic queue—which, incidentally, is a closed queueing network in series. We shall restrict ourselves mainly to Markovian systems as described earlier; that is, all holding times are exponential, all exogenous inputs are Poisson, and the routing probabilities r_{ij} are known and

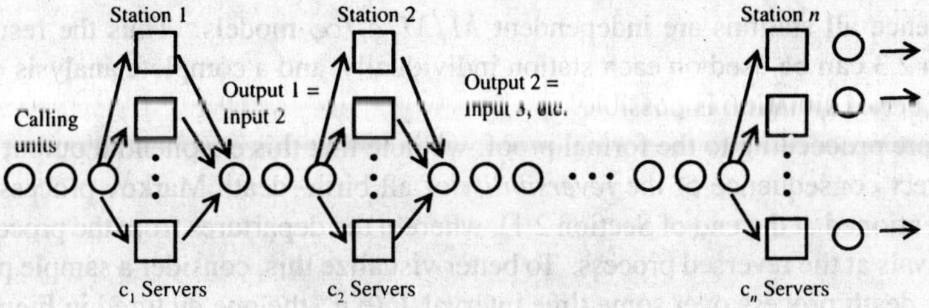

Figure 4.1 Series queue, infinite waiting room.

state independent (Jackson networks). We shall look briefly at some departures from the standard Jackson network assumptions, including the case where r_{ij} is allowed to be state dependent.

4.1 Series Queues

In this section we consider models in which there are a series of service stations through which each calling unit must progress prior to leaving the system. Some examples of such series queueing situations are a manufacturing or assembly-line process in which units must proceed through a series of work stations, each station performing a given task; a registration process (such as university registration) where the registrant must visit a series of desks (advisor, department chairperson, cashier, etc.); and a clinic physical examination procedure where the patient goes through a series of stages (lab tests, electrocardiogram, chest X-ray, etc.). In the following subsections several types of series queueing models are analyzed. (Analysis of *feed forward* queueing networks, i.e., networks of queues for which customers are not allowed to revisit previously visited nodes, is quite similar to that of the basic series queue first treated below.)

4.1.1 Queue Output

The first series model to be considered is a sequence of queues with no restriction on the waiting room's capacity between stations. Such a situation is pictured in Figure 4.1. We further assume that the calling units arrive according to a Poisson process, mean λ, and the service time of each server at station i $\quad (i = 1, 2, \ldots, n)$ is exponential with mean $1/\mu_i$. One can readily see that since there is no restriction on waiting between stations, each station can be analyzed separately as a single-stage (nonseries) queueing model.

The first station is an $M/M/c_1/\infty$ model. It is necessary to find the output distribution (distribution of times between successive departures) in order to find the input distribution (times between successive arrivals) to the next station. It turns out, rather surprisingly, that the departure-time distribution from an $M/M/c/\infty$ queue is identical to the interarrival-time distribution, namely, exponential with mean

$1/\lambda$; hence all stations are independent $M/M/c_i/\infty$ models. Thus the results of Section 2.3 can be used on each station individually, and a complete analysis of this type of series situation is possible.

Before proceeding to the formal proof, we note that this exponential output result is a direct consequence of the *reversibility* of all birth–death Markov processes (as first mentioned at the end of Section 2.1), wherein the departures from the process are the arrivals at the reversed process. To better visualize this, consider a sample path of a birth–death process over some time interval T (e.g., the one pictured in Figure 1.4 of Section 1.5.1). If we start at T and look backward, the departures become the arrivals and the arrivals become the departures, and the probabilistic characteristics of the backward sample path appear identical to those of the forward sample path. Hence the arrival and departure processes should be identical (this would not be the case for a sample path of an $M^{[2]}/M/1$ queue). Formally, in terms of the balance equations, a Markov chain is reversible if its steady-state probabilities and transition rates satisfy $\pi_i q_{ij} = \pi_j q_{ji}$.

We now proceed to verify the input–output identity with a constructive proof that utilizes a simple differential–difference argument (much like that used in the development of the birth–death process), which will show that, indeed, the interdeparture times are exponential with parameter λ. Consider an $M/M/c/\infty$ queue in steady state. Let $N(t)$ now represent the number of customers in the system at a time t after the last departure. Since we are considering steady state, we have

$$\Pr\{N(t) = n\} = p_n. \tag{4.1}$$

Furthermore, let T represent the random variable "time between successive departures" (interdeparture time), and

$$F_n(t) = \Pr\{N(t) = n \text{ and } T > t\}. \tag{4.2}$$

So $F_n(t)$ is the joint probability that there are n in the system at a time t after the last departure and that t is less than the interdeparture time; that is, another departure has not as yet occurred. The cumulative distribution of the random variable T, which will be denoted as $C(t)$, is given by

$$C(t) \equiv \Pr\{T \le t\} = 1 - \sum_{n=0}^{\infty} F_n(t), \tag{4.3}$$

since

$$\sum_{n=0}^{\infty} F_n(t) = \Pr\{T > t\}$$

is the marginal complementary cumulative distribution of T. To find $C(t)$, it is necessary to first find $F_n(t)$.

We can write the following difference equations concerning $F_n(t)$:

$$\begin{aligned} F_n(t+\Delta t) &= (1-\lambda\,\Delta t)(1-c\mu\,\Delta t)F_n(t)+\lambda\,\Delta t(1-c\mu\,\Delta t)F_{n-1}(t) \\ &\quad + o(\Delta t)(c \le n), \\ F_n(t+\Delta t) &= (1-\lambda\,\Delta t)(1-n\mu\,\Delta t)F_n(t)+\lambda\,\Delta t(1-n\mu\,\Delta t)F_{n-1}(t) \\ &\quad + o(\Delta t)(1 \le n \le c), \\ F_0(t+\Delta t) &= (1-\lambda\,\Delta t)F_0(t)+o(\Delta t). \end{aligned}$$

Moving $F_n(t)$ from the right side of each of the above equations, dividing by Δt, and taking the limit as $\Delta t \to 0$, we obtain the differential–difference equations as

$$\begin{aligned} \frac{dF_n(t)}{dt} &= -(\lambda+c\mu)F_n(t)+\lambda F_{n-1}(t) \quad (c \le n), \\ \frac{dF_n(t)}{dt} &= -(\lambda+n\mu)F_n(t)+\lambda F_{n-1}(t) \quad (1 \le n \le c), \\ \frac{dF_0(t)}{dt} &= -\lambda F_0(t). \end{aligned} \tag{4.4}$$

Using the boundary condition

$$F_n(0) \equiv \Pr\{N(0)=n \text{ and } T>0\} = \Pr\{N(0)=n\} = p_n \quad \text{[from (4.1)]}$$

and methodology similar to that of Section 1.7, where the Poisson process is derived, we find that the solution to (4.4) is (see Problem 4.1)

$$F_n(t) = p_n e^{-\lambda t}. \tag{4.5}$$

The reader can easily verify that (4.5) is a solution to (4.4) by substitution, recalling that for $M/M/c/\infty$ models,

$$p_{n+1} = \begin{cases} \dfrac{\lambda}{(n+1)\mu}p_n & (1 \le n \le c), \\ \dfrac{\lambda}{c\mu}p_n & (c \le n). \end{cases} \quad \text{[see (2.31)]}.$$

To obtain $C(t)$, the cumulative distribution of the interdeparture times, we use (4.5) in (4.3) to get

$$C(t) = 1-\sum_{n=0}^{\infty} p_n e^{-\lambda t} = 1-e^{-\lambda t}\sum_{n=0}^{\infty} p_n = 1-e^{-\lambda t}, \tag{4.6}$$

thus showing that the interdeparture times are exponential.

It is also true that the random variables $N(T)$ and T are independent and furthermore that successive interdeparture times are independent of each other (see Problem 4.2). This result was first proved by Burke (1956). So we see that the output distribution is identical to the input distribution and not at all affected by the exponential service mechanism. Intuitively, one would expect the means to be identical, since we are in steady state, so that the input rate must equal the output rate; but it is not quite so intuitive that the variances and indeed the distributions are identical. Nevertheless, it is true and proves extremely useful in analyzing series queues, where the initial input rate is Poisson, the service at all stations is exponential, and there is no restriction on queue size between stations.

We now illustrate a series queueing situation of the type above with an example.

EXAMPLE 4.1

Cary Meback, the president of a large Virginia supermarket chain, is experimenting with a new store design and has remodeled one of his stores as follows. Instead of the usual checkout-counter design, the store has been remodeled to include a checkout "lounge." As customers complete their shopping, they enter the lounge with their carts and, if all checkers are busy, receive a number. They then park their carts and take a seat. When a checker is free, the next number is called and the customer with that number enters the available checkout counter. The store has been enlarged so that for practical purposes, there is no limit on either the number of shoppers that can be in the food aisles or the number that can wait in the lounge, even during peak periods.

The management estimates that during peak hours customers arrive according to a Poisson process at a mean rate of 40/h and it takes a customer, on average, $\frac{3}{4}$ h to fill a shopping cart, the filling times being approximately exponentially distributed. Furthermore, the checkout times are also approximately exponentially distributed with a mean of 4 min, regardless of the particular checkout counter (during peak periods each counter has a cashier and bagger, hence the low mean checkout time). Meback wishes to know the following: (1) What is the minimum number of checkout counters required in operation during peak periods? (2) If it is decided to add one more than the minimum number of counters required in operation, what is the average waiting time in the lounge? How many people, on average, will be in the lounge? How many people, on average, will be in the entire supermarket?

This situation can be modeled by a two-station series queue. The first is the food portion of the supermarket. Since it is self-service and arrivals are Poisson, we have an $M/M/\infty$ model with $\lambda = 40$ and $\mu = \frac{4}{3}$. The second station is an $M/M/c$ model, since the output of an $M/M/\infty$ queue is identical to its input. Hence the input to the checkout lounge is Poisson with a mean of 40/h also. Since $c\mu > \lambda$ for steady-state convergence, the minimum number of checkout counters, c_m, must be greater than $\lambda/\mu = 40/15 \doteq 2.67$; hence c_m must be 3.

If it is decided to have four counters in operation, we have an $M/M/4$ model at the checkout stations with $\lambda = 40$ and $\mu = 15$. Meback desires to know W_q and L_q for the $M/M/4$ model, as well as the average total number in the supermarket, which is the sum of the L's for both models. Using (2.32) and (2.34) for an $M/M/c$ model, we get

$$p_0 = \left[\sum_{n=0}^{3} \frac{1}{n!}\left(\frac{8}{3}\right)^n + \frac{1}{4!}\left(\frac{8}{3}\right)^4 \left(\frac{4}{4-\frac{8}{3}}\right)\right]^{-1} \doteq 0.06$$

and

$$W_q = \frac{(\frac{8}{3})^4 15}{3!(60-40)^2}(0.06) \doteq 0.019 \text{ h} \doteq 1.14 \text{ min}.$$

To get L_q we use Little's formula and find

$$L_q = \lambda W_q \doteq 40(0.019) = 0.76,$$

so that, on average, less than one person will be waiting in the lounge for a checker to become free.

The total number of people in the system, on average, is the L for this $M/M/4$ model plus the L for the $M/M/\infty$ model. For the checkout station we get

$$L = \lambda W = \lambda\left(W_q + \frac{1}{\mu}\right) \doteq 40\left(0.019 + \frac{4}{60}\right) \doteq 3.44.$$

For the supermarket proper we have, from the $M/M/\infty$ model results, that $L = \lambda/\mu = 40/(\frac{4}{3}) = 30$. Hence the average number of customers in the store during peak hours is 33.44 if Meback decides on four checkout counters in operation. He might do well to perform similar calculations for three checkout counters operating to see how much the congestion increases (see Problem 4.3).

For series queues, therefore, as long as there are no capacity limitations between stations and the input is Poisson, results can be rather easily obtained. Furthermore, it can be shown (see Problem 4.4) that the joint probability that there are n_1 at station 1, n_2 at station 2, ..., and n_j at station j is merely the product $p_{n_1}p_{n_2}\cdots p_{n_j}$. This product-form type of result is quite typical of those available for Jackson networks, as we shall see in subsequent sections.

The analysis for series queues when there are limits on the capacity at a station (except for the case where the only limit is at the last station in a pure series flow situation and arriving customers who exceed the capacity are shunted out of the system—see Problem 4.5) is much more complex. This results from the blocking effect; that is, a station downstream comes up to capacity and thereby prevents any further processing at upstream stations that feed it. We treat some of these types of models in the next section.

4.1.2 Series Queues with Blocking

We consider first a simple sequential two-station, single-server-at-each-station model, where no queue is allowed to form at either station. If a customer is in station 2 and

Table 4.1 Possible System States

n_1, n_2	Description
0,0	System empty
1,0	Customer in process at 1 only
0,1	Customer in process at 2 only
1,1	Customers in process at 1 and 2
b,1	Customer in process at 2 and a customer finished at 1 but waiting for 2 to become available (i.e., system is blocked)

service is completed at station 1, the station-1 customers must wait there until the station-2 customer is completed; that is, the system is *blocked*. Arrivals at station 1 when the system is blocked are turned away. Also, if a customer is in process at station 1, then even if station 2 is empty, arriving customers are turned away, since the system is a sequential one; that is, all customers require service at 1 and then service at 2.

We wish to find the steady-state probability p_{n_1,n_2} of n_1 in the first station and n_2 in the second station. For this model, the possible states are given in Table 4.1.

Assuming arrivals at the system (station 1) are Poisson with parameter λ and service is exponential with parameters μ_1 and μ_2, respectively, the usual procedure leads to the steady-state equations for this multidimensional Markov chain:

$$\begin{aligned}
0 &= -\lambda p_{0,0} + \mu_2 p_{0,1},\\
0 &= -\mu_1 p_{1,0} + \mu_2 p_{1,1} + \lambda p_{0,0},\\
0 &= -(\lambda + \mu_2) p_{0,1} + \mu_1 p_{1,0} + \mu_2 p_{b,1},\\
0 &= -(\mu_1 + \mu_2) p_{1,1} + \lambda p_{0,1},\\
0 &= -\mu_2 p_{b,1} + \mu_1 p_{1,1}.
\end{aligned} \tag{4.7}$$

Using the boundary equation $\sum\sum p_{n_1,n_2} = 1$, we have six equations in five unknowns [there is some redundancy in (4.7); hence we can solve for the five steady-state probabilities]. Equation (4.7) can be used to get all probabilities in terms of $p_{0,0}$, and the boundary condition can be used to find $p_{0,0}$. If we let $\mu_1 = \mu_2$, the results are (see Problem 4.6)

$$\begin{aligned}
p_{1,0} &= \frac{\lambda(\lambda + 2\mu)}{2\mu^2} p_{0,0}, \quad p_{0,1} = \frac{\lambda}{\mu} p_{0,0}, \quad p_{1,1} = \frac{\lambda^2}{2\mu^2} p_{0,0},\\
p_{b,1} &= \frac{\lambda^2}{2\mu^2} p_{0,0}, \quad p_{0,0} = \frac{2\mu^2}{3\lambda^2 + 4\mu\lambda + 2\mu^2}.
\end{aligned} \tag{4.8}$$

It is easy to see how the problem expands if one allows limits other than zero on queue length or considers more stations. For example, if one customer is allowed to wait between stations, this results in seven state probabilities for which to solve, utilizing seven equations and a boundary condition (see Problem 4.7). The complexity results from having to write a balance equation for each possible system state. Conceptually, however, these types of series queueing situations can be attacked via the methodology presented earlier. For large numbers of equations, as long as we have a finite set, numerical techniques for solving these simultaneous equations can also be employed.

Hunt (1956) treated a modified series model using finite-difference operators to solve a two-station sequential series queue in which no waiting is allowed between stations, but where a queue with no limit is permitted in front of the first station. He obtained the steady-state probabilities for this model, the expected system size L, and the maximum allowable ρ (call it $\rho_{\max}$) for steady state to be assured. He also calculated $\rho_{\max}$ for some generalizations of this two-station model (infinite allowable queue in front of station 1) to three- and four-station systems with no waiting between stations, a two-station system with a capacity K allowed between stations, and a three-station system where $K = 2$ in between each of the stations. The interested reader is referred to Hunt (1956). A good general reference on networks of queues with blocking is Perros (1994).

4.2 Open Jackson Networks

We now treat a rather general network system that we previously described in the introduction to this chapter as the Jackson network, because of the landmark work done by Jackson (1957, 1963). To recapitulate, we consider a network of k service facilities (usually referred to as nodes). Customers can arrive from outside at any node according to a Poisson process. We will now represent the mean arrival rate to node i as γ_i (instead of the familiar λ_i) for reasons that will become clear shortly. All servers at node i work according to an exponential distribution with mean μ_i (so that all servers at a given node are identical). When customers complete service at node i, they go next to node j with probability r_{ij} (independent of the system state), $i = 1, 2, \ldots, k$. There is a probability r_{i0} that customers will leave the network at node i upon completion of service. There is no limit on queue capacity at any node; that is, we never have a blocked system or node.

Since we have a Markovian system, we can use our usual types of analyses to write the steady-state system equations. We first, however, must determine how to describe a system state. Since various numbers of customers can be at various nodes in the network, we desire the joint probability distribution for the number of customers at each node; that is, letting N_i be the random variable for the number of customers at node i in the steady state, we desire $\Pr\{N_1 = n_1, N_2 = n_2, \ldots, N_k = n_k\} \equiv p_{n_1,n_2,\ldots,n_k}$. From this joint probability distribution, we can obtain the marginal distribution for numbers of customers at a particular node by appropriately summing over the other nodes.

We shall use the method of stochastic balance to obtain the steady-state equations for this network. Rather than using the somewhat cumbersome k-component vector

Table 4.2 Simplified State Descriptors

State	Simplified Notation
$n_1, n_2, \ldots, n_i, n_j, \ldots, n_k$	$\bar{n}$
$n_1, n_2, \ldots, n_i + 1, n_j, \ldots, n_k$	$\bar{n}; i^+$
$n_1, n_2, \ldots, n_i - 1, n_j, \ldots, n_k$	$\bar{n}; i^-$
$n_1, n_2, \ldots, n_i + 1, n_j - 1, \ldots, n_k$	$\bar{n}; i^+ j^-$

$(n_1, n_2, \ldots, n_k)$ for describing a state, we employ the simplified notation given in Table 4.2.

Using stochastic balance for equating flow into state $\bar{n}$ to flow out of state $\bar{n}$, and assuming that $c_i = 1$ for all i (single-server nodes) and that $n_i \geq 1$ at each node (actually, the equation that results will also hold for $n_i = 0$ if we set terms with negative subscripts and terms containing μ_i for which the subscript $n_i = 0$ to zero), we obtain

$$\sum_{i=1}^{k} \gamma_i p_{\bar{n};i^-} + \sum_{j=1}^{k} \sum_{\substack{i=1 \\ (i \neq j)}}^{k} \mu_i r_{ij} p_{\bar{n};i^+j^-} + \sum_{i=1}^{k} \mu_i r_{i,0} p_{\bar{n};i^+} = \sum_{i=1}^{k} \mu_i (1 - r_{ii}) p_{\bar{n}} + \sum_{i=1}^{k} \gamma_i p_{\bar{n}}. \tag{4.9}$$

Jackson (1957, 1963) first showed that the solution to these steady-state balance equations is what has come to be generally called *product form*, where the joint probability distribution of system states can be written as

$$p_{\bar{n}} = C \rho_1^{n_1} \rho_2^{n_2} \cdots \rho_k^{n_k}.$$

We mention here that some authors consider a more restrictive definition of product form, namely, that the joint distribution is made up of the product of the marginal distributions at each of the nodes. We prefer the less restrictive definition wherein the constant C itself need not separate into a product, so that a product-form solution need not be a product of true marginals.

We will present Jackson's solution and then show that it satisfies (4.9). Let λ_i be the total mean flow rate into node i (from outside and from other nodes). Then, in order to satisfy flow balance at each node, we have the *traffic equations*

$$\boxed{\lambda_i = \gamma_i + \sum_{j=1}^{k} \lambda_j r_{ji}} \tag{4.10a}$$

or, in vector–matrix form,

$$\boxed{\boldsymbol{\lambda} = \boldsymbol{\gamma} + \boldsymbol{\lambda} \boldsymbol{R}.} \tag{4.10b}$$

We define ρ_i to be λ_i/μ_i for $i = 1, 2, \ldots, k$. Then, as Jackson showed, the steady-state solution to (4.9) is

$$p_{\bar{n}} \equiv p_{n_1,n_2,\ldots,n_k} = (1-\rho_1)\rho_1^{n_1}(1-\rho_2)\rho_2^{n_2}\cdots(1-\rho_k)\rho_k^{n_k}, \tag{4.11}$$

which, in this case, is a true product of marginal distributions. What this result says is that the network *acts as if* each node could be viewed as an independent $M/M/1$ queue, with parameters λ_i and μ_i, so that the joint probability distribution can be written as a product of marginal $M/M/1$'s. The reader should not be misled into believing that the network actually decomposes into individual $M/M/1$'s with the flow into each a true Poisson process with mean rate λ_i. In fact, it can be shown (see Disney, 1981) that, in general, the actual internal flow in these kinds of networks is *not* Poisson. Indeed, as long as there is any kind of *feedback* (i.e., customers can return to previously visited nodes), the internal flows are not Poisson. The surprising thing is that regardless of whether internal flows are really Poisson or not, (4.11) still holds and the network behaves as if its nodes were independent $M/M/1$'s.

The global balance equation (4.9) gives rise to a simple set of local balance equations in the same spirit as the classical birth–death process, namely, $\lambda_i p_{\bar{n};i^-} = \mu_i p_{\bar{n}}$. In other words, the expected rate at which the system goes from state $\bar{n}; i^-$ to $\bar{n}$ must equal the rate at which it goes in the reverse direction, $\bar{n}$ back to $\bar{n}; i^-$. It thus follows by linear difference methods that $p_{\bar{n}} = \rho_i^{n_i} p_{n_1,n_2,\ldots,0,\ldots,n_k}$. Therefore we can conclude that all the nodal marginal distributions are going to be *geometric* and that (4.11) is the likely form for the combined joint distribution.

To verify that (4.11) does indeed satisfy (4.9), we first show that $p_{\bar{n}} = C\rho_1^{n_1}\rho_2^{n_2}\cdots\rho_k^{n_k}$ satisfies (4.9) and then that C turns out to be $\Pi_{i=1}^k(1-\rho_i)$, in order to satisfy the summability-to-one criterion. We let $\Re^{\bar{n}} = \rho_1^{n_1}\rho_2^{n_2}\cdots\rho_k^{n_k}$, and plugging $p_{\bar{n}} = C\Re^{\bar{n}}$ into (4.9) gives

$$C\Re^{\bar{n}}\sum_{i=1}^k \frac{\gamma_i}{\rho_i} + C\Re^{\bar{n}}\sum_{j=1}^k\sum_{\substack{i=1\\(i\neq j)}}^k \mu_i r_{ij}\frac{\rho_i}{\rho_j} + C\Re^{\bar{n}}\sum_{i=1}^k \mu_i r_{i0}\rho_i$$

$$\stackrel{?}{=} C\Re^{\bar{n}}\sum_{i=1}^k \mu_i(1-r_{ii}) + C\Re^{\bar{n}}\sum_{i=1}^k \gamma_i.$$

Canceling out $C\Re^{\bar{n}}$, we have

$$\sum_{i=1}^k \frac{\gamma_i\mu_i}{\lambda_i} + \sum_{j=1}^k\sum_{\substack{i=1\\(i\neq j)}}^k \mu_i r_{ij}\frac{\lambda_i\mu_j}{\lambda_j\mu_i} + \sum_{i=1}^k \mu_i r_{i0}\frac{\lambda_i}{\mu_i} \stackrel{?}{=} \sum_{i=1}^k(\mu_i - \mu_i r_{ii} + \gamma_i).$$

From (4.10a) then,

$$\lambda_j = \gamma_j + \sum_{\substack{i=1\\(i\neq j)}}^k r_{ij}\lambda_i + r_{jj}\lambda_j \Rightarrow \sum_{\substack{i=1\\(i\neq j)}}^k r_{ij}\lambda_i = \lambda_j - \gamma_j - r_{jj}\lambda_j,$$

so that

$$\sum_{i=1}^{k} \frac{\gamma_i \mu_i}{\lambda_i} + \sum_{j=1}^{k} \frac{\mu_j}{\lambda_j}(\lambda_j - \gamma_j - r_{jj}\lambda_j) + \sum_{i=1}^{k} \mu_i r_{i0} \frac{\lambda_i}{\mu_i} \stackrel{?}{=} \sum_{i=1}^{k} (\mu_i - \mu_i r_{ii} + \gamma_i).$$

Changing the subscript on the second term on the left-hand side from j to i, we get

$$\sum_{i=1}^{k} \left(\frac{\gamma_i \mu_i}{\lambda_i} + \frac{\mu_i}{\lambda_i}(\lambda_i - \gamma_i - r_{ii}\lambda_i) + \lambda_i r_{i0} \right) \stackrel{?}{=} \sum_{i=1}^{k} (\mu_i - \mu_i r_{ii} + \gamma_i)$$

After moving through with the summation and canceling we have

$$\sum_{i=1}^{k} \lambda_i r_{i0} \stackrel{?}{=} \sum_{i=1}^{k} \gamma_i.$$

This is indeed true, since the left-hand side represents the total flow out of the network and the right-hand side represents the total flow in. For steady state, these must be equal.

To obtain ρ_i, we need to obtain λ_i from the traffic equations (4.10b), which are solved as $\boldsymbol{\lambda} = \boldsymbol{\gamma}(\boldsymbol{I} - \boldsymbol{R})^{-1}$. The invertibility of $\boldsymbol{I} - \boldsymbol{R}$ is assured as long as there is at least one node releasing its output to the outside and no node is totally absorbing.

Now to evaluate C, we have

$$\sum_{n_k=0}^{\infty} \cdots \sum_{n_2=0}^{\infty} \sum_{n_1=0}^{\infty} C \rho_1^{n_1} \rho_2^{n_2} \ldots \rho_k^{n_k} = 1.$$

Thus

$$C = \left(\sum_{n_k=0}^{\infty} \cdots \sum_{n_2=0}^{\infty} \sum_{n_1=0}^{\infty} \rho_1^{n_1} \rho_2^{n_2} \ldots \rho_k^{n_k} \right)^{-1} = \left(\sum_{n_k=0}^{\infty} \rho_k^{n_k} \ldots \sum_{n_2=0}^{\infty} \rho_2^{n_2} \sum_{n_1=0}^{\infty} \rho_1^{n_1} \right)^{-1}$$

$$= \left(\frac{1}{1-\rho_k} \cdots \frac{1}{1-\rho_2} \frac{1}{1-\rho_1} \right)^{-1} \quad (\rho_i < 1, i = 1, 2, \ldots, k).$$

Hence

$$C = \left(\frac{1}{(1-\rho_k) \cdots (1-\rho_2)(1-\rho_1)} \right)^{-1}$$

$$= \prod_{i=1}^{k} (1-\rho_i) \quad (\rho_i < 1, i = 1, 2, \ldots, k).$$

We can obtain expected measures rather easily for individual nodes, since $L_i = \rho_i/(1-\rho_i)$ and $W_i = L_i/\lambda_i$. This is so because of the product form of the solution for the joint probability distribution and again does not imply that the nodes are truly $M/M/1$ (which they may not be, although the system-size processes are indeed independent $M/M/1$'s, since the joint probability distribution is the product of the marginals). The expected total wait within the network of any customer before its final departure would be $W = \sum_i L_i / \sum_i \gamma_i$ (Little's formula for the entire network).

The above results for Jackson networks generalize easily to c-channel nodes (see Problem 4.9). Let c_i represent the number of servers at node i, each having exponential service time with parameter μ_i. Then (4.11) becomes (again we define λ_i/μ_i as r_i to be consistent with $M/M/c$ notation, but do not confuse with the double-subscripted r_{ij}, which are routing probabilities)

$$p_{\bar{n}} \equiv p_{n_1,n_2,\ldots,n_k} = \prod_{i=1}^{k} \frac{r_i^{n_i}}{a_i(n_i)} p_{0i} \quad (r_i \equiv \lambda_i/\mu_i), \tag{4.12}$$

where

$$a_i(n_i) = \begin{cases} n_i! & (n_i < c_i), \\ c_i^{n_i - c_i} c_i! & (n_i \geq c_i), \end{cases} \tag{4.13}$$

and p_{0i} is such that $\sum_{n_i=0}^{\infty} p_{0i} r_i^{n_i}/a_i(n_i) = 1$, which can be obtained from (2.32). Thus, again, what we have is a network that *acts as if* each node were an independent $M/M/c$.

With respect to waiting-time distributions, very little can be said. It is tempting, for example, to conclude that if we are interested in the unconditional waiting time at a particular node, since the node acts like an $M/M/c$, the waiting-time distribution should be the same as that for the $M/M/c$ model. But this is not necessarily true. A basic factor in developing $M/M/c$ waiting times was that the arrival-point steady-state probabilities $\{q_n\}$ were identical to the general-time steady-state probabilities $\{p_n\}$ because of the Poisson input. However, in a general Jackson network, as we have previously mentioned, flows are not necessarily truly Poisson, so that we cannot be sure that $q_n = p_n$, and, in fact, in general that is not true. What does appear to be the case is that the virtual waiting time (or work backlog, as it is sometimes called—see Section 2.2.4) at a node that requires use of the $\{p_n\}$ is the same as that of $M/M/c$, since the $\{p_n\}$ are identical. Hence equations such as (2.39) give virtual waiting times, but unless the network has a feedforward flow (i.e., are arborescent or treelike, indicating no direct or indirect feedback), which *is* truly Poisson, nothing can be concluded about actual waiting-time distributions, although the *mean* nodal values satisfy Little's formula.

Of even more interest would be customer waiting times to traverse portions of or the entire network (often referred to as *sojourn* times). Even less can be said here, due to the complicated correlation among node waiting times. Consider, for example (Disney, 1996), a simple feedback queue where we have a single $M/M/1$ node where, with probability p, a customer after being served returns to the end of the queue for additional service and with probability $1 - p$ leaves the system. Considering the sojourn time of a specific customer to be made up of (possibly) several passes through the queue, then it is clear that at the first pass, the time spent in the system is what would normally be spent in traversing an $M/M/1$ queue, but when the customer starts the second pass (assuming a feedback customer), ahead will be a number of other customers, some of whom were originally ahead of the customer during its first pass, while others may have arrived after the customer while it was making its first pass. The number of this latter kind of customers in the system depends on how long the subject customer spent going through the queue on the first pass. Its time to get through the queue the second time, then, depends on how many customers it finds ahead of it when it rejoins the queue, which depends on how long it spent in the system on the first pass; hence the sojourn times on these two passes are dependent.

Even with feedforward networks, sojourn times can be complex. While it is true in single-server queues that the total time, $T_i^{(n)}$, that the nth customer spends in queue plus service on a single pass through node i is a sum of IID exponential random variables, and that the total system sojourn time $T^{(n)} = T_1^{(n)} + \cdots + T_k^{(n)}$ is, in the limit, the sum of IID random variables (this result is due to Reich, 1957), it is not true in multiserver queues. For example, Burke (1969) showed that in a three-station series queue with the first and third stations having a single server but the middle station having multiple servers ($c_2 \geq 2$), in the limit, $T_1^{(n)}$ and $T_2^{(n)}$ are independent, and $T_2^{(n)}$ and $T_3^{(n)}$ are independent, but $T_1^{(n)}$ and $T_3^{(n)}$ are not independent. Simon and Foley (1979) considered a three-station queueing network with single servers at each station. Customers only enter the system at station 1 and exit the system at station 3, so that

$$\lambda_i = \begin{cases} \lambda & (i = 1), \\ 0 & (\text{elsewhere}). \end{cases}$$

However, this is not a series situation, since they allow for the possibility of bypassing, namely,

$$r_{i,j} = \begin{cases} p & (i = 1, j = 2), \\ 1 - p & (i = 1, j = 3), \\ 1 & (i = 2, j = 3; i = 3, j = 0), \\ 0 & (\text{elsewhere}). \end{cases}$$

They also showed that, in the limit, $T_1^{(n)}$ and $T_2^{(n)}$ are independent, and $T_2^{(n)}$ and $T_3^{(n)}$ are independent, but $T_1^{(n)}$ and $T_3^{(n)}$ are not independent. The implication of these results seems to be that even in the relatively well-behaved strict tandem (series) queue and feedforward types of networks, if there are multiple servers at stations other than the first or last so that customers can bypass one another, system sojourn times for successive customers are dependent, so that Reich's result holds only for single-server series systems. Thus the major culprits in complexity of sojourn times appear to be feedback and bypassing.

One is often interested, for networks, in output processes from individual nodes, especially since they influence input processes to other nodes. We saw in Section 4.1.1 that, for series situations, the output from each node is a Poisson process identical to the arrival process to the first node, so that the output from the last node (departure process out of the network) is identical to the arrival process (into the network). What now can be said concerning the more general Jackson networks? As the reader might suspect, not much is possible.

The survey by Disney (1981, 1996) summarizes key results. For single-server Jackson networks with an irreducible routing probability matrix $\boldsymbol{R} = \{r_{ij}\}$, and $\rho_i < 1, i = 1, 2, \ldots, k$ (every entering customer eventually leaves the network), Melamed (1979) showed that the departure process for nodes from which units could leave the network are Poisson and that the collection over all nodes that yield these Poisson departure processes are mutually independent. It obviously follows that the sum total of all departures from the network must be Poisson as well.

Furthermore, considering nodes with no feedback (there is no path a departing customer can follow that will eventually return to the same node, prior to exiting the network), the output process from this node is also Poisson. In nodes with feedback, one can think of two departing streams, one with customers who will either directly or eventually feed back, and the other with customers who will not. The nonfeedback stream is Poisson, but the feedback stream is not and, in fact, is not even a sequence of IID random variables. Disney et al. (1980) considered the single-node Poisson arrival process with direct feedback that was mentioned previously and showed that the total output process, feedback plus nonfeedback streams, is also not Poisson. It is not even a sequence of IID random variables; however, as Melamed has shown, the nonfeedback process is Poisson. Disney et al. (1980) conjecture that the feedback and nonfeedback processes are dependent, but no proof either way is known. Thus feedback causes problems with Jackson network flows, just as feedback and customer bypassing cause problems with sojourn times.

In summary, as long as there is no feedback, as in series or feedforward networks, flows between nodes and to the outside are truly Poisson. Feedback destroys Poisson flows, but Jackson's solution of (4.11) and (4.12) still holds.

While the results for departure processes and waiting and sojourn times are extremely complex and very little is really known other than the often counterintuitive results mentioned previously, the system-size results presented earlier are quite neat, and Jackson networks have been quite useful for modeling a variety of network situations in communications, computers, and repairable-item inventory control. Of course, the Poisson–exponential assumptions must hold, as well as state-independent routing probabilities and the absence of restrictions on waiting capacities.

■ EXAMPLE 4.2

The mid-Atlantic region of Hourfawlt Insurance Corporation has a three-node telephone system. Calls coming into the 800 number are Poisson, with a mean of 35/h. The caller gets one of two options: press 1 for claims service and press 2 for policy service. It is estimated that the caller's listening, decision, and button-pushing time is exponential with mean of 30 seconds. Only one call at a time can be processed, so that while a call is in process, any incoming calls that arrive are put in a queue with nice background music and a nice recorded message saying how important the call is and please wait (which we assume everyone does). Approximately 55% of the calls go to claims, and the remainder to policy service. The claims processing node has three parallel servers, and it is estimated that service times are exponential with mean 6 min (mostly, basic information is taken so that appropriate forms can be mailed out). The policy service node has seven parallel servers, again with exponential service times, and here the mean service time is 20 min. All buffers in front of the nodes can be assumed to hold as many calls as come into the queues. About 2% of the customers finishing at claims then go on to policy service, and about 1% of the customers finishing at policy service go to claims. It is desired to know the average queue sizes in front of each node and the total average time a customer spends in the system.

The routing matrix for the problem (calling claims node 2 and policy service node 3) is

$$\boldsymbol{R} = \begin{pmatrix} 0 & 0.55 & 0.45 \\ 0 & 0 & 0.02 \\ 0 & 0.01 & 0 \end{pmatrix},$$

and $\gamma_1 = 35/\text{h}, \gamma_2 = \gamma_3 = 0, c_1 = 1, \mu_1 = 120/\text{h}, c_2 = 3, \mu_2 = 10/\text{h}, c_3 = 7, \mu_3 = 3/\text{h}$.

First we must solve the traffic equations (4.10b):

$$(\boldsymbol{I} - \boldsymbol{R})^{-1} = \begin{pmatrix} 1 & 0.5546 & 0.4611 \\ 0 & 1.0002 & 0.02 \\ 0 & 0.01 & 1.0002 \end{pmatrix},$$

so that $\boldsymbol{\lambda} = \boldsymbol{\gamma}(\boldsymbol{I} - \boldsymbol{R})^{-1} = (35, 19.411, 16.138)$, which yields $r_1 = 35/120 = 0.292, r_2 = 19.411/10 = 1.941, r_3 = 16.132/3 = 5.379$. Now we can use the $M/M/c$ results of Section 2.3 to obtain L_q and L for each of the nodes. These turn out to be $L_{q1} = 0.120, L_{q2} = 0.765, L_{q3} = 1.402$ and $L_1 = 0.412, L_2 = 2.706, L_3 = 6.781$. The total system L is then $0.412 + 2.706 + 6.781 = 9.899$, and hence $W = 9.899/35 = 0.283$ h or approximately 17 min.

4.2.1 Open Jackson Networks with Multiple Customer Classes

It is a rather straightforward generalization to allow customers of different types, as reflected by a different routing matrix; that is, a customer of one type has different routing probabilities than a customer of another type. The essential modification is to first solve the traffic equations separately for each customer type and then add the resulting λ's. We will use a superscript to denote customer type, so that $\boldsymbol{R}^{(t)}$ is the routing probability matrix for a customer of type t $(t = 1, 2, \ldots, n)$. Solving (4.10b) yields a $\boldsymbol{\lambda}^{(t)}$ for each customer type t. We then obtain $\boldsymbol{\lambda} = \sum_t \boldsymbol{\lambda}^{(t)}$. We proceed as before to obtain the L_i for each of the nodes $(i = 1, 2, \ldots, k)$ using $M/M/c$ results. The average waiting time at each node (note that all customer types have the same average waiting time, since they have identical service-time distributions and wait in the same first-come, first-served queue) can be obtained by Little's formula as before. The same is true for the average system sojourn time. We can also obtain the average system size for customer type t, by simply weighting the node average total size by customer type t's relative flow rate, namely,

$$L_i^{(t)} = \frac{\lambda_i^{(t)}}{\lambda_i^{(1)} + \lambda_i^{(2)} \cdots \lambda_i^{(n)}} L_i.$$

EXAMPLE 4.3

Let us revisit Example 4.2. What was implicitly assumed in that example was that a caller going first to claims and then to policy service could go back to claims with probability 0.01. Furthermore, a customer going first to policy service and then to claims could go back again to policy service with probability 0.02. This probably isn't very realistic; that is, for this situation, customers really would not return to a previous node. We can get around this by calling customers who first go to claims type-1 customers and customers who first go to policy service type-2 customers. Then the two routing matrices are

$$\boldsymbol{R}^{(1)} = \begin{pmatrix} 0 & 1 & 0 \\ 0 & 0 & .02 \\ 0 & 0 & 0 \end{pmatrix}, \quad \boldsymbol{R}^{(2)} = \begin{pmatrix} 0 & 0 & 1 \\ 0 & 0 & 0 \\ 0 & .01 & 0 \end{pmatrix}.$$

Since 55% of the arrivals are type-1 customers, $\gamma_1^{(1)} = 19.25$, and 45% are type 2, so that $\gamma_1^{(2)} = 15.75$. Now solving the traffic equations separately gives $\lambda_1^{(1)} = 19.25, \lambda_2^{(1)} = 19.25, \lambda_3^{(1)} = 0.385$ and $\lambda_1^{(2)} = 15.75, \lambda_2^{(2)} = 0.1575, \lambda_3^{(2)} = 15.75$. Adding to get total flows gives $\boldsymbol{\lambda} = (35, 19.408, 16.135)$. Comparing this with Example 4.2, we see small differences in slightly lower flows in nodes 2 and 3 to account for the lack of recycling. The procedure now follows along as before, and using $M/M/c$ results we get $L_1 = 0.412, L_2 = 2.705$, and $L_3 = 6.777$, again with very slight differences from Example 4.2. The total system L is now 9.894, and the average system sojourn time is $9.894/35 = 0.283$ h or, again, about 17 min. We can also find the average number of each type of customer at each node:

$$L_1^{(1)} = \frac{19.25}{19.25 + 15.75} L_1 = 0.227, \quad L_2^{(1)} = \frac{19.25}{19.25 + 0.1575} L_2 = 2.683,$$

$$L_3^{(1)} = \frac{0.385}{0.385 + 15.75} L_3 = 0.162,$$

$$L_1^{(2)} = \frac{15.75}{19.25 + 15.75} L_1 = 0.185, \quad L_2^{(2)} = \frac{0.1575}{19.25 + 0.1575} L_2 = 0.022,$$

$$L_3^{(2)} = \frac{15.75}{0.385 + 15.75} L_3 = 6.616.$$

4.3 Closed Jackson Networks

If we set $\gamma_i = 0$ and $r_{i0} = 0$ for all i, we have a closed Jackson network, which is equivalent to a finite-source queue of, say, N items that continuously travel inside the network. If $c_i = 1$ for all i, we can get the steady-state flow-balance equations from (4.9), the open network model, by setting $\gamma_i = r_{i0} = 0$. This yields

$$\sum_{j=1}^{k}\sum_{\substack{i=1 \\ (i\neq j)}}^{k} \mu_i r_{ij} p_{\bar{n};i^+j^-} = \sum_{i=1}^{k} \mu_i(1-r_{ii})p_{\bar{n}}. \tag{4.14}$$

Since this is a special case of a general Jackson network, once more we have a product-form solution,

$$p_{\bar{n}} = C\rho_1^{n_1}\rho_2^{n_2}\cdots\rho_k^{n_k} \equiv C\Re^{\bar{n}}, \tag{4.15}$$

where $\rho_i = \lambda_i/\mu_i$ must satisfy the balance equations for flow at each node i, so that the flows into and out of node i must be equal. This yields the closed network equivalent to the traffic equations of (4.10a), namely,

$$\lambda_i = \mu_i\rho_i = \sum_{j=1}^{k} \lambda_j r_{ji} = \sum_{j=1}^{k} \mu_j r_{ji}\rho_j. \tag{4.16}$$

As for open networks, we assume that the routing matrix is irreducible and non-absorbing. But now, one of the equations of (4.16) is redundant because the sum of the λ_i is fixed. Thus we can arbitrarily set one ρ_i equal to 1 when solving $\mu_i\rho_i = \sum_{j=1}^{k}\mu_j r_{ji}\rho_j$. Problem 4.14 asks the reader to verify that (4.15) is a solution by substituting it into (4.14).

For this case, C does not "break apart" and must be evaluated by

$$\sum_{n_1+n_2+\cdots+n_k=N} C\rho_1^{n_1}\rho_2^{n_2}\cdots\rho_k^{n_k} = 1$$

$$\Rightarrow C = \left(\sum_{n_1+n_2+\cdots+n_k=N} \rho_1^{n_1}\rho_2^{n_2}\cdots\rho_k^{n_k}\right)^{-1},$$

where the sum is taken over all possible ways the N elements can be distributed among the k nodes. The constant C is often shown as $C(N)$ to emphasize that it is a function of the total population size N. Furthermore, the solution is often written in terms of $C^{-1}(N) \equiv G(N)$, so that

$$p_{n_1,n_2,\ldots,n_k} = \frac{1}{G(N)}\rho_1^{n_1}\rho_2^{n_2}\cdots\rho_k^{n_k},$$

where

$$G(N) = \sum_{n_1 n_2+\cdots+n_k=N} \rho_1^{n_1}\rho_2^{n_2}\cdots\rho_k^{n_k}.$$

Again, this closed network can easily be extended to c_i servers at node i (see Problem 4.15). The solution now becomes

$$p_{n_1,n_2,\ldots,n_k} = \frac{1}{G(N)}\prod_{i=1}^{k}\frac{\rho_i^{n_i}}{a_i(n_i)}, \tag{4.17}$$

where $a_i(n_i)$ is given by (4.13) and

$$G(N) = \sum_{n_1+n_2+\cdots+n_k=N} \prod_{i=1}^{k} \frac{\rho_i^{n_i}}{a_i(n_i)} \tag{4.18}$$

■ EXAMPLE 4.4

Two special-purpose machines are desired to be operational at all times. We call the operating node of this network node 1. The machines break down according to an exponential distribution with mean failure rate λ. Upon breakdown, a machine has a probability r_{12} that it can be repaired locally (node 2) by a single repairperson who works according to an exponential distribution with parameter μ_2. With probability $1 - r_{12}$ the machine must be repaired by a specialist (node 3), who also works according to an exponential distribution, but with mean rate μ_3. Further, after completing local service at node 2, there is a probability r_{23} that a machine will also require the special service (the probability of returning to operation from node 2 is then $1 - r_{23}$). After the special service (node 3), the unit always returns to operation ($r_{31} = 1$).

In solving this closed Jackson network, we first note that at node 1 the servers are the machines, so that $c_1 = 2$. Also, μ_1 becomes λ; that is, the mean service (or holding time) at node 1 is the mean time to failure of a machine. Thus the solution for the steady-state joint probability distribution is given by (4.17) as

$$p_{n_1,n_2,n_3} = \frac{1}{G(2)} \frac{\rho_1^{n_1}}{a_1(n_1)} \rho_2^{n_2} \rho_3^{n_3} \quad (n_i = 0, 1, 2, \quad i = 1, 2, 3),$$

where $a_1(n_1) = 1$ for $n_1 = 0, 1$, and $a_1(2) = 2$, and we must find ρ_i from (4.16). The routing probability matrix $\boldsymbol{R} = \{r_{ij}\}$ for nodes 1, 2, and 3 and is given by

$$\boldsymbol{R} = \begin{pmatrix} 0 & r_{12} & 1 - r_{12} \\ 1 - r_{23} & 0 & r_{23} \\ 1 & 0 & 0 \end{pmatrix}.$$

Using the above $\{r_{ij}\}$ in (4.16) yields

$$\begin{aligned} \lambda\rho_1 &= \mu_2(1 - r_{23})\rho_2 + \mu_3\rho_3, \\ \mu_2\rho_2 &= \lambda r_{12}\rho_1, \\ \mu_3\rho_3 &= \lambda(1 - r_{12})\rho_1 + \mu_2 r_{23}\rho_2. \end{aligned}$$

One equation in the set given by (4.16) is always redundant, so we can arbitrarily set one of the $\rho_i = 1$ [the constant $G(N)$ will account for the appropriate normalizing factor]. We choose to set $\rho_2 = 1$. Thus the solutions for the $\{\rho_i\}$ are

$$\rho_2 = 1, \qquad \rho_1 = \frac{\mu_2}{r_{12}\lambda},$$

$$\rho_3 = \frac{\lambda(1-r_{12})}{\mu_3}\frac{\mu_2}{\lambda r_{12}} + \frac{\mu_2}{\mu_3}r_{23} = \frac{\mu_2(1-r_{12}+r_{12}r_{23})}{r_{12}\mu_3},$$

and hence

$$p_{n_1,n_2,n_3} = \frac{1}{G(N)}\left(\frac{\mu_2}{r_{12}\lambda}\right)^{n_1}\frac{1}{a_1(n_1)}\left(\frac{\mu_2(1-r_{12}+r_{12}r_{23})}{r_{12}\mu_3}\right)^{n_3}$$

The normalizing constant $G(N)$ must be obtained by summing p_{n_1,n_2,n_3} over all cases for which $n_1+n_2+n_3=2$. There are actually six cases, namely, (2, 0, 0), (0, 2, 0), (0, 0, 2), (1, 1, 0), (1, 0, 1), and (0, 1, 1) for (n_1, n_2, n_3). To illustrate, assume $\lambda = 2, \mu_2 = 1, \mu_3 = 3, r_{12} = \frac{3}{4}$, and $r_{23} = \frac{1}{3}$. Then

$$p_{n_1,n_2,n_3} = \frac{1}{G(N)}\left(\frac{2}{3}\right)^{n_1}\frac{1}{a_1(n_1)}\left(\frac{2}{9}\right)^{n_3},$$

and we have

$$G(2) = \left(\frac{2}{3}\right)^2\cdot\frac{1}{2}+1+\left(\frac{2}{9}\right)^2+\frac{2}{3}+\frac{2}{3}\cdot\frac{2}{9}+\frac{2}{9} = \frac{187}{81} \doteq 2.3086.$$

Hence the steady-state probabilities are

$$p_{2,0,0} = \frac{(\frac{2}{3})^2(\frac{1}{2})}{2.3086} = 0.0962, \quad p_{0,2,0} = \frac{1}{2.3086} = 0.4332,$$

$$p_{0,0,2} = \frac{(\frac{2}{9})^2}{2.3086} = 0.0214, \quad p_{1,1,0} = \frac{\frac{2}{3}}{2.3086} = 0.2888,$$

$$p_{1,0,1} = \frac{(\frac{2}{3})(\frac{2}{9})}{2.3086} = 0.0642, \quad p_{0,1,1} = \frac{\frac{2}{9}}{2.3086} = 0.0962.$$

Thus in this particular situation only 9.62% of the time are both machines operating, with at least one machine available $0.0962 + 0.2888 + 0.0642 = 44.92\%$ of the time. It may be desirable to obtain more reliable machines (lower λ) or put on more repairpeople or have more machines installed, in order to have at least one available more of the time.

In the above example N was only 2 and k only 3, which made the calculation of $G(N)$ rather easy. For large N and k there are many possible ways to allocate the N customers among the k nodes, in fact, "$(N+k-1)$ choose N" ways (see Feller, 1957). Thus it would be highly advantageous to have an efficient algorithm to calculate $G(N)$. Buzen (1973) developed one, and we now present this most useful result for closed Jackson networks.

Let $f_i(n_i) = \rho_i^{n_i}/a_i(n_i)$. Then

$$G(N) = \sum_{n_1+n_2+\cdots+n_k=N}\prod_{i=1}^{k} f_i(n_i). \tag{4.19}$$

Buzen set up an auxiliary function

$$g_m(n) = \sum_{n_1+n_2+\cdots+n_m=n} \prod_{i=1}^{m} f_i(n_i). \tag{4.20}$$

Note that $g_m(n)$ would equal $G(N)$ if $k = m$ and $N = n$; that is, it is a normalizing constant for a system with n customers and m nodes. Also note that $G(N) = g_k(N)$. We can now set up a recursive scheme for calculating $G(N)$.

Consider $g_m(n)$ and suppose we fix $n_m = i$ first in the summation, so that we have

$$\begin{aligned} g_m(n) &= \sum_{i=0}^{n} \left(\sum_{n_1+n_2+\cdots+n_{m-1}+i=n} \prod_{j=1}^{m} f_j(n_j) \right) \\ &= \sum_{i=0}^{n} f_m(i) \left(\sum_{n_1+n_2+\cdots+n_{m-1}=n-i} \prod_{j=1}^{m-1} f_j(n_j) \right) \\ &= \sum_{i=0}^{n} f_m(i) g_{m-1}(n-i) \qquad (n = 0, 1, \ldots, N). \end{aligned} \tag{4.21}$$

Note that from (4.21) $g_1(n) = f_1(n)$ and $g_m(0) = 1$, so that we can use (4.21) recursively to calculate $G(N) = g_k(N)$. Furthermore, these functions aid in calculating marginal distributions as well. Suppose we desire the marginal distribution at node i, namely, $p_i(n) = \Pr\{N_i = n\}$. Let

$$S_i = n_1 + n_2 + \cdots + n_{i-1} + n_{i+1} + \cdots + n_k.$$

Then

$$\begin{aligned} p_i(n) = \sum_{S_i=N-n} p_{n_1,\ldots,n_k} &= \sum_{S_i=N-n} \frac{1}{G(N)} \prod_{i=1}^{k} f_i(N_i) \\ &= \frac{f_i(n)}{G(N)} \sum_{S_i=N-n} \prod_{\substack{j=1 \\ (j \neq i)}}^{k} f_i(n) \qquad (n = 0, 1, \ldots, N). \end{aligned}$$

This, however, is very cumbersome to compute. But for node k, the expression simplifies to

$$p_k(n) = \frac{f_k(n)}{G(N)} \sum_{S_k=N-n} \prod_{i=1}^{k-1} f_i(n) = \frac{f_k(n) g_{k-1}(N-n)}{G(N)} \qquad (n = 0, 1, \ldots, N), \tag{4.22}$$

and to find the other marginals $p_i(n), i \neq k$, Buzen (1973) suggests permuting the network to make the node i of interest $= k$. This of course requires resolving for some of the functions $g_m(n)$. Bruell and Balbo (1980) suggest an improved algorithm for obtaining $p_i(n), i \neq k$. In the next section, we present a method for the calculation of expected-value performance measures for closed Jackson networks, called *mean-value analysis*, which also can yield marginal probability distributions.

To illustrate, let us go back to Example 4.4. The factors $f_i(n_i)$ are

$$\begin{aligned}
&f_1(0) = 1, \quad f_1(1) = \tfrac{2}{3}, \quad f_1(2) = \tfrac{2}{9},\\
&f_2(0) = f_2(1) = f_2(2) = 1,\\
&f_3(0) = 1, \quad f_3(1) = \tfrac{2}{9}, \text{ and } f_3(2) = \tfrac{4}{81}.
\end{aligned}$$

The $g_i(n)$'s are

$$G(2) = g_3(2) = \sum_{i=0}^{2} f_3(i)g_2(2-i) = f_3(0)g_2(2) + f_3(1)g_2(1) + f_3(2)g_2(0)$$

and

$$\begin{aligned}
g_2(2) &= f_2(0)g_1(2) + f_2(1)g_1(1) + f_2(2)g_1(0),\\
g_2(1) &= f_2(0)g_1(1) + f_2(1)g_1(0),\\
g_2(0) &= f_2(0),
\end{aligned}$$

plus

$$g_1(0) = 1, \qquad g_1(1) = f_1(1) = \tfrac{2}{3}, \qquad g_1(2) = f_1(2) = \tfrac{2}{9}.$$

Thus we calculate from the bottom up to get

$$\begin{aligned}
g_0(0) &= 1,\\
g_2(1) &= 1 \times \tfrac{2}{3} + 1 \times 1 = \tfrac{5}{3}.\\
g_2(2) &= 1 \times \tfrac{2}{9} + 1 \times \tfrac{2}{3} + 1 \times 1 = \tfrac{17}{9},\\
g_3(2) &= G(2) = 1 \times \tfrac{17}{9} + \tfrac{2}{9} \times \tfrac{5}{3} + \tfrac{4}{81} \times 1 = \tfrac{187}{81} = 2.3086.
\end{aligned}$$

While Buzen's algorithm was not much easier for this simple example, it is quite efficient for large networks with large numbers of customers.

If now we desire the marginal distribution for the number of customers at node 3, say, we have from (4.22)

$$p_3(n) = \frac{f_3(n)g_2(2-n)}{G(2)} = \frac{(\frac{2}{9})^n g_2(2-n)}{2.3086},$$

so that

$$p_3(0) = \frac{1 \cdot g_2(2)}{2.3086} = .8182, \qquad p_3(1) = \frac{\frac{2}{9}g_2(1)}{2.3086} = .1604,$$

$$p_3(2) = \frac{(\frac{2}{9})^2 g_2(0)}{2.3086} = .0214,$$

which is the same answer one would get by summing appropriate joint probabilities, namely,

$$p_3(0) = p_{2,0,0} + p_{0,2,0} + p_{1,1,0} = .0962 + .4332 + .2888 = .8182,$$
$$p_3(1) = p_{0,1,1} + p_{1,0,1} = .0962 + .0642 = .1604,$$
$$p_3(2) = p_{0,0,2} = .0214.$$

Again, in large systems, making use of the already calculated functions $g_m(n)$ is considerably more efficient than summing over the joint distribution. A good general reference on computational algorithms for closed networks is Bruell and Balbo (1980).

4.3.1 Mean-Value Analysis

The previously described methods of analyzing closed Jackson queueing networks, which require computing the normalizing constant $G(N)$, are often referred to as "convolution procedures." Mean-value analysis is another approach that does not require evaluating $G(N)$. It is built on two basic principles (see Bruell and Balbo, 1980):

1. The queue length observed by an arriving customer is the same as the general-time queue length in a closed network with one less customer, that is, $q_n(N) = p_n(N-1)$.

2. Little's formula is applicable throughout the network.

The first principle allows us to write the average waiting time at a node in terms of the mean service time and average number in the system found by an arriving customer. Recall that, for the $M/M/1$ situation, it can easily be shown using (2.23) and (2.26) that $W = (1+L)/\mu$. What this implies is that the average time an arriving customer must wait is the average time to serve the queue size as seen by an arriving customer plus itself. For $M/M/c$, $q_n = p_n$, so no adjustment need be made for the fact that L is based on p_n and not q_n. For our closed network (we assume for the time being that all nodes have a single server) the equivalent equation becomes

$$W_i(N) = \frac{1 + L_i(N-1)}{\mu_i}, \tag{4.23}$$

where

$W_i(N)$ = mean waiting time at node i for a network containing N customers,

μ_i = mean service rate for the single server at node i,

$L_i(N-1)$ = mean number at node i in a network with $N-1$ customers.

The second principle, that of applying Little's formula throughout the network, allows us to write

$$L_i(N) = \lambda_i(N) W_i(N), \tag{4.24}$$

where $\lambda_i(N)$ is the throughput (arrival rate) for node i in an N-customer network. If we can find $\lambda_i(N)$, (4.23) and (4.24) give us a method for recursively calculating L_i and W_i, starting with an empty network [one with no customers, for which $L_i(0) = 0$ and $W_i(1) = 1/\mu_i$] and building up to the network of interest having N customers.

To be able to compute $\lambda_i(N)$, we note that if we let $D_i(N)$ represent the average delay per customer between successive visits to node i for a network with N customers, then by the laws of conservation we have $\lambda_i(N) = N/D_i(N)$. This merely states that the number of arrivals at node i per unit time must equal the total number of customers in the system divided by the mean time it takes each customer between successive visits to node i, and is a form of Little's formula applied to the entire network, since the expected number of customers in the system is exactly N.

To get $D_i(N)$, we return to the traffic equations (4.16). Letting $v_i = \mu_i \rho_i$ (we shall see shortly why we are not using λ_i), these equations become

$$v_i = \sum_{j=1}^{k} v_j r_{ji}. \tag{4.25}$$

Since one of the above equations is redundant, we can arbitrarily set one v_i (say, v_l) equal to 1 and solve for the others. The v_i then are *relative* throughputs through node i, that is, $v_i = \lambda_i/\lambda_l$, assuming v_l is the one on which we normalize. Now we can write $D_l(N) = \sum_{i=1}^{k} v_i W_i(N)$; that is, $D_l(N)$ is a weighted average of the average delays at each node, weighted by the relative throughputs (arrival rates) of each node to node l, or equivalently weighted by the expected number of visits to each node prior to returning to the "normalized" node, node l (note that v_i can also be interpreted as the expected number of visits to node i after leaving node l prior to returning to node l). For example, if we have a two-node network with $v_1 = 1$ and $v_2 = 2$, then since the arrival rate at node 2 is twice that at node 1, the expected number of visits to node 2 after leaving node 1 prior to returning to node 1 must be two.

We can now write the mean-value analysis (MVA) algorithm for finding $L_i(N)$ and $W_i(N)$ in a k-node, single-server-per-node network with routing probability matrix $\boldsymbol{R} = \{r_{ij}\}$ as follows:

(i) Solve the traffic equations (4.25), $v_i = \sum_{j=1}^{k} v_j r_{ji}$ $(i = 1, 2, \ldots, k)$, setting one of the v_j (say, v_l) equal to 1.

(ii) Initialize $L_i(0) = 0 \quad (i = 1, 2, \ldots, k)$.

(iii) For $n = 1$ to N, calculate

(a) $W_i(n) = \dfrac{1 + L_i(n-1)}{\mu_i} \quad (i = 1, 2, \ldots, k)$,

(b) $\lambda_l(n) = \dfrac{n}{\sum_{i=1}^{k} v_i W_i(n)} \quad$ (assume $v_l = 1$),

(c) $\lambda_i(n) = \lambda_l(n) v_i \quad (i = 1, 2, \ldots, k, i \neq l)$,

(d) $L_i(n) = \lambda_i(n) W_i(n) \quad (i = 1, 2, \ldots, k)$.

■ **EXAMPLE 4.5**

Consider Example 4.4, but now assume that there is only one machine, so that we now have single servers at each node. Solving (4.25) gives

$$(v_1, v_2, v_3) = (v_1, v_2, v_3)\begin{pmatrix} 0 & \frac{3}{4} & \frac{1}{4} \\ \frac{2}{3} & 0 & \frac{1}{3} \\ 1 & 0 & 0 \end{pmatrix},$$

or $v_1 = \frac{2}{3}v_2 + v_3, v_2 = \frac{3}{4}v_1, v_3 = \frac{1}{4}v_1 + \frac{1}{3}v_2$. Arbitrarily choosing $v_2 = 1$ ($l = 2$ here), we obtain $v_1 = \frac{4}{3}$ and $v_3 = \frac{2}{3}$.

Now $i = 1, 2, 3$ and $n = N = 1$. Applying step (iii) of the algorithm, we have for (a)

$$W_1(1) = \frac{1 + L_1(0)}{\lambda} = \frac{1}{\lambda} = \frac{1}{2} \quad \text{(note that } \mu_1 = \lambda \text{ for this example)},$$

$$W_2(1) = \frac{1 + L_2(0)}{\mu_2} = \frac{1}{\mu_2} = 1,$$

$$W_3(1) = \frac{1 + L_3(0)}{\mu_3} = \frac{1}{\mu_3} = \frac{1}{3};$$

for (b)

$$\lambda_2(1) = \frac{1}{\sum_{i=1}^{3} v_i W_i(1)} = \frac{1}{\frac{4}{3} \times \frac{1}{2} + 1 \times 1 + \frac{2}{3} \times \frac{1}{3}} = \frac{9}{17};$$

for (c)

$$\lambda_1(1) = v_i \lambda_2(1) = \tfrac{4}{3} \times \tfrac{9}{17} = \tfrac{12}{17},$$
$$\lambda_3(1) = v_3 \lambda_2(1) = \tfrac{2}{3} \times \tfrac{9}{17} = \tfrac{6}{17};$$

and for (d)

$$L_1(1) = \lambda_1(1) W_1(1) = \tfrac{12}{17} \times \tfrac{1}{2} = \tfrac{6}{17},$$
$$L_2(1) = \lambda_2(1) W_2(1) = \tfrac{9}{17} \times 1 = \tfrac{9}{17},$$
$$L_3(1) = \lambda_3(1) W_3(1) = \tfrac{6}{17} \times \tfrac{1}{3} = \tfrac{2}{17}.$$

Since we have only one machine ($N = 1$), we are finished. Had we had a spare machine, we would continue the algorithm for $n = 2$, by calculating $W_1(2) = [1 + L_1(1)]/\lambda = [1 + \frac{6}{17}]/2 = \frac{23}{24}$, and so on. We can continue for any number of spares, but we can have only one operating machine, since to accommodate multiple servers, we must modify the algorithm in step (iii)(a). We will present this a little later, but first, let us check our answer using the normalizing-constant method previously given.

To do this, we must solve the traffic equations (4.16), which gives us the same answer as for Example 4.4, since we have the same matrix $\boldsymbol{R}$, namely,

$\rho_1 = \frac{2}{3}, \rho_2 = 1$, and $\rho_3 = \frac{2}{9}$. Now we have from (4.15) that $p_{n_1,n_2,n_3} = [1/G(1)](\frac{2}{3})^{n_1}(\frac{2}{9})^{n_3}$ and $p_{100} = (\frac{2}{3})/G(1), p_{010} = 1/G(1), p_{001} = (\frac{2}{9})/G(1)$, so that $G(1) = (\frac{2}{3} + 1 + \frac{2}{9}) = \frac{17}{9}$. Thus $p_{100} = \frac{6}{17}, p_{010} = \frac{9}{17}, p_{001} = \frac{2}{17}$, and $L_1 = 0 \times \frac{11}{17} + 1 \times \frac{6}{17} = \frac{6}{17}, L_2 = \frac{9}{17}, L_3 = \frac{2}{17}$. Note here that we get the steady-state probabilities and have to calculate L_i using $\sum_{n=0}^{N} np_n$, whereas the MVA algorithm directly yields the L_i and W_i, but does not give us the steady-state probabilities.

It is, however, possible to get marginal steady-state probabilities also at each node by recursion, and in fact, we would add to the MVA algorithm a recursive relationship similar in spirit to that of $p_n = \rho p_{n-1}$ for $M/M/1$, the relation achieved from detailed (as opposed to global) stochastic balance (see Section 2.2.1),

$$p_i(n, N) = \frac{\lambda_i(N)}{\mu_i} p_i(n-1, N-1) \quad (n, N \geq 1), \tag{4.26}$$

where $p_i(n, N)$ is the marginal probability of n in an N-customer system at node i, and $p_i(0, 0) = 1$. For our example,

$$p_1(1,1) = \frac{\lambda_1(1)}{\lambda} p_1(0,0) = \frac{\lambda_1(1)}{\lambda} \times 1 = \frac{\frac{12}{17}}{2} = \frac{6}{17},$$

$$p_2(1,1) = \frac{\lambda_2(1)}{\mu_2} p_2(0,0) = \frac{\frac{9}{17}}{1} \times 1 = \frac{9}{17},$$

$$p_3(1,1) = \frac{\lambda_3(1)}{\mu_3} p_3(0,0) = \frac{\frac{6}{17}}{3} \times 1 = \frac{2}{17}.$$

So

$$p_1(0,1) = \frac{11}{17}, \qquad p_1(1,1) = \frac{6}{17},$$

$$p_2(0,1) = \frac{8}{17}, \qquad p_2(1,1) = \frac{9}{17},$$

$$p_3(0,1) = \frac{15}{17}, \qquad p_3(1,1) = \frac{2}{17}.$$

Checking with our normalizing constant solution,

$$p_1(0,1) = p_{010} + p_{001} = \frac{9}{17} + \frac{2}{17} = \frac{11}{17}, \qquad p_1(1,1) = \frac{6}{17},$$

$$p_2(0,1) = p_{100} + p_{001} = \frac{6}{17} + \frac{2}{17} = \frac{8}{17}, \qquad p_2(1,1) = \frac{9}{17},$$

$$p_3(0,1) = p_{100} + p_{010} = \frac{6}{17} + \frac{9}{17} = \frac{15}{17}, \qquad p_3(1,1) = \frac{2}{17}.$$

Thus it is relatively easy to obtain the marginal steady-state probabilities at each node using MVA.

We mentioned previously that the MVA algorithm must be modified in step (iii)(a) for multiple-server cases. Here, we do not have that the workload an arriving customer (inside observer) sees is $(1/\mu)(1 + L)$, since there are multiple servers who work simultaneously on reducing the customer queue. For this situation we have

$$W_i(n) = \frac{1}{\mu_i} + \frac{1}{c_i\mu_i}\sum_{j=c_i}^{n-1}(j - c_i + 1)p_i(j, n-1),$$

since if there are $j > c_i$ customers at node i when a customer arrives, it must wait until $j - c_i + 1$ are served at rate $c_i\mu_i$ to get into service. This may be simplified to

$$\begin{aligned} W_i(n) &= \frac{1}{c_i\mu_i}\left(c_i + \sum_{j=c_i}^{n-1} jp_i(j,n-1) - (c_i - 1)\sum_{j=c_i}^{n-1} p_i(j, n-1)\right) \\ &= \frac{1}{c_i\mu_i}\left[c_i + L_i(n-1) - \sum_{j=0}^{c_i-1} jp_i(j, n-1) - (c_i - 1)\right. \\ &\quad \left.\times\left(1 - \sum_{j=0}^{c_i-1} p_i(j, n-1)\right)\right] \\ &= \frac{1}{c_i\mu_i}\left(1 + L_i(n-1) + \sum_{j=0}^{c_i-2}(c_i - 1 - j)p_i(j, n-1)\right). \end{aligned}$$

Thus for the multiserver case, even if we are interested in only the W_i and L_i, we still need to calculate the marginal probabilities $p_i(j, n-1)$ for $j = 0, 1, \ldots, c_i - 2$. To calculate these recursively for the multiserver case we have, in the spirit of the $M/M/c$,

$$p_i(j, n) = \frac{\lambda_i(n)}{\alpha_i(j)\mu_i} p_i(j-1, n-1) \qquad (i \le j \le n-1),$$

where

$$\alpha_i(j) = \begin{cases} j & (j \le c_i), \\ c_i & (j \ge c_i). \end{cases} \tag{4.27}$$

Now we can modify the single-server MVA algorithm to allow for multiple servers:

(i) Solve the traffic equations (4.25) as done previously (note that these are the same regardless of the number of servers at a node).

(ii) Initialize for $i = 1, 2, \ldots, k$, $L_i(0) = 0$; $p_i(0,0) = 1$; $p_i(j, 0) = 0$, $(j \ne 0)$.

(iii) For $n = 1$ to N, calculate

(a) $W_i(n) = \dfrac{1}{c_i\mu_i}\left(1 + L_i(n-1) + \displaystyle\sum_{j=0}^{c_i-2}(c_i - 1 - j)p_i(j, n-1)\right)$

$(i = 1, 2, \ldots, k)$,

(b) $\lambda_l(n) = n/\sum_{i=1}^{k} v_i W_i(n)$ (assume $v_l = 1$),

(c) $\lambda_i(n) = \lambda_l(n)v_i \quad (i = 1, 2, \ldots, k;\, i \neq l)$,

(d) $L_i(n) = \lambda_i(n)W_i(n) \quad (i = 1, 2, \ldots, k)$,

(e) $p_i(j, n) = \dfrac{\lambda_i(n)}{\alpha_i(j)\mu_i} p_i(j-1, n-1) \quad (j = 1, 2, \ldots, n;\, i = 1, 2, \ldots, k)$.

We now illustrate this on Example 4.4. Initializing, we have $L_1(0) = L_2(0) = L_3(0) = 0, p_1(0,0) = p_2(0,0) = p_3(0,0) = 1$. The v_i's are the same as before, namely,

$$v_1 = \tfrac{4}{3}, \qquad v_2 = 1 \quad (l = 2), \qquad v_3 = \tfrac{2}{3}.$$

Now for the first iteration, steps (a) through (d) of the algorithm will produce identical results as those for the single-server case, since multiple servers are superfluous for a single-machine system. The results are, for (a),

$$W_1(1) = \tfrac{1}{2}, \qquad W_2(1) = 1, \qquad W_3(1) = \tfrac{1}{3};$$

for (b),

$$\lambda_2(1) = \tfrac{9}{17};$$

for (c),

$$\lambda_1(1) = \tfrac{12}{17}, \quad \lambda_3(1) = \tfrac{6}{17};$$

and for (d),

$$L_1(1) = \tfrac{6}{17}, \quad L_2(1) = \tfrac{9}{17}, \quad L_3(1) = \tfrac{2}{17}.$$

Step (e) yields the same marginal steady-state probabilities, namely,

$$\begin{aligned} p_1(1,1) &= \tfrac{6}{17}, & p_1(0,1) &= \tfrac{11}{17}, \\ p_2(1,1) &= \tfrac{9}{17}, & p_2(0,1) &= \tfrac{8}{17}, \\ p_3(1,1) &= \tfrac{2}{17}, & p_3(0,1) &= \tfrac{15}{17}. \end{aligned}$$

The second iteration of the algorithm gives, for (a),

$$\begin{aligned} W_1(2) &= \frac{1}{2\lambda}[1 + L_1(1) + (2-1)p_1(0,1)] = \tfrac{1}{4}[1 + \tfrac{6}{17} + \tfrac{11}{17}] = 0.5, \\ W_2(2) &= \frac{1}{\mu_2}[1 + L_2(1) + 0] = 1[1 + \tfrac{9}{17}] = \tfrac{26}{17} = 1.530, \\ W_3(2) &= \frac{1}{\mu_3}[1 + L_3(1) + 0] = \tfrac{1}{3}[1 + \tfrac{2}{17}] = 0.373; \end{aligned}$$

for (b),

$$\lambda_2(2) = \frac{2}{\sum_{i=1}^{3} v_i W_i(2)} = \frac{2}{\frac{4}{3}(0.5) + 1.53 + \frac{2}{3}(0.373)} = 0.818;$$

for (c),

$$\begin{aligned} \lambda_1(2) &= \lambda_2(2)v_1 = 0.818(\tfrac{4}{3}) = 1.091, \\ \lambda_3(2) &= \lambda_2(2)v_3 = 0.818(\tfrac{2}{3}) = 0.545; \end{aligned}$$

for (d),

$$L_1(2) = \lambda_1(2)W_1(2) = 0.546,$$
$$L_2(2) = \lambda_2(2)W_2(2) = 1.252,$$
$$L_3(2) = \lambda_3(2)W_3(2) = 0.203;$$

and for (e),

$$p_1(j,2) = \frac{\lambda_1(2)}{\alpha_1(j)\lambda} p_1(j-1,1) \quad (j = 1,2);$$

which yields

$$p_1(1,2) = \frac{1.091}{2} p_1(0,1) = 0.353,$$
$$p_1(2,2) = \frac{1.091}{4} p_1(1,1) = 0.096,$$
$$p_1(0,2) = 1 - 0.353 - 0.096 = 0.551,$$
$$p_2(j,2) = \frac{\lambda_2(2)}{\alpha_2(j)\mu_2} p_2(j-1,1) \quad (j = 1,2);$$

which yields

$$p_2(1,2) = 0.385, \quad p_2(2,2) = 0.433, \quad p_2(0,2) = 0.182,$$

and

$$p_3(j,2) = \frac{\lambda_3(2)}{\alpha_3(j)\mu_3} p_3(j-1,1) \quad (j = 1,2);$$

which yields

$$p_3(1,2) = 0.161, \quad p_3(2,2) = 0.021, \quad p_3(0,2) = 0.818.$$

Now, checking with the previous results for Example 4.4, we see that the marginal probabilities are

$$p_1(0) = p_{020} + p_{002} + p_{011} = 0.5508, \quad p_1(1) = p_{110} + p_{101} = 0.3530,$$
$$p_1(2) = p_{200} = 0.0962;$$
$$p_2(0) = p_{200} + p_{002} + p_{101} = 0.1818, \quad p_2(1) = p_{110} + p_{011} = 0.3850,$$
$$p_2(2) = p_{020} = 0.4332;$$
$$p_3(0) = p_{200} + p_{020} + p_{110} = 0.8182, \quad p_3(1) = p_{101} + p_{011} = 0.1604,$$
$$p_3(2) = p_{002} = 0.0214.$$

We mention here that we have not rigorously proved the recursive relationship on the marginal probabilities given by (4.26) and (4.27), but only argued intuitively based on $M/M/1$ and $M/M/c$ recursions.

We now prove (4.26) and note that the proof for (4.27) is similar when the multiserver factor $\alpha_i(n_i)$ is included. We let N_i represent the random variable "number of customers at node i (in the steady state)," so that the marginal probability distribution $p_i(n_i; N)$ is

$$\begin{aligned} p_i(n_i; N) &\equiv \Pr\{N_i = n_i | N \text{ customers in network}\} \\ &= \sum_{n_1+n_2+\cdots+n_{i-1}+n_{i+1}+\cdots+n_k=N-n_i} \frac{1}{G(N)} \rho_1^{n_1} \rho_2^{n_2} \cdots \rho_i^{n_i} \cdots \rho_k^{n_k}. \end{aligned}$$

The complementary marginal cumulative probability distribution is

$$\begin{aligned} &\bar{P}_i(n_i; N) \\ &\quad \equiv \Pr\{N_i \geq n_i | N \text{ customers in network}\} \\ &\quad = \sum_{j=n_i}^{\infty} \sum_{n_1+n_2+\cdots+n_{i-1}+n_{i+1}+\cdots+n_k=N-j} \frac{1}{G(N)} \rho_1^{n_1} \rho_2^{n_2} \cdots \rho_i^{j} \cdots \rho_k^{n_k} \\ &\quad = \sum_{j=n_i}^{\infty} \rho_i^j \sum_{n_1+n_2+\cdots+n_{i-1}+n_{i+1}+\cdots+n_k=N-j} \frac{1}{G(N)} \rho_1^{n_1} \rho_2^{n_2} \cdots \rho_{i-1}^{n_{i-1}} \rho_{i+1}^{n_{i+1}} \cdots \rho_k^{n_k} \\ &\quad = \sum_{j=n_i}^{\infty} \rho_i^j \frac{1}{G(N)} g_{k-1}(N-j) \quad [\text{from}(4.20)] \\ &\quad = \frac{\rho_i^{n_i}}{G(N)} \sum_{j=1}^{\infty} \rho_i^{j-n_i} g_{k-1}(N-j) = \frac{\rho_i^{n_i}}{G(N)} \sum_{l=0}^{\infty} \rho_i^l g_{k-1}(N-n_i-l) \\ &\quad = \frac{\rho_i^{n_i}}{G(N)} g_k(N-n_i) \quad [\text{from}(4.21)] \\ &\quad = \frac{\rho_i^{n_i}}{G(N)} G(N-n_i). \end{aligned}$$

Now

$$\begin{aligned} p_i(n_i; N) &= \bar{P}_i(n_i; N) - \bar{P}_i(n_i+1; N) \\ &= \frac{\rho_i^{n_i}}{G(N)} G(N-n_i) - \frac{\rho_i^{n_i+1}}{G(N)} G(N-n_i-1) \\ &= \frac{\rho_i^{n_i}}{G(N)} [G(N-n_i) - \rho_i G(N-n_i-1)]. \end{aligned}$$

Thus

$$\begin{aligned} &\frac{p_i(n_i; N)}{p_i(n_i-1; N-1)} \\ &\quad = \frac{\rho_i^{n_i}}{G(N)} \frac{G(N-1)}{\rho_i^{n_i-1}} \frac{G(N-n_i) - \rho_i G(N-n_i-1)}{G(N-1-n_i+1) - \rho_i G(N-1-n_i+1-1)} \\ &\quad = \frac{\rho_i G(N-1)}{G(N)} \frac{G(N-n_i) - \rho_i G(N-n_i-1)}{G(N-n_i) - \rho_i G(N-n_i-1)} = \frac{\rho_i G(N-1)}{G(N)}. \end{aligned}$$

Hence

$$p_i(n_i; N) = \frac{\rho_i G(N-1)}{G(N)} p_i(n_i - 1; N - 1).$$

But note that the throughput at node i is

$$\lambda_i(N) = \Pr\{\text{server busy at node } i\} \cdot \mu_i = \bar{P}(1; N)\mu_i = \frac{\rho_i}{G(N)} G(N-1)\mu_i,$$

so that

$$\frac{\rho_i G(N-1)}{G(N)} = \frac{\lambda_i(N)}{\mu_i},$$

and finally

$$p_i(n_i; N) = \frac{\lambda_i(N)}{\mu_i} p_i(n_i - 1; N - 1).$$

While all the methods [including "brute force" in calculating $G(N)$] are easy to employ for these small illustrative examples, Buzen's algorithm and MVA are computationally quite superior to "brute force," with respect to efficiency (storage and speed) and stability, for larger problems (larger k and larger N). Comparing Buzen and MVA, while they are both far superior to "brute force" for large problems, they nevertheless can still face some numerical difficulties for very large state-space problems emanating from real-world modeling. For a network made up of all single-server nodes for which we desire only mean waiting times and mean system sizes at each node, MVA is superior. However, if we have multiserver nodes or we desire marginal probability distributions at the nodes, then we must calculate the probabilities recursively for MVA, and it is not clear in these situations if MVA is really better than the Buzen procedure.

4.4 Cyclic Queues

If we consider a closed network of k nodes such that

$$r_{ij} = \begin{cases} 1 & (j = i+1, 1 \le i \le k-1), \\ 1 & (i = k, j = 1), \\ 0 & (\text{elsewhere}), \end{cases} \tag{4.28}$$

then we have a cyclic queue. A cyclic queue thus is a sort of series queue in a "circle," where the output of the last node feeds back to the first node. Since this is a special case of a closed queueing network, the results of the previous section apply. Hence, for single servers at each node, (4.15) and (4.16) apply, that is,

$$p_{n_1, n_2, \ldots, n_k} = C\rho_1^{n_1}\rho_2^{n_2} \cdots \rho_k^{n_k}, \tag{4.29}$$

with $\mu_i \rho_i = \sum_{j=1}^{k} \mu_j r_{ji} \rho_j$. Using (4.28) in the traffic equation results in

$$\mu_i \rho_i = \begin{cases} \mu_{i-1}\rho_{i-1} & (i = 2, 3, \ldots, k), \\ \mu_k \rho_k & (i = 1). \end{cases}$$

Thus we have

$$\rho_i = \begin{cases} (\mu_{i-1}/\mu_i)\rho_{i-1} & (i = 2, 3, \ldots, k), \\ (\mu_k/\mu_1)\rho_k & (i = 1). \end{cases} \tag{4.30}$$

From (4.30) we see that

$$\rho_2 = \frac{\mu_1}{\mu_2}\rho_1, \quad \rho_3 = \frac{\mu_2}{\mu_3}\rho_2 = \frac{\mu_1}{\mu_3}\rho_1, \quad \ldots, \quad \rho_{k-1} = \frac{\mu_1}{\mu_{k-1}}\rho_1, \quad \rho_k = \frac{\mu_1}{\mu_k}\rho_1.$$

Since one ρ can be set equal to one due to redundancy, we select $\rho_1 = 1$, and substituting into (4.29) we obtain

$$p_{n_1,\ldots,n_k} = \frac{1}{G(N)} \frac{\mu_1^{N-n_1}}{\mu_2^{n_2}\mu_3^{n_3}\cdots\mu_k^{n_k}}. \tag{4.31}$$

Again, $G(N)$ can be found by summing over all cases $n_1 + n_2 + \cdots + n_k = N$ or by Buzen's algorithm.

The multiple-server case can also be treated similarly, and we leave it as an exercise (see Problem 4.20). Of course, it is not necessary to develop special versions of (4.15) and (4.17); these results can be used as they are, with the appropriate $\{r_{ij}\}$ given by (4.28).

We have actually treated a cyclic queue previously in the text. The machine-repair problems of Section 2.8 are really two-node cyclic queues, with node 1 representing the machine "up" node, and the number of servers at this node being the number of machines desired operational, which we denoted by M. The presence of a queue at this node means that spare machines are on hand, while an idle server represents fewer than M machines operating. The mean "service" (holding) time at this node per customer (machine) is $1/\lambda$, where λ is the machine mean failure rate. Problem 4.19 asks the reader to show that (4.31) reduces to (2.58) when $M = c = 1$.

4.5 Extensions of Jackson Networks

Jackson networks have been extended in several ways. First, Jackson himself, in his 1963 paper, allowed state-dependent exogenous arrival processes and state-dependent internal service for open networks. The parameters of the exogenous "Poisson" arrival processes could depend on the total number of units in the network (a general birth process), while the service-time parameters at a given node could depend on the number of customers present at that node. The solution was also of product form [a more complex version of (4.17), as the general birth–death equation (2.3) is a more complex version of (2.31)]. The normalizing constant does not break apart as it did in the non-state-dependent case of Section 4.2, so that the nodes do not act like independent queues. Computation of the normalizing constant must be done similarly to that for closed networks, but is even more complicated, since the sum to one is over an infinite number of probabilities. A great deal of effort has been put forth in attempting to find efficient ways to obtain or approximate the normalizing constant. Unfortunately, no Buzen-type algorithm exists.

Another avenue of generalization of Jackson networks is to include travel times between nodes of the network. These, of course, could always be modeled as another node, but most often they are ample server nodes (no queueing for travel).

Posner and Bernholtz (1968) treated closed Jackson networks but allowed for ample-service travel-time "nodes" with general travel-time (holding-time) distributions. They showed that if one were interested in only marginal probability distributions of steady-state system sizes at sets of nodes exclusive of the travel-time nodes, the exact forms of the travel-time distributions did not matter—only their means. In fact, then, for any nodes in a Jackson network with ample service (number of servers equal to total number of units in the system), the forms of the service-time distributions do not explicitly enter (only the means) as long as the marginal distributions of interest do not include these nodes.

Another extension of Jackson networks we shall discuss (and probably the most significant) deals with extensions to multiple classes of customer networks, namely, multiclass Jackson networks, where, in addition to each class of customers having its own routing structure, each class also has its own mean arrival rate, and the mean service times at a node may depend on the particular customer type (class to which the customer belongs) as well.

Baskett et al. (1975) treated such multiclass Jackson networks and obtained product-form solutions for the following three queueing disciplines: (1) processor sharing (each customer gets a share of and is served simultaneously by a single server), (2) ample service, and (3) LCFS with preemptive-resume servicing. They allowed the network to be open for some classes of customers and closed for others. Customers may switch classes after finishing at a node according to a probability distribution; that is, there is a probability $r_{is;jt}$ that a customer of class s completing service at node i next goes to node j as a class t customer. Exogenous "Poisson" input can be state dependent (a general birth process), and service distributions can be of the phase type. Baskett et al. (1975) also considered c-server FCFS nodes, but for these, service times for *all* classes must be *IID exponential*; that is, for these nodes all customer types look alike, and service times are exponentially distributed.

Kelly's work (1975, 1976, 1979) considered multiple customer classes and set up a notational structure that allowed for unique class service times at multiserver FCFS nodes. In fact, his work is so general that it included "most" queueing disciplines [e.g., all those considered by Baskett et al. (1975); however, priority dependent on customer class is one it did not]. But a price must be paid for this generality in that the description of the state space becomes much more complex. Up to now, the state space could be described by a vector consisting of the number of each class of customer at each node, but now the state space must be described by a complete customer ordering, by type, at each node. Kelly considered, as well as exponential service time, Erlang service, which further expands the state-space descriptor to include service phase. Nevertheless, he proved that the solution is still of product form.

Kelly further conjectured that many of his results can be extended to include general service-time distributions. This conjecture was based on the fact that nonnegative probability distributions can be well approximated by finite mixtures of gamma distributions. Kelly's conjecture is proved by Barbour (1976). In Section 7.3, we will also consider general service distributions, but not using mixtures of gamma distributions. Rather, we will use general network-decomposition methods to approximate the performance of the network.

While the generalizations of Baskett and co-workers, Kelly, and Barbour are theoretically significant, obtaining computational results for these more general Jackson networks is another matter. Gross and Ince (1981) have applied Kelly's multiclass results to a closed network and obtained numerical solutions for an application in repairable-item inventory control (the machine-repair problem).

A great deal of effort has been expended in obtaining computational results for closed multiclass Jackson networks due to their use in modeling computer systems. The basic model generally considers a computer system with N terminals, one for each user logged on. While it is not strictly a closed system, since users log on and off, during busy periods one can assume all terminals are in use, so that there are always N customers (jobs) in the system. These can be at various stages in the system, such as "thinking" at the terminal, waiting in the queue to enter the central processing unit (CPU), being served by the CPU, waiting or in service at input/output stations, and so on. Multiple job classes are an important part of any such model. Bruell and Balbo (1980) provided a compendium of computing algorithms developed to treat such models.

Interest and research continue in the computational aspects of Jackson networks, particularly closed, multiclass networks, because of their immense importance for modeling a variety of systems of interest in the computer, communications, and logistics fields.

4.6 Non-Jackson Networks

The only non-Jackson networks we will consider here are those where the $\{r_{ij}\}$ are allowed to be state dependent. In many situations, customers have flexibility when leaving a node in deciding where to go next. For example, in an open network, if a customer has two more nodes to visit prior to departing the system, which node the customer goes to next can well depend on the relative congestion at the two nodes. Allowing this flexibility in the $\{r_{ij}\}$ destroys one of the Jackson assumptions, and hence the previous results of Section 4.2 for open networks and Section 4.3 for closed networks do not apply.

If, however, holding times at all nodes remain exponential and exogenous inputs (if any) remain Poisson, we can still model the network using Markov theory. A full Markov state-space analysis (a separate balance equation for each state) is required, however.

As an example, consider a closed network with three nodes and two customers. Suppose customers choose, for the node to visit next, that node with the fewest customers. If there is a tie, the customer chooses among the tied nodes with equal probability. We will further assume that customers will not directly feed back to the same node prior to visiting at least one other node.

We can develop the $\boldsymbol{Q}$ matrix for this network and solve the steady-state equations given by $\mathbf{0} = \boldsymbol{pQ}$, along with the boundary condition that all probabilities sum to one. Since in this example we have a finite number of customers, we have a finite state space and $\boldsymbol{Q}$ will be finite. Numerical solution techniques (see Section 8.1.1) can always be employed for finite-state-space problems.

The state space must again be described by a k (here $k = 3$) component vector (n_1, n_2, n_3). The six states for the $\boldsymbol{Q}$ matrix for this problem are, respectively, (2, 0, 0), (0, 2, 0), (0, 0, 2), (0, 1, 1), (1, 1, 0), (1, 0, 1), and the $\boldsymbol{Q}$ matrix is

$$\boldsymbol{Q} = \begin{bmatrix} * & 0 & 0 & 0 & \mu_1/2 & \mu_1/2 \\ 0 & * & 0 & \mu_2/2 & \mu_2/2 & 0 \\ 0 & 0 & * & \mu_3/2 & 0 & \mu_3/2 \\ 0 & 0 & 0 & * & \mu_3 & \mu_2 \\ 0 & 0 & 0 & \mu_1 & * & \mu_2 \\ 0 & 0 & 0 & \mu_1 & \mu_3 & * \end{bmatrix},$$

where * indicates the negative of the sum of the other elements in the row. Thus the steady-state equations are

$$0 = -\mu_1 p_{2,0,0}, \qquad 0 = -\mu_2 p_{0,2,0}, \qquad 0 = -\mu_3 p_{0,0,2},$$

$$0 = \frac{\mu_2}{2} p_{0,2,0} + \frac{\mu_3}{2} p_{0,0,2} - (\mu_3 + \mu_2) p_{0,1,1} + \mu_1 p_{1,1,0} + \mu_1 p_{1,0,1},$$

$$0 = \frac{\mu_1}{2} p_{2,0,0} + \frac{\mu_2}{2} p_{0,2,0} + \mu_3 p_{0,1,1} - (\mu_1 + \mu_2) p_{1,1,0} + \mu_3 p_{1,0,1},$$

$$0 = \frac{\mu_1}{2} p_{2,0,0} + \frac{\mu_3}{2} p_{0,0,2} + \mu_2 p_{0,1,1} + \mu_2 p_{1,1,0} - (\mu_1 + \mu_3) p_{1,0,1}.$$

We immediately see that for steady state $p_{2,0,0} = p_{0,2,0} = p_{0,0,2} = 0$, as we would expect from the facts that there are only two customers and that routing strategies are state dependent to avoid congestion. Therefore, for this model, we get the 3×3 set of equations

$$\begin{aligned} 0 &= -(\mu_3 + \mu_2) p_{0,1,1} + \mu_1 p_{1,1,0} + \mu_1 p_{1,0,1}, \\ 0 &= \mu_2 p_{0,1,1} - (\mu_1 + \mu_2) p_{1,1,0} + \mu_3 p_{1,0,1}, \\ 0 &= \mu_2 p_{0,1,1} + \mu_2 p_{1,1,0} - (\mu_1 + \mu_3) p_{1,0,1}, \end{aligned}$$

and of course $1 = p_{0,1,1} + p_{1,1,0} + p_{1,0,1}$, which can replace one of the above equations. If all the μ_i were equal, the probabilities would turn out to be equally likely, namely, the probability of finding any of the three nodes empty with one each at the other two is $\frac{1}{3}$. Even for closed networks, the number of equations can grow enormously, so that thousands or tens of thousands or even millions of equations might have to be solved. For example, a network with 50 customers and 10 nodes (not an unusually large network) could have as many as

$$\binom{N + k - 1}{N} = \binom{59}{50} \doteq 1.26 \times 10^{10}$$

equations, if all states are possible. Obviously, even for modern large-scale computers, this is a formidable task to say the least. This is why the product-form solutions of Jackson and of Gordon and Newell (1967) are so valuable.

PROBLEMS

4.1. Find the solution (4.5) to the equations (4.4) using the methodology of Section 1.7.

4.2. Show that the number in the system just after the last departure, $N(T)$, and the time between successive departures, T, are independent random variables. [*Hint*: Find the conditional distribution of $N(T)$ given $T = t$ from (4.5), and show that it does not depend on t.] Also show that successive interdeparture times are independent.

4.3. For Example 4.1 calculate the same performance measures using three checkout counters operating. If you were Meback's advisor, what would you recommend concerning the number of counters to have in operation?

4.4. For a two-station series queue (single server at each station) with Poisson input to the first with parameter λ, exponential service at each station with parameters μ_1 and μ_2, respectively, and no limit on number in system at either station, show that the steady-state probability that there are n_1 in the first-station system (queue and service) and n_2 in the second-station system is given by

$$p_{n_1,n_2} = p_{n_1}p_{n_2} = \rho_1^{n_1}\rho_2^{n_2}(1-\rho_1)(1-\rho_2).$$

[*Hint*: Find the steady-state difference equations, then show that $\rho_1^{n_1}\rho_2^{n_2}p_{0,0}$ is a solution to these equations by substitution, and then find $p_{0,0}$ by the boundary condition $\sum_{n_1=0}^{\infty}\sum_{n_2=0}^{\infty}p_{n_1,n_2} = 1$.]

4.5. Consider a three-station series queueing system (single server at each station) with Poisson input (parameter λ) and exponential service (parameters μ_1, μ_2, μ_3). There is no capacity limit on the queue in front of the first two stations, but at the third there is a limit of K allowed (including service). If K are in the third station, then any subsequent arrivals are shunted out of the system. Find the expected number in the system (all three stations) and the expected time spent in the system by a customer who completes all three stages of service.

4.6. Derive the results of (4.8) from (4.7) and the boundary condition.

4.7. Derive the steady-state difference equations for a sequential two-station, single-server system, with Poisson input (parameter λ) and exponential service (parameters μ_1, μ_2), where no queue is allowed in front of station 1 and at most one customer is allowed to wait between the stations. Blockage occurs when there is a customer waiting at station 2 and a customer completed at station 1. For the case $\mu_1 = \mu_2$, solve for the steady-state probabilities.

4.8. What are the eight system-state descriptors for a three-station series queueing model with blocking, infinite queue allowed in front of station 1, but no queues allowed at 2 or 3?

4.9. For an open Jackson network, generalize the steady-state balance equation set (4.9) to allow for c_i servers at each node, and show that the solution given by (4.12) satisfies (4.9). [*Hint*: Use the factor $a_i(n_i)$ in modifying (4.9).]

4.10. Consider a seven-node, open single-server Jackson network, where only nodes 2 and 4 get input from the outside (at rate 5/min). Nodes 1 and 2 have service rates of 85, nodes 3 and 4 have rates of 120, node 5 has a rate of 70, and nodes 6 and 7 have rates of 20 (all in units per minute). The routing matrix is given by

$$\begin{bmatrix} \frac{1}{3} & \frac{1}{4} & 0 & \frac{1}{4} & 0 & \frac{1}{6} & 0 \\ \frac{1}{3} & \frac{1}{4} & 0 & \frac{1}{3} & 0 & 0 & 0 \\ 0 & 0 & \frac{1}{3} & \frac{1}{3} & \frac{1}{3} & 0 & 0 \\ \frac{1}{3} & 0 & \frac{1}{3} & 0 & \frac{1}{3} & 0 & 0 \\ 0 & 0 & 0 & \frac{4}{5} & 0 & 0 & \frac{1}{6} \\ \frac{1}{6} & 0 & \frac{1}{6} & \frac{1}{6} & \frac{1}{6} & \frac{1}{6} & 0 \\ 0 & \frac{1}{6} & \frac{1}{6} & \frac{1}{6} & \frac{1}{6} & 0 & \frac{1}{6} \end{bmatrix}.$$

Find the average system size and mean line delay at each node.

4.11. Boobtube Boychicks, a TV repair facility that services many of the region's retail electronics stores' TV repair maintenance policies, receives sets to repair according to a Poisson process at an average rate of 9/h. All incoming sets are first looked at by a triage specialist, who determines whether the sets go to the facility's general repair station, require special attention and go to the expert repair station, or cannot be fixed locally and must be returned to the manufacturer and are sent to shipping. About 17% of all sets are sent directly to shipping to be returned to manufacturers. Of the 83% that are kept, 57% go to general repair, while 43% are sent to the experts. All repaired sets are sent to shipping; however, 5% of the sets received at the general repair station are sent back, unrepaired, to triage for redetermination. (*Note:* It is allowed for triage to send this set back again to general repair). Because of the varied nature of the problems and varying sizes of the sets, the exponential distribution is found to be an adequate representation for triage, repair (both general and expert), and shipping times. There is a single person at triage, taking a mean time of 6 min per set. There are three general repairpeople, each taking, on average, 35 min per set (including the ones they can't fix and send back to triage). Barry, Barney, Billy, and Bennie Boychick (quadruplets as it turns out) are the four experts manning the expert repair station and take, on average, 65 min to repair a set (they never fail to "fix 'em"). There are two shipping clerks, each of whom take an average time of 12.5 min to package the sets. Betty Boychick, Barry's wife and CEO of the company (who has an MS degree in OR from a major state university and teaches part-time at a local university) decides to ask her class to find the average number of sets at each node, the average time spent at each node, and the average time a set spends after it enters the receiving–triage station until it is packed and ready for delivery.

4.12. The Won Hung Rhee Chinese carry-out restaurant serves two dishes, chow mein and spareribs. There are two separate windows, one for chow mein and one for spareribs. Customers arrive according to a Poisson process with a mean rate of 20/h. Sixty percent go to the chow-mein window, and 40% go to the rib window. Twenty percent of those who come from the chow-mein window with their order go next to the rib window; the other 80% leave the restaurant. Ten percent of those who purchase ribs then go to the chow-mein window, and the other 90% leave. It takes on average 4 min to fill a chow-mein order and 5 min to fill a sparerib order, the service times being exponential. How many on average are in the restaurant? What is the average wait at each window? If a person wants both chow mein and ribs, how long, on average, does the person spend in the restaurant?

4.13. The diagram below represents a model of a call center that sells tickets for a local baseball team. Upon dialing, a customer first connects to an interactive voice response (IVR) system (node A, "If you would like to buy single-game tickets, press 1..."). There are two possible choices: (1) purchase single-game tickets or (2) purchase multigame ticket packages. (Customers who do not select any option effectively return to the IVR to start over.) Based on the selection, the customer is then transferred to either the single-game sales representatives (node B) or the multigame sales representative (node C). After selecting tickets, customers are transferred to another set of representatives (node D) to handle credit card payments. (From nodes B and C, customers may also choose to be transferred to the other set of agents to select more tickets before paying.) There is 1 representative at node B, 1 at node C, and 1 at node D. The IVR system can handle an arbitrarily large number of calls simultaneously. The service rates at each node are $\mu_A = 120$ per hour, $\mu_B = 30$ per hour, $\mu_C = 10$ per hour, and $\mu_D = 30$ per hour. The arrival rate to the center is $\gamma = 30$ calls per hour. Suppose all of the assumptions of an open Jackson network apply. The transition probabilities are shown in the diagram (probabilities that do not sum to 1 indicate the possibility of abandonment).

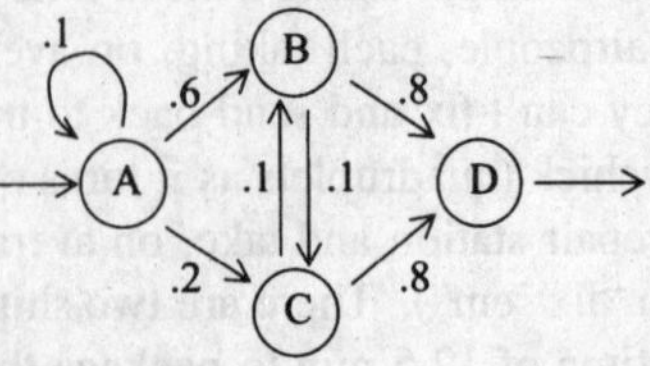

(a) Calculate the average number of customers in the system.
(b) Calculate the average time a customer spends in the system.

4.14. Verify, by substitution into (4.14), that (4.15) is a solution for single-server closed Jackson networks.

4.15. For a closed Jackson network, generalize the steady-state balance equation set (4.14) to allow for c_i servers at each node, and show that the solution given by (4.17) satisfies (4.14). [*Hint:* See the hint for Problem 4.9.]

4.16. Consider the same problem as Example 4.4, but suppose management decides to add a second server at node 2, identical to the one already there. Find the steady-state availabilities that (a) both machines are operating, and (b) at least one is operating.

4.17. Tonyexpress, a regional truck delivery service [one of its customers is Boobtube Boychicks (see Problem 4.11)], has a fleet of 50 trucks. Routine maintenance is done during off hours, so no trucks are unavailable for service because of routine maintenance. However, trucks do break down periodically and require repair. The breakdown process can be well approximated by a Poisson process, and repair records indicate a mean time to breakdown of 38 days. Sixty-eight percent of these breakdowns can be handled at Tonyexpress' own repair facility; the remainder must be handled by the manufacturer (the fleet consists of all the same make of trucks). Furthermore, about 7% of those sent to local repair must, after being worked on, be sent on to the manufacturer's repair facility. Repair times at the local repair facility are exponentially distributed with a mean of 2.75 days, and there are four repair bays, but only three operate at any given time, since there are only three repair crews available. Turnaround times for those sent to the factory have been found to be IID exponential random variables, with a mean of 10 days (since Tonyexpress is such a good customer, whenever a truck is received by the manufacturer, a mechanic is immediately put on the job). Since trucks are so well cared for, it is not unreasonable to assume that repaired trucks are as good as new. When trucks break down, Tonyexpress has a contract with TowsRus, a towing company with a very ample fleet of tow trucks. The mean time to pick up and tow a truck to the Tonyexpress facility is 0.15 days, and the mean time to tow a truck to the manufacturer (from the field or from Tonyexpress) is 0.75 days. Tony, the CEO of Tonyexpress, hires you to advise whether he should hire another repair crew. His major performance measure is truck availability—he is interested in the expected fraction of trucks available and the percentage of time that he has at least 45 trucks operational.

4.18. Find the average number of customers at each node and the node delay time for a closed network with 35 customers circulating between seven nodes using the switch matrix given by

$$\begin{bmatrix} \frac{1}{3} & \frac{1}{4} & 0 & \frac{1}{4} & 0 & \frac{1}{6} & 0 \\ \frac{1}{3} & \frac{1}{4} & 0 & \frac{1}{4} & \frac{1}{6} & 0 & 0 \\ 0 & 0 & \frac{1}{3} & \frac{1}{3} & \frac{1}{3} & 0 & 0 \\ \frac{1}{3} & 0 & \frac{1}{3} & 0 & \frac{1}{3} & 0 & 0 \\ 0 & 0 & 0 & \frac{5}{6} & 0 & 0 & \frac{1}{6} \\ \frac{1}{6} & \frac{1}{6} & \frac{1}{6} & \frac{1}{6} & \frac{1}{6} & \frac{1}{6} & 0 \\ 0 & \frac{1}{6} & \frac{1}{6} & \frac{1}{6} & \frac{1}{6} & \frac{1}{6} & \frac{1}{6} \end{bmatrix}$$

and assuming the same service rates as in Problem 4.10.

4.19. Show that for a single machine and a single repairperson, the solution obtained from (2.58) of Section 2.8 is the same as the solution obtained from (4.31).

4.20. Generalize (4.31) to handle multiple servers.

4.21. Solve Problem 2.47 by looking at it as a cyclic queue, that is, a special case of a closed queueing network.

4.22. Use the QtsPlus software to verify the solution to Example 4.5.

4.23. Using the QtsPlus software, (a) check the calculations for $p_3(n)$ of Example 4.4 given in the text section on Buzen's algorithm, and (b) calculate $p_2(n)$ and $p_1(n)$.

CHAPTER 5

GENERAL ARRIVAL OR SERVICE PATTERNS

This chapter considers queueing models involving general distributions. Because of the relaxation of the exponential assumption, the queues in this section cannot be modeled as continuous-time Markov chains. In particular, a Chapman–Kolmogorov analysis, as was done in the previous chapters, is no longer possible. However, for many of the models considered here, there is imbedded within the continuous-time non-Markov process a *discrete-time* Markov process, referred to as an *imbedded Markov chain* (see Section 1.9). For these types of models, we can employ some of the theory of Markov chains from Section 1.9 to analyze the queues.

The main sections of the chapter treat the cases of Poisson input with a general single server ($M/G/1$) and general input with a single exponential server ($G/M/1$).

5.1 General Service, Single Server ($M/G/1$)

This section considers a single-server queue with Poisson arrivals and a general service distribution. As usual, we assume that customers are served FCFS and that all service times and interarrival times are independent. Let λ be the arrival rate. Let S denote a random service time with a general distribution. Let $\mu = 1/\mathrm{E}[S]$ be the

Fundamentals of Queueing Theory, Fourth Edition.
By D. Gross, J. F. Shortle, J. M. Thompson, and C. M. Harris
Copyright © 2008 John Wiley & Sons, Inc.

service rate. We assume that $\rho \equiv \lambda/\mu < 1$. The analysis in this section considers the queue in steady state.

We begin by deriving a collection of formulas for expected-value measures of performance: W_q, W, L_q, and L. Formulas for these measures of service are typically referred to as Pollaczek–Khintchine (or PK) formulas. The approach is to derive one of these measures, and then to obtain the others using Little's formula and/or $W = W_q + \mathrm{E}[S]$, as we have done previously for other queues.

5.1.1 Expected-Value Measures of Effectiveness: The Pollaczek–Khintchine Formula

This section gives two derivations of expected-value measures for the $M/G/1$ queue. The first derivation obtains results by considering the system at times when customers arrive at the system. The second derivation obtains results by considering the system at times when customers depart from the system.

5.1.1.1 Derivation Using Arrival Times. Consider a customer arriving to the queue. Her delay is determined by the customers who are already in the system when she arrives. Specifically, there may be customers in the queue, and there may be a customer already in service.

We first consider the customers in the queue at the time of her arrival. Each customer who is in the queue ahead of her contributes, on average, $\mathrm{E}[S]$, to her delay. There are, on average, L_q customers in the queue when she arrives. Thus her average delay due to these customers is $L_q\mathrm{E}[S]$. (This logic requires the assumption of Poisson arrivals. The PASTA property, Section 1.7, implies that the average number of customers in the queue *as seen by an arriving customer* is the same as the time-average number of customers in queue, or L_q.)

Now, the customer who is in service (if there is any such customer) when she arrives contributes a different amount to her delay. This customer has completed some of his service already, so his contribution to her delay is his *remaining* service time, not his total service time. In general, these are not equal in expectation.

In summary, the average queue wait for the arriving customer is

$$W_q = L_q\mathrm{E}[S] + \Pr\{\text{server busy}\} \cdot \mathrm{E}[\text{residual service time} \mid \text{server busy}].$$

Using $L_q = \lambda W_q$ to eliminate L_q and then rearranging terms gives

$$W_q = \frac{\Pr\{\text{server busy}\} \cdot \mathrm{E}[\text{residual service time} \mid \text{server busy}]}{1-\rho}.$$

Now, $\Pr\{\text{server busy}\}$ is the probability that the arriving customer finds the server busy. By the PASTA property, this is the same as the fraction of time the server is busy, so $\Pr\{\text{server busy}\} = \rho$.

Thus it remains to find the expected residual service time, conditional on the arrival finding the server busy. It can be shown (Problem 5.7) that

$$\mathrm{E}[\text{residual service time} \mid \text{server busy}] = \frac{\mathrm{E}[S^2]}{2\,\mathrm{E}[S]} = \frac{1+C_B^2}{2}\,\mathrm{E}[S], \qquad (5.1)$$

where C_B^2 is the squared coefficient of variation of the service distribution, namely $\text{Var}[S]/\text{E}^2[S]$. This result is related to a standard result from renewal theory. The formula is sometimes called the average *excess* or average *residual* time of a renewal process. Intuitively, this is the average time until the end of a renewal cycle, as seen by an observer arriving to the process at a "random" time.

Note that (5.1) is typically greater than $\text{E}[S]/2$. That is, the expected remaining service time as seen by a customer arriving to a busy server is more than half of the expected time to service a customer. Equality is achieved only when $C_B^2 = 0$ or when the service distribution is deterministic. This is an example of the *inspection paradox* (e.g., Ross, 2007). The reason that the expected residual service time is more than what might be "intuitively" expected is that customers are more likely to arrive during long service intervals compared with short service intervals, and this increases the average above $\text{E}[S]/2$.

Combining the preceding results gives

$$\boxed{W_q = \frac{1 + C_B^2}{2} \cdot \frac{\rho}{1-\rho} \cdot \text{E}[S].} \tag{5.2}$$

This formula has three terms: a variability term, a utilization term, and a time-scale term. The first term $(1 + C_B^2)/2$ involves the squared coefficient of variation of the service distribution C_B^2. For exponential service, $C_B^2 = 1$, so $(1 + C_B^2)/2 = 1$. In this case, (5.2) reduces to the analogous formula for the $M/M/1$ queue (2.27). More specifically, (5.2) can be rewritten as

$$W_q = \frac{1 + C_B^2}{2} \cdot \{W_q \text{ for an analogous } M/M/1 \text{ queue}\}.$$

As the variability of the service distribution increases, the expected wait in queue increases. For large C_B^2, W_q is roughly linear in C_B^2.

The second term $\rho/(1-\rho)$ involves the queue utilization and increases to infinity as $\rho \to 1$. The last term $\text{E}[S]$ has units of time and can be thought of as a time-scale factor. Thus W_q is the product of two time quantities that are independent of the time scale chosen and the time-dependent quantity $\text{E}[S]$.

The formula for W_q is a powerful result. Only three parameters are needed to compute W_q: the arrival rate λ, the mean $\text{E}[S] = 1/\mu$ of the service distribution, and the SCV C_B^2 of the service distribution. (Equivalently, the second moment $\text{E}[S^2]$ or variance $\text{Var}[S]$ of the service distribution can be used in place of the SCV via the relations $C_B^2 = \text{Var}[S]/\text{E}^2[S]$ and $\text{Var}[S] = \text{E}[S^2] - \text{E}^2[S]$.) For a real system, information concerning the service mechanism is often readily available, so these parameters can easily be estimated.

Finally, other measures of effectiveness can easily be obtained from W_q. That is, L_q can be obtained from Little's formula $L_q = \lambda W_q$, W can be obtained from $W = W_q + 1/\mu$, and L can be obtained from either $L = \lambda W$ or $L = L_q + \rho$. Table 5.1 shows several different ways to express the results. The first column expresses the measures using the SCV of the service distribution C_B^2, the second column uses

Table 5.1 Measures of Effectiveness for the $M/G/1$ Queue

$$L_q = \frac{1+C_B^2}{2}\cdot\frac{\rho^2}{1-\rho} = \frac{\lambda^2\mathrm{E}[S^2]}{2(1-\rho)} = \frac{\rho^2+\lambda^2\sigma_B^2}{2(1-\rho)}$$

$$W_q = \frac{1+C_B^2}{2}\cdot\frac{\rho}{\mu-\lambda} = \frac{\lambda\mathrm{E}[S^2]}{2(1-\rho)} = \frac{\rho^2/\lambda+\lambda\sigma_B^2}{2(1-\rho)}$$

$$W = \frac{1+C_B^2}{2}\cdot\frac{\rho}{\mu-\lambda}+\frac{1}{\mu} = \frac{\lambda\mathrm{E}[S^2]}{2(1-\rho)}+\frac{1}{\mu} = \frac{\rho^2/\lambda+\lambda\sigma_B^2}{2(1-\rho)}+\frac{1}{\mu}$$

$$L = \frac{1+C_B^2}{2}\cdot\frac{\rho^2}{1-\rho}+\rho = \frac{\lambda^2\mathrm{E}[S^2]}{2(1-\rho)}+\rho = \frac{\rho^2+\lambda^2\sigma_B^2}{2(1-\rho)}+\rho$$

the second moment of the service distribution $\mathrm{E}[S^2]$, and the third column uses the variance of the service distribution σ_B^2. Collectively, the formulas are equivalent in view of Little's formula and the fundamental properties of the single-server queue. Thus each formula can be referred to as a "Pollaczek–Khintchine" (or PK) formula.

■ EXAMPLE 5.1

Check that the PK formulas for the $M/G/1$ queue are consistent with results for the $M/E_k/1$ queue (Section 3.3.3): The SCV of an E_k distribution is $\mathrm{Var}[S]/\mathrm{E}^2[S] = 1/k$; see (3.18) and (3.19). Thus (5.2) reduces to

$$W_q = \frac{1+1/k}{2}\cdot\frac{\rho}{1-\rho}\cdot\mathrm{E}[S] = \frac{k+1}{2k}\cdot\frac{\rho}{1-\rho}\cdot\mathrm{E}[S],$$

which is the same as (3.22). Similarly, we can obtain results for an $M/D/1$ queue by setting $k=\infty$ in the above equation or $C_B^2 = 0$ in (5.2).

■ EXAMPLE 5.2

Consider a single-server, Poisson-input queue with a mean arrival rate of 10 per hour. Currently, the server works according to an exponential distribution with mean service time of 5 min. Management has a training course that will result in an improvement (decrease) in the variance of the service time but a slight increase in the mean. After completion of the course, it is estimated that the mean service time will increase to 5.5 min but the standard deviation will decrease from 5 min (the exponential case) to 4 min. Management would like to know whether they should have the server undergo further training.

To answer the question, we compare L and W for each case, the first model being $M/M/1$ and the second being $M/G/1$. For $M/M/1$, we can either use

Table 5.2 Comparison of Models

	Present ($M/M/1$)	After Training ($M/G/1$)
L	5	8.625
W	30 min	51.75 min

the results of Section 2.2 or the PK formula (Table 5.1) with $\sigma_B = 1/\mu$; for the $M/G/1$, we use the PK results. The comparisons are presented in Table 5.2.

Thus it is not profitable to have the server "better" trained. In this example, the standard deviation of the service distribution decreases by 20% – twice as much as the increase of the service mean (10%). For this example, the performance is more sensitive to the mean of the service time than to the standard deviation. It is of interest to calculate the reduction in variance required to make up for the increase of 0.5 in the mean. We can do this by solving for σ_B^2 in the PK formula for L:

$$L = 5 = \rho + \frac{\rho^2 + \lambda^2\sigma_B^2}{2(1-\rho)},$$

where $\rho = \lambda/\mu = 10(\frac{11}{2})(\frac{1}{60}) = \frac{11}{12}$. This yields $\sigma_B^2 < 0$, which is not possible. What this means is that L is always greater than 5, even with a service-time variance of 0 (deterministic service times). The minimum value for L, achieved with deterministic service times, turns out to be

$$L = \rho + \frac{\rho^2}{2(1-\rho)} \doteq 6.0.$$

Problem 5.6 asks the reader to find the value of σ_B^2 required to yield the same L if the mean service time were increased to only 5.2 min after training.

5.1.1.2 *Derivation Using Departure Times.* Although we have already obtained formulas for various expected-value measures (Table 5.1), we now give a different derivation that considers the queue at times when customers *depart* from the queue. As we will see in the next section, considering the queue at departure points gives rise to a discrete-time Markov chain. This chain can be used to derive other system properties, such as the steady-state system-size probabilities.

Using this view of the queue, we now derive a formula for the expected system size L. Consider the queue immediately after a customer has departed from the system. Let X_n be the number of customers remaining in the system immediately after the nth customer departs (thus the departing customer is not included in the count). Let A_n be the number of customers who arrive during the service time of the

nth customer. Then, for all $n \geq 1$,

$$X_{n+1} = \begin{cases} X_n - 1 + A_{n+1} & (X_n \geq 1), \\ A_{n+1} & (X_n = 0). \end{cases} \tag{5.3}$$

This can be rewritten as

$$X_{n+1} = X_n - U(X_n) + A_{n+1}, \tag{5.4}$$

where U is the unit step function

$$U(X_n) = \begin{cases} 1 & (X_n > 0), \\ 0 & (X_n = 0). \end{cases}$$

Assuming that a steady-state solution exists and supposing that n is large enough so that the queue is in steady-state, we have $\mathrm{E}[X_{n+1}] = \mathrm{E}[X_n] = L^{(D)}$. $L^{(D)}$ denotes the expected steady-state system size *at departure points* (in contrast, L denotes the expected steady-state system size at arbitrary points in time). Taking the expectation of both sides of (5.4) yields

$$L^{(D)} = L^{(D)} - \mathrm{E}[U(X_n)] + E[A_{n+1}].$$

This implies that

$$\mathrm{E}[U(X_n)] = \mathrm{E}[A_{n+1}].$$

Now, $\mathrm{E}[A_{n+1}]$ can be computed by conditioning on the service time of the $(n+1)$st customer. Let S denote the random service time of this customer. Then,

$$\mathrm{E}[A_{n+1}] = \int_0^\infty \mathrm{E}[A_{n+1}|S = t]\, dB(t) = \int_0^\infty \lambda t\, dB(t) = \lambda \mathrm{E}[S] = \frac{\lambda}{\mu} = \rho.$$

The second equality follows since $\{A_{n+1}|S = t\}$ is a Poisson random variable with mean λt. Thus $\mathrm{E}[U(X_n)] = \mathrm{E}[A_{n+1}] = \rho$. Next, squaring (5.4) gives

$$X_{n+1}^2 = X_n^2 + U^2(X_n) + A_{n+1}^2 - 2X_nU(X_n) - 2A_{n+1}U(X_n) + 2A_{n+1}X_n.$$

Taking expected values and noting that $\mathrm{E}[X_{n+1}^2] = \mathrm{E}[X_n^2]$ gives

$$0 = \mathrm{E}[U^2(X_n)] + \mathrm{E}[A_{n+1}^2] - 2\mathrm{E}[X_nU(X_n)] - 2\mathrm{E}[A_{n+1}U(X_n)] + 2\mathrm{E}[A_{n+1}X_n].$$

Now, $U^2(X_n) = U(X_n)$ and $X_nU(X_n) = X_n$. Also, A_{n+1} is independent of X_n and $U(X_n)$. This is because the number of customers who arrive during the service time of the $(n+1)$st customer (A_{n+1}) is independent of an event that occurs earlier, namely, the number of customers remaining after the nth customer departs (X_n). This gives

$$0 = \mathrm{E}[U(X_n)] + \mathrm{E}[A_{n+1}^2] - 2\mathrm{E}[X_n] - 2\mathrm{E}[A_{n+1}]\mathrm{E}[U(X_n)] + 2\mathrm{E}[A_{n+1}]\mathrm{E}[X_n].$$

Now, we have shown that $\mathrm{E}[U(X_n)] = E[A_{n+1}] = \rho$, so this implies

$$0 = \rho + \mathrm{E}[A_{n+1}^2] - 2L^{(D)} - 2\rho^2 + 2\rho L^{(D)},$$

or

$$L^{(D)} = \frac{\rho - 2\rho^2 + \mathrm{E}[A_{n+1}^2]}{2(1-\rho)}. \tag{5.5}$$

Now,

$$\mathrm{E}[A_{n+1}^2] = \mathrm{Var}[A_{n+1}] + \mathrm{E}^2[A_{n+1}] = \mathrm{Var}[A_{n+1}] + \rho^2.$$

$\mathrm{Var}[A_{n+1}]$ can be computed by conditioning on the service time S of the $(n+1)$st customer and using the conditional variance formula (e.g., Ross, 2007, Proposition 3.1)

$$\mathrm{Var}[A_{n+1}] = \mathrm{E}[\mathrm{Var}[A_{n+1}|S]] + \mathrm{Var}[\mathrm{E}[A_{n+1}|S]].$$

Now, $\{A_{n+1}|S\}$ is a Poisson random variable with mean λS. The variance of this random variable is also λS. Therefore the above equation becomes

$$\mathrm{Var}[A_{n+1}] = \mathrm{E}[\lambda S] + \mathrm{Var}[\lambda S] = \rho + \lambda^2\sigma_B^2, \tag{5.6}$$

where σ_B^2 is the variance of the service-time distribution. Plugging (5.6) into (5.5) gives

$$L^{(D)} = \rho + \frac{\rho^2 + \lambda^2\sigma_B^2}{2(1-\rho)}. \tag{5.7}$$

It can be shown (see Section 5.1.3) that although $L^{(D)}$ is the expected steady-state system size *at departure points*, it is also equal to the expected steady-state system size at an arbitrary point in time, L. So we may write

$$L = \rho + \frac{\rho^2 + \lambda^2\sigma_B^2}{2(1-\rho)}.$$

This is equivalent to the previous result for W_q and was also given in Table 5.1.

5.1.2 Departure-Point System-Size Probabilities

This section treats the development of the steady-state system-size probabilities at departure points. Let π_n represent the probability of n in the system at a departure point (a point of time just slightly after a customer has completed service) after steady state is reached. The probabilities $\{\pi_n\}$ are not in general the same as the steady-state system-size probabilities $\{p_n\}$, which are valid for any arbitrary point of time after steady state is reached. For the $M/G/1$ model, however, it turns out that the set $\{\pi_n\}$ and the set $\{p_n\}$ are identical; this will be verified in a later section in this chapter.

We now show that the $M/G/1$ queue, viewed only at departure times, leads to an imbedded discrete-time Markov chain. (Viewing the queue only at *arrival times* does *not* yield a Markov chain; see Problem 5.2.) Let $t_1, t_2, t_3, \ldots$ be the sequence of departure times from the queue. Let $X_n \equiv X(t_n)$ be the number of customers left in the system immediately after the departure of the customer at time t_n. We previously obtained (5.3) and we repeat that result here:

$$X_{n+1} = \begin{cases} X_n - 1 + A_{n+1} & (X_n \geq 1), \\ A_{n+1} & (X_n = 0), \end{cases} \tag{5.8}$$

where A_{n+1} is the number of customers who arrive during the service time of the $(n+1)$st customer.

To show that $X_1, X_2, \ldots$ is a Markov chain, we must argue that future states of the chain depend only on the present state – more specifically, we must show that given the present state X_n, the future state X_{n+1} is independent of previous states X_{n-1}, $X_{n-2}, \ldots$. To see that this is so, observe from (5.8) that X_{n+1} depends on X_n and A_{n+1}. As long as A_{n+1} is independent of the past states $X_{n-1}, X_{n-2}, \ldots$, then $\{X_n\}$ is a Markov chain. This is true because A_{n+1}, the number of customer who arrive during the service time of the $(n+1)$st customer, depends on the length of this service time, but it does not depend on events that occurred earlier, namely, the queue size at earlier departure points, $X_{n-1}, X_{n-2}, \ldots$. Thus the imbedded discrete-time process $X_1, X_2, \ldots$ is a discrete-time Markov chain.

We now derive the transition probabilities for this Markov chain

$$p_{ij} \equiv \Pr\{X_{n+1} = j | X_n = i\}.$$

The transition probabilities depend on the distribution of the number of customers who arrive during a service time. Since this distribution does not depend on the index of the customer in service, we drop the subscript. Specifically, let S denote a random service time and A denote the random number of customers who arrive during this time. Define

$$k_i \equiv \Pr\{i \text{ arrivals during a service time}\} = \Pr\{A = i\}.$$

$\Pr\{A = i\}$ can be computed by conditioning on the length of the service time:

$$k_i = \Pr\{A = i\} = \int_0^\infty \Pr\{A = i | S = t\}\, dB(t).$$

[For generality, we use the Stieltjes integral here. In cases where the density function exists, $dB(t)$ can be replaced by $b(t)\,dt$, thus yielding the more familiar Riemann integral of elementary calculus.] Now, $\{A | S = t\}$ is a Poisson random variable with mean λt, so

$$\Pr\{A = i | S = t\} = \frac{e^{-\lambda t}(\lambda t)^i}{i!}.$$

Thus

$$k_i = \int_0^\infty \frac{e^{-\lambda t}(\lambda t)^i}{i!}\, dB(t). \tag{5.9}$$

Then, from (5.8),

$$\Pr\{X_{n+1} = j | X_n = i\} = \begin{cases} \Pr\{A = j - i + 1\} & (i \geq 1), \\ \Pr\{A = j\} & (i = 0). \end{cases}$$

In summary, we have the following transition matrix:

$$\boldsymbol{P} = \{p_{ij}\} = \begin{pmatrix} k_0 & k_1 & k_2 & k_3 & \cdots \\ k_0 & k_1 & k_2 & k_3 & \cdots \\ 0 & k_0 & k_1 & k_2 & \cdots \\ 0 & 0 & k_0 & k_1 & \cdots \\ 0 & 0 & 0 & k_0 & \cdots \\ \vdots & \vdots & \vdots & \vdots & \end{pmatrix}. \tag{5.10}$$

Assuming that steady state is achievable, the steady-state probability vector $\boldsymbol{\pi} = \{\pi_n\}$ can be found using the standard theory of Markov chains. In particular, $\{\pi_n\}$ is the solution to the stationary equation (see Section 1.9)

$$\boldsymbol{\pi P} = \boldsymbol{\pi}. \tag{5.11}$$

This yields (Problem 5.8)

$$\boxed{\pi_i = \pi_0 k_i + \sum_{j=1}^{i+1} \pi_j k_{i-j+1} \qquad (i = 0, 1, 2, \ldots).} \tag{5.12}$$

Now define the generating functions

$$\Pi(z) = \sum_{i=0}^{\infty} \pi_i z^i \quad \text{and} \quad K(z) = \sum_{i=0}^{\infty} k_i z^i \qquad (|z| \le 1). \tag{5.13}$$

Multiplying (5.12) by z^i, summing over i, and solving for $\Pi(z)$ yields (Problem 5.9)

$$\Pi(z) = \frac{\pi_0(1-z)K(z)}{K(z) - z}. \tag{5.14}$$

Using the fact that $\Pi(1) = 1$, along with L'Hôpital's rule, and realizing that $K(1) = 1$ and $K'(1) = \lambda(1/\mu)$, we find that

$$\pi_0 = 1 - \rho \qquad (\rho \equiv \lambda \mathrm{E}[\text{service time}]). \tag{5.15}$$

Hence

$$\boxed{\Pi(z) = \frac{(1-\rho)(1-z)K(z)}{K(z) - z}.} \tag{5.16}$$

Since by definition $\Pi'(1)$ is the expected system size, it is possible to derive the PK formula (see Problem 5.10) directly from (5.16).

Equation (5.16) is as far as we can go in obtaining the $\{\pi_n\}$ (which we show, in the next section, are equivalent to the $\{p_n\}$) without making assumptions as to the specific service-time distribution. Problem 5.11 asks the reader to verify that (5.16) reduces to the generating function for $M/M/1$ given by (2.15) when one assumes exponential service. We shall now present an example of an $M/G/1$ system with empirical service times.

■ EXAMPLE 5.3

To illustrate how the results above can be utilized on an *empirically* determined service distribution, consider the following situation. The Bearing Straight Corporation, a single-product company, makes a very specialized plastic bearing. The company is a high-volume producer, and the bearing undergoes a single machining operation, performed on a specialized machine. Because of the volume of sales, the company keeps a large number of machines in operation at all times (we may assume for practical purposes that the machine population is infinite).

Machines break down according to a Poisson process with a mean of 5/h. The company has a single repairman, and the machine characteristics are such that the breakdowns are due to one of two possible malfunctions. Depending on which of the malfunctions caused the breakdown, it takes the repairman either 9 or 12 min to repair a machine. Since he is an expert and the machines are identical, any variation in these service times is minuscule and can be ignored. The type of malfunction that causes a breakdown occurs at random, but it has been observed that one-third of the malfunctions require the 12-min repair time. The company wishes to know the probability that more than three machines will be down at any time.

The service-time mechanism can be viewed as a two-point discrete random variable that equals 9 min with probability 2/3 and 12 min with probability 1/3. The mean service time is $\mathrm{E}[S] = 10$ min and the variance is $\mathrm{Var}[S] = 2\ \mathrm{min}^2$.

If management were interested in only the expected number of machines down, it would be an easy matter to obtain it from the PK formula (Problem 5.12). However, to answer the question asked, it is necessary to find p_0, p_1, p_2, and p_3, since

$$\Pr\{\text{more than 3 machines down}\} = \sum_{i=4}^{\infty} p_i = 1 - \sum_{i=0}^{3} p_i.$$

The probability that no machines are down is

$$p_0 = \pi_0 = 1 - \rho = 1 - \frac{5}{6} = \frac{1}{6}.$$

To find the other $\{p_i\}$, we can apply (5.12) iteratively. However, in this example, we use a different method to illustrate the use of (5.16). The approach is to first calculate the generating function $\Pi(z)$ in (5.16) and then to expand it in a series to obtain $p_i = \pi_i (i = 1, 2, 3)$. To evaluate $\Pi(z)$, we must first find $K(z)$ from (5.13), which necessitates finding k_i as given in (5.9). Since we have a two-point discrete distribution for service, the Stieltjes integral reduces to a summation, and we have from (5.9) that

$$k_i = \frac{2}{3}\left(\frac{e^{-5(3/20)}[5(\frac{3}{20})]^i}{i!}\right) + \frac{1}{3}\left(\frac{e^{-5(1/5)}[5(\frac{1}{5})]^i}{i!}\right)$$
$$= \frac{2}{3i!}e^{-3/4}\left(\frac{3}{4}\right)^i + \frac{1}{3i!}e^{-1}.$$

Thus

$$K(z) = \frac{2}{3}e^{-3/4}\sum_{i=0}^{\infty}\frac{(3z/4)^i}{i!} + \frac{1}{3}e^{-1}\sum_{i=0}^{\infty}\frac{z^i}{i!}.$$

Although we can get $K(z)$ in closed form (since the sums are the infinite series expressions for $e^{3z/4}$ and e^z, respectively), we will keep it in power-series form, since we ultimately desire a power series expansion for $\Pi(z)$ in order to determine the individual probabilities $\{p_i\}$. To make $K(z)$ easier to work with, define

$$c_i = \frac{2}{3}e^{-3/4}\left(\frac{3}{4}\right)^i + \frac{1}{3}e^{-1}, \tag{5.17}$$

so that $K(z)$ can be written as

$$K(z) = \sum_{i=0}^{\infty}\frac{c_i z^i}{i!}.$$

Now from (5.16), we have

$$\Pi(z) = \frac{(1-\rho)(1-z)\sum_{i=0}^{\infty}\frac{c_i z^i}{i!}}{\left(\sum_{i=0}^{\infty}\frac{c_i z^i}{i!}\right) - z},$$

which gives

$$\Pi(z) = \frac{(1-\rho)\left(\sum_{i=0}^{\infty}\frac{c_i z^i}{i!} - \sum_{i=0}^{\infty}\frac{c_i z^{i+1}}{i!}\right)}{c_0 + (c_1 - 1)z + \sum_{i=2}^{\infty}\frac{c_i z^i}{i!}}$$
$$= \frac{(1-\rho)\left[1 + \sum_{i=1}^{\infty}\left(\frac{c_i}{c_0 i!} - \frac{c_{i-1}}{c_0(i-1)!}\right)z^i\right]}{1 + \frac{c_1 - 1}{c_0}z + \sum_{i=2}^{\infty}\frac{c_i z^i}{c_0 i!}}. \tag{5.18}$$

It is necessary to have (5.18) in terms of a power series in z and not the ratio of two power series. For our example, we require coefficients of terms up to and including z^3. These can be obtained by long division, carefully keeping track of the needed coefficients. However, it can be seen by long division that the ratio of two power series is itself a power series, namely,

$$\frac{1+\sum_{i=1}^{\infty} a_i z^i}{1+\sum_{i=1}^{\infty} b_i z^i} = \sum_{i=0}^{\infty} d_i z^i,$$

where d_i can be obtained recursively from

$$d_i = \begin{cases} a_i - \sum_{j=1}^{i} b_j\, d_{i-j} & (i = 1, 2, \ldots), \\ 1 & (i = 0). \end{cases}$$

Thus it is only necessary to obtain d_1, d_2, and d_3, which, when multiplied by $(1-\rho)$, give p_1, p_2, and p_3, respectively. In terms of the $\{c_i\}$, we get

$$\begin{aligned} p_1 &= (1-\rho)\left(\frac{1}{c_0} - 1\right), \\ p_2 &= (1-\rho)\left(\frac{1-c_1}{c_0} - 1\right)\frac{1}{c_0}, \\ p_3 &= (1-\rho)\frac{1}{c_0}\left[\frac{1-c_1}{c_0}\left(\frac{1-c_1}{c_0} - 1\right) - \frac{c_2}{2c_0}\right]. \end{aligned} \tag{5.19}$$

Finally, using (5.17) to evaluate c_0, c_1, and c_2, and then substituting into (5.19), yields

$$p_1 \doteq 0.2143, \qquad p_2 \doteq 0.1773, \qquad p_3 \doteq 0.1293$$

$$\Rightarrow \quad \Pr\{>3 \text{ machines down}\} = 1 - \sum_{n=0}^{3} p_n \doteq 0.3124.$$

One can generate as many p_i as desired by continuing the procedure further.

The results for the two-point service distribution can be generalized to a k-point service distribution using a similar analysis (see Greenberg, 1973). Let b_i, $i = 1, \cdots, k$, be the probability that the service time is t_i. Then, it can be shown (Problem 5.14) that

$$\Pi(z) = \frac{(1-\rho)\left[1 + \sum_{i=1}^{\infty}\left(\frac{c_i}{c_0 i!} - \frac{c_{i-1}}{c_0(i-1)!}\right) z^i\right]}{1 + \frac{c_1 - 1}{c_0} z + \sum_{i=2}^{\infty} \frac{c_i z^i}{c_0 i!}}. \tag{5.20}$$

This is identical to (5.18) except that the $\{c_i\}$ are different (see Problem 5.14). Using the same relationship for the quotient of two power series as given previously, the $\{d_i\}$ and hence the $\{p_i\}$ can be obtained and are identical to those given by (5.19), but with the new $\{c_i\}$.

EXAMPLE 5.4

The Bearing Straight Company of Example 5.3 has been able to tighten up its repair efforts (largely by increased automation) to the point where all repairs can now safely be assumed to take precisely 6 min (λ is still equal to $\frac{1}{12}$/ min). Thus we now have an $M/D/1$ problem, which we can completely solve by the direct use of (5.16).

Assuming that *all* service times are exactly $1/\mu$, we have from (5.9), (5.13), and (5.16) that

$$k_i = \frac{e^{-\rho}\rho^i}{i!} \quad (\rho = \lambda/\mu),$$

$$K(z) = \sum_{i=0}^{\infty} \frac{e^{-\rho}\rho^i}{i!} z^i = e^{-\rho(1-z)},$$

$$\Pi(z) = \frac{(1-\rho)(1-z)e^{-\rho(1-z)}}{e^{-\rho(1-z)} - z} = \frac{(1-\rho)(1-z)}{1 - ze^{\rho(1-z)}}. \tag{5.21}$$

Now, to obtain the individual probabilities, we expand (5.21) as a geometric series, giving

$$\begin{aligned}
\Pi(z) &= (1-\rho)(1-z)\sum_{k=0}^{\infty}\left[ze^{\rho(1-z)}\right]^k \\
&= (1-\rho)(1-z)\sum_{k=0}^{\infty} e^{k\rho(1-z)} z^k \\
&= (1-\rho)(1-z)\sum_{k=0}^{\infty} e^{k\rho}\sum_{m=0}^{\infty}\frac{(-k\rho z)^m}{m!} z^k \\
&= (1-\rho)(1-z)\sum_{k=0}^{\infty}\sum_{m=0}^{\infty} e^{k\rho}(-1)^m \frac{(k\rho)^m}{m!} z^{m+k} \\
&= (1-\rho)(1-z)\sum_{k=0}^{\infty}\sum_{n=k}^{\infty} e^{k\rho}(-1)^{n-k}\frac{(k\rho)^{n-k}}{(n-k)!} z^n.
\end{aligned}$$

We now change the order of summation [this can be verified by graphing the region of summation on the (k, n)-plane]:

$$\Pi(z) = (1-\rho)(1-z)\sum_{n=0}^{\infty}\sum_{k=0}^{n} e^{k\rho}(-1)^{n-k}\frac{(k\rho)^{n-k}}{(n-k)!} z^n.$$

Finally, we complete the multiplication by the factor $1 - z$ and find that

$$\Pi(z) = (1-\rho)\left(\sum_{n=0}^{\infty}\sum_{k=0}^{n} e^{k\rho}(-1)^{n-k}\frac{(k\rho)^{n-k}}{(n-k)!}z^n - \sum_{n=1}^{\infty}\sum_{k=0}^{n-1} e^{k\rho}(-1)^{n-k-1}\frac{(k\rho)^{n-k-1}}{(n-k-1)!}z^n\right). \tag{5.22}$$

We have a power series in z, so p_n is the coefficient in front of z^n. For $n = 0$, $p_0 = 1 - \rho$. For $n = 1$,

$$p_1 = (1-\rho)(e^{\rho} - 1).$$

For $n \geq 2$, the $k = 0$ term of both double summations in (5.22) is always 0 because of $k\rho$, so that both summations on k can start at $k = 1$. Hence we can write the final result as

$$p_n = (1-\rho) \times \left(\sum_{k=1}^{n} e^{k\rho}(-1)^{n-k}\frac{(k\rho)^{n-k}}{(n-k)!} - \sum_{k=1}^{n-1} e^{k\rho}(-1)^{n-k-1}\frac{(k\rho)^{n-k-1}}{(n-k-1)!}\right).$$

The mean system size follows from the PK formula with the variance set equal to 0 and $\rho = \frac{1}{2}$ as

$$L = \rho + \frac{\rho^2}{2(1-\rho)} = \frac{1}{2} + \frac{1}{4} = \frac{3}{4}.$$

5.1.3 Proof that $\pi_n = p_n$

We now show that π_n, the steady-state probability of n in the system at a departure point, is equal to p_n, the steady-state probability of n in the system at an arbitrary point in time. We begin by considering a specific realization of the actual process over a long interval $(0, T)$. Let $X(t)$ be the system size at time t. Let $A_n(t)$ be the number of unit upward jumps or crossings (arrivals) from state n occurring in $(0, t)$. Let $D_n(t)$ be the number of unit downward jumps (departures) to state n in $(0, t)$. Then, since arrivals occur singly and customers are served singly, we must have

$$|A_n(T) - D_n(T)| \leq 1. \tag{5.23}$$

Furthermore, the total number of departures, $D(T)$, relates to the total number of arrivals, $A(T)$, by

$$D(T) = A(T) + X(0) - X(T). \tag{5.24}$$

Now, the departure-point probabilities are

$$\pi_n = \lim_{T\to\infty} \frac{D_n(T)}{D(T)}. \tag{5.25}$$

When we add and subtract $A_n(T)$ from the numerator of (5.25) and use (5.24) in its denominator, we see that

$$\frac{D_n(T)}{D(T)} = \frac{A_n(T) + D_n(T) - A_n(T)}{A(T) + X(0) - X(T)}. \tag{5.26}$$

Since $X(0)$ is finite and $X(T)$ must be too because of the assumption of stationarity, it follows from (5.23), (5.26), and the fact that $A(T) \to \infty$ that

$$\lim_{T\to\infty} \frac{D_n(T)}{D(T)} = \lim_{T\to\infty} \frac{A_n(T)}{A(T)} \tag{5.27}$$

with probability one. Since the arrivals occur at the points of a Poisson process operating independently of the state of the process, we invoke the PASTA property that Poisson arrivals find time averages. Therefore the general-time probability p_n is identical to the arrival-point probability $q_n = \lim_{T\to\infty}[A_n(T)/A(T)]$, which is, in turn, equal to the departure-point probability from (5.27). Thus all three sets of probabilities are equal for the $M/G/1$ problem, and the desired result is shown.

5.1.4 Ergodic Theory

To find conditions for the existence of the steady state for the $M/G/1$ imbedded chain, we first obtain a sufficient condition from the theory of Markov chains and the previously shown equality of departure-point and general-time probabilities. Then we show that this sufficient condition is also necessary by the direct use of the generating function $\Pi(z)$ given by (5.14).

The Markov-chain proof relies heavily on two well-known theorems presented in Section 1.9 as Theorems 1.1 and 1.2. The former says that a discrete-time Markov chain has identical limiting and stationary distributions if it is irreducible, aperiodic, and positive recurrent, while the latter gives a sufficient condition for the positive recurrence of an irreducible and aperiodic chain.

The behavior of any single-channel queueing system is, at least, a function of the input parameters and the service-time distribution $B(t)$. We expect, as with the $M/M/1$ queue, that the existence of ergodicity depends on the value of the utilization factor ρ. The transition matrix $\boldsymbol{P}$ that characterizes the imbedded chain of the $M/G/1$ is given by (5.10) with

$$\begin{aligned} k_n &= \Pr\{n \text{ arrivals during a service time } S = t\} \\ &= \int_0^\infty \frac{e^{-\lambda t}(\lambda t)^n}{n!}\, dB(t), \end{aligned}$$

with $\int t\, dB(t) = \mathrm{E}[S]$. The problem is to determine a simple sufficient condition under which the $M/G/1$ has a steady state. Our past experience suggests that we try $\rho = \lambda \mathrm{E}[S] < 1$.

We show in the following that the $M/G/1$ imbedded chain is indeed irreducible, aperiodic, and positive recurrent when $\rho < 1$, and hence, by Theorem 1.1, possesses a long-run distribution under this condition. The chain is clearly irreducible, since any state can be reached from any other state. It can next be observed directly from the transition matrix that Pr{passing from state k to state k in one transition} $= p_{kk}^{(1)} > 0$ for all k, and therefore the period of each k, defined as the greatest common divisor of the set of all n such that $p_{kk}^{(n)} > 0$, is one. Hence the system is aperiodic. Neither irreducibility nor aperiodicity depends on ρ, but each is an inherent property of this chain. In fact, it is the positive recurrence that depends on the value of ρ. It thus remains for us to show that the chain is positive recurrent when $\rho < 1$, and we could then apply the theorem to infer the existence of the steady-state distribution.

To obtain the required result, we employ Theorem 1.2 of Section 1.9.6 and show, using Foster's method (Foster, 1953), that the necessary inequality has the required solution when $\rho < 1$. An educated guess at this required solution is

$$x_j = \frac{j}{1-\rho} \quad (j \geq 0).$$

We now show that this guess for x_j satisfies the condition for Theorem 1.2. From the matrix $\boldsymbol{P}$,

$$\begin{aligned}
\sum_{j=0}^{\infty} p_{ij}x_j &= \sum_{j=i-1}^{\infty} k_{j-i+1}\left(\frac{j}{1-\rho}\right) \\
&= \frac{k_0(i-1)}{1-\rho} + \frac{k_1 i}{1-\rho} + \frac{k_2(i+1)}{1-\rho} + \cdots \\
&= \frac{k_0(i-1)}{1-\rho} + \frac{k_1(i-1)}{1-\rho} + \frac{k_2(i-1)}{1-\rho} + \cdots + \frac{k_1}{1-\rho} + \frac{2k_2}{1-\rho} + \cdots \\
&= (i-1)\sum_{j=0}^{\infty}\frac{k_j}{1-\rho} + \sum_{j=1}^{\infty}\frac{jk_j}{1-\rho} \\
&= \frac{i-1}{1-\rho} + \sum_{j=1}^{\infty}\frac{jk_j}{1-\rho}.
\end{aligned}$$

Now,

$$\begin{aligned}
\sum_{j=1}^{\infty} jk_j &= \sum_{j=1}^{\infty} j\frac{1}{j!}\int_0^{\infty}(\lambda t)^j e^{-\lambda t}\,dB(t) \\
&= \int_0^{\infty} e^{-\lambda t}\left(\sum_{j=1}^{\infty}\frac{1}{(j-1)!}(\lambda t)^j\right) dB(t) \\
&= \int_0^{\infty} e^{-\lambda t}\lambda t e^{\lambda t}\,dB(t) \\
&= \int_0^{\infty}\lambda t\,dB(t) = \lambda \mathrm{E}[S] = \rho. \qquad (5.28)
\end{aligned}$$

So,

$$\sum_{j=0}^{\infty} p_{ij}x_j = \frac{i-1+\rho}{1-\rho} = x_i - 1 \quad (x_i \geq 0, \quad \text{since } 1-\rho > 0).$$

[This implies that the expected single-step displacement of the chain from state $i > 0$ is $\rho - 1 < 0$. This is because $\sum jp_{ij}$ is the mean destination state starting from i, and $\sum jp_{ij} = (1-\rho)\sum p_{ij}x_j = i - (1-\rho)$.]

Also,

$$\sum_{j=0}^{\infty} p_{0j}x_j = \sum_{j=0}^{\infty} k_j x_j = \sum_{j=0}^{\infty} \frac{jk_j}{1-\rho} = \frac{\rho}{1-\rho} < \infty.$$

Hence it follows that the chain possesses identical stationary and long-run distributions when $\rho < 1$.

The proof of the necessity of $\rho < 1$ for ergodicity arises directly from the existence of the generating function $\Pi(z)$ over the interval $|z| < 1$,

$$\Pi(z) = \frac{\pi_0(1-z)K(z)}{K(z)-z}.$$

We know that $\Pi(1)$ must be equal to one; hence

$$\begin{aligned} 1 &= \lim_{z\to 1} \Pi(z) \\ &= \pi_0 \frac{-K(1)}{K'(1)-1} \qquad \text{(using L'Hôpital's rule)} \\ &= \pi_0 \frac{-1}{\rho - 1} \qquad (\rho = \lambda \mathrm{E}[S]). \end{aligned}$$

But $\pi_0 > 0$ and therefore $\rho - 1 < 0$; thus $\rho < 1$ is necessary and sufficient for steady state.

5.1.5 Waiting Times

In this section we wish to present an assortment of important results concerning the delay times. It has already been shown that the average system wait is related to the average system size by Little's formula $W = L/\lambda$. A natural thing to require then is a possible relationship either between higher moments or between distribution functions [or equivalently between Laplace–Stieltjes transforms (LSTs)]. It turns out that this can be done with some extra effort.

To begin, we note that the stationary probability for the $M/G/1$ can always be written in terms of the waiting-time CDF as

$$p_n = \pi_n = \frac{1}{n!}\int_0^{\infty} (\lambda t)^n e^{-\lambda t}\, dW(t) \quad (n \geq 0),$$

since the system size under FCFS will equal n at an arbitrary departure point if there have been n (Poisson) arrivals during the departure's system wait. If we multiply this equation by z^n, sum on n, and define the usual generating function, then it is found that

$$P(z) = \sum_{n=0}^{\infty} p_n z^n = \int_0^\infty e^{-\lambda t} \sum_{n=0}^{\infty} \frac{(\lambda t z)^n}{n!} dW(t)$$

$$= \int_0^\infty e^{-\lambda t(1-z)} dW(t) = W^*[\lambda(1-z)], \qquad (5.29)$$

where $W^*(s)$ is the LST of $W(t)$. The succession of moments of system size and delay can now be easily related to each other by repeated differentiation of the equality $P(z) = W^*[\lambda(1-z)]$. We therefore have by the chain rule that

$$\frac{d^k P(z)}{dz^k} = (-1)^k \lambda^k \left. \frac{d^k W^*(u)}{du^k} \right|_{u=\lambda(1-z)}$$

$$= (-1)^k \lambda^k (-1)^k \mathrm{E}\left[T^k e^{-Tu}\right] |_{u=\lambda(1-z)}.$$

Hence if $L_{(k)}$ is used to denote the kth factorial moment of the system size and W_k the regular kth moment of the system waiting time, then

$$L_{(k)} = \left. \frac{d^k P(z)}{dz^k} \right|_{z=1} = \lambda^k W_k. \qquad (5.30)$$

This result provides a nice generalization of Little's formula, since the higher ordinary moments can be obtained from the factorial moments.

In the $M/M/1$ queue we were able to easily obtain a simple formula for the waiting-time distribution in terms of the service-time distribution (see Section 2.2.4 and Problem 2.5), namely,

$$W(t) = (1-\rho) \sum_{n=0}^{\infty} \rho^n B^{(n+1)}(t), \qquad (5.31)$$

where $B(t)$ is the exponential CDF and $B^{(n+1)}(t)$ is its $(n+1)$st convolution. The derivation of this result required the memoryless property of the exponential service, since the arrivals catch the server in the midst of a serving period with probability equal to ρ. However, we have now lost the memoryless property and therefore require an alternative approach to derive a comparable result for $M/G/1$.

To do so, we begin by deriving a simple relationship between the LSTs of the service and the waiting times, $B^*(s)$ and $W^*(s)$, respectively. From (5.29) we know that $P(z) = W^*[\lambda(1-z)]$, and from (5.16) that

$$P(z) = \Pi(z) = \frac{(1-\rho)(1-z)K(z)}{K(z)-z}.$$

But, from (5.13) and (5.9),

$$K(z) = \int_0^\infty e^{-\lambda t} \sum_{n=0}^\infty \frac{(\lambda t z)^n}{n!}\, dB(t)$$
$$= \int_0^\infty e^{-\lambda t(1-z)}\, dB(t) = B^*[\lambda(1-z)]. \tag{5.32}$$

Putting these three equations together, we find that

$$W^*[\lambda(1-z)] = \frac{(1-\rho)(1-z)B^*[\lambda(1-z)]}{B^*[\lambda(1-z)] - z},$$

or

$$\boxed{W^*(s) = \frac{(1-\rho)sB^*(s)}{s - \lambda[1 - B^*(s)]}.} \tag{5.33}$$

But, from the convolution property of transforms, $W^*(s) = W_q^*(s)B^*(s)$, since $T = T_q + S$. Thus

$$\boxed{W_q^*(s) = \frac{(1-\rho)s}{s - \lambda[1 - B^*(s)]}.} \tag{5.34}$$

Expanding the right side as a geometric series [since $(\lambda/s)[1 - B^*(s)] < 1$], gives

$$W_q^*(s) = (1-\rho)\sum_{n=0}^\infty \left(\frac{\lambda}{s}[1 - B^*(s)]\right)^n$$
$$= (1-\rho)\sum_{n=0}^\infty \left(\rho\frac{\mu}{s}[1 - B^*(s)]\right)^n.$$

It can be seen that $\mu[1 - B^*(s)]/s$ is the LST of the residual-service-time distribution

$$R(t) \equiv \mu \int_0^t [1 - B(x)]\, dx.$$

Intuitively, $R(t)$ is the CDF of the remaining service time of the customer being served at the time of an arbitrary arrival, given that the arrival occurs when the server is busy. The formula for $R(t)$ can be derived using renewal theory (e.g., Ross, 2007). It follows that

$$W_q^*(s) = (1-\rho)\sum_{n=0}^\infty [\rho R^*(s)]^n,$$

which yields, after term-by-term inversion utilizing the convolution property, a result surprisingly similar to (5.31), namely,

$$W_q(t) = (1-\rho)\sum_{n=0}^\infty \rho^n R^{(n)}(t). \tag{5.35}$$

This says that if time is reordered with this remaining service time as the fundamental unit, any arrival in the steady state finds n such time units of potential service in front of it with probability $(1-\rho)\rho^n$, giving a result remarkably like that for the $M/M/1$ queue. We can also use these results to derive an extension of the PK formula to relate iteratively the higher moments of the wait (call them $W_{q,k}$). First, rewrite the basic transform equation as

$$W_q^*(s)\{s-\lambda[1-B^*(s)]\}=(1-\rho)s.$$

Next, we take the kth derivative ($k>1$) of the previous equation, applying Leibniz's rule for the derivative of a product:

$$\sum_{i=0}^{k}\binom{k}{i}\left(\frac{d^{k-i}W_q^*(s)}{ds^{k-i}}\right)\left(\frac{d^i\{s-\lambda[1-B^*(s)]\}}{ds^i}\right)=\frac{d^k[(1-\rho)s]}{ds^k}.$$

We may assume $k>1$, since it is the higher moments we seek. Then the right-hand side of the above equation is zero, and thus

$$0=\frac{d^kW_q^*(s)}{ds^k}\{s-\lambda[1-B^*(s)]\}+k\frac{d^{k-1}W_q^*(s)}{ds^{k-1}}[1+\lambda B^{*\prime}(s)]$$
$$+\sum_{i=2}^{k}\binom{k}{i}\frac{d^{k-i}W_q^*(s)}{ds^{k-i}}\lambda\frac{d^iB^*(s)}{ds^i}.$$

Now set $s=0$, thus giving

$$0=k(-1)^{k-1}W_{q,k-1}(1-\rho)+\sum_{i=2}^{k}\binom{k}{i}(-1)^{k-i}W_{q,k-i}\mathrm{E}[S^i](-1)^i,$$

or

$$W_{q,k-1}=\frac{\lambda}{k(1-\rho)}\sum_{i=2}^{k}\binom{k}{i}W_{q,k-i}\mathrm{E}[S^i].$$

We may rewrite this slightly by letting $K=k-1$ and $j=i-1$, from which we get the more familiar form

$$W_{q,K}=\frac{\lambda}{1-\rho}\sum_{j=1}^{K}\binom{K}{j}W_{q,K-j}\frac{\mathrm{E}[S^{j+1}]}{j+1}.$$

■ **EXAMPLE 5.5**

Let us illustrate some of these results by going back to the Bearing Straight Corporation of Example 5.3. Bearing Straight has determined that it loses \$5000 per hour that a machine is down, and that an additional penalty must be incurred because of the possibility of an excessive number of machines being down. It is decided to cost this variability at \$10,000 × (standard deviation of customer delay). Under such a total-cost structure, what is the total cost of their policy, using the parameters indicated in Example 5.3 and assuming that repair labor is a sunk cost?

Problem 5.12 asks us to show that $L \doteq 2.96$. But we also need the variance of the system waits T, where we know using (5.30) that

$$\mathrm{E}[N(N-1)] \equiv L_{(2)} = \lambda^2 W_2.$$

Thus

$$\mathrm{Var}[T] = W_2 - W^2 = \frac{L_{(2)}}{\lambda^2} - \frac{L^2}{\lambda^2}.$$

To get $L_{(2)}$, the second derivative of $P(z)$ is found from (5.18) and then evaluated at $z = 1$ to be 14.50 (see Problem 5.19). Therefore $\mathrm{Var}[T] = (14.50 - 8.75)/25 = 0.23\ \mathrm{h}^2$. Thus the total cost of Bearing's policy computes as $C = 5000L + 10{,}000\sqrt{0.23} \doteq$ \$19,596/h.

5.1.6 Busy-Period Analysis

The determination of the distribution of the busy period for an $M/G/1$ queue is a somewhat more difficult matter than finding that of the $M/M/1$, particularly in view of the fact that the service-time CDF must be carried as an unknown. But it is not too much of a task to find the LST of the busy-period CDF, from which we can easily obtain any number of moments.

To begin, let $G(x)$ denote the CDF of the busy period X of an $M/G/1$ with service CDF $B(t)$. Then we condition X on the length of the first service time inaugurating the busy period. Since each arrival during that service time will contribute to the busy period by having arrivals come during its service time, we can look at each arrival during the first service time of the busy period as essentially generating its own busy period. Thus we can write

$$\begin{aligned} G(x) &= \int_0^x \Pr\{\text{busy period generated by all arrivals during } t \leq x - t \mid \\ &\qquad\qquad \text{first service time} = t\}\, dB(t) \\ &= \int_0^x \sum_{n=0}^{\infty} \frac{e^{-\lambda t}(\lambda t)^n}{n!} G^{(n)}(x-t)\, dB(t), \end{aligned} \tag{5.36}$$

where $G^{(n)}(x)$ is the n-fold convolution of $G(x)$. Next, let

$$G^*(s) = \int_0^\infty e^{-sx}\, dG(x)$$

be the LST of $G(x)$, and $B^*(s)$ be the LST of $B(t)$. Then, by taking the transform of both sides of (5.36), it is found that

$$G^*(s) = \int_0^\infty \int_0^x \sum_{n=0}^{\infty} e^{-sx} \frac{e^{-\lambda t}(\lambda t)^n}{n!} G^{(n)}(x-t)\, dB(t)\, dx.$$

Changing the order of integration gives

$$\begin{aligned} G^*(s) &= \int_0^\infty \int_t^\infty \sum_{n=0}^\infty e^{-sx} \frac{e^{-\lambda t}(\lambda t)^n}{n!} G^{(n)}(x-t)\,dx\,dB(t) \\ &= \int_0^\infty \sum_{n=0}^\infty \frac{e^{-\lambda t}(\lambda t)^n}{n!} \int_t^\infty e^{-sx} G^{(n)}(x-t)\,dx\,dB(t). \end{aligned}$$

Applying a change of variables $y = x - t$ to the inner integral gives

$$G^*(s) = \int_0^\infty \sum_{n=0}^\infty \frac{e^{-\lambda t}(\lambda t)^n}{n!} \left(e^{-st} \int_0^\infty e^{-sy} G^{(n)}(y)\,dy \right) dB(t).$$

By the convolution property,

$$\begin{aligned} G^*(s) &= \int_0^\infty \sum_{n=0}^\infty \frac{e^{-\lambda t}(\lambda t)^n}{n!} e^{-st} [G^*(s)]^n\,dB(t) \\ &= \int_0^\infty e^{-\lambda t} e^{\lambda t G^*(s)} e^{-st}\,dB(t) \\ &= B^*[s + \lambda - \lambda G^*(s)]. \end{aligned} \tag{5.37}$$

Hence the mean length of the busy period is found as

$$\mathrm{E}[X] = -\left.\frac{dG^*(s)}{ds}\right|_{s=0} \equiv -G^{*\prime}(0),$$

where

$$G^{*\prime}(s) = B^{*\prime}[s + \lambda - \lambda G^*(s)][1 - \lambda G^{*\prime}(s)].$$

Therefore

$$\mathrm{E}[X] = -B^{*\prime}[\lambda - \lambda G^*(0)][1 - \lambda G^{*\prime}(0)] = -B^{*\prime}(0) \cdot \{1 + \lambda \mathrm{E}[X]\},$$

or

$$\mathrm{E}[X] = -\frac{B^{*\prime}(0)}{1 + \lambda B^{*\prime}(0)}.$$

But $B^{*\prime}(0) = -1/\mu$; hence

$$\mathrm{E}[X] = \frac{1/\mu}{1 - \lambda/\mu} = \frac{1}{\mu - \lambda},$$

which, surprisingly, is exactly the same result we obtained earlier for $M/M/1$. However, we note that the proof given for $M/M/1$ in Section 2.12 is perfectly valid for the $M/G/1$ model, since no use is made of the exponentiality of service.

5.1.7 Finite $M/G/1$ Queues

The analysis of the finite-capacity $M/G/1/K$ queue proceeds in a way very similar to that of the unlimited-waiting-room case. Let us thus examine each of the main results of $M/G/1/\infty$ for applicability to $M/G/1/K$.

The PK formula will not hold now, since the expected number of (joined) arrivals during a service period must be conditioned on the system size. The best way to get the new result is directly from the steady-state probabilities, since there are now only a finite number of them.

The single-step transition matrix must here be truncated at $K-1$, so that

$$\boldsymbol{P} = \{p_{ij}\} = \begin{pmatrix} k_0 & k_1 & k_2 & \cdots & 1-\sum_{n=0}^{K-2} k_n \\ k_0 & k_1 & k_2 & \cdots & 1-\sum_{n=0}^{K-2} k_n \\ 0 & k_0 & k_1 & \cdots & 1-\sum_{n=0}^{K-3} k_n \\ 0 & 0 & k_0 & \cdots & 1-\sum_{n=0}^{K-4} k_n \\ \vdots & \vdots & \vdots & & \vdots \\ 0 & 0 & 0 & \cdots & 1-k_0 \end{pmatrix},$$

which implies that the stationary equation is

$$\pi_i = \begin{cases} \pi_0 k_i + \sum_{j=1}^{i+1} \pi_j k_{i-j+1} & (i = 0, 1, 2, \ldots, K-2), \\ 1 - \sum_{j=0}^{K-2} \pi_j & (i = K-1). \end{cases}$$

These K (consistent) equations in K unknowns can then be solved for all the probabilities, and the average system size at points of departure is thus given by $L = \sum_{i=0}^{K-1} i\pi_i$. (Note that the maximum state of the Markov chain is not K, since we are observing just after a departure. We assume $K > 1$ because otherwise the resultant model $M/G/1/1$ is just a special case of the $M/G/c/c$ to be discussed shortly in Section 5.2.2.)

The first portion of the stationary equation is identical to that of the unlimited $M/G/1$. Therefore the respective stationary probabilities $\{\pi_i\}$ for $M/G/1/K$ and $\{\pi_i^*\}$ for $M/G/1/\infty$ must be at worst proportional for $i \le K-1$; that is, $\pi_i = C\pi_i^*$, $i = 0, 1, \ldots, K-1$. The usual condition that the probabilities sum to one implies that $C = 1/\sum_{i=0}^{K-1} \pi_i^*$.

Furthermore, we note that the probability distribution for the system size encountered by an arrival will be different from $\{\pi_i\}$, since now the state space must be enlarged to include K. Let q'_n then denote the probability that an arriving customer finds a system with n customers. (Here we are speaking about the distribution of arriving customers whether or not they join the queue, as opposed to only those arrivals who join, denoted by q_n. The q'_n distribution is often of interest in its own right.) Now, if we go back to Section 5.1.3 and the proof that $\pi_n = p_n$ for $M/G/1$, it is noted in that proof, essentially (5.27), that the distribution of system sizes just prior to points of arrival (our $\{q_n\}$) is identical to the departure-point probabilities (here $\{\pi_n\}$) as long as arrivals occur singly and service is not in bulk. Such is also the case with q'_n, except that the state spaces are different. This difference is easily taken care of by first noting that (5.27) is really saying that

$$\begin{aligned}\pi_n &= \Pr\{\text{arrival finds } n|\text{customer does in fact join}\}\\ &= q_n = \frac{q'_n}{1-q'_K} \quad (0 \le n \le K-1).\end{aligned}$$

Therefore

$$q'_n = (1-q'_K)\pi_n \quad (0 \le n \le K-1).$$

To get q'_K now, we use an approach similar to one mentioned earlier for simple Markovian models in Section 2.5, where we equate the effective arrival rate with the effective departure rate; that is,

$$\lambda(1-q'_K) = \mu(1-p_0).$$

Therefore

$$\begin{aligned}q'_n &= \frac{(1-p_0)\pi_n}{\rho} \quad (0 \le n \le K-1),\\ q'_K &= \frac{\rho - 1 + p_0}{\rho}.\end{aligned}$$

But since the original arrival process is Poisson, $q'_n = p_n$ for all n. Thus

$$q'_0 = p_0 = \frac{(1-p_0)\pi_0}{\rho} \Rightarrow p_0 = \frac{\pi_0}{\pi_0+\rho},$$

and finally

$$q'_n = \frac{\pi_n}{\pi_0+\rho}.$$

5.1.8 Some Additional Results

In this section we present some assorted additional results for $M/G/1$ queues, including priorities, impatience, output, transience, finite source, and batching.

With respect to priorities, we have already obtained results for the general-service, nonpreemptive, many-priority $M/G/1$ queue; see Section 3.4.2. Specifically, we obtained the expected queue wait for each class of customer in (3.45).

With respect to impatience, one can easily introduce balking into the $M/G/1$ queue by prescribing a probability b that an arrival decides to actually join the system. Then the true input process becomes a (filtered) Poisson process with mean $b\lambda t$, and (5.9) thus has to be written as

$$k_n = \int_0^\infty \frac{e^{-b\lambda t}(b\lambda t)^n}{n!}\, dB(t).$$

But the rest of the analysis goes through parallel to that for the regular $M/G/1$ queue, with the probability of idleness, p_0, now equal to $1 - b\lambda/\mu$. For a more comprehensive treatment of impatience in $M/G/1$, the reader is referred to Rao (1968), who treats both balking and reneging.

As far as output is concerned, we have already shown in Chapter 4 that the steady-state $M/M/1$ has Poisson output, but we would now like to know whether there are any other $M/G/1/\infty$ queues that also possess this property. The answer is in fact no (except for the pathological case where service is 0 for all customers), and this follows from the simple observation that such queues are never reversible, so that their output processes cannot probabilistically match their inputs.

But what is the distribution of an arbitrary $M/G/1$ interdeparture time in the steady state? To derive this, define $C(t)$ as the CDF of the interdeparture times; then, for $B(t)$ equal to the service CDF, it follows that

$$\begin{aligned} C(t) &\equiv \Pr\{\text{interdeparture time} \le t\} \\ &= \Pr\{\text{system experienced no idleness during interdeparture period}) \\ &\quad \times \Pr\{\text{interdeparture time} \le t \mid \text{no idleness}\} \\ &\quad + \Pr\{\text{system experienced some idleness during interdeparture period}\} \\ &\quad \times \Pr\{\text{interdeparture time} \le t \mid \text{some idleness}\} \\ &= \rho B(t) + (1-\rho)\int_0^t B(t-u)\lambda e^{-\lambda u}\, du, \end{aligned}$$

since the length of an interdeparture period with idleness is the sum of the idle time and service time. Problem 5.25(b) asks the reader to show from the above equation that the exponentiality of $C(t)$ implies the exponentiality of $B(t)$.

Remember that the fact that the $M/M/1$ is the only $M/G/1$ with exponential output has serious negative implications for the solution of series models since we have shown that the output of a first stage will be exponential, which we would like it to be, only if it is $M/M/1$. However, small $M/G/1$ series problems can be handled numerically by appropriate utilization of the above formula for the CDF of the interdeparture times, $C(t)$.

By putting a capacity restriction on the $M/G/1$ queue at $K = 1$, it can be seen that such queues also have IID interdeparture times. This is because the successive departure epochs are identical to the busy cycles, which are found as the sums of each idle time paired with an adjacent service time.

To get any transient results for the $M/G/1$ queue, we appeal directly to the theory of Markov chains and the Chapman–Kolmogorov equation

$$p_j^{(m)} = \sum_k p_k^{(0)} p_{kj}^{(m)},$$

where $p_j^{(m)}$ is then the probability that the system state is in state j just after the mth customer has departed. The necessary matrix multiplications must be done with some caution, since we are dealing with an $\infty \times \infty$ matrix in the unlimited waiting room case. But this can indeed be done by carefully truncating the transition matrix at an appropriate point where the entries have become very small (e.g., Neuts, 1973).

To close this section, we make brief mention of two additional problem types, namely, finite-source and bulk queues. The finite-source $M/G/1$ is essentially the machine-repair problem with arbitrarily distributed repair times and has been solved in the literature, again using an imbedded-Markov chain approach. Any interested reader is referred to Takács (1962) for a fairly detailed discussion of these kinds of problems. The bulk-input $M/G/1$, denoted by $M^{[X]}/G/1$, and the bulk-service $M/G/1$, denoted by $M/G^{[Y]}/1$, can also be solved with the use of Markov chains. The bulk-input model is presented in the next section, but the bulk-service problem is quite a bit more messy and therefore is not treated in this book. However, the reader is referred to Prabhu (1965a) for the details of this latter model.

5.1.9 The Bulk-Input Queue ($M^{[X]}/G/1$)

The $M^{[X]}/G/1$ queueing system can be described in the following manner.

1. Customers arrive as a Poisson process with parameter λ in groups of random size C, where C has the distribution

$$\Pr\{C = n\} = c_n \quad (n > 1)$$

and the generating function (which will be assumed to exist) is

$$C(z) = \mathrm{E}[z^C] = \sum_{n=1}^{\infty} c_n z^n \quad (|z| \le 1).$$

The probability that a total of n customers arrive in an interval of length t is thus given by

$$p_n(t) = \sum_{k=0}^{n} e^{-\lambda t} \frac{(\lambda t)^k}{k!} c_n^{(k)} \quad (n \ge 0),$$

where $\{c_n^{(k)}\}$ is the k-fold convolution of $\{c_n\}$ with itself (i.e., the n arrivals form a compound Poisson process),

$$c_n^{(0)} \equiv \begin{cases} 1 & (n = 0), \\ 0 & (n > 0). \end{cases}$$

2. The customers are served singly by one server on a FCFS basis.

3. The service times of the succession of customers are IID random variables with CDF $B(t)$ and LST $B^*(s)$.

Let us make a slight change here from $M/G/1$ which will help us later: The imbedded chain we shall use will be generated by the points (therefore called regeneration points) at which either a departure occurs *or* an idle period is ended. This process will be called $\{X_n, n = 1, 2, \ldots | X_n =$ number of customers in the system immediately after the nth regeneration point$\}$, with transition matrix given by

$$\begin{pmatrix} 0 & c_1 & c_2 & \cdots \\ k_0 & k_1 & k_2 & \cdots \\ 0 & k_0 & k_1 & \cdots \\ 0 & 0 & k_0 & \cdots \\ \vdots & \vdots & \vdots & \end{pmatrix},$$

where

$$k_n = \Pr\{n \text{ arrivals during a full service period}\}$$

$$= \int_0^\infty p_n(t)\, dB(t) = \int_0^\infty \sum_{k=0}^{n} e^{-\lambda t} \frac{(\lambda t)^k}{k!} c_n^{(k)}\, dB(t) = p_{i,n+i-1} \quad (i > 0).$$

Application of Theorem 1.2 (Section 1.9.6) in a fashion similar to that of Section 5.1.4 for the $M/G/1$ queue shows that this chain is ergodic and hence possesses identical long-run and stationary distributions when

$$\rho \equiv \sum_{n=1}^{\infty} n k_n = \text{E[arrivals during a service time]} = \frac{\lambda \text{E}[C]}{\mu} < 1.$$

If the steady-state distribution $\{\pi_i\}$ is to exist for the chain then it is the solution of the system (for all $j \geq 0$)

$$\sum_{i=0}^{\infty} p_{ij}\pi_i = \pi_j \quad \text{and} \quad \sum_{i=0}^{\infty} \pi_i = 1.$$

(Keep in mind that these $\{\pi_i\}$ are slightly different from those that would have resulted had we restricted ourselves to departure points only, as we did for $M/G/1$.) From the transition matrix (as in Section 5.1.2),

$$\pi_j = c_j \pi_0 + \sum_{i=1}^{j+1} k_{j-i+1}\pi_i \quad (c_0 \equiv 0).$$

If the above stationary equation is multiplied by z^j and then summed on j, it is found that

$$\sum_{j=0}^{\infty} \pi_j z^j = \sum_{j=0}^{\infty} c_j \pi_0 z^j + \sum_{j=0}^{\infty} \sum_{i=1}^{j+1} k_{j-i+1}\pi_i z^j.$$

If the usual generating functions are defined as

$$\Pi(z) = \sum_{j=0}^{\infty} \pi_j z^j \quad \text{and} \quad K(z) = \sum_{j=0}^{\infty} k_j z^j,$$

then we find after reversing the order of summation in the final term that

$$\begin{aligned}\Pi(z) &= \pi_0 C(z) + \frac{K(z)}{z} \sum_{i=1}^{\infty} \pi_i z^i \\ &= \pi_0 C(z) + \frac{K(z)}{z}[\Pi(z) - \pi_0],\end{aligned}$$

or

$$\Pi(z) = \frac{\pi_0[K(z) - zC(z)]}{K(z) - z}.$$

Furthermore, it can be shown for $|z| \leq 1$ that

$$\begin{aligned}K(z) &= \sum_{j=0}^{\infty} \int_0^{\infty} \sum_{k=0}^{j} e^{-\lambda t} \frac{(\lambda t)^k}{k!} c_j^{(k)} \, dB(t) z^j \\ &= \int_0^{\infty} e^{-\lambda t} \sum_{k=0}^{j} \frac{(\lambda t)^k}{k!} [C(z)]^k \, dB(t) \\ &= \int_0^{\infty} e^{-\lambda t + \lambda t C(z)} \, dB(t) = B^*[\lambda - \lambda C(z)].\end{aligned}$$

These results have all been derived without much difficulty in a manner similar to the approach for $M/G/1$. However, there now exists a problem that we have not faced before, namely, that the results derived for the imbedded Markov chain do not directly apply to the total general-time stochastic process, $\{X(t), t \geq 0 | X(t) =$ number in the system at time $t\}$. In order to relate the general-time steady-state probabilities $\{p_n\}$ to $\{\pi_n\}$, we must appeal to some results from semi-Markov processes, which are presented in Chapter 6, Section 6.4; a reference on this subject for anyone with further interest in this material is Heyman and Sobel (1982).

5.1.10 Departure-Point State Dependence, Decomposition, and Server Vacations

In Section 2.9 we treated queues with state dependences such that the mean service rate was a function of the number of customers in the system. Whenever the number in the system changed (arrival or departure), the mean service rate would itself change accordingly. But in many situations, it might not be possible to change the service rate at any time a new arrival may come, but rather only upon the initiation of a service (or, almost equivalently, at the conclusion of a service time). For example, in many cases where the service is a human–machine operation and the machine is capable of running at various speeds, the operator would set the speed only prior to the actual commencement of service. Once service had begun, the speed could not be changed until that service was completed, for to do otherwise would necessitate stopping work to alter the speed setting, and then restarting and/or repositioning the work. Furthermore, stopping the operation prior to completion might damage the unit. This type of situation, where the mean service rate can be adjusted only prior to commencing service or at a customer departure point, is what we refer to as departure-point state dependence and is considered in this section.

We assume the state-dependent service mechanism is as follows. Let $B_i(t)$ be the service-time CDF of a customer who enters service after the most recent departure left i customers behind, and

$$\begin{aligned} k_{ni} &= \Pr\{n \text{ arrivals during a service time} \mid \\ &\qquad\qquad i \text{ in the system at latest departure}\} \\ &= \int_0^\infty e^{-\lambda t} \frac{(\lambda t)^n}{n!}\, dB_i(t). \end{aligned} \tag{5.38}$$

Then the transition matrix is given as

$$\boldsymbol{P} = \{p_{ij}\} = \begin{pmatrix} k_{00} & k_{10} & k_{20} & k_{30} & \cdots \\ k_{01} & k_{11} & k_{21} & k_{31} & \cdots \\ 0 & k_{02} & k_{12} & k_{22} & \cdots \\ 0 & 0 & k_{03} & k_{13} & \cdots \\ \vdots & \vdots & \vdots & \vdots & \end{pmatrix}. \tag{5.39}$$

A sufficient condition for the existence of a steady-state solution (see Crabill, 1968) is

$$\limsup\{\rho_j \equiv \lambda \mathrm{E}[S_j] < 1\}, \tag{5.40}$$

which says that all but a finite number of the ρ_j must be less than 1. Thus assuming this condition is met, we can find the steady-state probability distribution by solving the usual stationary equation $\boldsymbol{\pi P} = \boldsymbol{\pi}$. Although this gives the departure-point state probabilities, we have shown in Section 5.1.3 for non-state-dependent service that these are equivalent to the general-time probabilities. The addition of state dependence of service does not alter the proof in any way, so that here, also, the departure-point and general-time state probabilities are equivalent.

Thus the use of the stationary equation (5.11) results in

$$p_j = \pi_j = \pi_0 k_{j,0} + \pi_1 k_{j,1} + \pi_2 k_{j-1,2} + \pi_3 k_{j-2,3} + \cdots + \pi_{j+1} k_{0,j+1} \quad (j \ge 0). \tag{5.41}$$

Again, we define the generating functions

$$\Pi(z) = \sum_{j=0}^{\infty} \pi_j z^j \quad \text{and} \quad K_i(z) = \sum_{j=0}^{\infty} k_{ji} z^j.$$

Then multiplying both sides of (5.41) by z^j and summing over all j, we get

$$\Pi(z) = \pi_0 K_0(z) + \pi_1 K_1(z) + \pi_2 z K_2(z) + \cdots + \pi_{j+1} z^j K_{j+1}(z) + \cdots. \tag{5.42}$$

This is as far as we are able to proceed in general. No closed-form expression for $\Pi(z)$ in terms of the $K_i(z)$ is obtainable, so we now present a specific case for $B_i(t)$ to illustrate the typical solution procedure and give another specific case as an exercise (see Problem 5.24).

Consider the case where customers beginning a busy period get exceptional service at a rate μ_0 unequal to the rate μ offered to everybody else. Thus we have

$$B_n(t) = \begin{cases} 1 - e^{-\mu_0 t} & (n = 0), \\ 1 - e^{-\mu t} & (n > 0). \end{cases}$$

From (5.38), it follows that

$$\begin{aligned} k_{n0} &= \int_0^\infty \frac{e^{-\lambda t}(\lambda t)^n}{n!} \mu_0 e^{-\mu_0 t}\, dt = \frac{\mu_0 \lambda^n}{(\lambda + \mu_0)^{n+1}}, \\ k_n &= \int_0^\infty \frac{e^{-\lambda t}(\lambda t)^n}{n!} \mu e^{-\mu t}\, dt = \frac{\mu \lambda^n}{(\lambda + \mu)^{n+1}}. \end{aligned} \tag{5.43}$$

Now using (5.42), with $K(z) \equiv K_i(z)$ for $i > 0$, we thus have that

$$\Pi(z) = \pi_0 K_0(z) + \frac{K(z)}{z}[\Pi(z) - \pi_0],$$

so that

$$\Pi(z) = \frac{\pi_0[zK_0(z) - K(z)]}{z - K(z)}. \tag{5.44}$$

To get π_0, we take the limit of $\Pi(z)$ as $z \to 1$, recognizing that $K(1) = K_0(1) = 1$, $K_0'(1) = \rho_0 = \lambda/\mu_0$, and $K'(1) = \rho = \lambda/\mu$, and then use L'Hôpital's rule to get

$$1 = \frac{\pi_0(1 + \rho_0 - \rho)}{1 - \rho} \quad \Rightarrow \quad \pi_0 = \frac{1 - \rho}{1 + \rho_0 - \rho}.$$

The remaining probabilities may be found by iteration on the stationary equation (5.41), and we find after repeated calculations and verification by induction that [see Problem 5.24(b)]

$$\pi_n = \left(\frac{\rho_0^n}{(\rho_0 + 1)^n} + \sum_{k=0}^{n-1} \frac{\rho_0^{n-k} \rho^{k+1}}{(\rho_0 + 1)^{n-k}} \right) \pi_0.$$

Now, particularly interesting things happen to the above state-dependent service model when the probability generating function $K_0(z)$ can be expressed as the product $K(z)D(z)$, where $D(z)$ is also a probability generating function defined over the nonnegative integers (see Harris and Marchal, 1988). Then it follows from (5.44) that

$$\Pi(z) = \frac{\pi_0 K(z)[1 - zD(z)]}{K(z) - z}.$$

What happens now is that an appropriate rewriting leads to a *decomposition* of $\Pi(z)$ itself into the product of two generating functions as

$$\Pi(z) = \frac{\pi_0(1-z)K(z)}{K(z)-z} \times \frac{1-zD(z)}{1-z}. \tag{5.45}$$

We immediately notice that the first factor on the right-hand side is the precise probability generating function for the steady-state system size of the $M/G/1$, as given in (5.14) (although the exact value of the $M/G/1$'s π_0 will not be the same as that of the state-dependent model). Now, it turns out that the second factor can also be algebraically shown to be a legitimate probability generating function. Hence it follows from the product decomposition of $\Pi(z)$ in (5.45) that the queue's system sizes are the sum of two random variables defined on the nonnegative integers, the first coming from an ordinary $M/G/1$, while the second is associated with a random variable having generating function proportional to $[1-zD(z)]/[1-z]$.

We can connect the above development to a particular type of *server-vacation* queue. Suppose that, after each busy period is concluded, the server takes a vacation away from its work. If upon return from vacation the server finds that the system is still empty, it goes away again, and so on. Therefore it follows that k_{i0} is the probability that $i+1$ arrivals occur during the service time combined with the final vacation, with probability generating function

$$K_0(z) = K(z) \times \frac{K_V(z)}{z}, \tag{5.46}$$

where $K_V(z)$ is the generating function for the number of arrivals during an arbitrary vacation time. Thus the role of the $D(z)$ of (5.45) is played by the quotient $K_V(z)/z$, and the second factor on the right-hand side is now $[1-K_V(z)]/[1-z]$. By analogy with the LST of the residual service time of Section 5.1.5, as we show later, this latter quotient is proportional to the generating function of the number of arrivals coming in during a residual vacation time. So, taken altogether, we are claiming that the system state of the vacation queue is the sum of the size of an ordinary, nonvacation $M/G/1$ and the number of arrivals coming in during a random residual vacation time.

Now we verify the claim that $[1-K_V(z)]/[1-z]$ is proportional to the generating function of the number of arrivals coming in during a residual vacation time, where $K_V(z)$ is the generating function of (5.13) based on a k_n value derived using the vacation distribution function $V(t)$ in place of the service-time CDF $B(t)$ in (5.9). From Section 5.1.5, the residual distribution function associated with the vacation CDF $V(t)$ is given as

$$R_V(t) = \frac{1}{\overline{v}} \int_0^t [1-V(x)]\,dx,$$

where $\overline{v}$ is the mean vacation length. Then, from the results for the ordinary $M/G/1$ queue in Section 5.1.5, including (5.32) and the material leading to (5.35), we see that

$$K_V(z) = V^*[\lambda(1-z)] \quad \text{and} \quad R_V^*(s) = \frac{1-V^*(s)}{\overline{v}s}.$$

Therefore the probability generating function for the number of arrivals during a residual vacation time may be written as

$$K_{R-V}(z) = \frac{1 - V^*[\lambda(1-z)]}{\overline{v}\lambda(1-z)} = \frac{1 - K_V(z)}{\overline{v}\lambda(1-z)}.$$

The result is thus shown.

■ **EXAMPLE 5.6**

The Hugh Borum Company has a production process that involves drilling holes into castings. The interarrival times of the castings at the single drill press were found to be governed closely by an exponential distribution with mean 4 min, while there was such a variety of hole dimensions that service times also appeared to follow exponentiality, with mean 3 min. However, the very high speed at which the press operated necessitated a cooling-off period, which was done whenever the queue emptied, for a fixed amount of time equal to 2 min. Management wanted to know what penalty they paid in lengthening the average system size because of the breather that the equipment had to take.

Since we have shown in (5.45) that the generating function of the system size is the product of two distinct generating functions, it follows that the system size is the sum of two independent random variables, with the first coming from an ordinary $M/G/1$ ($G = M$ in this problem) and the second equal to the number of arrivals during a residual vacation time. Thus it follows that the mean system size is the sum of the $M/M/1$ mean $\rho/(1-\rho) = 3$ and the mean number expected to arrive during a residual vacation time. Since the vacation time is the constant 2 min, it follows that the mean remaining vacation time at a random point is merely $2/2 = 1$ min. Thus the required answer for the Hugh Borum management is $\lambda \times 1 = 0.25$ customers, a value with which they are quite comfortable in view of the base average of 3 customers.

5.1.11 Level-Crossing Methods

This section gives a brief introduction to level-crossing methods. We apply level crossing to derive the steady-state PDF of queue wait in several variants of the $M/G/1$ queue. Level crossing is useful for analyzing state-dependent queues (Brill, 2008).

To illustrate basic concepts, consider a stable $M/G/1$ queue having Poisson arrivals with rate λ, general service times denoted by the random variable S, and traffic intensity $\rho = \lambda \mathrm{E}[S] < 1$.

Let $W_q(t)$ denote the *virtual* wait of the queue at the instant $t \geq 0$. The virtual wait is the waiting time that a virtual customer *would* experience if such a customer arrived at time t. Figure 5.1 shows a sample path of $W_q(t)$. At actual customer arrival points, the virtual wait $W_q(t)$ increases by a positive jump. The value of the jump is equal to the service time of the arriving customer. This is because a virtual customer who arrives just after an actual customer experiences a delay in queue equal

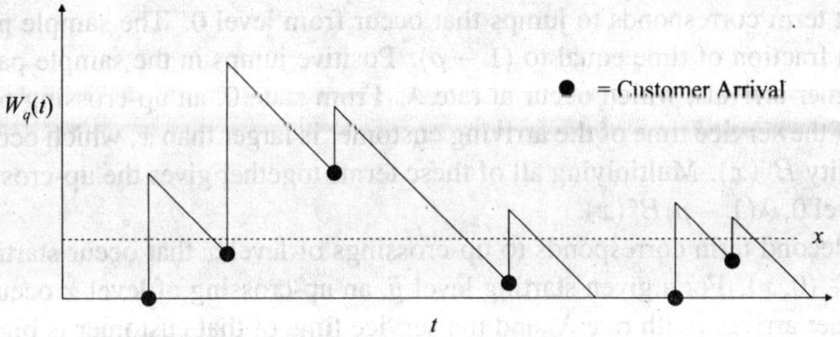

Figure 5.1 Sample path for virtual queue wait.

to the delay in queue of the arriving customer plus the service time of that customer. At nonarrival points, the virtual wait decreases continuously with time, since the customer in service is getting closer to completing service.

Let $W_{q,n}$ denote the *actual* queue wait of the nth arrival. Because of Poisson arrivals, the steady-state distributions of the virtual queue wait and actual queue wait are identical in the $M/G/1$ queue (Takács, 1962). That is,

$$\lim_{t\to\infty} \Pr\{W_q(t) \le x\} = \lim_{n\to\infty} \Pr\{W_{q,n} \le x\}.$$

Let x denote a horizontal level of the virtual wait process. A sample path makes a *crossing* of level x when the sample path down-crosses or up-crosses the level at a point in time. For the $M/G/1$ queue, down-crossings occur at instants when the virtual wait decays below the level x. Up-crossings occur via jumps at instants when an actual customer arrives to the queue. (A slight exception occurs at the level $x = 0$. Since the sample path cannot go below 0, we say that a down-crossing occurs when the system enters state 0 from above and then stays there. An up-crossing occurs whenever the system leaves state 0 due to a customer arrival.)

The basic idea in level crossing is that the long-term rates of up-crossings and down-crossings must be equal. Since every up-crossing is followed by a down-crossing (assuming a stable queue), the long-run rates are equal. The basic approach in level crossing, then, is to determine the rates of up-crossings and down-crossings for an arbitrary level $x \ge 0$ and to equate these rates. Drawing a typical sample path is usually helpful in this process.

For the $M/G/1$ queue, a down-crossing occurs when $W_q(t)$ is just above the level x. The fraction of time that $W_q(t) \in [x, x + \Delta x]$ is approximately $w_q(x)\Delta x$, where $w_q(x)$ is the PDF of the steady-state distribution of the queue wait (where the PDF exists). That is, for $x > 0$,

$$w_q(x) \equiv \frac{d}{dx}\left(\lim_{t\to\infty} \Pr\{W_q(t) \le x\}\right) = \frac{d}{dx}\left(\lim_{n\to\infty} \Pr\{W_{q,n} \le x\}\right).$$

Thus the down-crossing rate is simply $w_q(x)$. The up-crossing rate for level x is determined by the distribution of jump sizes. The up-crossing rate of level x is

$$\lambda(1-\rho)B^c(x) + \lambda \int_0^x B^c(x-y) w_q(y)\, dy.$$

The first term corresponds to jumps that occur from level 0. The sample path is in state 0 a fraction of time equal to $(1-\rho)$. Positive jumps in the sample path occur at customer arrivals, which occur at rate λ. From state 0, an up-crossing of state x occurs if the service time of the arriving customer is larger than x, which occurs with probability $B^c(x)$. Multiplying all of these terms together gives the up-crossing rate from level 0, $\lambda(1-\rho)B^c(x)$.

The second term corresponds to up-crossings of level x that occur starting from level $y \in (0, x)$. For a given starting level y, an up-crossing of level x occurs when a customer arrives (with rate λ) and the service time of that customer is bigger than $(x-y)$, which occurs with probability $B^c(x-y)$. Since the virtual queue wait is at level y according to the PDF $w_q(y)$, the final result is the integral of these terms over $y \in (0, x)$.

Setting the down-crossing rate equal to the up-crossing rate gives the following integral equation:

$$w_q(x) = \lambda(1-\rho)B^c(x) + \lambda \int_0^x B^c(x-y)w_q(y)\,dy \qquad x > 0. \tag{5.47}$$

This integral equation can be solved numerically for practically all $B(t)$. Specifically, the right side of (5.47) depends on $w_q(y)$ for $y \in (0, x)$. The basic approach is to successively build up sample values of $w_q(y)$ for $y \in (0, x)$. These values can then be used to approximate the integral of $w_q(x + \Delta x)$ in the next iteration. Technically, $w_q(x)$ does not exist for $x = 0$, but we can start the procedure by letting $w_q(0) = \lambda(1-\rho)$, which is $\lim_{x\to 0} w_q(x)$ in (5.47) [assuming $B^c(0) = 1$].

Now, we already have derived (in principle) the steady-state queue-wait distribution for the $M/G/1$ queue. However, this was done in terms of Laplace–Stieltjes transforms; see (5.34). In other words, we derived the LST of the queue-wait, but not the actual distribution itself. Obtaining the actual distribution requires inversion of a Laplace transform, which can be done numerically using techniques discussed in Section 8.2.

In summary, both (5.47) and (5.34) provide implicit solutions for the waiting-time distribution. Equation (5.47) provides an integral equation which can be solved numerically. Equation (5.34) provides a solution for the transform that can then be inverted numerically. Finally, (5.47) can be used to derive (5.34) (Problem 5.26). We now apply level-crossing techniques to queues not studied yet.

■ **EXAMPLE 5.7**

This example considers an $M/G/1$ queue with waiting-time-dependent service. We suppose that the service distribution varies depending on the waiting time process. We have previously considered models where the service distribution depends on the number in the system (Sections 2.9 and 5.1.10), so this is a slight variation on that theme. Such a queue can be used to model effects where the server increases the service rate in response to higher delays in the queue. More specifically, let T_q denote the random wait in queue for a customer in steady state. Suppose that the random service time S for this customer has the following CDF:

$$B_y(x) \equiv \Pr\{S \le x | T_q = y\}.$$

The level-crossing equation for this queue is similar to (5.47), but with the wait-dependent service distribution replacing the usual service distribution.

$$w_q(x) = \lambda p_0 B_0^c(x) + \lambda \int_0^x B_y^c(x-y) w_q(y) dy, \qquad x > 0,$$

where p_0 is the fraction of time the system is empty.

A special case of this model is where the first customer of every busy period receives specialized service. For example, perhaps there is a start-up time to begin service after an idle period. Let S_0 be a random service time of a customer arriving to an empty system and let $B_0(t)$ be its CDF. Let S_1 be a random service time of a customer arriving to a nonempty system and let $B_1(t)$ be its CDF. Then the previous equation reduces to

$$w_q(x) = \lambda p_0 B_0^c(x) + \lambda \int_0^x B_1^c(x-y) w_q(y) dy, \qquad x > 0.$$

The value p_0 can be found by integrating with respect to x:

$$\int_0^\infty w_q(x)\, dx = \int_0^\infty \lambda p_0 B_0^c(x)\, dx + \int_0^\infty \int_0^x \lambda B_1^c(x-y) w_q(y) dy\, dx.$$

Using the fact that $\int_0^\infty w_q(x)\, dx = 1 - p_0$ gives

$$\begin{aligned} 1 - p_0 &= \lambda p_0 E[S_0] + \int_0^\infty \int_y^\infty \lambda B_1^c(x-y) w_q(y) dx\, dy \\ &= \lambda p_0 \mathrm{E}[S_0] + \int_0^\infty \lambda w_q(y) \mathrm{E}[S_1]\, dy \\ &= \lambda p_0 \mathrm{E}[S_0] + \lambda(1-p_0)\mathrm{E}[S_1]. \end{aligned}$$

Thus,

$$p_0 = \frac{1-\rho_1}{1-\rho_1+\rho_0},$$

where $\rho_0 = \lambda \mathrm{E}[S_0]$ and $\rho_1 = \lambda \mathrm{E}[S_1]$. Given the value for p_0, the complete PDF $w_q(x)$ can be computed via numerical integration. In a similar manner, multiplying the level crossing equation by x and integrating for $x \in (0, \infty)$ gives a modified version of the PK formula

$$W_q = \frac{\lambda p_0 \mathrm{E}[S_0^2] + \lambda(1-p_0)\mathrm{E}[S_1^2]}{2(1-\rho_1)}.$$

■ **EXAMPLE 5.8**

This example considers an $M/G/1$ queue with balking. A customer, upon arriving to the queue, can decide to wait in line and receive full service or can decide to leave (balk). The decision of whether to balk or join the queue is based on the virtual wait of the system at the time of the arrival. This is slightly different from the discussion in Section 5.1.8, which considers balking based on the number in the system. The balking decision here is given by the following function:

$$R(y) \equiv \Pr\{\text{arriving customer leaves} \mid \text{required wait in queue is } y\}.$$

If $R(y) = 0$, then the model reduces to the standard $M/G/1$ queue. We assume that there is no balking upon arrival to an empty system, so $R(0) = 0$. Furthermore, we assume that $R(y)$ is a nondecreasing function. This model implicitly assumes that customers know exactly how long they will wait upon arriving to the system. In many practical settings, an arriving customer can count the number of people waiting in line, but does not know the exact service times of those customers in advance, so cannot predict the exact wait in queue. The level-crossing equation for $w_q(x)$ is

$$w_q(x) = \lambda p_0 B^c(x) + \lambda \int_0^x B^c(x - y) R^c(y) w_q(y) dy, \qquad x > 0,$$

where $R^c(y) \equiv 1 - R(y)$. As before, the left side is the down-crossing rate of level x. The right side is the up-crossing rate of level x, where the first term corresponds to jumps starting from level $y = 0$ and the second term corresponds to jumps starting from level $y \in (0, x)$. When the system is empty, each arrival generates a jump starting at level 0 having CDF $B(\cdot)$. When the virtual queue wait is at level $y > 0$, each arrival generates a jump having CDF $B(\cdot)$ with probability $R^c(y)$. There is no jump if an arrival balks, and this occurs with probability $R(y)$.

5.2 General Service, Multiserver ($M/G/c/\cdot$, $M/G/\infty$)

We begin here with an immediate disadvantage from the point of view of being able to derive necessary results, since $M/G/c/\infty$ and the $M/G/c/K$ loss-delay system do not possess imbedded Markov chains in the usual sense. This is so at departure points because the number of arrivals during any interdeparture period is dependent on more than just the system size at the immediate departure predecessor, due to the presence of multiple servers. There are, however, some special $M/G/c$'s that do possess enough structure to get fairly complete results, including, of course, $M/M/c$. Another such example is the $M/D/c$, which will be taken up in detail in Chapter 6. What we wish then to do in this section is to try to get some general results for both $M/G/c/\infty$ and $M/G/c/c$, which can easily be applied by merely specifying G.

5.2.1 Some Results for $M/G/c/\infty$

For $M/G/c/\infty$, the main general result that may be found is a line version of the relationship between the kth factorial moment of system size and the regular kth moment of system delay given by (5.30), namely,

$$L_{(k)} = \lambda^k W_k.$$

The proof of this result follows.

To begin, we know from our earlier work on waiting times for $M/G/1$ (Section 5.1.5) that

$$\pi_n = \Pr\{n \text{ in system just after a departure}\} = \frac{1}{n!}\int_0^\infty (\lambda t)^n e^{-\lambda t}\, dW(t).$$

But this equation is also valid in modified form for $M/G/c$ if we consider everything in terms of the queue and not the system. Then it is true that

$$\pi_n^q \equiv \Pr\{n \text{ in queue just after a departure}\} = \frac{1}{n!}\int_0^\infty (\lambda t)^n e^{-\lambda t}\, dW_q(t),$$

with the mean queue length at departure points $L_q^{(D)}$ given by

$$L_q^{(D)} = \sum_{n=1}^{\infty} n\pi_n^q = \int_0^\infty \lambda t\, dW_q(t) = \lambda W_q,$$

which is Little's formula. Now suppose we use $L_{q(k)}^{(D)}$ to denote the kth factorial moment of the departure-point queue size. Then

$$\begin{aligned} L_{q(k)}^{(D)} &= \sum_{n=1}^{\infty} n(n-1)\cdots(n-k+1)\pi_n^q \\ &= \int_0^\infty dW_q(t) \sum_{n=1}^{\infty} \frac{n(n-1)\cdots(n-k+1)(\lambda t)^n e^{-\lambda t}}{n!}. \end{aligned}$$

But the summand is nothing more than the kth factorial moment of the Poisson, which can be shown to be $(\lambda t)^k$. Hence

$$L_{q(k)}^{(D)} = \lambda^k W_{q,k}, \tag{5.48}$$

where $W_{q,k}$ is the ordinary kth moment of the line waiting time. This is now a generalization of (5.30) to line waits for an $M/G/c$.

5.2.2 The $M/G/\infty$ Queue and $M/G/c/c$ Loss System

We begin this section with a derivation of two key results for the $M/G/\infty$ model, namely, the transient distribution for the number of customers in the system at time t

(as we did for $M/M/\infty$), and the transient distribution for the number of customers who have completed service by time t, that is, the departure counting process. To start, let the overall system-size process be called $N(t)$, the departure process $Y(t)$, and the input process $X(t) = Y(t) + N(t)$. By the laws of conditional probability, we find that

$$\Pr\{N(t) = n\} = \sum_{i=n}^{\infty} \Pr\{N(t) = n | X(t) = i\} \frac{e^{-\lambda t}(\lambda t)^i}{i!},$$

since the input is Poisson. The probability that a customer who arrives at time x will still be present at time t is given by $1 - B(t - x)$, $B(u)$ being the service-time CDF. Hence it follows that the probability that an arbitrary one of these customers is still in service is given by

$$q_t = \int_0^t \Pr\{\text{service time} > t - x | \text{arrival at time } x\} \Pr\{\text{arrival at } x\}\, dx.$$

Since the input is Poisson, $\Pr\{\text{arrival at } x\}$ is uniform on (0, t); hence it is 1/t, from (1.16). Thus

$$q_t = \frac{1}{t} \int_0^t [1 - B(t - x)]\, dx = \frac{1}{t} \int_0^t [1 - B(x)]\, dx$$

and is independent of any other arrival. Therefore by the binomial law,

$$\Pr\{N(t) = n | X(t) = i\} = \binom{i}{n} q_t^n (1 - q_t)^{i-n} \quad (n \geq 0),$$

and the transient distribution is

$$\begin{aligned} \Pr\{N(t) = n\} &= \sum_{i=n}^{\infty} \binom{i}{n} \frac{q_t^n (1 - q_t)^{i-n} e^{-\lambda t} (\lambda t)^i}{i!} \\ &= \frac{(\lambda q_t t)^n e^{-\lambda t}}{n!} \sum_{i=n}^{\infty} \frac{[\lambda t(1 - q_t)]^{i-n}}{(i - n)!} \\ &= \frac{(\lambda q_t t)^n e^{-\lambda t} e^{\lambda t - \lambda q_t t}}{n!} = \frac{(\lambda q_t t)^n e^{-\lambda q_t t}}{n!}, \end{aligned}$$

namely, nonhomogeneous Poisson with mean $\lambda q_t t$.

To derive the equilibrium solution, take the limit as $t \to \infty$ of this transient answer. It is thereby found that

$$\lim_{t \to \infty} (\lambda q_t t) = \lambda \int_0^\infty [1 - B(x)]\, dx = \frac{\lambda}{\mu},$$

and hence the equilibrium solution is Poisson with mean $\lambda \mathrm{E}[S] = \lambda/\mu$. We make special note of this result and its similarity to the steady-state probabilities we derived in Chapter 2 for the $M/M/\infty$—the importance of this observation will become more apparent later in our discussion of the $M/G/c/c$ problem.

The distribution of the departure-counting process $Y(t)$ can be found by exactly the same argument as above, using $1 - q_t = \int_0^t B(x)dx/t$ instead of q_t. The result, as expected, is

$$\Pr\{Y(t) = n\} = \frac{[\lambda(1 - q_t)t]^n e^{-\lambda(1-q_t)t}}{n!}.$$

In the limit as $t \to \infty$, we see that q_t goes to zero, and thus the interdeparture process is Poisson in the steady state, which is precisely the same as the arrival process.

Note how this output result compares with our discussion in Chapter 4, where we showed that the output of an $M/M/c$ is Poisson for any value of c. Furthermore, in Section 5.1.8 we pointed out the converse result that $M/M/1$ is the only $M/G/1$ with Poisson output [Problem 5.25(b)]. More generally, we can say that $M/M/c$ is the only $M/G/c$ with Poisson output, although we must now, as a result of the calculations of this section, add the caveat that c must be finite.

Now, getting back to $M/G/c/c$, we wish to look into a result that we quoted earlier in Section 2.6. It is the surprising fact that the steady-state system-size distribution given by (2.39), namely, the truncated Poisson

$$p_n = \frac{(\lambda/\mu)^n/n!}{\sum_{i=0}^{c}(\lambda/\mu)^i/i!} \qquad (0 \le n \le c),$$

is valid for *any* $M/G/c/c$, independent of the form of G. The specific value from this for p_c, as noted in Section 2.6, is called Erlang's loss or B formula, and, as noted just above in this section, the result extends to $M/G/\infty$, where $p_n = e^{-\lambda/\mu}(\lambda/\mu)^n/n!$ for any form of G. We now sketch a proof of the general assertion for $M/G/c/c$.

The $c = 1$ case (i.e., $M/G/1/1$) is very simple, since we have essentially already noted this fact with the observation (from Section 2.5) that $p_0 = 1 - \rho_{\text{eff}} = 1 - p_1$ for any $G/G/1/1$ queue. Since $\rho_{\text{eff}} = \lambda(1 - p_1)/\mu$, it follows that

$$p_1 = \frac{\lambda/\mu}{1 + \lambda/\mu}, \quad p_0 = \frac{1}{1 + \lambda/\mu},$$

which is precisely the result needed.

For the general problem $c > 1$, the formula follows from a set of more complex observations involving reversibility, product-form solutions, and multidimensional Markov processes, much as was done in Chapter 4 for the modeling of Jackson, product-form networks. In order to invoke these theories, it is critical to define a Markov process in the context of the $M/G/c/c$, since, as we have already noted, such systems are not Markovian by definition. This is done by expanding the model state space from n to the multidimensional vector $(n, u_1, u_2, \ldots, u_c)$, where $0 \le u_1 \le u_2 \le \cdots \le u_c$ are the c ordered service ages (i.e., completed service times so far, ranked smallest to largest, recognizing that $u_1, \ldots, u_{c-n}$ will be zero when the system state is n). That such a vector is Markovian can be seen from the fact that its future state is clearly a function of only its current position, with change over infinitesimal intervals depending on the fact that the instantaneous probability of a service completion when the service age is u depends solely on u. Then, altogether, it turns out that this augmented process is *reversible*—we have already had a major hint that this is true from our earlier observation that the output process of an $M/G/\infty$ is Poisson. More detailed discussions of the precise nature of the reversibility can be found in Ross (1996) and Wolff (1989).

From the reversibility, then, comes the critical observation that the limiting joint distribution of $(n, u_1, u_2, \ldots, u_c)$ has the proportional product form

$$p_n(u_1, u_2, \ldots, u_c) = Ca_n\bar{B}(u_1)\bar{B}(u_2)\cdots\bar{B}(u_c)/n!, \tag{5.49}$$

where $u_1, \ldots, u_{c-n}$ are zero, C is proportional to the zero-state probability, a_n is proportional to the probability that n servers are busy, $B(u_i)$ is the service-time distribution function [thus $\bar{B}(u)$ is the probability that a service time is at least equal to u], and $n!$ accounts for all of the rearrangements possible for the n order statistics within the $\{u_1, \ldots, u_c\}$ corresponding to the n busy servers. From the boundary conditions of the problem and the Poisson input assumption, we see that

$$\begin{aligned} C \equiv p_n(0,0,\ldots,0) &= \lambda p_{n-1}(0,0,\ldots,0) = \lambda[\lambda p_{n-2}(0,0,\ldots,0)] \\ &= \cdots = \lambda^n p_0(0,0,\ldots,0) = \lambda^n p_0. \end{aligned}$$

Thus it follows from (5.49) that this is a completely separable product form, since we know from Section 5.1.5 that $\mu\bar{B}(u)$ is a legitimate density function (viz., of the residual service time) and each such term will independently integrate out to 1. The resultant marginal system-size probability function is

$$p_n = p_0\frac{(\lambda/\mu)^n}{n!}.$$

It then follows that

$$p_0 = \left(\sum_{n=0}^{c}\frac{(\lambda/\mu)^n}{n!}\right)^{-1}$$

and

$$p_n = \frac{(\lambda/\mu)^n/n!}{\sum_{i=0}^{c}(\lambda/\mu)^i/i!} \qquad (0 \le n \le c),$$

which is precisely what we got for $M/M/c/c$ in Chapter 2.

Note that the fact that the steady-state probabilities for the $M/G/c/c$ are insensitive to the choice of G means that these probabilities will always satisfy the $M/M/c/c$ birth–death recursion

$$\lambda p_n = (n+1)\mu p_{n+1} \qquad (n = 0, \ldots, c-1).$$

It also turns out that we can retain the insensitivity with respect to the service distribution G even when the arrival process is generalized from a Poisson to a state-dependent birth process with rate λ_n depending on the system state (Wolff, 1989). In this case, the left-hand side of the birth–death recursion is merely changed to $\lambda_n p_n$.

We summarize in Table 5.3 the results we have just discussed on the insensitivity of $M/G/c/K$ models to the form of the service-time distribution G.

Table 5.3 Insensitivity Results for $M/G/c/K$ Models

$M/G/c/c$ vs. $M/M/c/c$	Steady-state probabilities and output process independent of form of G
$M/G/\infty$ vs. $M/M/\infty$	Steady-state probabilities and output process independent of form of G
$M/G/c/\infty$ vs. $M/M/c/\infty$	Output processes equal if and only if $G \equiv M$

5.3 General Input ($G/M/1$, $G/M/c$)

In this section, we consider queues where service times are assumed to be exponential and no specific assumption is made concerning the arrival pattern other than that successive interarrival times are IID. For this case, results can be obtained for c parallel servers using an analysis similar to that for the $c = 1$ case with a slight increase in complexity in certain calculations. We first consider $c = 1$ and then generalize to c servers.

5.3.1 Single-Server $G/M/1$ Queue

We first examine a single-server queue. Service times are assumed to be exponential with mean $1/\mu$. Interarrival times follow a general distribution and are assumed to be IID. The queue is assumed to be operating in steady state.

As with the $M/G/1$ queue, an imbedded-Markov-chain approach is used to obtain results. Following the type of analysis used in Section 5.1, we now consider the system at *arrival* times (for the $M/G/1$ queue, we considered the system at *departure* times). Let X_n denote the number of customers in the system just prior to the arrival of the nth customer. Then,

$$X_{n+1} = X_n + 1 - B_n \quad (B_n \le X_n + 1, \quad X_n \ge 0),$$

where B_n is the number of customers served during the interarrival time $T^{(n)}$ between the nth and $(n+1)$st arrivals. Since the interarrival times are assumed to be IID, the random variable $T^{(n)}$ can be denoted by T, and we denote its CDF by $A(t)$. Furthermore, the random variable B_n does not depend on the past history of the queue, given the current number in the system X_n at the time of the nth arrival. Thus $\{X_0, X_1, X_2, \ldots\}$ is a Markov chain.

Define the following probabilities:

$$b_k \equiv \int_0^\infty \frac{e^{-\mu t}(\mu t)^k}{k!}\, dA(t). \tag{5.50}$$

The interpretation is that b_k is the probability that there are exactly k service completions between two consecutive arrivals (given that there are at least k in the system just prior to the first arrival). That is, $b_k = \Pr\{B_n = k | X_n \geq k\}$.

Then, the imbedded, single-step transition probability matrix for the Markov chain $\{X_0, X_1, X_2, \ldots\}$ is

$$\mathbf{P} = \{p_{ij}\} = \begin{pmatrix} 1 - b_0 & b_0 & 0 & 0 & 0 & \cdots \\ 1 - \sum_{k=0}^{1} b_k & b_1 & b_0 & 0 & 0 & \cdots \\ 1 - \sum_{k=0}^{2} b_k & b_2 & b_1 & b_0 & 0 & \cdots \\ \vdots & \vdots & \vdots & \vdots & \vdots & \end{pmatrix}. \tag{5.51}$$

Assuming that a steady-state solution exists (this will be taken up later) and denoting the probability vector that an arrival finds n in the system by $\boldsymbol{q} = \{q_n\}, n = 0, 1, 2, \ldots$, we have the usual stationary equations

$$\boldsymbol{q}\mathbf{P} = \boldsymbol{q} \quad \text{and} \quad \boldsymbol{q}\boldsymbol{e} = \mathbf{1}, \tag{5.52}$$

which yield

$$\begin{aligned} q_i &= \sum_{k=0}^{\infty} q_{i+k-1} b_k \quad (i \geq 1), \\ q_0 &= \sum_{j=0}^{\infty} q_j \left(1 - \sum_{k=0}^{j} b_k\right). \end{aligned} \tag{5.53}$$

A major difference between (5.53) and its counterpart for the $M/G/1$ queue, (5.12), is that the equations of (5.53) have an infinite summation, whereas the equations of (5.12) have a finite summation. It turns out that this works to our advantage, and we now employ the method of operators on (5.53). (Problem 5.31 asks the reader to derive the same results using generating functions.)

Letting $Dq_i = q_{i+1}$, we find for $i \geq 1$ that (5.53) can be written as

$$q_i - (q_{i-1}b_0 + q_i b_1 + q_{i+1} b_2 + \cdots) = 0,$$

so that

$$q_{i-1}(D - b_0 - Db_1 - D^2 b_2 - D^3 b_3 - \cdots) = 0.$$

The characteristic equation for this difference equation is

$$z - b_0 - zb_1 - z^2 b_2 - z^3 b_3 - \cdots = 0,$$

which can be written as

$$\sum_{n=0}^{\infty} b_n z^n = z. \tag{5.54}$$

Since b_n is a probability, the term on the left is merely the probability generating function of $\{b_n\}$, which we call $\beta(z)$. Then, (5.54) becomes

$$\boxed{\beta(z) \equiv \sum_{n=0}^{\infty} b_n z^n = z.} \tag{5.55}$$

Analogous to (5.32), it can be shown that $\beta(z) = A^*[\mu(1-z)]$, where $A^*(z)$ is the LST of the interarrival-time CDF (see Problem 5.31), so that (5.55) may also be written as

$$\boxed{z = A^*[\mu(1-z)].} \tag{5.56}$$

The goal is to find solutions of the characteristic equation which can then be used to determine $\{q_i\}$. We shortly prove that (5.55) has exactly one real root in $(0,1)$ (assuming $\lambda/\mu < 1$) and that there are no other roots with absolute value less than 1. Denoting this root by r_0, we therefore have that

$$q_i = Cr_0^i \quad (i \ge 0). \tag{5.57}$$

The constant C, as usual, is determined from the condition that the probabilities sum to one. This implies that $C = 1 - r_0$.

To show that there is one and only one positive root of (5.55) in $(0,1)$, we consider the two sides of the equation separately, as

$$y = \beta(z) \quad \text{and} \quad y = z. \tag{5.58}$$

First, we observe that

$$0 < \beta(0) = b_0 < 1 \quad \text{and} \quad \beta(1) = \sum_{n=0}^{\infty} b_n = 1.$$

We can also easily show that $\beta(z)$ is both monotonically nondecreasing and convex, because

$$\beta'(z) = \sum_{n=1}^{\infty} nb_n z^{n-1} \ge 0,$$

$$\beta''(z) = \sum_{n=2}^{\infty} n(n-1)b_n z^{n-2} \ge 0.$$

Furthermore, since the service times are exponential, each b_n is strictly positive. That is, $b_n > 0$ for $n \ge 0$, so $\beta(z)$ is strictly convex. Thus there are two possible cases for the graphs of $y = \beta(z)$ and $y = z$ as shown in Figure 5.2. Either there are no intersections in $(0,1)$, or there is exactly one intersection in $(0,1)$. The latter case occurs when

$$\beta'(1) = \text{E[number served during interarrival time]} = \frac{\mu}{\lambda} > 1.$$

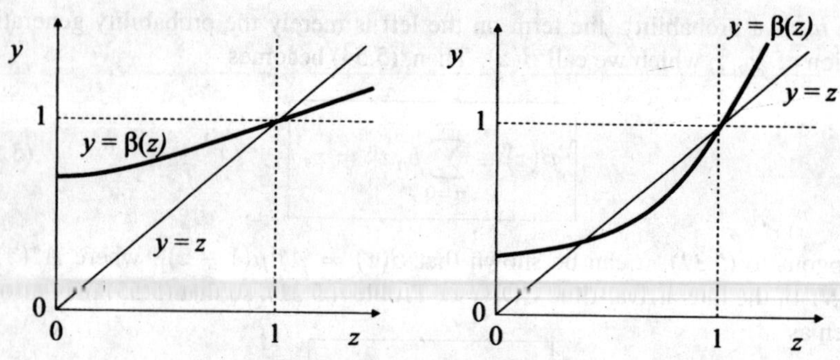

Figure 5.2 Plot of Equation (5.58).

That is, when $\lambda/\mu < 1$, there is exactly one root r_0 of (5.55) in $(0, 1)$.

We have not yet shown that this is the only *complex* root with absolute value less than one. To do this, we use Rouché's theorem (see Section 2.11.2). Assume that $\beta'(1) = 1/\rho > 1$. Let $f(z) \equiv -z$ and $g(z) \equiv \beta(z)$. Because $g(1) = 1$ and $g'(1) > 1$, then $g(1 - \epsilon) < 1 - \epsilon$ for small enough $\epsilon > 0$. Consider the set z such that $|z| = 1 - \epsilon$. By the triangle inequality,

$$|g(z)| \leq \sum_{n=0}^{\infty} b_n |z|^n = g(1 - \epsilon) < 1 - \epsilon = |f(z)|.$$

By Rouché's theorem, $f(z) = -z$ and $f(z) + g(z) = -z + \beta(z)$ have the same number of roots within the contour $|z| = 1 - \epsilon$. Since ϵ can be made arbitrarily small, there is exactly one complex root of $z = \beta(z)$ whose absolute value is less than one. Thus it must be the real root r_0 found earlier.

Finding the root r_0 generally involves numerical procedures, but it is readily obtainable. For example, the method of successive substitution,

$$z^{(k+1)} = \beta(z^{(k)}) \quad (k = 0, 1, 2, \ldots, 0 < z^{(0)} < 1), \tag{5.59}$$

is guaranteed to converge because of the shape of $\beta(z)$. More on the numerical aspects of the problem will follow later.

When these results are put together, the steady-state arrival-point distribution is

$$\boxed{q_n = (1 - r_0)\, r_0^n \quad (n \geq 0, \rho < 1).} \tag{5.60}$$

It is informative to note the analogy between (5.60) and the $M/M/1$ steady-state probability given by $p_n = (1 - \rho)\rho^n$. We can therefore use all the expected-value measures of effectiveness results obtained for $M/M/1$ by merely replacing ρ with r_0. However, note that q_n is the steady-state probability of n in the system just prior to an arrival and not the general-time steady-state probability p_n, so that the expected-value measures apply only at arrival points. Unlike the $M/G/1$ model, it is not true here that $q_n = p_n$. In fact, it turns out that the equality holds for the current model if, and only if, the arrivals are Poisson; that is, $q_n = p_n$ for $G/M/1$ if, and only if, $G = M$. However, q_n and p_n can be related; more will be said about this in Chapter 6, Section 6.3.

In light of these circumstances, we use a superscript (A) to denote the fact that a particular measure of effectiveness is taken relative to arrival points only, and thus write from (5.60) that

$$L^{(A)} = \frac{r_0}{1-r_0} \quad \text{and} \quad L_q^{(A)} = \frac{r_0^2}{1-r_0}. \tag{5.61}$$

The line-delay and system-waiting-time distribution functions, $W_q(t)$ and $W(t)$, can also be obtained from the $M/M/1$ with r_0 replacing ρ, that is,

$$\begin{aligned} W_q(t) &= 1 - r_0 e^{-\mu(1-r_0)t} \quad (t \geq 0), \\ W(t) &= 1 - e^{-\mu(1-r_0)t} \quad (t \geq 0), \end{aligned} \tag{5.62}$$

with mean values given by

$$W_q = \frac{r_0}{\mu(1-r_0)} \quad \text{and} \quad W = \frac{1}{\mu(1-r_0)}. \tag{5.63}$$

These results refer to the distribution of the waiting times as observed by customers arriving to the system. This is in contrast to the distribution of the *virtual* waiting time, corresponding to the waiting times that would be observed for virtual customers arriving at random points in time rather than at the actual customer arrival points. If one refers back to the development of $W_q(t)$ and $W(t)$ for the $M/M/1$ queue (Section 2.2.5), the development of (2.28) depended on the fact that $q_n = p_n$, which was justified by Poisson arrivals (PASTA). Here, $q_n \neq p_n$, so the virtual and actual waiting-time distributions are different.

■ **EXAMPLE 5.9**

Suppose we know in a single-server queueing situation that the service time is exponential with mean μ but have no theoretical basis for assuming the input to be Poisson or Erlang. From past history, we have determined a k-point probability distribution for interarrival times, so that

$$\Pr\{\text{interarrival time} = t_i\} = a(t_i) = a_i \quad (1 \leq i \leq k).$$

We must first determine the root r_0 from (5.56), written here as

$$z = A^*[\mu(1-z)] = \sum_{i=1}^{k} a_i e^{-\mu t_i(1-z)}.$$

Table 5.4 Interarrival Distribution

t (min)	$a(t)$	$A(t)$
2	0.2	0.2
3	0.7	0.9
4	0.1	1.0

Table 5.5 Successive Substitution Steps

k	$z^{(k)}$	$\beta(z^{(k)})$
1	0.500	0.489
2	0.489	0.481
3	0.481	0.476
4	0.476	0.472
5	0.472	0.470
6	0.470	0.468
7	0.468	0.467
8	0.467	0.467

To illustrate numerically, consider the case where the interarrival-time distribution is given in Table 5.4 and $1/\mu = 2$ min. We must solve for the root of

$$A^*(z) = \beta(z) = 0.2e^{-(1-z)} + 0.7e^{-1.5(1-z)} + 0.1e^{-2(1-z)} = z, \tag{5.64}$$

which is a relatively easy equation to solve. We shall use successive substitution on (5.64), beginning from $z^{(0)} = 0.5$, and the resulting sequence of values is as shown in Table 5.5. We see that it took only eight iterations for the process to converge to three decimal places. With r_0 thus estimated as 0.467, we now compute the measures of effectiveness as

$$q_n \doteq 0.533(0.467)^n \quad (n \geq 0),$$

$$L^{(A)} = \frac{r_0}{1-r_0} \doteq 0.876, \quad L_q^{(A)} = \frac{r_0^2}{1-r_0} \doteq 0.409,$$

$$W_q = \frac{r_0}{\mu(1-r_0)} \doteq 1.752 \text{ min}, \quad W = W_q + 1/\mu \doteq 3.752 \text{ min}.$$

(Remember that Little's formula cannot be used to relate mean waiting times with mean queue sizes measured at arrival points.)

Thus we see from this exercise that it is not difficult to obtain results for empirical distributions. This was true also for the $M/G/1$ as illustrated by

Example 5.3. It turns out that this is quite a useful result, since any probability distribution can be approximated by a finite discrete distribution of k points, through the use of an approximating histogram.

5.3.2 Multiserver $G/M/c$ Queue

When we move to generalize to c servers, much of the derivation remains the same, with the major exception of the value of b_n and its effect on the imbedded matrix and the rootfinding problem. The mean system service rate is now going to be either $n\mu$ or $c\mu$, depending on the state, so that b_n will now depend on both i and j. This leads to a very different-looking transition matrix for the Markov chain, and its derivation follows.

To begin, we note that it is still true that $p_{ij} = 0$ for all $j > i+1$. Now, for $j \le i+1$ but $\ge c$, the system serves at the mean rate $c\mu$, since all servers are busy; so, as with $G/M/1$,

$$p_{ij} = b_{i+1-j} \quad (c \le j \le i+1),$$

although here

$$b_n = \int_0^\infty \frac{e^{-c\mu t}(c\mu t)^n}{n!}\, dA(t). \tag{5.65}$$

As a result, the transition probability matrix for the multiserver problem has the same kind of layout from its cth column (starting the count from the 0th one) on out to the right as the $G/M/1$ does from its 1st column on out to the right, as seen in (5.51). Furthermore, it turns out that the rootfinding problem remains the same except that now the number of servers must enter the calculation as

$$\beta(z) = A^*[c\mu(1-z)] = z.$$

The two cases that cause some difficulties are $(i \ge c, j < c)$, for which the system service rate varies from $c\mu$ down to $j\mu$, and $(i < c, j < c)$, for which the system service rate varies from $i\mu$ down to $j\mu$. Thus there are now c columns replacing the first one of (5.51), numbered 0 to $c-1$.

Consider now the case when $j \le i+1 \le c$. Here, everyone is being served, and the probability that anyone has completed service by a time t is the CDF of an individual server, namely, $1-e^{-\mu t}$. To go from i to j, there must be $i+1-j$ service completions by time t; hence, using the binomial distribution, we have

$$p_{ij} = \int_0^\infty \binom{i+1}{i+1-j} (1-e^{-\mu t})^{i+1-j} e^{-\mu t j}\, dA(t) \quad (j \le i+1 \le c). \tag{5.66}$$

Lastly, it remains to consider the case $i+1 > c > j$. Here, the system starts out with all servers busy, since $i \ge c$, and sometime during the interarrival time T, servers start to become idle, until finally only j servers are busy. Let us assume that at a time V after the arrival comes $(0 < V < T)$, it goes into service (all prior customers have left), with $H(v)$ the CDF of V. Thus to get from state i to j in time T, we must have $c-j$ service completions from V to T, or during a time interval of length $T - V$. Using the binomial distribution again and realizing that service time is memoryless, we have

$$p_{ij} = \int_0^\infty \int_0^t \binom{c}{c-j} (1 - e^{-\mu(t-v)})^{c-j} e^{-\mu(t-v)j} \, dH(v) \, dA(t). \tag{5.67}$$

The random variable V is merely the time until $i-c+1$ people have been served with all c servers working, which is the $(i-c+1)$-fold convolution of the exponential distribution with parameter $c\mu$. Hence $h(v) = dH(v)/dv$ is Erlang type $i - c + 1$, namely,

$$h(v) = \frac{c\mu(c\mu v)^{i-c} e^{-c\mu v}}{(i-c)!}.$$

Substituting for $dH(v)$ gives us for $j < c < i+1$ that

$$p_{ij} = \binom{c}{c-j} \frac{(c\mu)^{i-c+1}}{(i-c)!} \int_0^\infty \int_0^t (1-e^{-\mu(t-v)})^{c-j} e^{-\mu(t-v)j} v^{i-c} e^{-c\mu v} \, dv \, dA(t). \tag{5.68}$$

Now, the stationary equation is

$$q_j = \sum_{i=0}^{\infty} p_{ij} q_i \quad (j \geq 0),$$

where the $\{p_{ij}\}$ are as given throughout the preceding discussion. However, for $j \geq c$, we have

$$\begin{aligned} q_j &= \sum_{i=0}^{j-2} 0 \cdot q_i + \sum_{i=j-1}^{\infty} b_{i+1-j} q_i \\ &= \sum_{k=0}^{\infty} b_k q_{j+k-1} \quad (j \geq c). \end{aligned} \tag{5.69}$$

Equation (5.69) is identical to the first line of (5.53), and hence, using analyses like those for $c = 1$, we have

$$q_j = C r_0^j \quad (j \geq c), \tag{5.70}$$

where r_0 is the root of $\beta(z) = A^*[c\mu(1-z)] = z$.

The constant C and the $q_j (j = 0, 1, \ldots, c-1)$ must be determined from the boundary condition $\sum_{i=0}^{\infty} q_j = 1$ and the first $c-1$ of the stationary equations, using the various formulas for the $\{p_{ij}\}$ given above. This is not particularly easy to do, since the $c+1$ equations in $c+1$ unknowns are all infinite summations. We can, however, get an expression for C in terms of $q_1, q_2, \ldots, q_c$ and r_0, and then develop a recursive relation among the q_j. The procedure is as follows.

The boundary condition yields

$$1 = \sum_{j=0}^{\infty} q_j = \sum_{j=0}^{c-1} q_j + \sum_{j=c}^{\infty} C r_0^j.$$

Hence

$$C = \frac{1-\sum_{j=0}^{c-1} q_j}{\sum_{j=c}^{\infty} r_0^j} = \frac{1-\sum_{j=0}^{c-1} q_j}{r_0^c(1-r_0)^{-1}}. \tag{5.71}$$

Now, a recursive relation for q_j when $j < c$ can be obtained:

$$\begin{aligned} q_j &= \sum_{i=0}^{\infty} p_{ij}q_i = \sum_{i=0}^{c-1} p_{ij}q_i + \sum_{i=c}^{\infty} p_{ij}Cr_0^i \\ &= \sum_{i=j-1}^{c-1} p_{ij}q_i + C\sum_{i=c}^{\infty} p_{ij}r_0^i \quad (1 \le j \le c-1), \end{aligned}$$

since $p_{ij} = 0$ for $j > i+1$. Then, rewriting, we have

$$q_{j-1} = \frac{q_j - \sum_{i=j}^{c-1} p_{ij}q_i - C\sum_{i=c}^{\infty} p_{ij}r_0^i}{p_{j-1,j}} \quad (1 \le j \le c-1).$$

Dividing through by C and letting $q_j' = q_j/C$ gives

$$q_{j-1}' = \frac{q_j' - \sum_{i=j}^{c-1} p_{ij}q_i' - \sum_{i=c}^{\infty} p_{ij}r_0^i}{p_{j-1,j}} \quad (1 \le j \le c-1). \tag{5.72}$$

From the stationary equation, we may also write that

$$q_c = \sum_{i=0}^{\infty} p_{ic}q_i = \sum_{i=c-1}^{c} p_{ic}q_i + \sum_{i=c+1}^{\infty} b_{i+1-c}Cr_0^i.$$

Hence

$$q_{c-1} = \frac{(1-p_{cc})q_c - C\sum_{i=c+1}^{\infty} b_{i+1-c}r_0^i}{p_{c-1,c}}$$

and

$$\begin{aligned} q_{c-1}' &= \frac{(1-p_{cc})q_c' - \sum_{i=c+1}^{\infty} b_{i+1-c}r_0^i}{p_{c-1,c}} \\ &= \frac{(1-b_1)q_c' - \sum_{i=c+1}^{\infty} b_{i+1-c}r_0^i}{b_0}. \end{aligned}$$

But we also have, using (5.70), that $q_c' = r_0^c$, so that q_{c-1}' can be determined using the earlier formulas for b_n and p_{ij}. Then $q_{c-1}', q_{c-2}', \ldots, q_0'$ can be obtained by repeated use of (5.72). Now writing (5.71) in terms of q_i' finally gives

$$C = \frac{1 - C\sum_{j=0}^{c-1} q_j'}{r_0^c(1-r_0)^{-1}} = \left(\sum_{j=0}^{c-1} q_j' + \frac{r_0^c}{1-r_0}\right)^{-1} \tag{5.73}$$

Although we have just presented a complete derivation of a very specific approach to obtaining the initial c arrival-point probabilities for the $G/M/c$ model, namely, $\{q_j, j = 0, 1, 2, \ldots, c-1\}$, there is a computational recursion due to Takács (1962) which is better suited to the kind of spreadsheet computations we have developed for our software. Its derivation, however, is extremely long, so we just present the steps of this process, precisely as we have implemented them for the $G/M/c$ queues presented in the software.

We let r_0 be the unique real solution in (0, 1) of $z = A^*[c\mu(1-z)]$, where A^* is the LST of the interarrival distribution. Then the Takács algorithm is used to define and calculate, in sequence:

$$A_j^* \equiv A^*(j\mu) \quad \text{for} \quad j = 0, 1, 2, \ldots, c,$$

$$C_j \equiv \frac{A_1^*}{1-A_1^*} \cdot \frac{A_2^*}{1-A_2^*} \cdots \frac{A_j^*}{1-A_j^*} \quad \text{for} \quad j = 1, 2, \ldots, c, \quad C_0 = 1$$

and

$$D_j \equiv \sum_{k=j+1}^{c} \binom{c}{k} \frac{c(1-A_k^*) - k}{[C_k(1-A_k^*)][c(1-r_0) - k]} \quad \text{for} \quad j = 0, 1, 2, \ldots, c-1,$$

$$M \equiv \left(\frac{1}{1-r_0} + D_0 \right)^{-1}.$$

Then it can be proved that

$$q_j = \begin{cases} \displaystyle\sum_{i=j}^{c-1} (-1)^{i-j} \binom{i}{j} M C_i D_i & (j = 0, 1, \ldots, c-1), \\ M r_0^{j-c} & (j \geq c). \end{cases}$$

It follows by comparing the above with (5.70) that the constant $M = q_c$.

To determine the line-delay distribution function $W_q(t)$, we recognize that there should be a direct analogy back to the derivation of $W_q(t)$ for the $M/M/c$ in Section 2.3, leading to (2.39). Since both $G/M/c$ and $M/M/c$ have exponential service, the only difference in the respective delay distribution derivations would be the values used for the arrival-point probabilities. For the $G/M/c$, we need to use the probabilities $\{q_j, j \geq c\}$. By rewriting (2.31) as

$$W_q(t) = 1 - \frac{p_c}{1-\rho} e^{-c\mu(1-\rho)t},$$

we are able to conclude that the line-delay distribution of the $G/M/c$ is given for $t \geq 0$ by

$$\boxed{W_q(t) = 1 - \frac{q_c}{1-r_0} e^{-c\mu(1-r_0)t} = 1 - \frac{C r_0^c}{1-r_0} e^{-c\mu(1-r_0)t}.} \tag{5.74}$$

The mean delay is easily seen to be

$$W_q = \frac{q_o}{c\mu(1-r_0)^2} = \frac{Cr_0^c}{c\mu(1-r_0)^2}.$$

Just as in $M/G/1$, there are numerous other $G/M/1$-type problems we might want to consider, such as busy periods, $G/M/1/K$, impatience, priorities, output, and transience. Due to space limitations, we are not able to pursue any of these topics at length, and shall instead make a few comments on each and indicate a number of references.

Cohen (1982) is probably the most comprehensive reference for nearly all of these problems. Of course, specific references in the open literature may be better for particular subjects. For example, it is not difficult to get the expected length of the busy period for any $G/M/1$ queue, and there is a very nice approach to this in Ross (2007) using renewal theory. We supply some of the details of this argument in the next chapter in the context of the $G/G/1$ queue.

The approach to the truncation of the queue would be very similar to that described early in Section 5.1.7 for the imbedded chain of $M/G/1$, while impatience could be nicely introduced by permitting some departures from the system to occur before customers reach service. This can essentially be accomplished by changing the parameter of the exponential service to $\mu + r$, r now being the probability of such a renege. If, in addition, we desire to make reneges functions of queue size, as they probably should be, then we have a problem with state-dependent departures, such that

$$\begin{aligned} b_{mn} &= \Pr\{m \text{ services during an interarrival time} \mid \\ &\qquad\qquad n \text{ in system at latest departure}\} \\ &= \int_0^\infty \frac{e^{-[\mu+r(n)]t}\{[\mu + r(n)]t\}^m}{m!}\,dA(t), \end{aligned}$$

where $r(n)$ is defined to be the reneging rate during interarrival periods which began with n in the system. The analysis would then proceed in a way similar to the departure-point state dependence of Section 5.1.10.

As far as priorities are concerned, when the assumption of Poisson inputs is relaxed, it becomes very difficult to obtain any results. One possible way of approaching the problem, suggested by Jaiswal (1968), is to use the technique of supplementary variables. However, one supplementary variable is required for each priority, hence at least two for any such problem. Even the near-Markovian assumption of Erlang input is messy, although some work has appeared on this problem for a small number of priorities.

For output, we have already indirectly obtained some results. It was noted in our discussion of series queues that the limiting output of an $M/G/1$ is Poisson if, and only if, G is exponential. Likewise, it can be shown that the limiting output $G/M/1$ is Poisson if, and only if, G is exponential (see Problem 5.40).

Bulk services ($G/M^{[Y]}/1$) can be handled in a way comparable to the manner in which we solve the $M^{[X]}/G/1$ problem in Section 5.1.9. Some results are also possible for these extensions of the basic $G/M/1$ models when considering c-channels.

Finally, we close with a few comments about transient analysis. As for $M/G/1$, we have to again appeal to the CK equation

$$p_j^{(m)} = \sum_k p_k^{(0)} p_{kj}^{(m)},$$

where $p_j^{(m)}$ is the probability that the system state is j just before the mth customer has arrived. The necessary matrix multiplications must be done with some caution, since we are still dealing with an $\infty \times \infty$ matrix when there is unlimited waiting room. But this can be done (albeit carefully) by truncating the transition matrix at an appropriate point (see Neuts, 1973).

PROBLEMS

5.1. Calculate the imbedded transition probabilities for an $M/G/1$ where service is uniformly distributed on (a, b).

5.2. Let X_n be the number of customers in an $M/G/1$ queue just prior to the *arrival* of the nth customer. Explain why $X_1, X_2, \ldots$ is *not* a discrete-time Markov chain.

5.3. Do Problem 3.24 without making any assumptions concerning the service-time distribution, that is, utilizing only the mean and variance of the service-time data.

5.4. (a) Find L, L_q, W_q, and W for an $M/G/1$ queue with service times that are beta-distributed.

(b) Find L, L_q, W_q, and W for an $M/G/1$ queue with service times that follow a type-2 Erlang distribution and where the arrival rate is $\lambda = 1/3$.

5.5. Derive the waiting-time distribution $W(t)$ and the queue-waiting-time distribution $W_q(t)$ for an $M/M/1$ queue using the formulas (5.33) and (5.34) for an $M/G/1$ queue.

5.6. For Example 5.2, find the σ_B^2 necessary to yield $L = 5$ if the mean service time after training increases to 5.2 min.

5.7. (a) Consider a customer arriving to an $M/G/1$ queue in steady state. Under the condition that the arriving customer finds the server busy, show that at the time of the arrival

$$\text{E[residual service time} \mid \text{server busy]} = \frac{\text{E}[S^2]}{2\,\text{E}[S]},$$

where S is a random service time. [*Hint:* Relate this problem to the average residual time of a renewal process.]

(b) Similarly, for an arbitrary arrival (without conditioning on the status of the server), show that at the time of the arrival

$$\text{E[residual service time]} = \frac{\lambda \text{E}[S^3]}{2}.$$

5.8. Verify (5.12) for $i = 0, 1, 2, 3$.

5.9. Derive the generating function $\Pi(z)$ for $M/G/1$ as given by (5.14).

5.10. From (5.16), derive the PK formula for L in Table 5.1 by using the fact that $L = \Pi'(1)$. [*Hint*: Use l'Hôpital's rule twice.]

5.11. Use (5.16) for the $M/G/1$ queue with G assumed exponential and show that it reduces to the generating function of $M/M/1$ given by (2.15).

5.12. Calculate L for Example 5.3.

5.13. A certain assembly-line operation is assumed to be of the $M/G/1$ type, with input rate 5/h and service times with mean 9 min and variance 90 min^2. Find L, L_q, W, and W_q. Is the operation improved or degraded if the service times are forced to be exponential with the same mean?

5.14. Derive the generating function $\Pi(z)$ in (5.20) for the $M/G/1$ k-point service-time model. Also, obtain the form of the constants c_i in (5.20).

5.15. Customers arrive at an ATM machine according to a Poisson process with rate $\lambda = 60$/h. The following transaction times are observed (in seconds): 28, 71, 70, 70, 51, 62, 36, 25, 35, 87, 69, 27, 56, 25, 36.

(a) Would an $M/M/1$ queue be an appropriate model for this system (why or why not)?

(b) Estimate the average number of people waiting in line at the ATM.

(c) The bank wishes to keep the average line length (number in queue) less than or equal to one. What is the average transaction time needed to achieve this goal (assuming the variance of the transaction times is held constant)?

5.16. Consider an $M/D/1$ queue with service time equal to b time units. Suppose that the system size is measured when time is a multiple of b, and let X_n denote the system size at time $t = n \cdot b$. Show that the stochastic process $\{X_n, n = 0, 1, 2, \ldots\}$ is a Markov chain, and find its transition matrix.

5.17. Determine the lowest value of the Erlang parameter k which will allow you to approximate the mean system waiting time of an $M/D/1$ queue to no worse than 0.5% by an $M/E_k/1$ when $\lambda = 4$ and $\mu = 4.5$.

5.18. Find the LST of the queue-waiting-time distribution for the $M/E_2/1$ queue using (5.34). Then invert the result using partial fractions.

5.19. Verify the computation for Example 5.5 that $P''(1) = 14.50$.

5.20. Find the third and fourth ordinary moments of the system delay for the Bearing Straight problem in Example 5.3, given that the third and fourth moments of the system size are 149.2 and 1670.6, respectively.

5.21. Find the variance of the $M/G/1$ busy period from the Laplace–Stieltjes transform of its CDF.

5.22. Consider a single-server queue to which customers arrive according to a Poisson process with parameter $\lambda = 0.04/\text{min}$ and where the service times of all customers are fixed at 10 min. When there are three units in line, the system becomes saturated and all additional arrivals are turned away. The instants of departure give rise to an imbedded Markov chain with states 0, 1, 2, and 3. Find the one-step transition matrix of this chain and the resultant stationary distribution. Then compare this answer with the result you would have gotten without truncation.

5.23. Given the two-point service distribution of Example 5.3, find the output CDF for that $M/G/1$ queue.

5.24. Find the steady-state probabilities for an $M/G/1$ state-dependent queue where

$$B_i(t) = \begin{cases} 1 - e^{-\mu_1 t} & (i = 1), \\ 1 - e^{-\mu t} & (i > 1). \end{cases}$$

That is, the service distribution is exponential with mean μ_1 if there is no queue when the customer begins service; the service distribution is exponential with mean μ if there is a queue when the customer begins service.

5.25. **(a)** Prove that the distribution of system sizes just prior to arrivals is equal to that after departures in any *G/G/c* queue.

(b) Use the result of Section 5.1.8 for the output CDF $C(t)$ of the inter-departure process of the $M/G/1$ to show that the $M/M/1$ is the only $M/G/1$ with Poisson output.

5.26. Use the $M/G/1$ level-crossing equation (5.47) for the steady-state queue-wait PDF $w_q(t)$ to derive the corresponding transform equation (5.34).

(a) First show that (5.47) can be written as

$$w_q(x) = \lambda W^c(x) - \lambda W_q^c(x).$$

(b) Then, multiply by e^{-sx} and integrate to obtain an equation involving transforms. Use properties of transforms to complete the proof.

5.27. Derive the stationary system-size probabilities for the $M/G/2/2$ queue using only Little's formula and fundamental steady-state identities.

5.28. The Mutual Exclusive Life Insurance Company (MELIC) is building a new headquarters in downtown Burbank. The telephone company wishes to determine the number of lines to feed into the building to assure MELIC of no more than 5% loss in calls due to busy circuits. Find the number of lines when it is estimated that the calling stream is Poisson with mean 100/h throughout the day and that the mean call duration is 2 min.

5.29. When an AIDS case is first diagnosed in the United States by a physician, it is required that a report be filed (i.e., serviced) on the case with the Centers for Disease Control in Atlanta, Georgia. It takes a random amount of time (with an expected value of approximately 3 months) for a doctor to finish the report and send it into the CDC (distribution unknown). An OR team has analyzed the report arrival stream and believes that new patients come to the doctors all over the nation as a Poisson process. If 50,000 new reports are completed each year, what is the mean number of reports in process by doctors at any one instant (assuming steady state)?

5.30. Derive the equivalent to Erlang's loss formula (2.52) for the case in which the input source is *finite* of size M with arrival rate proportional to the remaining source size. Assume that service times are general and that there are c servers. (The resultant answer is an example of a so-called *Engset* formula.)

5.31. Derive (5.55) using the method of generating functions on (5.53), $i \geq 1$. Then show that $\beta(z) = A^*[\mu(1-z)]$.

5.32. **(a)** Find the solution r_0, in closed form, of $\beta(z) = z$, the generating-function equation for a $G/M/1$ queue, when the interarrival times follow a *hyperexponential* distribution with density function

$$a(t) = q\lambda_1 e^{-\lambda_1 t} + (1-q)\lambda_2 e^{-\lambda_2 t}.$$

(b) Find the steady-state waiting-time distribution for a $G/M/1$ queue with $\mu = 8$ and interarrival density function

$$a(t) = (0.3)3e^{-3t} + (0.7)10e^{-10t}.$$

5.33. Show that if $G = M$, the root r_0 of (5.55) is ρ, and that (5.60) yields the familiar $M/M/1$ result.

5.34. **(a)** Prove that r_0 is always greater than $e^{-1/\rho}$.

(b) You are observing two different $G/M/1$ queues, and know that one interarrival CDF (say, A_1) is everywhere larger than the second, A_2. Show that $\beta_1(z) \leq \beta_2(z)$, that $r_0^{(1)} \leq r_0^{(2)}$, and thus that $\sum_{i=0}^{n} q_i^{(1)} \geq \sum_{i=0}^{n} q_i^{(2)}$.

5.35. Find $\beta(z)$ for $G/M/1$ in the event that G is (a) deterministic, and (b) Erlang type 2. Then, under the assumption that the mean input rate λ is 3 and that the mean service time is $\frac{1}{5}$, find the steady-state arrival-point distribution for each of the two cases.

5.36. Consider a $G/M/1$ system whose solution depends on the real root of the nonlinear equation

$$z = \frac{12}{31 - z} + \frac{3}{5}e^{-10(1-z)/3}.$$

Use successive substitution to find the root rounded to three decimal places. Then determine the line-delay CDF under the assumption that the service rate is 1.

5.37. Consider a $G/M/1$ queue in steady state, where the interarrival times are uniformly distributed from 0 to 6 minutes and the expected service time is 2 minutes.

(a) Find the average queue length (as found by an arriving customer).

(b) Find the average time spent in queue for each customer.

(c) Find the probability that a customer's waiting time in queue is greater than 3 minutes.

5.38. Suppose in Example 5.9 that there are two servers, each working at a mean rate of 0.25/min with their times being exponential. Find the arrival-point steady-state system-size probabilities, the expected system size and queue length at arrival points, and the expected time in queue and in system.

5.39. Suppose it is known in a $G/M/1$ queue that some customers are reneging. Specifically, assume that the reneging pattern is such that

$$\Pr\{\text{one customer reneges in } (t, t + \Delta t)\} = r\Delta t + o(\Delta t).$$

The reneging pattern is the same regardless of the number of customers in the system (assuming there is at least one customer in the system). Also, it is assumed that the reneging customer can be the customer in service. Find the resultant imbedded Markov chain.

5.40. (a) Show for an $M/G/1$ queue that stationary system waiting times are exponential if, and only if, $G = M$.

(b) Show that the stationary output of a $G/M/1$ queue is Poisson if, and only if, G is exponential. [*Hint:* The idle time in an arbitrary interdeparture period (called the *virtual idle time*) has a CDF given by $F(u) = A(u) + \int_u^\infty e^{-\mu(1-r_0)(t-u)} dA(t)$. Use the fact that each departure time is a sum of a virtual idle time and a service time.]

5.41. Determine the lowest value of the Erlang parameter k which will allow you to approximate the mean line delay of a $D/M/c$ queue to no worse than 2.0% by an $E_k/M/c$ when $\lambda = 4$, $\mu = 1.5$, and $c = 3$.

5.42. Arrivals occur to a $G/M/4$ system from either of two sources. Any particular interarrival time is twice as likely to come from the first source as the second, and will be exponentially distributed with parameter $\lambda_1 = 0.5$ per unit time when coming from the first source and exponential with parameter $\lambda_2 = 0.25$ when coming from the second. It is further known that the mean service time is 9 time units. Find the mean system size and expected system waiting time.

5.43. The arrival stream described in Problem 5.42 is now observed to come equally likely from the two sources. Furthermore, suppose that the four servers have been merged into a single server whose service time is exponentially distributed with a mean of 2.25. Compute the queue-waiting time beyond which only 5% of the arriving customer delays in queue will fall.

CHAPTER 6

GENERAL MODELS AND THEORETICAL TOPICS

In this chapter we provide some assorted additional results. As a rule, these results were not included earlier because the models were either not Markovian or inappropriate for the discussions in Chapter 5 dealing with $M/G/1$ and $G/M/c$. Most of this new material follows in a logical way from previous material in the sense that it ties up some loose theoretical ends and should also help to provide a more complete picture of the kinds of models that may occur in real life.

6.1 $G/E_k/1$, $G^{[k]}/M/1$, and $G/PH_k/1$

Recall from Chapter 5 that the waiting-time distribution function for the $G/M/1$ queue requires the single real root on $(0, 1)$ for the characteristic equation

$$z = A^*[\mu(1 - z)] = \beta(z),$$

where A^* is the Laplace–Stieltjes transform (LST) of the interarrival times, and $\beta(z)$ is the probability generating function (PGF) of the number of service completions during interarrival times, with $\rho = \lambda/\mu < 1$. The PGF (defined and analytic at least on the complex unit circle) is easily shown to be monotone nondecreasing and

Fundamentals of Queueing Theory, Fourth Edition.
By D. Gross, J. F. Shortle, J. M. Thompson, and C. M. Harris
Copyright © 2008 John Wiley & Sons, Inc.

convex for real z, and thus the root is readily obtainable. For example, the well-known method of successive substitution is guaranteed to converge from any nonnegative starting point less than 1 because of the shape of $\beta(z)$.

For the $G/E_k/1$ queue, the problem becomes more interesting and potentially quite useful. Let λ and μ be the average arrival and service rates, respectively (i.e., the rate of a service phase is $k\mu$). Here the roots need to be generally located and then found from within the interior of the unit circle for

$$z^k = A^*[k\mu(1-z)] = \beta(z), \tag{6.1}$$

where $A^*(\cdot)$ is once more the LST of the interarrival distribution function, and $\beta(z)$ is now the PGF of the number of *phases* completed during an interarrival period (rather than the number of service completions, as defined for the $G/M/1$ queue). An example of the form of (6.1) is the characteristic equation of the $M/E_k/1$ queue,

$$k\mu z^{k+1} - (\lambda + k\mu)z^k + \lambda = 0,$$

which can be obtained directly from the rate-balance equations in (3.25) or from (6.1) with $A^*(s) = \lambda/(s+\lambda)$.

Clearly, one solution to (6.1) is $z = 1$. By Rouché's theorem, it can be shown that there are k others strictly inside the unit circle $|z| = 1$ when the traffic intensity $\lambda/\mu < 1$. We now attempt to find their precise locations. The following result from Chaudhry et al. (1990) provides an important sufficient condition under which the roots are distinct and offers some key information on determining their location. The result can be stated as follows:

One of the k roots of the characteristic equation of the $G/E_k/1$ model (or, equivalently, the $G^{[k]}/M/1$) is real and in $(0,1)$ for all values of k, and there is a second real root in $(-1,0)$ only when k is even. In addition, the other roots ($k-1$ for k odd, $k-2$ for k even) are distinct if the Laplace–Stieltjes transform $A^*(s)$ of the interarrival-time distribution can be written as $[A_1^*(s)]^k$, where $A_1^*(s)$ is itself a legitimate LST.

The proof of this assertion is straightforward. For simplicity, let us use (6.1) in the form $z^k = \beta(z)$. Then by a geometric argument essentially the same as that for the $G/M/1$ used in Figure 5.2, it follows that there exists a unique *real* root in $(0, 1)$ for *all* k when

$$\beta'(1) > \left.\frac{dz^k}{dz}\right|_{z=1} = k.$$

Now, $\beta'(1)$ is the expected number of phase completions during an interarrival period, so $\beta'(1) = k\mu/\lambda$. Thus, the above condition is equivalent to $\lambda/\mu < 1$. When k is even, there is an additional real root in $(-1,0)$ with a smaller modulus than the positive root. This is because z^k is a symmetric function and $0 < \beta(-z) < \beta(z)$ for $z \in (0,1)$.

To show the distinctness of the roots, let us first write z as $re^{i\theta}$. Then (6.1) can be rewritten as

$$r^k e^{i\theta k} = A^*[k\mu(1-re^{i\theta})]e^{2\pi ni} = \beta(re^{i\theta})e^{2\pi ni}, \tag{6.2}$$

where n is an integer. The extra exponential factor equals 1 and is added in preparation for taking the complex kth root. Taking the kth root gives

$$re^{i\theta} = \beta_1(re^{i\theta})e^{2\pi ni/k}, \tag{6.3}$$

where $\beta_1(z) = [\beta(z)]^{1/k}$ is a unique and analytic PGF. The existence of $\beta_1(z)$ comes from the assumption that $A^*(s)$ can be written as $[A_1^*(s)]^k$ (which is the property of *infinite divisibility*).

Thus (6.3) is a different equation for each value of $n = 1, \ldots, k$. Then, by an argument virtually identical to that used to show that there is one characteristic root for the $G/M/1$ inside the unit circle, we can show that the above equation has a unique root strictly inside the unit circle for each value of $n = 1, \ldots, k$, provided $\lambda/\mu < 1$. Finally, no two of these k roots can be equal, since that would imply the equality of their respective roots of unity, which is a contradiction.

The sufficient condition of this theorem can be weakened by asking that $\beta(z)$ be nonzero but not necessarily infinitely divisible. The kth root function is then analytic, and a proof can be devised (see Chaudhry et al., 1990) using Rouché's theorem and not requiring that the function $\beta(z)$ be a legitimate probability generating function.

Unfortunately, there are some $\beta(z)$ associated with $G/E_k/1$ models that have zeros inside the unit circle. However, we can feel comfortable knowing that a good number of queues encountered in practice will have infinitely divisible interarrival-time distributions, since the exponential and Erlang are infinitely divisible and distributions built up from them by convolutions (such as generalized Erlang) will be also. For all other distributions, one should always first try to determine whether $\beta(z)$ is ever zero, for if not, the kth-root approach of the infinitely divisible case will work. If there is a zero of $\beta(z)$ in the unit disk and it is isolated (as it will be in most cases), it is fairly easy to program any numerical procedure to go around the difficulties. Furthermore, there are examples for which $\beta(z)$ has zeros, but where the characteristic equations still have distinct roots. Because of the distinctness of the roots inside the unit circle, we see that the waiting-time distribution function of Section 1.4 for the line delays will be a linear but possibly nonconvex combination of negative exponential functions with possibly complex parameters.

The algorithm we recommend for this problem is successive substitution, although this time we are to work in the complex domain and must thus solve k different problems as we move n from 1 to k. A priori, we know that each of these problems has a distinct, complex-valued solution.

■ EXAMPLE 6.1

To illustrate this approach to rootfinding, consider an $E_3/E_3/1$ problem, with $\lambda = 1$ and $\mu = 4/3$ ($\rho = \lambda/\mu = 3/4$). Then, $A^*(s) = [3\lambda/(3\lambda + s)]^3$, and the resultant problem is to find the roots of

$$z^3 = A^*[3\mu(1-z)] = \left(\frac{3\lambda}{3\lambda + 3\mu(1-z)}\right)^3 = \frac{27}{(7-4z)^3}.$$

Table 6.1 Successive Substitution on (6.5)

m	$z^{(m)}$	Right-Hand Side
1	$(-.500, -.866)$	$(-.242, -.196)$
2	$(-.242, -.196)$	$(-.218, -.305)$
3	$(-.218, -.305)$	$(-.236, -.294)$
4	$(-.236, -.294)$	$(-.233, -.293)$
5	$(-.233, -.293)$	$(-.233, -.293)$

The problem is equivalent to finding the roots of a sixth-degree polynomial, with one root at 1, one real root in $(0, 1)$, two complex conjugate roots with absolute values less than 1, and two possibly complex roots with absolute values more than 1. Rather than turning to a polynomial rootfinder, we shall use successive substitution on (6.3), which for this example is

$$z = \frac{3}{7-4z} e^{2\pi n i/3} \quad (n = 1, 2, 3). \tag{6.4}$$

First, for $n = 3$, we get the real root on (0, 1), and the solution is easily found by the quadratic formula to be 0.75, since (6.4) simplifies in this case to $(4z - 3)(1 - z) = 0$.

Now, let $n = 2$ in (6.4), so that we are to solve

$$z = \frac{3}{7-4z} e^{4\pi i/3},$$

where $\exp(4\pi i/3) = \cos(4\pi/3) + i \sin(4\pi/3) = -0.5 - (\sqrt{3}/2)i$, which we use as the starting point for iterations defined by

$$z^{(m+1)} = \frac{3}{7-4z^{(m)}} e^{4\pi i/3}. \tag{6.5}$$

The results are displayed in Table 6.1, where we note a very rapid convergence. The roots in question are therefore $-0.233 \pm 0.293i$, each of which is clearly less than 1 in absolute value. Thus, in total analogy with the $G/M/1$ model, we may write the steady-state arrival-point probabilities as

$$q_n = C_1(0.75)^n + C_2(-0.233 - 0.293i)^n + C_3(-0.233 + 0.293i)^n,$$

where C_3 must equal the complex conjugate of C_2 in order for the complex arithmetic to lead to real-valued answers. Furthermore, it follows that the line-delay distribution function has the form

$$W_q(t) = 1 - [K_1 e^{-0.25\mu t} - K_2 e^{-(1.233+0.293i)\mu t} - \overline{K}_2 e^{-(1.233-0.293i)\mu t}],$$

where the complex constants (with positive real parts) need to be found from boundary conditions.

At this point, we are temporarily going to put aside the question of how to find the remaining constants in these solutions. These steps turn out to be more clearly explained by two alternative methods for solving this problem, and discussions of these follow in Sections 6.1.1 and 6.1.2.

6.1.1 Matrix Geometric Solutions

By use of a clever device largely due to Neuts (1981), the fundamental arrival-point argument used for the $G/M/1$ may be extended to the $G/PH_k/1$. The idea is to establish a transition matrix whose form is similar to (5.51) except that the entries become matrices. To show how this results, consider a $G/E_2/1$ system. Denoting the state of the system in the usual way for Erlang service $[(n,i) = n$ in system, customer in phase i of service], we have the transition matrix for the imbedded discrete parameter chain at arrival epochs as

$$
\boldsymbol{P} = \begin{array}{c} \\ 0 \\ 1,2 \\ 1,1 \\ 2,2 \\ 2,1 \\ 3,2 \\ 3,1 \\ 4,2 \\ 4,1 \\ \vdots \end{array}
\begin{array}{c}
\begin{array}{cccccccccccc} 0 & 1,2 & 1,1 & 2,2 & 2,1 & 3,2 & 3,1 & 4,2 & 4,1 & 5,2 & \cdots \end{array} \\
\left(\begin{array}{cccccccccccc}
1-\Sigma b_j & b_0 & b_1 & 0 & \cdots & & & & & & \\
1-\Sigma b_j & b_2 & b_3 & b_0 & b_1 & 0 & \cdots & & & & \\
1-\Sigma b_j & b_1 & b_2 & 0 & b_0 & 0 & \cdots & & & & \\
1-\Sigma b_j & b_4 & b_5 & b_2 & b_3 & b_0 & b_1 & 0 & \cdots & & \\
1-\Sigma b_j & b_3 & b_4 & b_1 & b_2 & 0 & b_0 & 0 & \cdots & & \\
1-\Sigma b_j & b_6 & b_7 & b_4 & b_5 & b_2 & b_3 & b_0 & b_1 & 0 & \cdots \\
1-\Sigma b_j & b_5 & b_6 & b_3 & b_4 & b_1 & b_2 & 0 & b_0 & 0 & \cdots \\
1-\Sigma b_j & b_8 & b_9 & b_6 & b_7 & b_4 & b_5 & b_2 & b_3 & b_0 & \cdots \\
1-\Sigma b_j & b_7 & b_8 & b_5 & b_6 & b_3 & b_4 & b_1 & b_2 & 0 & \cdots \\
\vdots & \vdots & \vdots & \vdots & \vdots & \vdots & \vdots & \vdots & \vdots & \vdots &
\end{array}\right)
\end{array}
$$

where now $b_n = \Pr\{n$ phases of service completed during an interarrival time$\}$.

Denoting by $\boldsymbol{B}_i$ the matrix

$$
\begin{bmatrix} b_{2i} & b_{2i+1} \\ b_{2i-1} & b_{2i} \end{bmatrix},
$$

by $\boldsymbol{B}_{01}$ the vector (b_0, b_1), by B_{00} the scalar $1 - b_0 - b_1$, and by $\boldsymbol{B}_{i0}$ the column vector

$$
\begin{pmatrix} 1-\sum b_j \\ 1-\sum b_j \end{pmatrix},
$$

this can be rewritten as follows:

$$
\boldsymbol{P} = \begin{array}{c} \\ 0 \\ 1 \\ 2 \\ 3 \\ \vdots \end{array}
\begin{array}{c}
\begin{array}{ccccccc} 0 & 1 & 2 & 3 & 4 & 5 & \cdots \end{array} \\
\left(\begin{array}{ccccccc}
\boldsymbol{B}_{00} & \boldsymbol{B}_{01} & \boldsymbol{0} & \cdots & & & \\
\boldsymbol{B}_{10} & \boldsymbol{B}_1 & \boldsymbol{B}_0 & \boldsymbol{0} & \cdots & & \\
\boldsymbol{B}_{20} & \boldsymbol{B}_2 & \boldsymbol{B}_1 & \boldsymbol{B}_0 & \boldsymbol{0} & \cdots & \\
\boldsymbol{B}_{30} & \boldsymbol{B}_3 & \boldsymbol{B}_2 & \boldsymbol{B}_1 & \boldsymbol{B}_0 & \boldsymbol{0} & \cdots \\
\vdots & \vdots & \vdots & \vdots & \vdots & &
\end{array}\right)
\end{array}
\tag{6.6}
$$

The $\boldsymbol{P}$ matrix of (6.6) looks very similar to the $\boldsymbol{P}$ matrix of the $G/M/1$ queue given in (5.51), but with 2×2 matrices replacing scalars. A slight variation exists in the first row with its 1×2 matrix and the first column with its 2×1 matrices. This same pattern would exist for E_k service with k replacing 2; that is, the $\boldsymbol{B}_i$ would be $k \times k$ matrices, and so on. Furthermore, a similar kind of structural pattern can be obtained for PH_k service. Differences exist only in the first row and first few columns at most.

Neuts further showed that the similarity does not end here, but that the solution is totally analogous to that for the $G/M/1$ and leads to what Neuts has called a matrix-geometric solution. We of course wish to find the invariant, nonnegative probability vector $\boldsymbol{q}$ satisfying $\boldsymbol{qP} = \boldsymbol{q}$ and $\boldsymbol{qe} = \boldsymbol{1}$. If we partition the vector $\boldsymbol{q}$ into a sequence of contiguously laid-out vectors $q_0, \boldsymbol{q}_1, \ldots$, where $\boldsymbol{q}_j$ is k-dimensional, $j \geq 1$, and q_0 is of dimension one, then, as in (5.53), the stationary equations become

$$\boldsymbol{q}_j = \sum_{i=0}^{\infty} \boldsymbol{q}_{j+i-1}\boldsymbol{B}_i \qquad (j > 1),$$

$$(q_0, \boldsymbol{q}_1) = \sum_{i=0}^{\infty} \boldsymbol{q}_i\boldsymbol{A}_i \qquad [\boldsymbol{A}_0 = (B_{00}, \boldsymbol{B}_{01}), \ \ \boldsymbol{A}_i = (\boldsymbol{B}_{i0}, \boldsymbol{B}_i), i \geq 1]$$

with $\sum \boldsymbol{q}_i\boldsymbol{e} = \boldsymbol{1}$.

The form of the solution to this turns out to be $\boldsymbol{q}_n = \boldsymbol{CR}^n$, where $\boldsymbol{R}$ is a special nonnegative and irreducible $k \times k$ matrix. It is in fact the minimal nonnegative solution (in the sense that all its entries are individually the smallest in the set of solving matrices $\boldsymbol{Z}$) to the matrix equation

$$\boldsymbol{Z} = \sum_{i=0}^{\infty} \boldsymbol{Z}^i\boldsymbol{B}_i = \boldsymbol{B}(\boldsymbol{Z}). \tag{6.7}$$

This last relationship is the matrix equivalent of the fundamental generating function equation for the $G/M/1$ given by (5.55). The answer is unique for $\rho < 1$, and the actual computation of $\boldsymbol{R}$ must be numerical. One method of finding the matrix $\boldsymbol{R}$ is a numerical iterative procedure motivated by the process of successive substitution we used for getting the root of (5.55). This time, we write

$$\boldsymbol{Z}^{(k+1)} = \boldsymbol{B}(\boldsymbol{Z}^{(k)}) \qquad (k = 0, 1, 2, \ldots).$$

The reader is referred to Neuts (1981) and Kao (1991) for further details on this extension of $G/M/1$ to $G/PH_k/1$. This current discussion also carries through to $G/PH_k/c$, although the size of the state space increases rapidly, making computations formidable unless c and k are small. Since the Erlang type k is a special type of phase-type distribution, we note that the matrix-geometric approach presents an alternative solution method to the complex-plane rootfinding procedure introduced earlier in this chapter for the $G/E_k/1$ and $G/E_k/c$ problems.

The matrix geometric $\{\boldsymbol{q}_n\}$ determined for $\boldsymbol{PH}$ service as described above can be used to find line waiting-time distribution functions, by setting up an appropriate continuous-parameter Markov chain with an absorbing state and finding the time to absorption, much as we did in Section 3.3.2. The line-delay distribution of an arriving customer can be obtained beginning from conditional waiting-time probabilities, given that an arrival finds n in the system, and then summing over all states. Each conditional problem requires the simultaneous solution of a system of linear differential equations in the unknown conditional waiting-time distributions, which are then multiplied by their respective $\boldsymbol{q}_n$. Thus obtaining waiting-time distributions in this fashion requires solving large numbers of differential equations using carefully chosen numerical procedures (see Neuts, 1981).

6.1.2 Quasi-Birth–Death Processes

An interesting special case of the $G/PH_k/1$ queue occurs when the input is Poisson. The infinitesimal generator of such a queue has a matrix structure composed of tridiagonal elements that are themselves matrices. This is a matrix analog of the standard birth–death process. This structure arises because the allowable transitions from state (n, i) (i.e., n in system, customer in phase i of service) are to the set $\{(n, i-1), (n+1, i), (n-1, k)\}$, where k is the "first" stage of service, with successive service stages labeled with decreasing numbers.

An illustrative $M/PH_2/1$ generator matrix $\boldsymbol{Q}$ with mean service rates μ_1 and μ_2 is as follows, analogous to that provided for the $M/E_2/1$ of Section 6.1.1:

$$\boldsymbol{Q} = \begin{array}{c} \\ 0 \\ 1,2 \\ 1,1 \\ 2,2 \\ 2,1 \\ 3,2 \\ 3,1 \\ \vdots \end{array} \begin{array}{c} \begin{array}{cccccccc} 0 & 1,2 & 1,1 & 2,2 & 2,1 & 3,2 & 3,1 & \end{array} \\ \left(\begin{array}{cccccccc} -\Sigma b_0 & \lambda & 0 & 0 & \cdots & & & \\ 0 & -\Sigma b_1 & \mu_2 & \lambda & 0 & 0 & \cdots & \\ \mu_1 & 0 & -\Sigma b_2 & 0 & \lambda & 0 & 0 & \cdots \\ 0 & 0 & 0 & -\Sigma b_3 & \mu_2 & \lambda & 0 & \cdots \\ 0 & \mu_1 & 0 & 0 & -\Sigma b_4 & 0 & \lambda & \cdots \\ 0 & 0 & 0 & 0 & 0 & -\Sigma b_5 & \mu_2 & \cdots \\ 0 & 0 & 0 & \mu_1 & 0 & 0 & -\Sigma b_6 & \cdots \\ \vdots & \vdots & \vdots & \vdots & \vdots & \vdots & \vdots & \end{array}\right) \end{array}$$

where $\sum b_i$ denotes the sum of the quantities in row i. When this matrix is rewritten in block form, we find that

$$\boldsymbol{Q} = \begin{array}{c} \\ 0 \\ 1 \\ 2 \\ 3 \\ \vdots \end{array} \begin{array}{c} \begin{array}{cccccc} 0 & 1 & 2 & 3 & 4 & 5\,\cdots \end{array} \\ \left(\begin{array}{cccccc} \boldsymbol{B}_{00} & \boldsymbol{B}_{01} & \boldsymbol{0} & \cdots & & \\ \boldsymbol{B}_{10} & \boldsymbol{B}_1 & \boldsymbol{B}_0 & \boldsymbol{0} & \cdots & \\ \boldsymbol{0} & \boldsymbol{B}_2 & \boldsymbol{B}_1 & \boldsymbol{B}_0 & \boldsymbol{0} & \cdots \\ \boldsymbol{0} & \boldsymbol{0} & \boldsymbol{B}_2 & \boldsymbol{B}_1 & \boldsymbol{B}_0 & \boldsymbol{0}\,\cdots \\ \vdots & \vdots & \vdots & \vdots & \vdots & \vdots \end{array}\right) \end{array},$$

where

$$\boldsymbol{B}_{00} = [-\textstyle\sum b_0] = [-\lambda], \quad \boldsymbol{B}_{01} = [\lambda \ \ 0], \quad \boldsymbol{B}_{10} = [0 \ \ \mu_1],$$

$$\boldsymbol{B}_0 = \begin{bmatrix} \lambda & 0 \\ 0 & \lambda \end{bmatrix}, \quad \boldsymbol{B}_1 = \begin{bmatrix} -\sum b_i & \mu_2 \\ 0 & -\sum b_i \end{bmatrix}, \quad \boldsymbol{B}_2 = \begin{bmatrix} 0 & 0 \\ \mu_1 & 0 \end{bmatrix}.$$

Note that this block-matrix representation of the Q matrix has a tridiagonal structure reminiscent of the usual birth–death generator seen back in Section 1.9.3; thus the name *quasi-birth–death* (QBD) *process*.

After formulating the Q values for a specific problem, the solution can be determined analytically via the techniques mentioned in Section 6.1.1 or computed through successive substitution in the equation $\mathbf{0} = \pi Q$. The QBD approach is useful in analyzing priority queues, as in the previously mentioned work of Miller (1981) in Section 3.4.1, or in problems like those with one server and two queues, with priority for the longer line. Such problems are not $M/PH_k/1$ queues, but they have multiple classes of Poisson arrivals coupled with exponential service that leads to a QBD form.

6.2 General Input, General Service ($G/G/1$)

Although nearly completely devoid of specific structure, we are nonetheless able to get some results for single-server queues with general (i.e., arbitrary) input and service. The major things we are able to do follow from an integral equation of the Wiener–Hopf type for the stationary distribution of the waiting time in queue of an arbitrary customer. This equation is largely due to Lindley (1952) and goes under his name. Further details on the $G/G/1$ may be found in the literature (e.g., Cohen, 1982), but a good portion of it would be beyond the level of this text.

We begin by observing that the relationship between the line waiting times, $W_q^{(n)}$ and $W_q^{(n+1)}$, of the nth and $(n+1)$st customers given in (1.5) is valid for the arbitrary $G/G/1$ problem. This recursion is given by

$$W_q^{(n+1)} = \begin{cases} W_q^{(n)} + S^{(n)} - T^{(n)} & (W_q^{(n)} + S^{(n)} - T^{(n)} > 0), \\ 0 & (W_q^{(n)} + S^{(n)} - T^{(n)} \leq 0), \end{cases}$$

or

$$W_q^{(n+1)} = \max(0, W_q^{(n)} + S^{(n)} - T^{(n)}),$$

where $S^{(n)}$ is the service time of the nth customer and $T^{(n)}$ is the time between the arrivals of the two customers. We can immediately note that the stochastic process $\{W_q^{(n)}, n = 0, 1, 2, \ldots\}$ is a discrete-time Markov process, since the behavior of $W_q^{(n+1)}$ is only a function of the stochastically determined value of $W_q^{(n)}$ and is independent of prior waiting-time history.

Now, from basic probability arguments, we have

$$\begin{aligned} W_q^{(n+1)}(t) &\equiv \Pr\{[\text{line delay } W_q^{(n+1)} \text{ of } (n+1)\text{st customer}] \leq t\} \\ &= \Pr\{W_q^{(n+1)} = 0\} + \Pr\{0 < W_q^{(n+1)} \leq t\} \\ &= \Pr\{W_q^{(n)} + S^{(n)} - T^{(n)} \leq 0\} + \Pr\{0 < W_q^{(n)} + S^{(n)} - T^{(n)} \leq t\} \\ &= \Pr\{W_q^{(n)} + S^{(n)} - T^{(n)} \leq t\}. \end{aligned}$$

If we now define the random variable $U^{(n)} \equiv S^{(n)} - T^{(n)}$ with CDF $U^{(n)}(x)$, then, making use of the convolution formula,

$$W_q^{(n+1)}(t) = \int_{-\infty}^{t} W_q^{(n)}(t-x)\,dU^{(n)}(x) \qquad (0 \le t < \infty).$$

In the steady state ($\rho < 1$), the two waiting-time CDFs must be identical; hence using $W_q(t)$ to denote the stationary delay distribution, we find *Lindley's equation* as

$$\begin{aligned} W_q(t) &= \begin{cases} \int_{-\infty}^{t} W_q(t-x)\,dU(x) & (0 \le t < \infty), \\ 0 & (t < 0) \end{cases} \\ &= -\int_{0}^{\infty} W_q(y)\,dU(t-y) \quad (0 \le t < \infty), \end{aligned} \tag{6.8}$$

where $U(x)$ is the equilibrium $U^{(n)}(x)$ and is given by the convolution of S and $-T$:

$$U(x) = \int_{\max(0,x)}^{\infty} B(y)\,dA(y-x). \tag{6.9}$$

The usual approach to the solution of a Wiener–Hopf integral equation such as (6.8) (see Feller, 1971) begins with the definition of a new function as

$$W_q^-(t) \equiv \begin{cases} \int_{-\infty}^{t} W_q(t-x)\,dU(x) & (t < 0), \\ 0 & (t \ge 0). \end{cases} \tag{6.10}$$

Hence it follows from (6.8) that

$$W_q^-(t) + W_q(t) = \int_{-\infty}^{t} W_q(t-x)\,dU(x) \qquad (-\infty < t < \infty). \tag{6.11}$$

Note that $W_q^-(t)$ is the portion of the CDF associated with the negative values of $W_q^{(n)} + S - T$ when there is idle time between the nth and the $(n+1)$st customer.

It turns out to be easiest to try to obtain $W_q(t), t > 0$ (i.e., the positive part), since $W_q(t)$ is not continuous at 0, but has a jump equal to the arrival-point probability q_0, so that $W_q(0) = q_0$. Start this by denoting the *two-sided* Laplace transforms (LTs) of $W_q(t)$ and $W_q^-(t)$ as

$$\overline{W}_q(s) = \int_{-\infty}^{\infty} e^{-st} W_q(t)\,dt = \int_{0}^{\infty} e^{-st} W_q(t)\,dt$$

and

$$\overline{W}_q^-(s) = \int_{-\infty}^{\infty} e^{-st} W_q^-(t)\,dt = \int_{-\infty}^{0} e^{-st} W_q^-(t)\,dt.$$

In addition, we shall use $U^*(s)$ as the (two-sided) LST of $U(t)$.

We then take the two-sided Laplace transform of both sides of (6.11). The transform of the right-hand side is found to be

$$\mathcal{L}_2\left\{\int_{-\infty}^{t} W_q(t-x)\,dU(x)\right\} = \int_{-\infty}^{\infty}\int_{-\infty}^{t} e^{-(t-x)s} W_q(t-x)e^{-sx}\,dU(x)\,dt.$$

Since $W_q(t-x) = 0$ for $x \geq t$, we can write

$$\begin{aligned}\mathcal{L}_2 &= \int_{-\infty}^{\infty}\int_{-\infty}^{\infty} e^{-(t-x)s} W_q(t-x)e^{-sx}\,dU(x)\,dt \\ &= \left(\int_{-\infty}^{\infty} e^{-su} W_q(u)\,du\right)\left(\int_{-\infty}^{\infty} e^{-sx}\,dU(x)\right) \\ &= \overline{W}_q(s)U^*(s).\end{aligned}$$

But U is the CDF of the difference of the interarrival and service times and hence by the convolution property must have (two-sided) LST equal to the product of the interarrival transform $A^*(s)$ evaluated at $-s$ and the service transform $B^*(s)$, since $A(t)$ and $B(t)$ are both zero for $t < 0$. Hence $U^*(s) = A^*(-s)B^*(s)$, and thus from (6.11),

$$\begin{aligned}\overline{W}_q^-(s) + \overline{W}_q(s) &= \overline{W}_q(s)A^*(-s)B^*(s) \\ \Rightarrow \quad \overline{W}_q(s) &= \frac{\overline{W}_q^-(s)}{A^*(-s)B^*(s)-1}.\end{aligned} \tag{6.12}$$

Therefore, given any pair $\{A(t), B(t)\}$ for the $G/G/1$, we can theoretically find the Laplace transform of the line delay. The determination of $\overline{W}_q^-(s)$ is the primary difficulty in this computation, often requiring advanced concepts from the theory of complex variables.

To show how this all may work, let us consider the $M/M/1$ problem and then check the answer against our earlier result from Chapter 2. Here

$$B(t) = 1 - e^{-\mu t}, \quad B^*(s) = \frac{\mu}{\mu+s},$$
$$A(t) = 1 - e^{-\lambda t},$$

and

$$A^*(-s) = \frac{\lambda}{\lambda - s}.$$

So, from (6.9),

$$U(x) = \begin{cases} \displaystyle\int_0^\infty (1 - e^{-\mu y})\lambda e^{-\lambda(y-x)}\, dy & (x < 0), \\ \displaystyle\int_x^\infty (1 - e^{-\mu y})\lambda e^{-\lambda(y-x)}\, dy & (x \geq 0), \end{cases}$$

$$= \begin{cases} \dfrac{\mu e^{\lambda x}}{\lambda + \mu} & (x < 0), \\ 1 - \dfrac{\lambda e^{-\mu x}}{\lambda + \mu} & (x \geq 0). \end{cases} \tag{6.13}$$

Thus

$$\begin{aligned} W_q^-(t) &= \int_{-\infty}^t W_q(t - x)\, dU(x) \quad (t < 0) \\ &= \frac{\lambda\mu}{\lambda + \mu} \int_{-\infty}^t W_q(t - x) e^{\lambda x}\, dx \quad (t < 0). \end{aligned}$$

Letting $u = t - x$ yields

$$\begin{aligned} W_q^-(t) &= \frac{\lambda\mu}{\lambda + \mu} \int_0^\infty W_q(u) e^{-\lambda(u-t)}\, du \\ &= \frac{\lambda\mu e^{\lambda t}}{\lambda + \mu} \int_0^\infty W_q(u) e^{-\lambda u}\, du \\ &= \frac{\lambda\mu e^{\lambda t} \overline{W}_q(\lambda)}{\lambda + \mu}. \end{aligned} \tag{6.14}$$

Now $\overline{W}_q(\lambda)$ may easily be found from some of our earlier work. In Section 5.2.1, we noted that

$$\begin{aligned} \pi_n^q &= \Pr\{n \text{ in queue just after a departure}\} \\ &= \frac{1}{n!} \int_0^\infty (\lambda t)^n e^{-\lambda t}\, dW_q(t) \end{aligned}$$

for any $M/G/c$. Hence if $G = M$ and $c = 1$, we find that

$$\pi_0^q = \int_0^\infty e^{-\lambda t}\, dW_q(t).$$

Integration by parts then gives

$$\pi_0^q = e^{-\lambda t} W_q(t)|_0^\infty + \lambda \int_0^\infty e^{-\lambda t} W_q(t)\, dt.$$

But $\lim_{t\to\infty} e^{-\lambda t} = 0$, and since we are only concerned computationally with $W_q(t)$ for $t > 0$, let us make $W_q(0) = 0$ for the time being to simplify the analysis henceforth. In the end, we shall simply set $W_q(0) = p_0$, since it is true for all $M/G/1$ (which always satisfy Little's formula) that $W_q(0) = p_0 = 1 - \lambda/\mu$. Therefore

$$\pi_0^q = \lambda \overline{W}_q(\lambda).$$

We also know that π_0^q must be equal to $p_0 + p_1 = (1-\rho)(1+\rho)$, since this is an $M/M/1$. Hence

$$\overline{W}_q(\lambda) = \frac{(1-\rho)(1+\rho)}{\lambda},$$

and, from (6.14),

$$W_q^-(t) = \frac{e^{\lambda t}(1-\rho)(1+\rho)}{1+\rho} = e^{\lambda t}(1-\rho),$$

with transform

$$\overline{W}_q^-(s) = \frac{1-\rho}{\lambda - s}.$$

Putting everything together, we find, using (6.12), that

$$\begin{aligned}\overline{W}_q(s) &= \frac{(1-\rho)/(\lambda - s)}{\lambda\mu/[(\lambda - s)(\mu + s)] - 1} \\ &= \frac{(1-\rho)(\mu + s)}{s(\mu - \lambda + s)} \\ &= \frac{1-\rho}{s} + \frac{\lambda(1-\rho)}{s(\mu - \lambda + s)},\end{aligned}$$

which inverts to

$$\begin{aligned}W_q(t) &= 1 - \rho + \frac{\lambda(1-\rho)(1 - e^{-(\mu-\lambda)t})}{\mu - \lambda} \\ &= 1 - \rho e^{-\mu(1-\rho)t} \quad (t > 0).\end{aligned}$$

Now, realizing that $W_q(0)$ equals $p_0 = 1 - \rho$, the result is thus the same as obtained in (2.28).

We illustrate the use of Lindley's equation by the following example.

EXAMPLE 6.2

Our friend the hair-salon operator, H. R. Cutt, has decided that she would like to find the distribution of the line waits her customers undergo. She realizes that under the new priority rules she recently designated (see Example 3.11, Section 3.4.1), service times may be either of two possibilities: exponential with mean 5 min (for the trims, used one-third of the time), and exponential with mean 12.5 min (for the others). The arrival stream remains Poisson with parameter $\lambda = 5/\text{h}$. If we assume that there are no priorities, then Cutt's system is an $M/G/1$ queue, with service times given to be the mixed exponential (i.e., the weighted average of two the exponentials) having

$$\begin{aligned} b(t) &= \tfrac{1}{3}\tfrac{1}{5}e^{-t/5} + \tfrac{2}{3}\tfrac{2}{25}e^{-2t/25}, \\ B(t) &= 1 - \left(\tfrac{1}{3}e^{-t/5} + \tfrac{2}{3}e^{-2t/25}\right), \\ B^*(s) &= \frac{1}{15s+3} + \frac{4}{75s+6}. \end{aligned}$$

So, from (6.10), noting that $a(t) = \frac{1}{12}e^{-t/12}$, we find that

$$W_q^-(t) = \int_{-\infty}^{t} W_q(t-x)\,dU(x) \quad (t<0),$$

where, from (6.9),

$$U(x) = \begin{cases} \displaystyle\int_0^\infty \left(1 - \frac{e^{-y/5} + 2e^{-2y/25}}{3}\right) \frac{e^{-(y-x)/12}}{12}\,dy & (x<0), \\ \displaystyle\int_x^\infty \left(1 - \frac{e^{-y/5} + 2e^{-2y/25}}{3}\right) \frac{e^{-(y-x)/12}}{12}\,dy & (x \geq 0), \end{cases}$$

$$= \begin{cases} e^{x/12}\left(1 - \dfrac{\frac{1}{36}}{\frac{1}{12}+\frac{1}{5}} - \dfrac{\frac{1}{18}}{\frac{1}{12}+\frac{2}{25}}\right) & (x<0), \\ 1 - \dfrac{\frac{1}{36}e^{x/5}}{\frac{1}{12}+\frac{1}{5}} - \dfrac{\frac{1}{18}e^{-2x/25}}{\frac{1}{12}+\frac{2}{25}} & (x \geq 0). \end{cases}$$

Thus

$$\begin{aligned} W_q^-(t) &= \frac{1}{12} \times \frac{468}{833} \int_{-\infty}^{t} W_q(t-x)e^{x/12}\,dx \quad (t<0) \\ &= \tfrac{39}{833}e^{t/12}\overline{W}_q\left(\tfrac{1}{12}\right). \end{aligned}$$

We must therefore next find $\overline{W}_q(\lambda)$, $\lambda = \frac{1}{12}$. Since the result we quoted earlier in this section from Section 5.2.1 regarding π_n^q was valid for any $M/G/1$, it is certainly true in this current case. Hence, again,

$$\overline{W}_q(\lambda) = \frac{\pi_0^q}{\lambda}.$$

But here π_0^q will clearly be different from its value for $M/M/1$. We know that $\pi_0^q = \pi_0 + \pi_1$, where π_0 and π_1 refer to the departure-point probabilities of 0 and 1 in the system, respectively. Recalling the analysis of Chapter 5, we have

$\pi_0 = 1 - \rho$, and from (5.12), $\pi_1 = \pi_0(1-k_0)/k_0$. But

$$k_0 = \int_0^\infty e^{-\lambda t}\, dB(t)$$
$$= \int_0^\infty e^{-t/12}\left(\frac{e^{-t/5}}{15} + \frac{4e^{-2t/25}}{75}\right) dt = \frac{468}{833}.$$

Since

$$\rho = \tfrac{1}{12}\left[\tfrac{1}{3}(5) + \tfrac{2}{3}\left(\tfrac{25}{2}\right)\right] = \tfrac{5}{6},$$

we get $\pi_0 = \frac{1}{6}$ and

$$\pi_1 = \tfrac{1}{6}\left(\tfrac{365}{468}\right) = \tfrac{365}{2808}.$$

So

$$\pi_0^q \doteq 0.297 \quad \text{and} \quad \overline{W}_q(\lambda) \doteq \frac{0.297}{\frac{1}{12}} = 3.564.$$

Therefore

$$W_q^-(t) \doteq \tfrac{39}{833}(3.564)e^{t/12} \quad \text{and} \quad \overline{W}_q^-(s) \doteq \frac{2.00}{1-12s}.$$

Since $A^*(-s) = 1/(1-12s)$, we now have, from (6.12), that

$$\overline{W}_q(s) \doteq \frac{\frac{2.00}{1-12s}}{\frac{1}{1-12s}\left(\frac{1}{3+15s} + \frac{4}{6+75s}\right) - 1}$$
$$= \frac{2.00(18 + 315s + 1125s^2)}{36s + 2655s^2 + 13,500s^3}$$
$$\doteq \frac{1}{s} - \frac{0.010}{s+0.18} - \frac{0.82}{s+0.015}.$$

Finally, inversion of the LT yields

$$W_q(t) \doteq 1 - 0.01e^{-0.18t} - 0.82e^{-0.015t}.$$

6.2.1 $GE_j/GE_k/1$

The $G/G/1$ problem can be greatly simplified if it can be assumed that the interarrival and service distributions can be expressed individually as convolutions of independent and not necessarily identical exponential random variables. Such a form is called the *generalized Erlang*, and of course the regular Erlang is just a special case. This is not a particularly strong restriction, since it can be shown that any CDF can be approximated to almost any degree of accuracy by such a convolution, using an argument similar to that used to show the completeness of the polynomials in function space.

So we may write that

$$A^*(s) = \prod_{i=1}^{j} \frac{\lambda_i}{\lambda_i + s}, \quad B^*(s) = \prod_{i=1}^{k} \frac{\mu_i}{\mu_i + s}.$$

Thus from (6.12),

$$\begin{aligned} \overline{W}_q(s) &= \frac{\overline{W}_q^-(s)}{\prod_{i=1}^{j} \frac{\lambda_i}{\lambda_i - s} \prod_{i=1}^{k} \frac{\mu_i}{\mu_i + s} - 1} \\ &= \frac{\overline{W}_q^-(s) \prod_{i=1}^{j} (\lambda_i - s) \prod_{i=1}^{k} (\mu_i + s)}{\prod_{i=1}^{j} \lambda_i \prod_{i=1}^{k} \mu_i - \prod_{i=1}^{j} (\lambda_i - s) \prod_{i=1}^{k} (\mu_i + s)}. \end{aligned} \tag{6.15}$$

The denominator of $\overline{W}_q(s)$ is clearly a polynomial of degree $j + k = n$; its roots will be denoted by $s_1, \ldots, s_n$, where s_1 is easily seen to be 0. Furthermore, it can be shown from the form of the polynomial and Rouché's theorem that there are exactly $j - 1$ roots, $s_2, \ldots, s_j$, whose real parts are positive, and thus k roots, $s_{j+1}, \ldots, s_n$, with negative real parts. Thus we must be able to write that

$$\begin{aligned} &\prod_{i=1}^{j} \lambda_i \prod_{i=1}^{k} \mu_i - \prod_{i=1}^{j} (\lambda_i - s) \prod_{i=1}^{k} (\mu_i + s) \\ &\qquad = s(s - s_2) \cdots (s - s_j)(s - s_{j+1}) \cdots (s - s_n). \end{aligned}$$

Hence, letting $z_i = s_{j+i}$,

$$\overline{W}_q(s) = \frac{\overline{W}_q^-(s) \prod_{i=1}^{j} (\lambda_i - s) \prod_{i=1}^{k} (\mu_i + s)}{s \prod_{i=2}^{j} (s - s_i) \prod_{i=1}^{k} (s - z_i)}.$$

But the numerator must also have the roots $s_2, \ldots, s_j$ to preserve analyticity for $\text{Re}(s) > 0$ and hence may be rewritten as $Cf(s)\Pi_{i=2}^{j}(s - s_i)$, where C is a constant to be found later. So now, after cancellation,

$$\overline{W}_q(s) = \frac{Cf(s)}{s\Pi_{i=1}^{k}(s - z_i)}.$$

The last key step is to show that $f(s)$ is itself also a polynomial. This can, in fact, be done again using concepts from the theory of complex variables. The final form of $\overline{W}_q(s)$ is therefore determined by finding $Cf(s)$. It turns out that $f(s)$ cannot have any roots with positive real parts and, in fact, has the roots $-\mu_1, \ldots, -\mu_k$. Hence $\overline{W}_q(s)$ may be written as

$$\overline{W}_q(s) = \frac{C\Pi_{i=1}^{k}(s + \mu_i)}{s\Pi_{i=1}^{k}(s - z_i)}.$$

To get C now we note that

$$\begin{aligned}\mathcal{L}\{W_q'(t)\} &= s\overline{W}_q(s) - W_q(0)\\ &= C\prod_{i=1}^{k}\frac{s+\mu_i}{s-z_i} - q_0.\end{aligned}$$

But the value of the transform of $W_q'(t)$ as $s \to 0$ is equal to $1 - q_0$, since there is a jump in the CDF $W_q(t)$ equal to q_0 at the origin. Thus

$$1 - q_0 = C\prod_{i=1}^{k}\frac{\mu_i}{-z_i} - q_0 \Rightarrow C = \prod_{i=1}^{k}\frac{-z_i}{\mu_i}.$$

So

$$\overline{W}_q(s) = \frac{\Pi_{i=1}^{k}(-z_i/\mu_i)(s+\mu_i)}{s\Pi_{i=1}^{k}(s-z_i)}. \tag{6.16}$$

The above result is an extremely useful simplification, since it now puts $\overline{W}_q(s)$ into an easily invertible form. A partial-fraction expansion is then performed (assuming distinct z_i) to give the result

$$\overline{W}_q(s) = \frac{1}{s} - \sum_{i=1}^{k}\frac{C_i}{s-z_i},$$

which inverts to the generalized exponential mixture

$$W_q(t) = 1 - \sum_{i=1}^{k} C_i e^{z_i t},$$

where the $\{z_i\}$ have *negative* real parts and the values of the C_i would be determined in the usual way from the partial-fraction expansion.

While conceptually this is a simplification, in practice it is sometimes difficult to estimate the parameters $\lambda_i (i = 1, 2, \ldots, j)$ and $\mu_i (i = 1, 2, \ldots, k)$, since j and k generally must be large in order for this method to be accurate. In the event that the $\{\mu_i\}$ are all equal to begin with, the model becomes a $G/E_k/1$, which therefore has "mixed-exponential" waiting times for any interarrival-time distribution. Recognize also that each generalized Erlang (GE_k) is of phase type and further that any GE_k with distinct parameters may be written as a linear combination of exponentials with possibly negative mixing parameters.

As a check, let us verify these results for $M/M/1$. From (6.15), we need the roots of the polynomial denominator of (6.15),

$$0 = \lambda\mu - (\lambda - s)(\mu + s) = s^2 - (\lambda - \mu)s$$

with negative real parts. There is clearly one, and it is $z_1 = \lambda - \mu$. Hence, from (6.16),

$$\overline{W}_q(s) = \frac{(1-\rho)(s+\mu)}{s(s-\lambda+\mu)},$$

which completely agrees with the result obtained earlier.

EXAMPLE 6.3

Ms. W. A. R. Mup of the Phil R. Upp Company of Example 3.8 would like to know the percentage of her customers who wait more than 4 h for oil delivery. She has already computed the average delay as 2 h but is now particularly concerned that too many customers are excessively delayed. We know that $\lambda = \frac{6}{5}$/h and $\mu = \frac{3}{2}$/h for this $M/E_2/1$ queue, and the results of this section now permit us to obtain the necessary service-time distribution $W_q(t)$.

Since there are two identical exponential stages in the service process, we see that $\mu_1 = \mu_2 = 2\mu = 3$/h. Hence the LST of the waiting times is computed from (6.15) as

$$\overline{W}_q(s) = \frac{(s+3)^2 \Pi_{i=1}^2 z_i}{3^2 s \Pi_{i=1}^2 (s - z_i)};$$

the quantities z_1 and z_2 are the roots with negative real parts of the polynomial denominator of (6.15),

$$\begin{aligned} 0 &= \lambda\mu_1\mu_2 - (\lambda - s)(\mu_1 + s)(\mu_2 + s) \\ &= s(5s^2 + 24s + 9). \end{aligned}$$

Both roots of the quadratic factor are found to be real, with values -4.39 and -0.410. Hence

$$\begin{aligned} \overline{W}_q(s) &= \frac{(s+3)^2}{5s(s+4.39)(s+0.410)} \\ &= \frac{1}{s} + \frac{0.0221}{s+4.39} - \frac{0.8221}{s+0.410}. \end{aligned}$$

Therefore

$$\begin{aligned} W_q(t) &= 1 + 0.0221e^{-4.39t} - 0.8221e^{-0.410t} \\ &\Rightarrow \Pr\{T_q > 4\} = 0.1595. \end{aligned}$$

6.2.2 Discrete $G/G/1$

From the foregoing, it should be readily apparent that actual results for waiting times using Lindley's equation when interarrival and service times are continuous are often difficult to obtain. This is so primarily because of the complexity in obtaining $W_q^-(t)$. However, if the interarrival and service times are discrete (recall that any continuous distribution can be approximated by a k-point distribution), then (6.8) can be used iteratively [since its right-hand side becomes a sum and we know that $\sum w_q(t_i) = 1$] to obtain the values of $W_q(t)$ at all realizable values of the discrete random variable t.

Since the state space is discrete here, the Markovian waiting-time process $\{W_q^{(n)}\}$ now clearly becomes a Markov chain. To illustrate briefly, let us assume that interarrival and service times can take on only two values each, namely, (a_1, a_2) and

(b_1, b_2), respectively. Then $U = S - T$ can have at most four values—let them be $U_1 = -2, U_2 = -1, U_3 = 0$, and $U_4 = 1$. Such a system is ergodic whenever $E[U] < 0$, which will be true if we assign equal probabilities of $\frac{1}{4}$ to these values. Let us denote the steady-state probability of a wait of j as w_j; it can be found by solving the stationary equation $w_j = \sum_i w_i p_{ij}$.

Given the above values, the transition probabilities are all derived simply as

$$p_{00} = \Pr\{U = -2, -1 \text{ or } 0\} = \frac{3}{4}, \qquad p_{01} = \Pr\{U = 1\} = \frac{1}{4},$$
$$p_{10} = \Pr\{U = -2, \text{ or } -1\} = \frac{1}{2}, \qquad p_{11} = \Pr\{U = 0\} = \frac{1}{4},$$
$$p_{12} = \Pr\{U = 1\} = \frac{1}{4},$$

and, for $i > 1$,

$$p_{i,i-2} = p_{i,i-1} = p_{i,i} = p_{i,i+1} = \frac{1}{4}.$$

Thus the $\{w_j\}$ are found as the solution to

$$w_0 = \frac{3}{4}w_0 + \frac{1}{2}w_1 + \frac{1}{4}w_2,$$
$$w_j = \sum_{i=j-1}^{j+2} \frac{w_i}{4} \quad (j \geq 1),$$
$$1 = \sum_{j=0}^{\infty} w_j.$$

A more complete expansion of these ideas will be presented in Chapter 7 as a means of finding approximate $G/G/1$ solutions.

6.3 Poisson Input, Constant Service, Multiserver ($M/D/c$)

In the event that an $M/G/c$ is found to have deterministic service, there is sufficient special structure available to permit the obtaining of the stationary probability generating function for the distribution of the queue size, something that is not possible generally for $M/G/c$. The approach we use is fairly typical, and a similar one may be found in Saaty (1961), which is essentially originally due to Crommelin (1932).

We begin by rescaling time so that the constant service time (say, $b = 1/\mu$) is now the basic unit of time; then λ becomes λ/b and μ becomes 1. For ease of notation, let us henceforth use P_n to denote the CDF of the system size in the steady state. We are then able to observe (in the spirit of Problem 5.16) that the queueing process is indeed Markovian and therefore

$$
\begin{aligned}
p_0 &= \Pr\{c \text{ or less in system at arbitrary instant in steady state}\} \\
&\quad \times \Pr\{0 \text{ arrivals in subsequent unit of time}\}, \\
p_1 &= \Pr\{c \text{ or less in system}\}\Pr\{1 \text{ arrival}\} \\
&\quad + \Pr\{c+1 \text{ in system}\}\Pr\{0 \text{ arrivals}\}, \\
p_2 &= \Pr\{c \text{ or less in system}\}\Pr\{2 \text{ arrivals}\} \\
&\quad + \Pr\{c+1 \text{ in system}\}\Pr\{1 \text{ arrival}\} \\
&\quad + \Pr\{c+2 \text{ in system}\}\Pr\{0 \text{ arrivals}\}, \\
&\;\;\vdots \\
p_n &= \Pr\{c \text{ or less in system}\}\Pr\{n \text{ arrivals}\} \\
&\quad + \Pr\{c+1 \text{ in system}\}\Pr\{n-1 \text{ arrivals}\} \\
&\quad + \cdots + \Pr\{c+n \text{ in system}\}\Pr\{0 \text{ arrivals}\},
\end{aligned}
$$

or

$$
\begin{aligned}
p_0 &= P_c e^{-\lambda}, \\
p_1 &= P_c \lambda e^{-\lambda} + p_{c+1} e^{-\lambda}, \\
p_2 &= \frac{P_c \lambda^2 e^{-\lambda}}{2} + p_{c+1}\lambda e^{-\lambda} + p_{c+2} e^{-\lambda}, \\
&\;\;\vdots \\
p_n &= \frac{P_c \lambda^n e^{-\lambda}}{n!} + \frac{p_{c+1}\lambda^{n-1} e^{-\lambda}}{(n-1)!} + \cdots + p_{c+n} e^{-\lambda}.
\end{aligned}
\tag{6.17}
$$

When we define the usual generating function $P(z) \equiv \sum_{n=0}^{\infty} p_n z^n$, (6.17) leads, after multiplying the ith row by z^i and then summing over all rows, to (see Problem 6.9)

$$
P(z) = \frac{\sum_{n=0}^{c} p_n z^n - P_c z^c}{1 - z^c e^{\lambda(1-z)}} = \frac{\sum_{n=0}^{c-1} p_n (z^n - z^c)}{1 - z^c e^{\lambda(1-z)}}. \tag{6.18}
$$

Our first step is to prove that the poles of (6.18) are distinct. This is done by showing that no value that makes the denominator of (6.18) vanish can do the same for its derivative [see Problem 6.10(a)]. Then to get rid of the c unknown probabilities in the numerator of (6.18), we invoke the usual arguments, employing Rouché's theorem and the fact that $P(1) = 1$. Since $P(z)$ is analytic and bounded within the unit circle, all the zeros of the denominator within and on the unit circle must also make the numerator vanish. Rouché's theorem will tell us that there are $c - 1$ zeros of the numerator inside $|z| = 1$, while the cth is clearly $z = 1$. Since the numerator is a polynomial of degree c, it may be written as

$$
N(z) = K(z-1)(z-z_1)\cdots(z-z_{c-1}),
$$

where $1, z_1, z_2, \ldots, z_{c-1}$ are the coincident roots of the denominator and numerator. To get K, we use the fact that $P(1) = 1$. By L'Hôpital's rule,

$$1 = \lim_{z \to 1} P(z) = \frac{K(1-z_1)\cdots(1-z_{c-1})}{\lambda - c} \Rightarrow K = \frac{\lambda - c}{(1-z_1)\cdots(z-z_{c-1})}.$$

Hence

$$P(z) = \frac{\lambda - c}{(1-z_1)\cdots(1-z_{c-1})} \frac{(z-1)(z-z_1)\cdots(z-z_{c-1})}{1 - z^c e^{\lambda(1-z)}}, \tag{6.19}$$

The roots $\{z_i, i = 1, \ldots, c-1\}$ can be obtained exactly like those of the $G/E_k/1$ model from

$$z^c = e^{-\lambda(1-z)}.$$

The first step in obtaining the $\{p_n\}$ is to find p_0 by evaluating the generating function of (6.19) at $z = 0$, from which we find that

$$p_0 = \frac{(c-\lambda)(-1)^{c-1}\Pi_{i=1}^{c-1} z_i}{\Pi_{i=1}^{c-1}(1-z_i)} \quad (c \geq 2). \tag{6.20}$$

Next, we get $\{p_1, \ldots, p_{c-1}\}$ from a $(c-1) \times (c-1)$ complex-valued linear system of equations created from the numerator of (6.18) set equal to zero at each of the $c-1$ roots $\{z_i\}$. The resultant matrix of coefficients has (i, j) entry given by

$$a_{ij} = z_i^j - z_i^c \quad (i, j = 1, \ldots, c-1)$$

and right-hand-side elements equal to

$$b_i = p_0(z_i^c - 1) \quad (i = 1, \ldots, c-1).$$

The remaining probabilities are found by recursion on (6.17), first obtaining p_c from $p_0 = P_c e^{-\lambda}$ as

$$p_c = (e^\lambda - 1)p_0 - \sum_{i=1}^{c-1} p_i.$$

Various other measures of effectiveness may also be obtained for this model. The reader is referred to Saaty (1961) and/or Crommelin (1932) for the computation of the probability of no delay and the waiting-time distribution, as well as some approximate methods for computing measures of effectiveness that are useful when extreme accuracy is not required.

6.4 Semi-Markov and Markov Renewal Processes in Queueing

In this section we treat a discrete-valued stochastic process that transits from state to state according to a Markov chain, but in which the time required to make each transition may be a random variable that is a function of both the "to" and "from" states. In the special cases where the transition times are independent exponential random variables with parameters dependent only on the "from" state, the semi-Markov process reduces to a continuous-parameter Markov chain. We denote the general semi-Markov process (SMP) as $\{X(t)|X(t) = \text{state of the process at time } t \geq 0\}$, with the state space equal to the nonnegative integers. There are numerous references for semi-Markov processes such as Ross (1996, 2007) and Heyman and Sobel (1982).

Define $Q_{ij}(t)$ to be the joint conditional probability that, starting in state i, the next transition is to state j in an amount of time less than or equal to t. Let $G_i(t) = \sum_{j=0}^{\infty} Q_{ij}(t)$. This is the CDF of the time spent in state i before a transition, regardless of the destination j. Let $\{X_n|n = 0, 1, 2, \ldots\}$ be the imbedded Markov chain. The transition probabilities of the imbedded chain are $p_{ij} = Q_{ij}(\infty)$. In the SMP, the conditional distribution of the time to transition from i to j, given that the next transition is to j, is

$$F_{ij}(t) = \frac{Q_{ij}(t)}{p_{ij}}, \qquad (p_{ij} > 0). \tag{6.21}$$

Let $\{R_i(t)|t > 0\}$ be the number of transitions into state i that occur during (0, t). Let $\boldsymbol{R}(t)$ be the vector whose ith component is $R_i(t)$. The stochastic process $\boldsymbol{R}(t)$ is called a Markov renewal process (MRP). The Markov renewal process and the semi-Markov process have slightly different definitions. The MRP $\boldsymbol{R}(t)$ is a counting process that keeps track of the total number of visits to each state, while the SMP $X(t)$ records the system state at each point in time. For all intents and purposes, one always determines the other.

■ EXAMPLE 6.4

For an $M/M/1$ queue, let $N(t)$ be the number of customers in the system at time t. Then $\{N(t)\}$ is a semi-Markov process with

$$p_{01} = 1, \qquad F_{01}(t) = 1 - e^{-\lambda t}, \qquad Q_{01}(t) = p_{01}F_{01}(t),$$

$$p_{i,i-1} = \frac{\mu}{\lambda+\mu}, \quad F_{i,i-1}(t) = 1 - e^{-(\lambda+\mu)t}, \quad Q_{i,i-1}(t) = p_{i,i-1}F_{i,i-1}(t),$$

$$p_{i,i+1} = \frac{\lambda}{\lambda+\mu}, \quad F_{i,i+1}(t) = 1 - e^{-(\lambda+\mu)t}, \quad Q_{i,i+1}(t) = p_{i,i+1}F_{i,i+1}(t),$$

where $i \geq 1$. In fact, any Markov chain, whether in continuous time or discrete time, is an SMP as well. A Markov chain in discrete time is an SMP whose transition times are the same constant, while a Markov chain in continuous time is an SMP who transition times are all exponential.

■ EXAMPLE 6.5

For an $M/G/1$ queue, the system size $N(t)$ is neither a Markov process nor a semi-Markov process. But the following process is an SMP: Let $\{X(t)\}$

be the number of customers left behind by the most recent departure. This is often called an *imbedded* SMP because it lies within the total process. Note that $\{X(t)\}$ is slightly different than the number of customers $\{X_n\}$ left behind by the nth departure (defined in Section 5.1.1). $\{X_n\}$ is an imbedded Markov chain that occurs in discrete time, while $\{X(t)\}$ is an imbedded SMP that occurs in continuous time.

The main use of SMPs in queueing is as a means to quantify systems that are non-Markovian, but that either are semi-Markovian or possess an imbedded SMP. The existence of semi-Markovian structure is quite advantageous and often easily permits the obtaining of some of the fundamental relationships required in queueing. To derive some of the key SMP results, we need the following notation:

$$H_{ij}(t) = \text{CDF of time until first transition into } j \text{ beginning from } i,$$

$$m_{ij} = \text{mean first passage time from } i \text{ to } j = \int_0^\infty t\,dH_{ij}(t),$$

$$m_i = \text{mean time spent in state } i \text{ during each visit} = \int_0^\infty t\,dG_i(t),$$

$$\eta_{ij} = \text{mean time spent in state } i \text{ before going to } j = \int_0^\infty t\,dF_{ij}(t).$$

From the definition of $G_i(t)$ given previously and (6.21), we are able to immediately deduce the intuitive result that

$$m_i = \sum_{j=0}^{\infty} p_{ij}\eta_{ij}.$$

The key results that we desire are those that determine the limiting probabilities of the SMP and then relate these to the general-time queueing process in the event that the SMP is imbedded and does not possess the same distribution as the general process. These relationships are all fairly intuitive but of course require proof. A reference for the direct SMP results is Ross (1996). Fabens (1961) gives a presentation of the relationship between the imbedded SMP and the general-time process. The key points of interest to us are presented as follows, much of it built up from the material of Section 1.9.

1. If state i communicates with state j in the imbedded Markov chain of an SMP, if $F_{ij}(t)$ is not lattice (i.e., is not discrete with all of its outcomes multiples of one of the values it takes), and if $m_{ij} < \infty$, then

$$\begin{aligned} v_j &\equiv \text{the steady-state probability that the SMP is in state } j \\ &\qquad \text{given that it started in state } i \\ &= \lim_{t\to\infty} \Pr\{X(t) = j | X(0) = i\} = \frac{m_j}{m_{jj}}. \end{aligned}$$

2. If the underlying Markov chain of an SMP is irreducible and positive recurrent, if $m_j < \infty$ for all j, and if π_j is the stationary probability of j in the imbedded Markov chain, then

$$v_j = \frac{\pi_j m_j}{\sum_{i=0}^{\infty} \pi_i m_i}.$$

3. Let $\delta(t)$ be the time that has elapsed since the most recent transition (looking back from time t). If $\{X(t)\}$ is an aperiodic SMP with $m_i < \infty$,

$$\lim_{t\to\infty} \Pr\{\delta(t) \le u | X(t) = i\} = \int_0^u \frac{1 - G_i(x)}{m_i}\, dx \equiv R_i(u).$$

   In other words, $R_i(u)$ is the CDF of the time since the last transition, given that the system is in state i in steady state.

4. The general-time state probabilities p_n of the total process are related to the imbedded SMP as follows:

$$p_n = \sum_i v_i \int_0^{\infty} \Pr\{\text{system moves from state } i \text{ to } n \text{ in time } t\}\, dR_i(t).$$

We now proceed to illustrate the use of the above results on the $G/M/1$ queue, whose analysis we were not quite able to finish in Chapter 5. With this current theory, we are able to obtain the general-time state probabilities p_n starting from the geometric arrival-point distribution $q_n = (1 - r_0)r_0^n$.

For the $G/M/1$ queue, recall that the imbedded Markov chain $\{X_n\}$ measures the system size just before an arrival. Let $X(t)$ be the system size just before the most recent arrival. This is an imbedded SMP. The time spent in state i of the SMP has the same CDF as the interarrival distribution, namely, $A(t)$, with mean $1/\lambda$, since the transitions of the SMP occur precisely at the arrival points. From the second property, it follows that the imbedded SMP has steady-state probabilities

$$v_n = \frac{q_n/\lambda}{\sum_{j=0}^{\infty} q_j/\lambda} = q_n = (1 - r_0)r_0^n, \quad (n \ge 0).$$

In other words, the fraction of time spent in state n in the SMP (v_n) is the same as the stationary probability of state n in the imbedded Markov chain (q_n). These are equal because the mean time spent in each state n of the SMP is the same constant $(1/\lambda)$ for every n.

Also, because the times between the transition points of the SMP are IID, the CDF of the time since the last transition is independent of the current state. In other words, $R_i(t)$ (from the third property) is independent of i, namely,

$$R_i(t) = \lambda \int_0^t [1 - A(x)]\, dx \equiv R(t).$$

From the fourth property,

$$p_n = \sum_{i=n-1}^{\infty} v_i \int_0^\infty \Pr\{i-n+1 \text{ departures in } t\}\, \lambda[1-A(t)]\,dt \quad (n>0),$$

$$p_0 = \sum_{i=0}^{\infty} v_i \int_0^\infty \Pr\{\text{at least } i+1 \text{ departures in } t\}\, \lambda[1-A(t)]\,dt.$$

We already know that p_0 must equal $1-\lambda/\mu = 1-\rho$, since we have long ago shown that this is true for all $G/G/1$ queues. Therefore we focus on $n>0$, and it follows from the above that

$$p_n = \lambda \sum_{i=n-1}^{\infty} (1-r_0) r_0^i \int_0^\infty \frac{e^{-\mu t}(\mu t)^{i-n+1}}{(i-n+1)!}[1-A(t)]\,dt \quad (n>0).$$

Letting $j = i-n+1$, we have

$$\begin{aligned} p_n &= \lambda(1-r_0) r_0^{n-1} \int_0^\infty e^{-\mu t}[1-A(t)] \sum_{j=0}^{\infty} \frac{(r_0 \mu t)^j}{j!}\,dt \\ &= \lambda(1-r_0) r_0^{n-1} \int_0^\infty e^{-\mu t(1-r_0)}[1-A(t)]\,dt. \end{aligned}$$

Integration by parts yields

$$p_n = \frac{\lambda}{\mu} r_0^{n-1}\left[1 - \int_0^\infty e^{-\mu t(1-r_0)}\,dA(t)\right].$$

But, from (5.55) and Problem 5.31,

$$\int_0^\infty e^{-\mu t(1-r_0)}\,dA(t) = \beta(r_0) = r_0.$$

Hence for $n>0$,

$$p_n = \frac{\lambda}{\mu} r_0^{n-1}(1-r_0) = \frac{\lambda q_{n-1}}{\mu} \quad (n>0). \tag{6.22}$$

We have now shown that the *general-time* probabilities p_n for all $G/M/1$ queues are of the familiar geometric pattern, which we have shown earlier for the $M/M/1$ and $E_k/M/1$ systems.

For completeness, we provide without proof the comparable results for the $G/M/c$ queue:

$$p_0 = 1 - \frac{\lambda}{c\mu} - \frac{\lambda}{\mu}\sum_{j=1}^{c-1} q_{j-1}\left(\frac{1}{j} - \frac{1}{c}\right),$$

$$p_n = \frac{\lambda q_{n-1}}{n\mu} \quad (1 \le n < c),$$

$$p_n = \frac{\lambda q_{n-1}}{c\mu} \quad (n \ge c).$$

6.5 Other Queue Disciplines

As we mentioned early in Chapter 1, Section 1.2.3, there are many possible approaches to the selection from the queue of customers to be served, and certainly FCFS is not the only choice available. We have already considered some priority models in Chapter 3, and mentioned in Chapter 1 at least two other important possibilities, namely, random selection for service (RSS) and last come, first served (LCFS). We might even consider a third to be general discipline (GD), that is, no particular pattern specified. We shall obtain results in this section for some of our earlier FCFS models under the first two of these three possible variations. To do any more would be quite time consuming, and it would not be in the interest of the reader to be bogged down in such detail.

We should note that the system state probabilities do not change when the discipline is modified from FCFS to another. As observed before, the proof of Little's formula remains unchanged, and thus the average waiting time is the same. But there will indeed be changes in the waiting-time distribution, and, of course, this implies that any higher moment generalization of Little's formula is, in general, no longer applicable. For the sake of illustration we shall now derive the waiting-time distribution for the $M/M/c$ under the RSS discipline and for the $M/G/1$ under LCFS, two results that are thought to be quite typical.

We begin with a discussion of the waiting times for the $M/M/c$ when service is in random order. In the usual way let us define $W_q(t)$ as the CDF of the line delay, and then it may be written that

$$W_q(t) = 1 - \sum_{j=0}^{\infty} p_{c+j} \tilde{W}_q(t|j) \quad (t \geq 0),$$

where $\tilde{W}_q(t|j)$ represents the probability that the delay undergone by an arbitrary arrival who joined when $c+j$ were in the system is more than t. But, from the results of Section 2.3, (2.31), where we see that $p_{c+j} = p_c \rho^j$, we may rewrite the above equation as

$$W_q(t) = 1 - p_c \sum_{j=0}^{\infty} \rho^j \tilde{W}_q(t|j).$$

To calculate $\tilde{W}_q(t|j)$ we observe that the waiting times depend not only on the number of customers found in line by an arbitrary arrival, but also on the number who arrive afterward. We shall then derive a differential–difference equation for $\tilde{W}_q(t|j)$ by considering this Markov process over the time intervals $(0, \Delta t)$ and $(\Delta t, t + \Delta t)$, and evaluating the appropriate CK equation. There are three possible ways to have a waiting time greater than $t + \Delta t$, given that $c + j$ were in the system upon arrival: (1) the interval $(\Delta t, t + \Delta t)$ passes without any change; (2) there is another arrival in $(0, \Delta t)$ [thus bringing the system size, not counting the first arrival, up to $c + j + 1$], and then the remaining waiting time is greater than t, given that $c + j + 1$ were in the system at the instant of arrival; and (3) there is a service completion in $(0, \Delta t)$ [thus leaving $c + j - 1$ of the originals], the subject customer is not the one selected for service, and the line wait now exceeds t, given that $c+j-1$ were in the system. Thus

$$\tilde{W}_q(t+\Delta t|j) = [1-(\lambda+c\mu)\Delta t]\tilde{W}_q(t|j) + \lambda\,\Delta t\tilde{W}_q(t|j+1) + \frac{j}{j+1}c\mu\Delta t\tilde{W}_q(t|j-1) + o(\Delta t) \quad [j \geq 0, \quad \tilde{W}_q(t|-1) \equiv 0]. \tag{6.23}$$

The usual algebra on (6.23) leads to

$$\frac{d\tilde{W}_q(t|j)}{dt} = -(\lambda+c\mu)\tilde{W}_q(t|j) + \lambda\tilde{W}_q(t|j+1) + \frac{j}{j+1}c\mu\tilde{W}_q(t|j-1) \quad (j \geq 0) \tag{6.24}$$

with $\tilde{W}_q(t|-1) \equiv 0$ and $\tilde{W}_q(0|j) = 1$ for all j. Equation (6.24) is somewhat complicated, but can be examined using the following approach. Assume that $\tilde{W}_q(t|j)$ has the Maclaurin series representation

$$\tilde{W}_q(t|j) = \sum_{n=0}^{\infty} \frac{\tilde{W}_q^{(n)}(0|j)t^n}{n!} \quad (j = 0, 1, \ldots),$$

with $\tilde{W}_q^{(0)}(0|j) \equiv 1$. We thus get

$$W_q(t) = 1 - p_c \sum_{j=0}^{\infty} \rho^j \sum_{n=0}^{\infty} \frac{\tilde{W}_q^{(n)}(0|j)t^n}{n!}. \tag{6.25}$$

Then the derivatives can be directly determined by the successive differentiation of the original recurrence relation given by (6.24). For example,

$$\begin{aligned}
\tilde{W}_q^{(1)}(0|j) &= -(\lambda+c\mu)\tilde{W}_q^{(0)}(0|j) + \lambda\tilde{W}_q^{(0)}(0|j+1) + \frac{j}{j+1}c\mu\tilde{W}_q^{(0)}(0|j-1) \\
&= -(\lambda+c\mu) + \lambda + \frac{jc\mu}{j+1} = \frac{-c\mu}{j+1}, \\
\tilde{W}_q^{(2)}(0|j) &= -(\lambda+c\mu)\tilde{W}_q^{(1)}(0|j) + \lambda\tilde{W}_q^{(1)}(0|j+1) + \frac{j}{j+1}c\mu\tilde{W}_q^{(1)}(0|j-1) \\
&= \frac{(\lambda+c\mu)c\mu}{j+1} - \frac{\lambda c\mu}{j+2} - \frac{(c\mu)^2}{j+1},
\end{aligned}$$

and so on. Putting everything together into (6.25) gives a final series representation for $W_q(t)$. The ordinary moments of the line delay may be found by the appropriate manipulation of $W_q(t)$ and its complement [this kind of manipulation is well detailed by Parzen (1960)].

For the other model of this section, let us now consider $M/G/1/\infty/\text{LCFS}$. In this case, the waiting time of an arriving customer who comes when the server is busy is leaving $c + j - 1$ of the originals], the subject customer is not the one selected for service, and the line wait now exceeds t, given that $c+j-1$ were in the system. Thus

$$\begin{aligned}\tilde{W}_q(t+\Delta t|j) = [1-(\lambda+c\mu)\Delta t]\tilde{W}_q(t|j) + \lambda\,\Delta t\tilde{W}_q(t|j+1)\\ + \frac{j}{j+1}c\mu\Delta t\tilde{W}_q(t|j-1) + o(\Delta t)\\ [j \geq 0, \quad \tilde{W}_q(t|-1) \equiv 0].\end{aligned} \tag{6.23}$$

The usual algebra on (6.23) leads to

$$\begin{aligned}\frac{d\tilde{W}_q(t|j)}{dt} = -(\lambda+c\mu)\tilde{W}_q(t|j) + \lambda\tilde{W}_q(t|j+1)\\ + \frac{j}{j+1}c\mu\tilde{W}_q(t|j-1) \quad (j \geq 0)\end{aligned} \tag{6.24}$$

with $\tilde{W}_q(t|-1) \equiv 0$ and $\tilde{W}_q(0|j) = 1$ for all j. Equation (6.24) is somewhat complicated, but can be examined using the following approach. Assume that $\tilde{W}_q(t|j)$ has the Maclaurin series representation

$$\tilde{W}_q(t|j) = \sum_{n=0}^{\infty}\frac{\tilde{W}_q^{(n)}(0|j)t^n}{n!} \quad (j = 0, 1, \ldots),$$

with $\tilde{W}_q^{(0)}(0|j) \equiv 1$. We thus get

$$W_q(t) = 1 - p_c\sum_{j=0}^{\infty}\rho^j\sum_{n=0}^{\infty}\frac{\tilde{W}_q^{(n)}(0|j)t^n}{n!}. \tag{6.25}$$

Then the derivatives can be directly determined by the successive differentiation of the original recurrence relation given by (6.24). For example,

$$\begin{aligned}\tilde{W}_q^{(1)}(0|j) &= -(\lambda+c\mu)\tilde{W}_q^{(0)}(0|j) + \lambda\tilde{W}_q^{(0)}(0|j+1) + \frac{j}{j+1}c\mu\tilde{W}_q^{(0)}(0|j-1)\\ &= -(\lambda+c\mu) + \lambda + \frac{jc\mu}{j+1} = \frac{-c\mu}{j+1},\\ \tilde{W}_q^{(2)}(0|j) &= -(\lambda+c\mu)\tilde{W}_q^{(1)}(0|j) + \lambda\tilde{W}_q^{(1)}(0|j+1) + \frac{j}{j+1}c\mu\tilde{W}_q^{(1)}(0|j-1)\\ &= \frac{(\lambda+c\mu)c\mu}{j+1} - \frac{\lambda c\mu}{j+2} - \frac{(c\mu)^2}{j+1},\end{aligned}$$

and so on. Putting everything together into (6.25) gives a final series representation for $W_q(t)$. The ordinary moments of the line delay may be found by the appropriate manipulation of $W_q(t)$ and its complement [this kind of manipulation is well detailed by Parzen (1960)].

For the other model of this section, let us now consider $M/G/1/\infty$/LCFS. In this case, the waiting time of an arriving customer who comes when the server is busy is

6.5 Other Queue Disciplines

As we mentioned early in Chapter 1, Section 1.2.3, there are many possible approaches to the selection from the queue of customers to be served, and certainly FCFS is not the only choice available. We have already considered some priority models in Chapter 3, and mentioned in Chapter 1 at least two other important possibilities, namely, random selection for service (RSS) and last come, first served (LCFS). We might even consider a third to be general discipline (GD), that is, no particular pattern specified. We shall obtain results in this section for some of our earlier FCFS models under the first two of these three possible variations. To do any more would be quite time consuming, and it would not be in the interest of the reader to be bogged down in such detail.

We should note that the system state probabilities do not change when the discipline is modified from FCFS to another. As observed before, the proof of Little's formula remains unchanged, and thus the average waiting time is the same. But there will indeed be changes in the waiting-time distribution, and, of course, this implies that any higher moment generalization of Little's formula is, in general, no longer applicable. For the sake of illustration we shall now derive the waiting-time distribution for the $M/M/c$ under the RSS discipline and for the $M/G/1$ under LCFS, two results that are thought to be quite typical.

We begin with a discussion of the waiting times for the $M/M/c$ when service is in random order. In the usual way let us define $W_q(t)$ as the CDF of the line delay, and then it may be written that

$$W_q(t) = 1 - \sum_{j=0}^{\infty} p_{c+j} \tilde{W}_q(t|j) \quad (t \geq 0),$$

where $\tilde{W}_q(t|j)$ represents the probability that the delay undergone by an arbitrary arrival who joined when $c+j$ were in the system is more than t. But, from the results of Section 2.3, (2.31), where we see that $p_{c+j} = p_c\rho^j$, we may rewrite the above equation as

$$W_q(t) = 1 - p_c \sum_{j=0}^{\infty} \rho^j \tilde{W}_q(t|j).$$

To calculate $\tilde{W}_q(t|j)$ we observe that the waiting times depend not only on the number of customers found in line by an arbitrary arrival, but also on the number who arrive afterward. We shall then derive a differential–difference equation for $\tilde{W}_q(t|j)$ by considering this Markov process over the time intervals $(0, \Delta t)$ and $(\Delta t, t + \Delta t)$, and evaluating the appropriate CK equation. There are three possible ways to have a waiting time greater than $t + \Delta t$, given that $c + j$ were in the system upon arrival: (1) the interval $(\Delta t, t + \Delta t)$ passes without any change; (2) there is another arrival in $(0, \Delta t)$ [thus bringing the system size, not counting the first arrival, up to $c + j + 1$], and then the remaining waiting time is greater than t, given that $c + j + 1$ were in the system at the instant of arrival; and (3) there is a service completion in $(0, \Delta t)$ [thus the sum of the duration of time from his instant of arrival, T_A, to the first subsequent service completion, at T_S, and the length of the total busy period generated by the $n \geq 0$ other customers who arrive in (T_A, T_S).

To find $W_q(t)$ let us first consider the joint probability that n customers arrive in (T_A, T_S) and $T_S - T_A \leq x$. Then, for $R(t)$ equal to the CDF of the remaining service time (see Section 5.1.5),

$$\Pr\{n \text{ arrivals } \in (T_A, T_S) \text{ and } T_S - T_A \leq x\} = \int_0^x \frac{(\lambda t)^n e^{-\lambda t}}{n!}\, dR(t),$$

$$R(t) = \mu \int_0^t [1 - B(x)]dx.$$

Since $\pi_0 = 1 - \rho$, it is found by an argument similar to that used for the $M/G/1$ busy period that

$$\begin{aligned} W_q(t) = 1 - \rho + \Pr\{\text{system busy}\} \\ \times \sum_n \Pr\{n \text{ customers arrive } \in (T_A, T_S), T_S - T_A \leq t - x, \\ \text{and total busy period generated by these arrivals is } x\}, \end{aligned} \tag{6.26}$$

because under the LCFS discipline the most recent arrival in the system is the one chosen for service when the server has just had a completion. But the busy-period distribution is exactly the same as that derived in the FCFS case in Section 5.1.6, since the sum total of customer service times is unchanged by the discipline [it will be denoted henceforth by $G(x)$, with $G^{(n)}(x)$ used for its n-fold convolution]. Therefore (6.26) may be rewritten as

$$W_q(t) = 1 - \rho + \rho \sum_{n=0}^{\infty} \int_0^t \int_0^{t-x} \frac{(\lambda u)^n e^{-\lambda u}}{n!}\, dR(u)\, dG^{(n)}(x).$$

A change of the order of integration then gives

$$\begin{aligned} W_q(t) &= 1 - \rho + \rho \sum_{n=0}^{\infty} \int_0^t \frac{(\lambda u)^n e^{-\lambda u}}{n!}\, dR(u) \int_0^{t-u} dG^{(n)}(x) \\ &= 1 - \rho + \rho \sum_{n=0}^{\infty} \int_0^t \frac{(\lambda u)^n e^{-\lambda u}}{n!} G^{(n)}(t-u)\, dR(u). \end{aligned} \tag{6.27}$$

Although this form of $W_q(t)$ is adequate for some purposes, it turns out to be possible to refine the result further to make it free of $G(t)$. The final version as given in Cooper (1981) is

$$W_q(t) = 1 - \rho + \lambda \sum_{n=1}^{\infty} \int_0^t \frac{(\lambda u)^{n-1} e^{-\lambda u}}{n!} [1 - B^{(n)}(u)]\, du, \tag{6.28}$$

where $B^{(n)}(t)$ is the n-fold convolution of the service-time CDF.

There are, of course, numerous other models with various disciplines, but, as mentioned earlier, it is felt that this discussion should suffice for this text. It is to be emphasized, however, that each discipline must be approached in a unique manner, and any interested reader is referred to the literature, one reference again being Cooper (1981).

6.5.1 Conservation

Little's formula and global and local stochastic balance are examples of conservation laws (see Sections 1.5.1 and 2.2.1). The general idea of conservation is that the expected change of a state function is zero over any finite (including infinitesimal) span of time picked at random in the steady state. These sorts of results play a particularly vital role in the modeling of systems with priorities (see Section 3.4).

For example, Little's result says that the expected change of aggregate waiting time of customers in line or system is zero during a randomly chosen finite time interval. We would similarly find that the time needed to clear out the system is unchanged, on average, over time intervals. As a further illustration, the fact that $\rho = 1 - p_0$ for all $G/G/1$ queues expresses the fact that the expected net change in the number in system over a random time interval is zero, since the equation is equivalent to $\lambda = \mu(1 - p_0)$.

Another previous result that made use of a conservation argument is the proof that $\pi_n = p_n$ for $M/G/1$ queues given in Section 5.1.3. Actually, contained within that proof is a more general result that $q_n = \pi_n$ for any stationary $G/G/c$ queue. To see this, reconsider (5.28) in Section 5.1.3. The right-hand side is q_n, while the left-hand side is π_n. Furthermore, nothing in the proof up to this point assumed anything about the number of servers or distributions for arrival or service times, so that these results hold for $G/G/c$. The final step of the $p_n = \pi_n$ proof required only Poisson arrivals, so that $p_n = \pi_n$ thus holds for all $M/G/c$ systems.

We can redo this argument in a slightly different fashion following the logic used in Krakowski (1974), and similar logic used by us previously in Chapter 2 when deriving q_n for the finite queueing models (e.g., see Section 2.5):

$$\begin{aligned} q_n &\equiv \Pr\{n \text{ in system}|\text{arrival about to occur}\} \\ &= \lim_{\Delta t \to 0} \frac{\Pr\{n \text{ in system and arrival within } (t, t + \Delta t)\}}{\Pr\{\text{arrival within } (t, t + \Delta t)\}}. \end{aligned}$$

Similarly,

$$\pi_n = \lim_{\Delta t \to 0} \frac{\Pr\{n + 1 \text{ in system and a departure within}(t, t + \Delta t)\}}{\Pr\{\text{a departure within } (t, t + \Delta t)\}}.$$

For systems in steady state, the numerators of q_n and π_n must be equal, since the limiting transition rate (as $\Delta t \to 0$) from n to $n + 1$ must be equal to the limiting transition rate from $n + 1$ to n (which is what the respective numerators represent). Furthermore, the denominators are also equal for steady-state systems, since the arrival rate must equal the departure rate; hence $q_n = \pi_n$ for all $G/G/c$ systems.

Now if input is Poisson, then the numerator of q_n is [ignoring $o(\Delta t)$ terms] $p_n \lambda \Delta t$ and the denominator is $\lambda \Delta t$, so that

$$q_n = \lim_{\Delta t \to 0} \frac{p_n \lambda \Delta t}{\lambda \Delta t} = p_n.$$

From the above argument, it is also easy to get the relationship between q_n and p_n for $G/M/1$. The limiting transition rate for going from state $n+1$ to n is μp_{n+1} and since for steady state this must equal the limiting transition rate of going from n to $n+1$, we see that the numerator of q_n for the above equation must equal $p_{n+1} \mu \Delta t$ [again ignoring $o(\Delta t)$ terms]. The arrival rate must equal the departure rate, which is $\mu(1 - p_0) = \mu\rho$, so that the denominator is $\mu\rho \Delta t$ and we have

$$q_n = \lim_{\Delta t \to 0} \frac{p_{n+1} \mu \Delta t}{\mu\rho \Delta t} = \frac{\mu p_{n+1}}{\lambda}.$$

This is the same result as obtained in Section 6.4 using semi-Markov processes, since $q_n = r_0^n(1 - r_0)$, r_0 being the root of the generating function equation $\beta(r) = r$ (see Section 5.3.1), and hence $p_{n+1} = \rho r_0^n(1 - r_0)$ as in Section 6.4. For further general discussion on the topic of conservation, we refer the reader to Krakowski (1973, 1974). See also the related work on level crossing in Section 5.1.11 and Brill (2008).

To see the role that this concept plays in priority queues, let us first define a work-conserving queue discipline to be one where the service need of each customer is unaltered by the queue discipline and where it is also true that the server is not forced to be idle with customers waiting. The relationship between virtual waits, backlog or workload (first introduced in Section 2.2.4, with a typical evolving pattern of workload shown in Figure 6.1) in a priority system, and the customer service and waiting times is crucial to understanding how such systems operate. It is certainly fairly clear how the virtual wait relates to the more usual system parameters for nonpriority, FCFS $G/G/1$ queues, for example. If we separate the steady-state mean virtual delay V into the mean delay caused by those in queue (call it V_q) and the mean residual service time R, then it follows that

$$V_q = \frac{L_q}{\mu} = \frac{\lambda W_q}{\mu}.$$

When the input is Poisson, we see from Problem 5.7 that

$$V = V_q + R = \frac{\lambda}{\mu} W_q + \frac{\lambda}{2} \mathrm{E}[S^2].$$

Hence, using the PK formula for this case and simplifying, we find that

$$\begin{aligned} V &= \frac{\lambda}{\mu} \frac{\rho^2 + \lambda^2 \mathrm{Var}[S]}{2\lambda(1 - \lambda/\mu)} + \frac{\lambda(\mathrm{Var}[S] + 1/\mu^2)}{2} \\ &= \frac{\rho^2 + \lambda^2 \mathrm{Var}[S]}{2\lambda(1 - \lambda/\mu)}, \end{aligned}$$

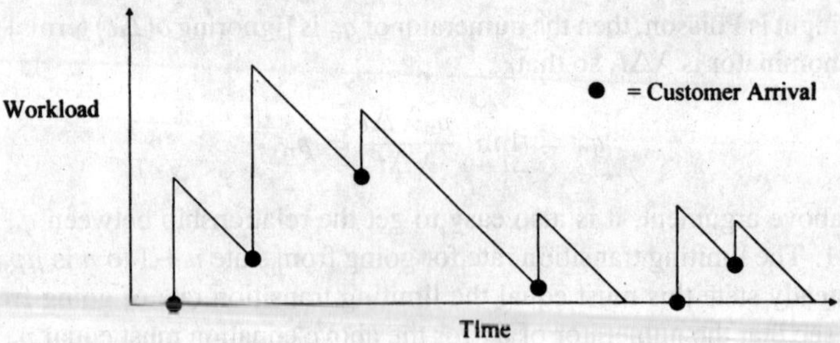

Figure 6.1 Sample workload seen by a $G/G/1$ queue.

which is precisely W_q, as it should be.

But it is of special interest to see how these latter results may differ from those for systems with priorities, since then an arrival's delay is not simply the remaining time necessary to complete service of those in front. A major quantity for an arbitrary system is the product of a customer's service time and line delay, since this represents the total contribution to the remaining workload over the full extent of the customer's waiting time. Furthermore, the contribution of a given customer's service time to V is $S^2/2$ (as the average value of 0 through S). When we assume that the discipline is work-conserving and nonpreemptive (Poisson input or otherwise), it follows that $R = \lambda S^2/2$ and that $\mathrm{E}[S \cdot T_q] = V_q/\lambda$.

We note a number of consequences of this argument for the $M/M/1$ and $M/G/1$ queues. For any stationary $M/M/1/GD$, we see that

$$\mathrm{E}[S \cdot T_q] = \frac{V_q}{\lambda} = \frac{V - R}{\lambda} = \frac{W_q - \lambda \mathrm{E}[S^2]/2}{\lambda} = \frac{\rho}{\mu^2(1-\rho)}.$$

This is also true for the preemptive case in light of the lack of memory of the exponential. For the $M/G/1/GD$, we find that

$$\mathrm{E}[S \cdot T_q] = \frac{V - R}{\lambda} = \frac{W_q - \lambda \mathrm{E}[S^2]/2}{\lambda} = \frac{\rho \mathrm{E}[S^2]}{2(1-\rho)}.$$

6.6 Design and Control of Queues

Models have, on occasion, been classified into two general types—descriptive and prescriptive. Descriptive models are models that *describe* some current real-world situation, while prescriptive models (also often called normative) are models that *prescribe* what the real-world situation should be, that is, the optimal behavior at which to aim.

Most of the queueing models presented thus far are descriptive, in that for given types of arrival and service patterns, and specified queue discipline and configuration, the state probabilities and expected-value measures of effectiveness that describe the system are obtained. This type of model does not attempt to prescribe any action (such as to put on another server or to change from FCFS to priority), but merely represents the current state of affairs.

On the other hand, consider a resource allocation problem, for example, using linear programming to determine how much of each type of product to make for a certain sales period. This model indicates the best setting of the variables, that is, how much of each product *should* be made, within the limitations of the resources needed to produce the products. This, then, is a prescriptive model, for it prescribes the optimal course of action to follow. There has been much work on prescriptive queueing models, and this effort is generally referred to under the title of *design and control of queues*.

Design and control models are those dealing with what the optimal system parameters (in this context, they are actually variables) should be (e.g., the optimal mean service rate, the optimal number of channels). Just which and how many parameters are to be optimized depends entirely on the system being modeled, that is, the particular set of parameters that are actually subject to control.

Generally, the controllable parameters are the service pattern (μ and/or its distribution), number of channels, and queue discipline, or some combination of these. Occasionally, one may even have control over arrivals in that certain arrivals can be shunted to certain servers, truncation can be instituted, the number of servers removed, increased or decreased, and so on. The arrival rate can even be influenced by the levying of tolls. In many cases, however, some "design" parameters are beyond control or perhaps limited to a few possibilities. For example, physical space may prevent increasing the number of channels, and the workers may prevent any proposed decreases. Similar types of effects could fix or limit other potentially controllable parameters.

It is not always easy to make a distinction between those models classified as design and those classified as control. Generally, design models are *static* in nature; that is, cost or profit functions are superimposed on classical descriptive models, so that λ, μ, c, or some combination of these parameters can be optimized. Control models are more dynamic in nature and usually deal with determining the optimal control policy. For example, one may be operating an $M/M/c$ queue and desire to find the optimal c that balances customer waits against server idleness. Since it has already been determined that the form of the policy is always to have c servers, it remains to determine the optimal c by using the descriptive waiting-time information and superimposing a cost function that is to be minimized with respect to c. This is a design problem.

On the other hand, perhaps we desire to know if a single-variable (c) policy is optimal, that is, perhaps we should have a (c_1, c_2) policy where when the queue size exceeds N, we shift from c_1 to c_2 servers, and when the queue size falls back to N, we shift back to c_1 servers. Or perhaps the best policy is (c_1, c_2, N_1, N_2), where, when the queue builds up to N_1, a shift from c_1 to c_2 servers is effected, but we shift back to c_1 only when the queue falls to a different level $N_2 (N_2 < N_1)$. This type of problem is a control problem.

The methodology is vastly different between design and control problems, and perhaps this is the easiest way to classify a problem as to whether it is design or control. For design problems, basic probabilistic queueing results are used to build objective (cost, profit, time, etc.) functions to be minimized or maximized, often subject to constraints (which are also functions of the probabilistic queueing measures). Classical optimization techniques (differential calculus; linear, nonlinear, or integer programming) are then applied. For control problems, techniques such as dynamic programming (or variations such as value or policy iteration) are used. These latter techniques all fall under the category of *Markov decision problems* (MDPs).

As a simple illustration, consider an $M/M/c/K$ problem where a fixed fee per customer is charged but the cost of serving a customer depends on the total time the customer spends in the system. We desire to find the optimal truncation value $K(1 \leq K < \infty)$ that maximizes the expected profit rate. Letting R be the fee charged per customer served and C be the cost incurred per customer per unit time, we have the classic optimization problem

$$\max_K Z = R\lambda(1 - p_k) - C\lambda(1 - p_k)W, \tag{6.29}$$

where p_K and W are given by (2.45) and (2.48), respectively. We have an integer programming problem to solve, which can be done by stepping K from 1 and finding the K that yields the maximum profit, since it can be shown that Z is concave in K (Rue and Rosenshine, 1981).

In the above design problem, we assumed the "optimal" policy was a single-value K. It might have been that the policy should have been "turn customers away when queue size reaches K_1, but do not let customers enter again until queue size drops to $K_2 (K_1 > K_2)$." Actually, a single K value is optimal for this cost structure (Rue and Rosenshine, 1981), which can be proved by value iteration. We would classify this latter type of analysis as a control problem, as opposed to the design problem, where it was predecided that the policy form was a single-value K and the objective function Z was then maximized using classical optimization techniques.

In this section, we shall concentrate mostly on design problems (without getting too much into the details of the necessary optimization procedures), for which much of the appropriate probabilistic analysis has already been done in prior sections of this book.

6.6.1 Queueing Design Problems

Although it was not explicitly pointed out, we have already introduced the ideas of prescriptive models. The reader is specifically referred to Examples 2.10, 3.3, and 5.5 and to Problems 2.9, 2.10, 2.24, 2.25, 2.33, 2.34, 2.35, 2.41, 2.50, and 2.51, for some illustrations of prescriptive analyses.

One of the earliest uses of queueing design models in the literature is described in Brigham (1955). He was concerned with the optimum number of clerks to place behind tool-crib service counters in plants belonging to the Boeing Airplane Company. From observed data, Brigham inferred Poisson arrivals and exponential service, so that an $M/M/c$ model could be used. Costing both clerk idle time and customer waiting time, he presented curves showing the optimal number of clerks (optimal c) as a function of λ/μ and the ratio of customer waiting cost to clerk idle cost.

Morse (1958) considered optimizing the mean service rate μ for an $M/M/1$ problem framed in terms of ships arriving at a harbor with a single dock. He desired the value of μ that would minimize costs proportional to it and the mean customer wait W. This is easily done by taking the derivative of the total-cost expression with respect to μ and equating it to zero. Morse also treated an $M/M/1/K$ problem with a service cost proportional to μ and a cost due to lost customers. The problem was actually framed as a profit model (a lost customer detracts from the profit), and again differential calculus was employed to find the optimal μ.

Yet another model considered by Morse dealt with finding an optimal μ when there is customer impatience. The greater μ, the less impatience on the part of a customer. Again, a profit per customer and a service cost proportional to μ were considered (see Problem 6.24). Morse also considered a model dealing with optimizing the number of channels in an $M/M/c/c$ (no queue allowed to form). The cost of service was now assumed proportional to c rather than μ, and again lost customers were accounted for economically by a profit-per-customer term. His solution was in the form of a graph where, for a given λ, μ, and ratio of service cost to profit per customer, the optimal c could be found.

Finally, Morse considered several machine-repair problems, with a cost proportional to μ and a machine profitability dependent on the machine's uptime. For the latter model, he also built in a preventive maintenance function.

A second reference on economic models in queueing falling into the category of design discussed here is Hillier and Lieberman (1995, Chapter 16). They have considered three general classes of models, the first dealing with optimizing c, the second with optimizing μ and c, and the third with optimizing λ and c. Mention is also made of optimization with respect to λ, c, and μ. Many of the previously mentioned models of Morse turn out to be special cases of one of the classes above. We present here a few of the examples from Hillier and Lieberman and refer the reader to this source for further cases.

We first consider an $M/M/c$ model with unknown c and μ. Assume that there is a cost per server per unit service of C_S and a cost of waiting per unit time for each customer of C_W. Then the expected cost rate (expected cost per unit time) is given by

$$\mathrm{E}[C] = cC_S\mu + C_W L. \tag{6.30}$$

Considering first the optimal value of c, we see that the first term is independent of c for a fixed value of ρ. Furthermore, Morse showed that L is increasing in c for fixed ρ; thus it follows that $\mathrm{E}[C]$ is minimized when $c = 1$. So we now consider finding the optimal value of μ for an $M/M/1$ queue.

From (6.30),

$$\mathrm{E}[C] = C_S\mu + C_W L = C_S\mu + C_W\frac{\lambda}{\mu - \lambda}.$$

Taking the derivative $d\mathrm{E}[C]/d\mu$ and setting it equal to zero yields

$$0 = C_S - \frac{\lambda C_W}{(\mu^* - \lambda)^2} \Rightarrow \mu^* = \lambda + \sqrt{\frac{\lambda C_W}{C_S}}.$$

Looking at the sign of the second derivative confirms that μ^* minimizes $\mathrm{E}[C]$.

Hillier and Lieberman have pointed out an interesting interpretation of the results. If a service channel consists of a crew, such that the mean service rate is proportional to crew size, it is better to have one large crew ($M/M/1$) whose size corresponds to μ^* than several smaller crews (or $M/M/c$). Clearly, given an individual service rate μ, using μ^* to obtain the crew size (μ^*/μ) may not yield an integer value. However, the expected-cost function is convex, so that checking the integer values on either side of μ^*/μ will give the optimum crew size. All of this, of course, is true only as long as the assumptions of crew service rate being proportional to crew size and cost functions being linear are valid. Stidham (1970) extended this model by relaxing the exponential assumptions on interarrival and service times and also considered nonlinear cost functions. He showed that even for these relaxed assumptions, $c = 1$ is generally optimal.

Another interesting model treated by Hillier and Lieberman deals with finding optimal λ (optimal assignment of arrivals to servers) and c for a given μ. The situation considered is one in which a population (such as employees of a building) must be provided with a certain service facility (such as a rest room). The problem is to determine what proportion of the population to assign to each facility (or equivalently the number of facilities) and the optimal number of channels for each facility (in the rest-room example, a channel would correspond to a stall). To simplify the analysis, it is assumed that λ and c are the same for all facilities. Thus if λ_p is the mean arrival rate for the entire population, we need to find the optimal λ, say, λ^* (optimal mean number of arrivals to assign to a facility). The optimal number of facilities then would be λ_p/λ^*.

Using C_S to denote the marginal cost of a server per unit time and C_f to denote a fixed cost per facility per unit time, we desire to minimize $\mathrm{E}[C]$ with respect to c and λ (or c and n), where

$$\mathrm{E}[C] = (C_f + cC_S)n + nC_W L$$
$$\text{subject to } n = \lambda_p/\lambda.$$

Hillier and Lieberman have shown under quite general conditions ($C_f \geq 0$, $C_S \geq 0$, and $L = \lambda W$) that the optimal solution is to make $\lambda^* = \lambda_p$, or equivalently $n = 1$, that is, to provide only a single service facility. The problem then reduces to finding the optimal value for c that minimizes

$$\mathrm{E}[C] = cC_S + C_W L,$$

a problem discussed by Hillier and Lieberman and by Brigham (1955).

It is interesting to consider further that $n = 1$ was shown to be optimal, regardless of the particular situation. For example, in a large skyscraper, this model would say to have only one gigantic rest room. This is obviously absurd, and the reason, as Hillier and Lieberman point out, is that travel time to the facility is not considered in the model. If travel time is insignificant (which might be the case in certain situations), then a single facility would be optimal. When travel time is explicitly considered, that is not always the case, depending on the amount of travel time involved. Hillier and Lieberman present a variety of travel-time models with examples, and again, it is recommended that the interested reader consult this source.

Most of the attention so far has been in designing optimal service facilities (c or μ), the above being one exception, since λ was also of concern. In the introduction to Section 6.6, we also considered a design model involving the arrival process with the objective function given by (6.29). It is interesting to return briefly to this model and reconsider the cost criteria of (6.29). It was desired there to find the K that maximized (6.29), namely, the average profit rate *to the system*. This is referred to in the design and control literature as *social optimization*; that is, it maximizes benefits to the entire system rather than the gain of an individual customer. We can reformulate this problem from an individual customer's standpoint by viewing R as a reward given directly to each customer served and C as a direct cost charged to the customer for each unit of time the customer spends in the system. This gives rise to the objective function

$$Z = R - \frac{C(N+1)}{\mu},$$

where N is the number of customers an arrival finds in the system. The customer serves its own self-interest and joins the queue only when $Z \geq 0$ or when $N \leq R\mu/C - 1$. If every customer uses this strategy, an $M/M/1/K$ queue results, where $K = \lfloor R\mu/C \rfloor$ and $\lfloor x \rfloor$ is the greatest integer less than or equal to x. Rue and Rosenshine (1981) show that the optimal K for this system is less than or equal to the K coming from the individual optimal criterion. The difference between social and individual optimization comes up frequently in control problems involving the arrival-rate process.

We conclude with an example dealing with design in a queueing network.

■ EXAMPLE 6.6

This example is from Gross et al. (1983). We consider a three-node closed Jackson network. The first node U represents the population of operating units. There are M servers at node U, representing the desired M operating units. A queue at node U indicates spare units on hand, while idle servers represent a population operating below the desired M level. The second node B represents local repair, and the third node D represents remote repair. There are c_B and c_D servers at nodes B and D.

The time to failure for each unit is exponential with rate λ. Repair times are exponential with mean rates μ_B and μ_D for local and remote. When a unit fails, it can be repaired locally with probability α and must be repaired at the remote site with probability $1 - \alpha$. A machine repaired locally returns to operational status with probability $1 - \beta$ and must be subsequently sent to the remote site with probability β. All units repaired at the remote site return to operational status.

The problem is to find the optimal numbers of spares and the optimal number of servers at each site to satisfy a constraint on the availability of units. Specifically, we desire to

$$\underset{y,\, c_b,\, c_D}{\text{minimize}} \quad Z = k_S y + k_B c_B + k_D c_D$$

$$\text{subject to} \quad \sum_{n=M}^{M+y} p_n \geq A,$$

where

$$\begin{aligned}
p_n &= \text{steady-state probability that } n \text{ units are operational,}\\
M &= \text{operating population size,}\\
A &= \text{minimum fraction of time } M \text{ units are to be operational,}\\
y &= \text{number of spares,}\\
c_B, c_D &= \text{number of servers at local and remote sites,}\\
k_B, k_D &= \text{annual operating cost per local and remote server,}\\
k_S &= \text{annual operating cost per spare unit.}
\end{aligned}$$

The variables y, c_B, and c_D are the decision variables to be determined by an optimization algorithm, with the steady-state probabilities $\{p_n\}$ determined using the closed Jackson network theory from Chapter 4.

Holding times at all nodes are assumed to be independent exponentially distributed random variables. At node U, the holding time is the time to failure of a component, with the mean failure rate denoted by $\lambda \equiv \mu_U$. At nodes B and D, the holding times are repair times, and the mean repair rates are denoted by μ_B and μ_D, respectively. Let $N \equiv M + y$. From (4.17) we have

$$p_{n_U, n_B, n_D} = \frac{1}{G(N)} \frac{\rho_U^{n_U}}{a_U(n_U)} \frac{\rho_B^{n_B}}{a_B(n_B)} \frac{\rho_D^{n_D}}{a_D(n_D)},$$

where $n_U + n_B + n_D = N$ and

$$a_i(n) = \begin{cases} n! & (n < c_i, \quad i = U, B, D), \\ c_i^{n-c_i} c_i! & (n \geq c_i, \quad i = U, B, D). \end{cases}$$

The routing-probability matrix is

$$\boldsymbol{R} = \{r_{ij}\} = \begin{matrix} U \\ B \\ D \end{matrix} \overset{\begin{matrix} U & \quad B & \quad D \end{matrix}}{\begin{pmatrix} 0 & \alpha & 1-\alpha \\ 1-\beta & 0 & \beta \\ 1 & 0 & 0 \end{pmatrix}}.$$

Now, we substitute $\{r_{ij}\}$ into (4.14) and use Buzen's algorithm to calculate the constant $G(N)$. Once the joint probabilities are obtained, we calculate the marginal probabilities $\{p_{n_U}\}$ required for the constraint, again using Buzen's algorithm.

The resultant probability distribution is a function of the decision variables y, c_B, and c_D. The distribution exhibits certain monotonicity properties in relation to these variables, which play a crucial role in developing an implicit enumeration optimization algorithm. We do not dwell on the optimization routine here. However, we do present results for the specific problem given below:

$$\begin{aligned} &\underset{y, c_b, c_D}{\text{minimize}} \quad Z = 20y + 8c_B + 10c_D \\ &\text{subject to} \quad \sum_{n_U=M}^{M+y} p_{n_U} \geq .90 \quad (A_1), \qquad\qquad (6.31) \\ &\qquad\qquad\quad \sum_{n_U=.9M}^{M+y} p_{n_U} \geq .98 \quad (A_2). \end{aligned}$$

The parameters are: $\alpha = 0.5, \beta = 0.5, u = 1, \mu_B = \mu_D = 5$. This example involves two availability constraints. The first constraint requires the entire (M) population to be operating 90% of the time. The second constraint requires 90% of the population to be operating 98% of the time.

The optimization algorithm is quite efficient for these small problems. The efficiency of the computations relies not only on the implicit enumeration scheme, but also on the efficiency of calculations of the probabilities, which in turn depends on the efficiency of Buzen's algorithm. Table 6.2 shows sample results for this problem. In the last line of the table, the problem formulation uses constraint A_1, but not A_2. The more constraints that are imposed, the more efficient the implicit enumeration scheme is. Thus the most demanding run of (6.31) is the last one.

In any problem like this, because of the complexity of the probabilities and the requirement of integer values for the decision variables, a search technique such as implicit enumeration is usually required; that is, a set of decision variables must first be specified, and then the constraints and objective function are evaluated. Implicit enumeration schemes appear to be well suited to these types of problems.

6.6.2 Queueing Control Problems

As previously mentioned, the focus in control problems is on determining the form of the optimal policy. Often, researchers are interested in determining when stationary policies (such as those treated in the previous section) are truly optimal. Also, although it seems to a somewhat lesser extent, there is some concern about actually determining the optimal values of the control parameters.

Table 6.2 Sample Results

M	c_B^*	c_D^*	y^*	Z^*	A_1	A_2	Calls to Buzen's Algorithm
5	2	2	3	96	.938	.982	25
10	3	3	5	154	.926	.988	38
20	4	6	8	252	.907	.989	66
30	6	8	11	348	.904	—	137

The methodology used generally involves an analysis of the functional equation of dynamic programming arising from an objective function that minimizes (maximizes) total discounted cost (profit) streams or average cost (profit) rates over either a finite or an infinite planning horizon. Since this type of analysis is rather different from anything we have thus far encountered in the text, we merely present to the reader some general notions and point to appropriate references.

The earliest efforts involving queueing control in the literature appear to date back to Romani (1957) and Moder and Phillips (1962). The variable under control was the number of servers in an $M/M/c$ queue, and while the form of the policy was prechosen so that their analyses would be technically categorized under our previous heading of static design, they did form the nucleus for the sizable research effort on service-rate control that followed. Romani considered a policy where, if the queue builds up to a certain critical value, additional servers are added as new arrivals come, thus preventing the queue from ever exceeding the critical value. Servers are then removed when they finish service and no one is waiting in the queue.

Moder and Phillips (1962) modified the Romani model in that there are a certain number of servers c_1 always available. If the size of the queue exceeds a critical value M_1, additional servers are added as arrivals enter in a manner similar to the Romani model, except here there is a limit, c_2, to how many servers can be added. Furthermore, in the Moder–Phillips model, the added servers are removed when they finish serving and the queue has fallen below another critical value M_2.

The usual measures of effectiveness, such as idle time, mean queue lengths, and mean wait, are derived, as well as the mean number of channel starts. No explicit cost functions or optimizations are considered, nor is any attempt made to prove the chosen policy to be optimal, so that these models not only fall into the static design category, but also are in reality more descriptive than prescriptive, although the measures of effectiveness are compared for various values of certain parameters such as the switch points (critical values of queue size for which servers are added and removed).

Yadin and Naor (1967) further generalized the Moder–Phillips model by assuming that the service rate can be varied at any time and is under the control of the decisionmaker. The class of policies considered is of the following form. Denoting the feasible service capacities by $\mu_0, \mu_1, \ldots, \mu_k, \ldots$, where $\mu_{k+1} > \mu_k$ and $\mu_0 = 0$, the policy is stated as "whenever system size reaches a value R_k (from below) and service capacity equals μ_{k-1}, the latter is increased to μ_k; whenever system size drops to S_k (from above) and service capacity is μ_{k+1}, the latter is decreased to μ_k." The $\{R_k\} = \{R_1, R_2, \ldots, R_k, \ldots\}$ and $\{S_k\} = \{S_0, S_1, \ldots, S_k, \ldots\}$ are vectors of integers, ordered by $R_{k+1} > R_k, S_{k+1} > S_k, R_{k+1} > S_k, S_0 = 0$, and are the policy parameters; that is, a specific set of values for $\{R_k\}$ and $\{S_k\}$ yields a specific decision rule (policy) from the class of policies described above within the quotation marks. Input is assumed to be Poisson and service exponential, so the queueing model is essentially $M/M/1$ with state-dependent service. Note that this type of $M/M/1$ model with state-dependent service can also represent a situation where additional channels are added and removed as the state of the system changes.

Given the class of policies above, the authors derive the steady-state probabilities, the expected system size, and expected number of rate switches per unit time. While they do not specifically superimpose any cost functions in order to compare various policies of the class studied, they do discuss the problem in general terms. Given a feasible set $\{\mu_k\}$, a cost structure made up of costs proportional to customer wait, service rate, and number of rate switches, and sets $\{R_k\}$ and $\{S_k\}$ that represent a particular policy, the expected cost of the policy can be computed. Thus various policies (different sets $\{R_k\}, \{S_k\}$) can be compared. They further point out the extreme difficulty, however, of trying to find the optimal policy, that is, finding the particular sets $\{R_k\}$ and $\{S_k\}$ that minimize expected cost.

Gebhard (1967) considered two particular service-rate-switching policies for situations with two possible service rates, also under the assumption of Poisson input and exponential service (state-dependent $M/M/1$). He refers to the first as single-level control, and the policy is "whenever the system size is $\leq N_1$, use rate μ_1; otherwise, use rate μ_2." The second policy considered, called bilevel hysteretic control, can be stated as "when the system size reaches a value N_2 from below, switch to rate μ_2; when the system size drops to a value N_1 from above, switch back to rate μ_1." The term *hysteretic* (which was also used by Yadin and Naor) stems from the control loop that can be seen in a plot of system size versus service rate. Gebhard actually compared his two policies for specific cost functions (including both service and queueing costs) after deriving the steady-state probabilities, expected system size, and expected rate of service switching.

As the Yadin–Naor and Gebhard papers appeared simultaneously, it was not specifically pointed out that the Gebhard policies were special cases of the Yadin–Naor policies. The single-level control policy is the case for which, in the Yadin–Naor notation, $R_1 = 1, R_2 = N_1 + 1$, and $S_1 = N_1$; and the bilevel hysteretic control is the case for which $R_1 = 1, R_2 = N_2$, and $S_1 = N_1$.

Heyman (1968) considered an $M/G/1$ state-dependent model where there are two possible service rates, one being zero (the server is turned off) and the other being μ (the server is turned on). He included a server startup cost, a server shutdown cost, a cost per unit time when the server is running, and a customer waiting cost. Heyman was able to *prove* that the form of the optimal policy is "turn the server on when there are n customers in the system, and turn the server off when the system is empty." Heyman examined various combinations of cases involving discounting or not discounting costs over time and a finite or infinite planning horizon.

This paper is the first of those mentioned so far in this subsection that properly fit the category of queueing control, since the emphasis is on the determination of what the optimal form of the class of policies should be. Heyman also considered the problem of determining the optimal n for the various combinations of cases mentioned earlier (infinite horizon with and without discounting, and finite horizon), although this is not the prime thrust of his work.

Sobel (1969) looked at the same problem as Heyman, namely, starting and stopping service but generalized the results to $G/G/1$, as well as assumed a more general cost structure. Considering only the criterion of average cost rate (undiscounted) over an infinite horizon, he showed that almost any type of stationary policy is equivalent to one that is a slight generalization of Heyman's policy, and furthermore, that it is optimal under a wide class of cost functions. The policy form is "provide no service (server turned off) if system size is m or less; when the system size increases to $M(M > m)$, turn the server on and continue serving until the system size again drops to m." He refers to these as (M, m) policies, and one can readily see that Heyman's is a proper subset of Sobel's class of policies, namely, (M, m) policies where $m = 0$. Sobel's results are strictly qualitative in that he is interested in showing that almost any type of policy imaginable for this kind of problem falls into his class of (M, m) policies. He also shows conditions under which (M, m) policies are optimal but does not deal with determining, for specific costs, the optimum values of M and m. (For a complete discussion of this and other related material, see Heyman and Sobel 1984.)

The reader having some familiarity with inventory theory will notice the similarity of the (M, m) policies to the classical (S, s) inventory policies. Sobel does, in one section of his paper, show the applicability of these (M, m) policies to a production inventory system, namely, that the rule becomes "if inventory is at least as high as M, do not produce until inventory drops to m; at that time, start to produce and continue until the inventory level reaches M once again."

Sobel (1974) gives a survey of work on control of service rates, and the reader desiring to delve further into the topic is referred to that source. Another reference on service control problems is Bengtsson (1983). Much of this material is again covered in Heyman and Sobel (1984).

While the early control work centered on the control of service parameter(s), and work continues in this area, much of the later work has involved controlling the arrival process. Forms of the optimal policies for both social and individual optimization and comparisons of the optimal policies for these cases have been a major focus. Stidham (1982) and Rue and Rosenshine (1981) are recommended references for readers interested in delving further into this area. General survey papers on design and control are Crabill et al. (1977) and Stidham and Prabhu (1974). More recently, Serfozo (1981) and Serfozo and Lu (1984) have studied Markovian queues with simultaneous control of arrival and service rates and derived conditions for the existence of natural monotone optimal policies.

A somewhat different type of control problem involves determining the optimal stopping time for operating a transient queueing system, that is, determining when to shut down the queue ("close up shop"). The cost trade-offs here generally involve loss of revenue versus incurring overtime costs. The methodology for treating these problems is different from the dynamic programming techniques used for the arrival or service control problems, and we refer the interested reader to Prabhu (1974). An additional, useful, and more up-to-date reference on control problems in queueing, and more generally on Markov decision processes, is Puterman (1991).

6.7 Statistical Inference in Queueing

The role of statistics (as contrasted with probability) in queueing analyses is focused on the estimation of arrival and service parameters and/or distributions from observed data. Since one must, in practice, often use observable data to decide on what arrival and service patterns of a queueing system actually are, it is extremely important to utilize the data to the fullest extent possible. Since there are also many statistical problems associated with the use of discrete-event simulation modeling in queueing analyses, we have divided the coverage of the broad topic of inference between this section and the later one on simulation in Section 8.3. In the current section, we examine only those statistical questions related to what might be called *a posteriori* analysis; that is, data are collected on the behavior of a queueing problem, and we wish to make inferences about the underlying interarrival and service structure giving rise to these observations. We contrast this with the *a priori* analysis necessary for selecting the appropriate interarrival and/or service distributions to input to a queueing model (analytical or simulation analyses) or for the isolated examination of data on interarrival and/or service times. These latter problems are to be addressed in Section 8.3; of course, we recognize the overall importance of distribution selection and estimation throughout all of queueing theory.

Known statistical procedures can help in making the best use of existing data, or in determining what and how much new data should be taken. That this is an important facet of real queueing studies is readily seen, since the output from a queueing model can be no better than its input. Furthermore, when statistical procedures are used to estimate input parameters (say, from data), the output measures of effectiveness actually become random variables, and it is often of interest to obtain confidence statements concerning them. The primary references for the material to follow are a survey paper by Cox (1965) as well as Clarke (1957), Billingsley (1961), Wolff (1965), and Harris (1974).

As already stated, it is our intention in this section to describe the problems of statistical inference which arise when it is assumed that a specific model form is applicable. This in turn leads to a further subdivision into problems of estimation, hypothesis testing, and confidence statements, though all of course are clearly related.

The initial step in any statistical procedure is a determination of the availability of sample information. The method that is chosen for estimation and the form of the estimators depend very much on the completeness of the monitoring process. It is one thing if one can observe a system fully over a certain period and is thus able to record the instants of service for each customer, but quite a different problem to have such incomplete information as only the queue size at each departure.

The earliest work on the statistics of queues did, in fact, assume that the subject queue was fully observed over a period of time and therefore that complete information was available in the form of the arrival instants and the points of the beginning and end of the service of each customer. As would then be expected, the queue was assumed to be a Markov chain in continuous time. Clarke (1957) began a sequence of papers on this and related topics by obtaining the maximum-likelihood estimators for the arrival and service parameters of an $M/M/1$ queue, in addition to the variance–covariance matrix for the two statistics. This work was followed shortly thereafter by a similar exposition for $M/M/\infty$ by Beneš (1957). Clarke's work was fundamental, and therefore a brief discussion of it follows.

We recall that the class known as maximum-likelihood estimates (MLEs) can be obtained as follows. First, we form the likelihood L as a function of the (one or more) model parameters equal to the joint density of the observed sample. Then, the MLEs of the parameters are the values maximizing L. Remember that it is often mathematically easier to find the maximum of the natural logarithm of L, which is generally written as $\mathcal{L}$.

To describe Clarke's work, begin by assuming that a stationary $M/M/1$ queue is being observed, with unknown mean arrival rate λ and mean service rate μ ($\rho \equiv \lambda/\mu$). We suppose that the queue begins operation with no customers present and then consider both the conditional likelihood given $N(0) = n_0$ and the likelihood ignoring the initial condition. It is clear that times between transitions are exponential, with mean $1/(\lambda+\mu)$ when the zero state is not occupied and mean $1/\lambda$ when the zero state is occupied. All jumps are upward from zero, while jumps upward from nonzero states occur with probability $\lambda/(\lambda+\mu)$ and jumps downward with probability $\mu/(\lambda+\mu)$, all independent of previous queue history.

Let us further assume that we are observing the system for a fixed amount of time t, where t is sufficiently large to guarantee some appropriate number of observations and must be chosen independently of the arrival and service processes, so that the sampling interval is independent of λ and μ. Let us then use t_e and t_b to denote, respectively, the amounts of time the system is empty and busy ($t_b = t - t_e$). In addition, let n_c denote the number of customers who have been served, n_{ae} the number of arrivals to an empty system, and n_{ab} the number to a busy system, with the total number of arrivals $n_a = n_{ae} + n_{ab}$. This is essentially Clarke's notation, and it is most convenient to use in this situation.

The likelihood function is then made up of components that are formed from the following kinds of information:

1. intervals of length x_b spent in a nonzero state and ending in an arrival or departure;

2. intervals of length x_e spent in the zero state and ending in an arrival;
3. the very last (unended) interval (length x_l) of observation;

4. arrivals at a busy system;
5. departures;
6. the initial number of customers, n_0.

The contributions to the likelihood of each of these are

1. $(\lambda+\mu)e^{-(\lambda+\mu)x_b}$;
2. $\lambda e^{-\lambda x_e}$;
3. $e^{-\lambda x_l}$ or $e^{-(\lambda+\mu)x_l}$;
4. $\lambda/(\lambda+\mu)$;
5. $\mu/(\lambda+\mu)$;
6. $\Pr\{N(0)=0\}$.

Since $\sum x_b = t_b, \sum x_e = t_e$, and $n_{ae}+n_{ab}=n_a$, the likelihood is found to be (with the unended time x_l properly included in either t_b or t_e)

$$\begin{aligned} L(\lambda,\mu) &= (\lambda+\mu)^{n_c+n_{ab}} \exp\left(-(\lambda+\mu)\sum x_b\right)\lambda^{n_{ae}}\exp\left(-\lambda\sum x_e\right) \\ &\quad\times\left(\frac{\lambda}{\lambda+\mu}\right)^{n_{ab}}\left(\frac{\mu}{\lambda+\mu}\right)^{n_c}\Pr\{n_0\} \\ &= e^{-(\lambda+\mu)t_b}\lambda^{n_a}e^{-\lambda t_e}\mu^{n_c}\Pr\{n_0\}. \end{aligned}$$

The log-likelihood function $\mathcal{L}$ corresponding to the above L is given by

$$\mathcal{L}(\lambda,\mu) = -\lambda t - \mu t_b + n_a \ln\lambda + n_c\ln\mu + \ln\Pr\{n_0\}. \tag{6.32}$$

In the event that the queue is in equilibrium, the initial size may be ignored, and then the MLEs $\hat{\lambda}$ and $\hat{\mu}$ ($\hat{\rho}=\hat{\lambda}/\hat{\mu}$) will be given by the solution to

$$\frac{\partial\mathcal{L}}{\partial\lambda}=0,\quad \frac{\partial\mathcal{L}}{\partial\mu}=0.$$

In this case,

$$\frac{\partial\mathcal{L}}{\partial\lambda}=-t+\frac{n_a}{\lambda},\quad \frac{\partial\mathcal{L}}{\partial\mu}=-t_b+\frac{n_c}{\mu}.$$

Thus

$$\hat{\lambda}=\frac{n_a}{t},\quad \hat{\mu}=\frac{n_c}{t_b}. \tag{6.33}$$

Note that this is what one would obtain by observing individual interarrival and service times and taking their sample averages (ignoring the last unended interval); we later show in Section 8.3 that these are also the MLEs.

It must be that $\hat{\rho}=\hat{\lambda}/\hat{\mu}<1$, since equilibrium has been assumed; but if this condition is violated, then we must assume that $\hat{\lambda}/\hat{\mu}\approx 1$ and minimize $\mathcal{L}(\lambda,\mu)+$

$\theta(\lambda, \mu)$, where θ is a Lagrange multiplier, and obtain as the common estimator $\hat{\lambda} = \hat{\mu} = (n_a + n_c)/(t + t_b)$.

If we now instead assume that $N(0)$ cannot be ignored, then in order to obtain any meaningful results, some assumption must be made regarding the distribution of this initial size. If ρ is known to be less than one, then by choosing $\Pr\{N(0) = n_0\}$ to be $\rho^{n_0}(1-\rho)$ we would immediately place ourselves in the steady state. On the other hand, we may want to do the estimation under the assumption that ρ can indeed be greater than one. But then the suggested geometric distribution for system size would not be appropriate. In this case, an alternative approach must be tried, and the choice is somewhat arbitrary.

For illustrative purposes, let us now use $\Pr\{n_0\} = \rho^{n_0}(1-\rho)$, in which case (6.32) becomes

$$\mathcal{L}(\lambda, \mu) = -\lambda t - \mu t_b + n_a \ln \lambda + n_c \ln \mu + n_0(\ln \lambda - \ln \mu) + \ln\left(\frac{1-\lambda}{\mu}\right).$$

Then

$$\frac{\partial \mathcal{L}}{\partial \lambda} = -t + \frac{n_a}{\lambda} + \frac{n_0}{\lambda} - \frac{1}{\mu - \lambda},$$
$$\frac{\partial \mathcal{L}}{\partial \mu} = -t_b + \frac{n_c}{\mu} - \frac{n_0}{\mu} + \frac{\lambda}{\mu(\mu - \lambda)}.$$

The estimators $\hat{\lambda}$ and $\hat{\mu}$ are thus the solution to

$$0 = -t + \frac{n_a + n_0}{\hat{\lambda}} - \frac{1}{\hat{\mu} - \hat{\lambda}},$$
$$0 = -t_b + \frac{n_c - n_0}{\hat{\mu}} + \frac{\hat{\lambda}}{\hat{\mu}(\hat{\mu} - \hat{\lambda})}. \tag{6.34}$$

The first of these equations simplifies to

$$\hat{\mu} - \hat{\lambda} = \frac{\hat{\lambda}}{n_a + n_0 - \hat{\lambda} t}.$$

Then eliminating $\hat{\mu}$ from the second equation gives a quadratic in $\hat{\lambda}$, which would then be used to obtain two values of $\hat{\lambda}$. Any negative value obtained is rejected, and for the remaining values of $\hat{\lambda}$, the corresponding value of $\hat{\mu}$ is obtained. In addition, one would reject any $(\hat{\lambda}, \hat{\mu})$ pair for which $\hat{\mu} \leq 0$ or $\hat{\lambda}/\hat{\mu} > 1$. If both solutions are valid, then the one that maximizes the likelihood function is kept. If neither solution is valid and it is positiveness that is violated, let the violating parameter be equal to a small positive ϵ; otherwise, let $\hat{\lambda} = \hat{\mu}$.

Another approach, incorporating the initial state, is to adjust (6.33) by subtracting the initial system size minus the estimated mean equilibrium system size divided by the observing time from $\hat{\mu}$, since the effect of the difference of n_0 from the steady-state mean is then removed. The reverse is true of $\hat{\lambda}$; that is, this quantity is added.

This gives the approximations

$$\hat{\lambda} \approx \frac{n_a}{t} + \frac{n_0 - (n_a t_b/n_c t)/(1 - n_a t_b/n_c t)}{t},$$
$$\hat{\mu} \approx \frac{n_c}{t_b} + \frac{n_0 - (n_a t_b/n_c t)/(1 - n_a t_b/n_c t)}{t_b}, \tag{6.35}$$

where $(n_a t_b/n_c t)/(1 - n_a t_b/n_0 t)$ is an estimate of L, namely, $\hat{L} = \hat{\rho}/(1 - \hat{\rho})$. It should be noted that all the estimators presented here for λ and μ are indeed consistent.

Cox (1965) and Lilliefors (1966) have considered the problem of finding confidence intervals for the actual $M/M/1$ traffic intensity from the MLEs given by (6.33). Since the individual interarrival times are IID exponential random variables, the quantity t is Erlang type n_a with mean n_a/λ; hence λt is Erlang type n_a with mean n_a. Likewise, the quantity μt_b is Erlang type n_c with mean n_c. If the sampling stopping rule is carefully specified to guarantee independence of n_a and n_c, then the distribution for the ratio $t_b/t\rho$ is given as $F_{2n_c,2n_a}(t_b/t\rho)$, where $F_{a,b}(x)$ is the usual F distribution with degrees of freedom a and b. (The twos in the degrees of freedom enter when the Erlang distributions are converted to χ^2 distributions, the ratio of which then yields the F distribution.) But, from (6.33), $t_b/t\rho = (n_c/n_a)\hat{\rho}/\rho$, and thus confidence intervals can readily be found for ρ by the direct use of the F distribution, with the upper $1 - \alpha$ confidence limit ρ_u found from

$$\frac{n_c \hat{\rho}}{n_a \rho_u} = F_{2n_c,2n_a}(\alpha/2),$$

and lower $1 - \alpha$ confidence limit ρ_l found from

$$\frac{n_c \hat{\rho}}{n_a \rho_l} = F_{2n_c,2n_a}(1 - \alpha/2).$$

In addition, confidence intervals can be found for any of the usual measures of effectiveness which are functions of ρ.

EXAMPLE 6.7

As a simple illustration, let us consider the following problem. Observations are made of an $M/M/1$ queue, and it is noted at time $t = 400$ h that all of 60 arrivals have been duly served and departed. Of the 400 h of observation, the server was actually busy for a total of 300 h. Let us then find a 95% confidence interval for the traffic intensity ρ.

By the previous discussion, $\hat{\lambda} = 60/400 = \frac{3}{20}$ and $\hat{\mu} = 60/300 = \frac{1}{5}$, so that $\hat{\rho} = \frac{3}{4}$. Furthermore, we are interested in confidence intervals at a level $\alpha = 0.05$. Then the appropriate upper and lower limits for degrees of freedom 120 and 120 are found to be approximately

$$\frac{n_c \hat{\rho}}{n_a \rho_u} \doteq 0.70 \quad \Rightarrow \quad \rho_u \doteq \frac{\hat{\rho}}{0.70} \doteq 1.07$$

and

$$\frac{n_c \hat{\rho}}{n_a \rho_l} \doteq 1.43 \quad \Rightarrow \quad \rho_l \doteq \frac{\hat{\rho}}{1.43} \doteq 0.52.$$

We therefore conclude with 95% confidence that ρ will fall in the interval (0.52, 1.07).

These same kinds of ideas can be nicely extended to exponential queues with many servers, and to cases with Erlang input and/or service. In addition, Billingsley (1961) has made a detailed study of likelihood estimation for Markov chains in continuous time, including limit theory and hypothesis testing, and then these results were used and extended to obtain results for birth–death queueing models by Wolff (1965).

Suppose that we now wish to apply the likelihood procedure to an $M/G/1$ queue. The approach is similar except that the loss of memorylessness alters the likelihood function, since no use can be made of data that distinguish between empty and busy intervals, and there are now four components of the likelihood, namely:

1. interarrival intervals of length x, which are exponential, with contribution $\lambda e^{-\lambda x}$;

2. service times of duration x for the n_c completed customers, with contribution $b(x)$;

3. time spent in service (say, x_l) by the very last customer, with contribution $1 - B(x_l)$;

4. the initial number of customers.

Hence the likelihood may be written as

$$L(\lambda, \mu) = e^{-\lambda(t - x_l)} \lambda^{n_a} \left(\prod_{i=1}^{n_c} b(x_i) \right) [1 - B(x_l)] \Pr\{n_0\},$$

and the log likelihood as

$$\mathcal{L}(\lambda, \mu) = n_a \ln \lambda - \lambda(t - x_l) + \sum_{i=1}^{n_c} \ln b(x_i) + \ln[1 - B(x_l)] + \ln \Pr\{n_0\}.$$

Then derivatives are taken in the usual way, and the procedure follows that of the $M/M/1$ thereafter.

EXAMPLE 6.8

Let us find the MLEs for an $M/E_2/1$ queue with mean arrival rate λ and mean service time $2/\mu$. From the above, since $b(t) = \mu^2 t e^{-\mu t}$, the log likelihood may be written as

$$\mathcal{L}(\lambda,\mu) = n_a \ln\lambda - \lambda(t - x_l) + \sum_{i=1}^{n_c}(2\ln\mu + \ln x_i - \mu x_i) + \ln\left(1 - \int_0^{x_l}\mu^2 t e^{-\mu t}\,dt\right) + \ln\Pr\{n_0\}.$$

But the integral of an Erlang may be rewritten in terms of a Poisson sum [see (1.15)] as

$$\int_0^{x_l}\mu^2 t e^{-\mu t}\,dt = 1 - e^{-\mu x_l}\sum_{i=0}^{1}\frac{(\mu x_l)^i}{i!};$$

hence

$$\mathcal{L}(\lambda,\mu) = n_a\ln\lambda - \lambda(t - x_l) + 2n_c\ln\mu + \sum_{i=1}^{n_c}\ln x_i - \mu\sum_{i=1}^{n_c}x_i + \ln(e^{-\mu x_l} + \mu x_l e^{-\mu x_l}) + \ln\Pr\{n_0\}.$$

The partial derivatives can be computed to be

$$\frac{\partial\mathcal{L}}{\partial\lambda} = \frac{n_a}{\lambda} - (t - x_l) + \frac{\partial\ln\Pr\{n_0\}}{\partial\lambda},$$

$$\frac{\partial\mathcal{L}}{\partial\mu} = \frac{2n_c}{\mu} - \sum_{i=1}^{n_c}x_i - x_l + \frac{x_l}{1+\mu x_l} + \frac{\partial\ln\Pr\{n_0\}}{\partial\mu}.$$

If now the initial state is assumed to be chosen free of λ and μ, then the MLEs for λ and μ are found by equating the partial derivatives to zero as

$$\hat{\lambda} = \frac{n_a}{t - x_l},$$

and the appropriate solution $\hat{\mu}$ to the quadratic equation

$$x_l\left(x_l + \sum_{i=2}^{n_c}x_i\right)\hat{\mu}^2 + \left(\sum_{i=1}^{n_c}x_i - 2n_c x_l\right)\hat{\mu} - 2n_c = 0.$$

Thus far we have assumed that we were dealing with simple processes subject to complete information. But suppose now that certain kinds of information are just not available. For example, suppose we observe only the stationary output of an $M/G/1$ (G known) queue and are then asked to estimate the mean service and interarrival times. Under the assumption that the stream is in equilibrium, the mean interarrival time $1/\lambda$ must equal the long-term arithmetic mean of the interdeparture times. If the mean of the departure process is denoted by $\bar{d}$, then the maximum-likelihood estimator of λ is $\hat{\lambda} = 1/\bar{d}$.

If service is exponential, then we know that the limiting distribution of output is the same as that of the input. Hence no inference is possible about the mean service time. But if service is assumed to be other than exponential, then the estimation is possible. The CDF of the interdeparture process of any $M/G/1$ queue is given by (recalling that $p_0 = 1 - \lambda/\mu$)

$$C(t) = \frac{\lambda}{\mu}B(t) + \left(1 - \frac{\lambda}{\mu}\right)\int_0^t B(t-x)\lambda e^{-\lambda x}\,dx, \tag{6.36}$$

where the last term is the convolution of service and interarrival-time CDFs. In the special case where the service time is constant, (6.36) reduces to

$$C(t) = \begin{cases} 0 & (t < 1/\mu), \\ \lambda/\mu & (t = 1/\mu), \\ \lambda/\mu + (1 - \lambda/\mu)(1 - e^{-\lambda(t-1/\mu)}) & (t > 1/\mu). \end{cases}$$

Since there is nonzero probability associated with the point $t = 1/\mu$, we may directly obtain our estimate by equating $1/\mu$ with the minimum observed interdeparture time.

On the other hand, if the distribution is other than exponential or deterministic, an approach is to take LSTs of (6.36), yielding

$$C^*(s) = \frac{(1 + s/\mu)B^*(s)}{1 + s/\lambda}, \tag{6.37}$$

where $B^*(s)$ is the LST of the service-time CDF, whose form is known, but not the values of its parameters. Then the moments of $C(t)$ and $B(t)$ may be directly related by the successive differentiation of (6.37), using enough equations to determine all parameters of $B(t)$. However, there still exists a small problem: successive interdeparture times will be correlated, and this correlation must be considered when calculating moments from data. For example, if enough data are present, data spread sufficiently far apart may be considered to form an approximately random sample so that formulas based on uncorrelated observations can be used. It would be advisable, nevertheless, to test for lack of correlation by computing the sample correlation coefficient between successive observations before making any definitive statements. We have asked the reader to solve a problem of this type at the end of the chapter as Problem 6.32. There is also a discussion of this and related questions in Cox (1965).

Suppose we slightly modify the previous problem and instead now consider an $M/G/1$ with the form of G known and observations now made on both the input and output. The analysis might then proceed in a very similar manner, except that, since we are observing ordered instants of arrival and departure, the successive customer waiting times are now available. The relevant relationship between the transforms of $B(t)$ and the system-wait CDF $W(t)$ is [see (5.33)]

$$W^*(s) = \frac{(1 - \lambda/\mu)sB^*(s)}{s - \lambda + \lambda B^*(s)}. \tag{6.38}$$

Again, problems of autocorrelation surface. But assuming that these are taken into account as previously discussed, we know that in the event $B(t)$ is exponential,

$$\left.\frac{dW^*(s)}{ds}\right|_{s=0} = -\frac{1}{\mu - \lambda}.$$

Hence

$$\hat{W} = \frac{1}{\hat{\mu} - \hat{\lambda}} \quad \Rightarrow \quad \hat{\mu} = \hat{\lambda} + \frac{1}{\hat{W}}.$$

In fact, for any one-parameter service distribution, we may directly appeal to the PK formula [found via the first derivatives of (6.38) or from the results of Section 5.1], namely,

$$W = \frac{1}{\mu} + \frac{(\lambda/\mu)^2 + \lambda^2\sigma_S^2}{2\lambda(1 - \lambda/\mu)}.$$

In the deterministic case, for example, we find that

$$W = \frac{1}{\mu} + \frac{\lambda/\mu^2}{2(1 - \lambda/\mu)} = \frac{2 - \lambda/\mu}{2\mu(1 - \lambda/\mu)}.$$

Therefore we may write that

$$\hat{W} = \frac{2 - \hat{\lambda}/\hat{\mu}}{2\hat{\mu}(1 - \hat{\lambda}/\hat{\mu})},$$

or that $\hat{\mu}$ is the appropriate solution to the quadratic

$$2\hat{W}\hat{\mu}^2 - 2(\hat{W}\hat{\lambda} + 1)\hat{\mu} + \hat{\lambda} = 0,$$

where $\hat{\lambda}$ is found as before from $\overline{d}$. Again, we leave the Erlang case as a problem (see Problem 6.33).

As for $M/G/\infty$, we have shown previously in Section 5.2.2 that both output and system size are nonhomogeneous Poisson streams. Hence we should be able to estimate parameters by converting all the observable processes to Poisson processes, from which it is easy to obtain appropriate estimators.

PROBLEMS

6.1. Reformulate the $M/E_2/1$ rootfinding problem of Section 3.3.4 to the type required in Section 6.1, and then verify the answers from Section 3.3.4.

6.2. Find the general form of $W_q(t)$ for the $D/E_k/1$.

6.3. You have learned that the Laplace transform of the distribution function of the system waiting time in a $G/G/1$ system is

$$\overline{W}(s) = \frac{1}{s} - \frac{3s^2 + 22s + 36}{3(s^2 + 6s + 8)(s + 3)}.$$

Find $W(t)$ and its mean.

6.4. The Bearing Straight Corporation of Example 5.3 now finds that its machines do not break down as a Poisson process, but rather as the two-point distribution given below:

Interarrival Time	Probability	$A(t)$
9	$\frac{2}{3}$	$\frac{2}{3}$
18	$\frac{1}{3}$	1

Given that the service has the same two-point distribution as it did earlier, use Lindley's equation to find the line-delay CDF.

6.5. Use the approach indicated for $M/G/1$ waiting times in Section 5.1.5 to verify the waiting-time result of Example 6.2.

6.6. Consider a $G/G/1$ queue whose service times are uniformly distributed on (0, 1), with $\lambda = 1$. A specific arrival in steady state encounters 12 people in the system when it arrives. What is its mean wait in queue?

6.7. Find the probability mass functions for the arrival counts associated with deterministic and Erlang input streams.

6.8. Show that the virtual idle time for a $G/M/1$ has CDF

$$F(u) = A(u) + \int_u^\infty e^{-\mu(1-r_0)(t-u)}\, dA(t).$$

6.9. Verify the derivation of (6.18) from (6.17).

6.10. **(a)** Show that the poles of (6.18) are distinct.
(b) Fully solve an $M/D/2$ queue with $\lambda = \mu = 1$.

6.11. Use the generating function for the state probabilities of Problem 6.10(b) to find the variance of the line wait under the same assumptions.

6.12. Consider an $M^{[X]}/M/1$ queue with the following modification: If there are two or more customers waiting when the server is free, then the next two customers are served together; if there is one customer waiting when the server is free, then this customer is served by itself. The service time is exponential with mean $1/\mu$ regardless of the number being served. Find the single-step transition probabilities p_{ij} of the imbedded Markov chain.

6.13. Repeat Problem 6.12 for the two-channel exponential–exponential machine-repair problem, and also find the general-time stationary probabilities.

6.14. Consider the $G/M/3$ queue with the same interarrival distribution as in Problem 6.4, and mean service rate $\mu = 0.035$. Find the general-time probabilities.

6.15. Find the general-time probability distribution associated with the queue in Example 5.9, as well as the mean system and queue sizes.

6.16. Find the general-time probability distribution associated with the queue in Problem 5.32(b), as well as the mean system and queue sizes.

6.17. For Problem 5.36, use the results of Section 6.4 to get general-time probabilities.

6.18. Find the general-time probability distribution associated with the queue in Problem 5.38.

6.19. Find p_0, p_1, L, and W for a $G/G/1/1$ queue with $\lambda = \mu = 1$.

6.20. Consider an $M/M/1/\infty$ queue for which $\lambda = 1$ and $\mu = 3$. Every customer going through the system pays an amount \$15, but costs the system \$6 per unit time it spends in the system.

(a) What is the average profit rate of this system?

(b) A bright young OR analyst from a prestigious middle Atlantic state university tells management that the profit can be increased by shutting off the queue at a certain point, that is, by preventing customers from entering whenever the queue gets to a certain size. Do you agree, and if so, what is the optimal point at which to prevent arrivals from joining?

6.21. Redo Problem 2.51 for C_1 = \$24, C_2 = \$138, and downtime cost = \$10/h.

6.22. Consider a two-server system, where each server is capable of working at two speeds. The service times are exponential with mean rate μ_1 or μ_2. When there are $k(> 2)$ customers in the system, the mean rate of both servers switches from μ_1 to μ_2. Suppose that the cost per operating hour at low speed is \$50 and at high speed is \$220, and the cost of waiting time (time spent in the system) per customer is \$10/h. The arrival rate is Poisson with mean 20/h, while the service rates μ_1 and μ_2 are 7.5 and 15/h, respectively. Use the results of Problem 2.52 to find the optimal k. Compare the solution with that of Problem 2.51.

6.23. Consider an $M/M/1$ queue with a three-state service rate as explained in Problem 2.53. Suppose that management policy sets the upper switch point k_2 at 5. Using the results of Problem 2.53, find the optimal k_1 (the lower switch point) for a system with mean arrival rate of 40, μ_1 of 20, μ_2 of 40, and μ of 50. Assume that the waiting cost per customer is \$10/h and the service costs per hour are \$100, \$150, and \$250, respectively.

6.24. Suppose we have an $M/M/1$ queue with customer balking. Furthermore, suppose the balking function b_n is given as $e^{-n/\mu}$. where μ is the mean service rate (see Section 2.10.1). It is known that the salary of the server depends on his or her skill, so that the marginal cost of providing service is proportional to the rate μ and, in fact, is estimated as $\$1.50\mu$/h. Arrivals are Poisson with a mean rate of 10/h. The profit per customer served is estimated at \$75. Find the optimal value of μ.

6.25. Consider a tool crib where the manager believes that costs are only associated with idle time, that is, the time that the clerks are idle (waiting for mechanics to come) and the time that the mechanics are idle (waiting in the queue for a clerk to become available). Using costs \$30 per hour idle per clerk and \$70 per hour idle per mechanic, with $1/\lambda = 50$ s, $1/\mu = 60$ s, find the optimal number of clerks. Comment.

6.26. Consider an $M/M/1$ queue where the mean service rate μ is under the control of management. There is a customer waiting cost per unit time spent in the system of C_W and a service cost that is proportional to the *square* of the mean service rate, the constant of proportionality being C_S. Find the optimal value of μ in terms of λ, C_S, and C_W. Solve when $\lambda = 10, C_S = 2$, and $C_W = 20$.

6.27. Find the quadratic in $\hat{\lambda}$ that arises in the simultaneous solution of (6.34), and then solve.

6.28. Get maximum-likelihood estimators for λ and μ in an $M/M/1$ queue, first ignoring the initial state and then with an initial state of 4 in steady state, from the data $t = 150$, $t_b = 100$, $n_a = 16$, and $n_c = 12$.

6.29. Under the assumption that n_0 is always fixed at 0, find the MLEs for λ, μ, and ρ in an $M/M/1$ queue from the data $t = 150$, $t_b = 100$, $n_a = 16$, and $n_c = 8$.

6.30. Give a 95% confidence interval for the ρ in Problem 6.29.

6.31. Find the formulas for the MLEs of λ and μ in an $M/E_3/1$.

6.32. Use (6.37) to find the estimators of the service parameter from the output of an $M/E_2/1$ queue.

6.33. Use (6.38) and subsequent results to find the estimator of the service parameter ($\hat{\mu}$) from the system waiting times and arrival times of the $M/E_2/1$ queue.

6.34. The observed interdeparture times of an $M/G/1$ queue are:

0.6, 2.4, 1.0, 1.1, 0.2, 0.2, 0.2, 2.7, 1.5, 0.3, 0.6, 0.9, 0.5, 0.2, 0.5, 3.8, 0.4, 0.1, 1.5, 1.4, 0.7, 0.5, 0.1, 0.6, 1.6, 1.5, 0.5, 0.8, 1.3, 0.4, 0.7, 2.4, 2.4, 0.3, 0.8, 0.9, 1.5, 0.3, 1.2, 1.0, 0.6, 0.1, 0.4, 0.3, 2.5, 3.5, 0.8, 0.6, 9.5, 1.6.

Test to determine whether this stream is exponential and therefore whether the queue is $M/M/1$.

CHAPTER 7

BOUNDS AND APPROXIMATIONS

For many queueing systems, direct analytical results are not available. One way to proceed is to develop bounds or approximations for such systems. Bounds are useful in that they provide worst-case or best-case performance metrics. For example, with an upper bound, a safe choice can be made for the number of servers so that congestion does not exceed a desired threshold. Furthermore, with bounds, approximations can be developed based on the bounds. For example, a simple approximation is to take the average of an upper bound and a lower bound. Of course, more sophisticated approximations can be developed by using information about the performance of the bounds. Approximations can also be developed from other theoretical principles – for example, by analyzing a separate queueing system whose behavior is similar to the original, but which can be analyzed more easily.

The technical presentation of this chapter is divided into three parts. First, we offer a derivation of some popular upper and lower bounds for expected line waits (and therefore also for the mean queue sizes) of both steady-state $G/G/1$ and $G/G/c$ queues. Then we move on to the derivation of some commonly used approximations. A number of these are developed directly from the bounds presented in the first part. Other approximations are developed in an independent fashion. Finally, we conclude this chapter with the application of approximation methods to queueing networks.

Fundamentals of Queueing Theory, Fourth Edition.
By D. Gross, J. F. Shortle, J. M. Thompson, and C. M. Harris
Copyright © 2008 John Wiley & Sons, Inc.

7.1 Bounds

In the following section, we first present upper and lower bounds for the mean line delay of a steady-state $G/G/1$ queue as a function only of the first and second moments of its interarrival and service times. This is followed by the derivation of a further lower bound when the full forms of the interarrival and service-time distribution functions are known—but possibly too complicated to analyze in one of the usual ways. In Section 7.1.3 we then present some comparable results for the general multiserver queue.

7.1.1 Basic Relationships for Single-Server Queues

Before obtaining bounds, we first need to derive several basic relationships for stationary $G/G/1$ queues with $\rho < 1$. The relationships involve moments of the interarrival times, service times, idle periods, waiting times, and interdeparture times. Much of this development parallels results from Kingman (1962b) and Marshall (1968).

The starting point is the iterative equation for the line delays (see Section 6.2),

$$W_q^{(n+1)} = \max(0, W_q^{(n)} + U^{(n)}), \tag{7.1}$$

where $U^{(n)} \equiv S^{(n)} - T^{(n)}$, $S^{(n)}$ is the service time of the nth customer and $T^{(n)}$ is the interarrival time between the nth and $(n+1)$st customers. Let

$$X^{(n)} = -\min(0, W_q^{(n)} + U^{(n)}). \tag{7.2}$$

This is the time between the departure of the nth customer and the start of service of the $(n+1)$st customer. Then,

$$W_q^{(n+1)} - X^{(n)} = W_q^{(n)} + U^{(n)}. \tag{7.3}$$

To see this, if $W_q^{(n)} + U^{(n)}$ is positive, then $W_q^{(n+1)} = W_q^{(n)} + U^{(n)}$ from (7.1) and $-X^{(n)} = 0$ from (7.2). On the other hand, if $W_q^{(n)} + U^{(n)}$ is negative, then $W_q^{(n+1)} = 0$ and $-X^{(n)} = W_q^{(n)} + U^{(n)}$. Either way, $W_q^{(n+1)} - X^{(n)}$ equals $W_q^{(n)} + U^{(n)}$.

Now, for a stationary queue, $\mathrm{E}[W_q^{(n+1)}] = \mathrm{E}[W_q^{(n)}]$. Taking expectations of (7.3) gives

$$\mathrm{E}[X] = -\mathrm{E}[U] = \frac{1}{\lambda} - \frac{1}{\mu}. \tag{7.4}$$

Since X is the time between a customer departure and the next start of service (in steady-state), we also have

$$\begin{aligned}\mathrm{E}[X] &= \Pr\{\text{system found empty by an arrival}\} \cdot \mathrm{E}[\text{length of idle period}] \\ &= q_0 \mathrm{E}[I],\end{aligned}$$

where $\{q_n\}$ denotes the arrival point probabilities and I denotes the length of an idle period. This gives

$$\mathrm{E}[I] = \frac{\mathrm{E}[X]}{q_0} = -\frac{\mathrm{E}[U]}{q_0} = \frac{1/\lambda - 1/\mu}{q_0}. \tag{7.5}$$

The next result is a formula for the expected wait for a stable $G/G/1$ queue in terms of the first and second moments of U and I. Squaring both sides of (7.3) gives

$$(W_q^{(n+1)})^2 - 2W_q^{(n+1)}X^{(n)} + (X^{(n)})^2 = (W_q^{(n)})^2 + 2W_q^{(n)}U^{(n)} + (U^{(n)})^2. \tag{7.6}$$

From (7.1) and (7.2), $W_q^{(n+1)}X^{(n)} = 0$, since one or the other of the two factors must be zero. Also, $W_q^{(n)}$ and $U^{(n)}$ are independent. Then, taking expectations of (7.6) and using $\mathrm{E}[(W_q^{(n+1)})^2] = \mathrm{E}[(W_q^{(n)})^2]$ from stationarity gives

$$\mathrm{E}[X^2] = 2W_q\mathrm{E}[U] + \mathrm{E}[U^2],$$

or

$$W_q = \frac{\mathrm{E}[X^2] - \mathrm{E}[U^2]}{2\mathrm{E}[U]}. \tag{7.7}$$

Now, $\mathrm{E}[X^2] = \Pr\{\text{system found empty by an arrival}\} \cdot \mathrm{E}[(\text{length of idle period})^2]$. Hence,

$$W_q = \frac{q_0\mathrm{E}[I^2] - \mathrm{E}[U^2]}{2\mathrm{E}[U]}.$$

From (7.5), we have that $\mathrm{E}[U] = -q_0\mathrm{E}[I]$. Combining this with the previous result gives

$$\boxed{W_q = -\frac{\mathrm{E}[I^2]}{2\mathrm{E}[I]} - \frac{\mathrm{E}[U^2]}{2\mathrm{E}[U]}.} \tag{7.8}$$

A similar expression can be found for the variance of the wait in queue by cubing (7.3) (see Marshall, 1968). In the case of Poisson arrivals, (7.8) reduces to the PK formula (Problem 7.2).

Before using these results to derive bounds for the $G/G/1$ queue, we derive a similar result for the departure process. This result will be used in Section 7.3. Let $D^{(n)}$ be the time between the successive departures of the nth and $(n+1)$st customers. Then,

$$D^{(n)} \equiv S^{(n+1)} + X^{(n)}.$$

To see this relationship, $X^{(n)}$ is the time between the departure of the nth customer and the start of service of the $(n+1)$st customer. Adding the service time of the $(n+1)$st customer gives the time between the departures of the two customers. Now, $S^{(n+1)}$ and $X^{(n)}$ are independent, so

$$\mathrm{Var}[D^{(n)}] = \mathrm{Var}[S^{(n+1)}] + \mathrm{Var}[X^{(n)}]. \tag{7.9}$$

From (7.3), and since $W_q^{(n)}$, $S^{(n)}$, and $T^{(n)}$ are independent,

$$\begin{aligned}\text{Var}[W_q^{(n+1)} - X^{(n)}] &= \text{Var}[W_q^{(n)} + U^{(n)}] \\ &= \text{Var}[W_q^{(n)} + S^{(n)} - T^{(n)}] \\ &= \text{Var}[W_q^{(n)}] + \text{Var}[S^{(n)}] + \text{Var}[T^{(n)}]. \end{aligned} \tag{7.10}$$

Using basic properties of variance and covariance,

$$\begin{aligned}&\text{Var}[W_q^{(n+1)} - X^{(n)}] \\ &= \text{Var}[W_q^{(n+1)}] + \text{Var}[X^{(n)}] - 2\text{Cov}[W_q^{(n+1)} X^{(n)}] \\ &= \text{Var}[W_q^{(n+1)}] + \text{Var}[X^{(n)}] - 2(\text{E}[W_q^{(n+1)} X^{(n)}] - \text{E}[W_q^{(n+1)}]\text{E}[X^{(n)}]) \\ &= \text{Var}[W_q^{(n+1)}] + \text{Var}[X^{(n)}] + 2\text{E}[W_q^{(n+1)}]\text{E}[X^{(n)}]. \end{aligned} \tag{7.11}$$

The last equality follows since $W_q^{(n+1)} X^{(n)} = 0$, as discussed before. Combining (7.10) and (7.11), letting $n \to \infty$, and using $\text{Var}[W_q^{(n+1)}] = \text{Var}[W_q^{(n)}]$ for a stationary queue gives

$$\text{Var}[S] + \text{Var}[T] = \text{Var}[X] + 2W_q\text{E}[X].$$

Solving for $\text{Var}[X]$ and substituting into (7.9) (with $n \to \infty$) gives

$$\text{Var}[D] = \text{Var}[S] + \text{Var}[S] + \text{Var}[T] - 2W_q\text{E}[X].$$

This can be rewritten as

$$\boxed{\text{Var}[D] = 2\sigma_B^2 + \sigma_A^2 - 2W_q\left(\frac{1}{\lambda} - \frac{1}{\mu}\right),} \tag{7.12}$$

where D is a random interdeparture time in steady state, $\sigma_B^2 = \text{Var}[S]$, $\sigma_A^2 = \text{Var}[T]$, $W_q = \text{E}[W_q^{(n+1)}]$, and $\text{E}[X]$ comes from (7.4).

7.1.2 Bounds for Single-Server Queues

We now apply some of these relationships to derive bounds that are valid for all stationary $G/G/1$ queues, with $\rho < 1$. Again, we follow work by Kingman (1962b) and Marshall (1968).

The first is a lower bound on the mean idle time. From (7.5), since $q_0 \le 1$, we have $\text{E}[I] \ge 1/\lambda - 1/\mu$, where equality is achieved for the $D/D/1$ queue. We can use this inequality to derive an upper bound for W_q. Begin by rewriting (7.8) as

$$W_q = \frac{-(\text{Var}[I] + \text{E}^2[I])}{2\text{E}[I]} - \frac{\text{E}[U^2]}{2\text{E}[U]}.$$

Since Var[I] is nonnegative,

$$W_q \leq \frac{-\mathrm{E}^2[I]}{2\mathrm{E}[I]} - \frac{\mathrm{E}[U^2]}{2\mathrm{E}[U]} = \frac{1}{2}\left(-\mathrm{E}[I] - \frac{\mathrm{E}[U^2]}{\mathrm{E}[U]}\right) = \frac{1}{2}\left(\frac{\mathrm{E}[U]}{q_0} - \frac{\mathrm{E}[U^2]}{\mathrm{E}[U]}\right),$$

where the last equality follows from (7.5). Since $\mathrm{E}[U] = 1/\mu - 1/\lambda < 0$ and $q_0 \leq 1$, which implies that $\mathrm{E}[U]/q_0 \leq \mathrm{E}[U]$, we see that

$$\begin{aligned} W_q &\leq \frac{1}{2}\left(\mathrm{E}[U] - \frac{\mathrm{E}[U^2]}{\mathrm{E}[U]}\right) = \frac{1}{2}\left(\frac{\mathrm{E}^2[U] - \mathrm{E}[U^2]}{\mathrm{E}[U]}\right) = \frac{1}{2}\left(\frac{-\mathrm{Var}[U]}{\mathrm{E}[U]}\right) \\ &= \frac{1}{2}\left(\frac{\mathrm{Var}[S] + \mathrm{Var}[T]}{1/\lambda - 1/\mu}\right), \end{aligned}$$

which can be rewritten as

$$\boxed{W_q \leq \frac{\lambda(\sigma_A^2 + \sigma_B^2)}{2(1-\rho)}.} \tag{7.13}$$

For a similar lower bound, we refer to the work of Marchal (1978). From (7.7), we can bound W_q from below if we can find a lower bound for $\mathrm{E}[X^2]$. To do so, we recognize from (7.2) that the random variable X is stochastically smaller than the interarrival time variable T, since X is either 0 or $T - (W_q + S)$. (We say that X is stochastically smaller than T if $\Pr\{X \leq x\} \geq \Pr\{T \leq x\}$ for all x, and we write this as $X \leq_{\mathrm{st}} T$.) It then follows that $\mathrm{E}[X^2] \leq \mathrm{E}[T^2]$. Thus from (7.7) we have

$$\begin{aligned} W_q &\geq \frac{\mathrm{E}[T^2] - \mathrm{E}[U^2]}{2\mathrm{E}[U]} \\ &= \frac{\mathrm{Var}[T] + \mathrm{E}^2[T] - \mathrm{Var}[U] - \mathrm{E}^2[U]}{2\mathrm{E}[U]} \\ &= \frac{\mathrm{Var}[T] + \mathrm{E}^2[T] - \mathrm{Var}[T] - \mathrm{Var}[S] - \mathrm{E}^2[S-T]}{2\mathrm{E}[U]} \\ &= \frac{1/\lambda^2 - \sigma_B^2 - 1/\lambda^2 - 1/\mu^2 + 2/(\mu\lambda)}{2(1/\mu - 1/\lambda)}. \end{aligned}$$

Finally, this simplifies to

$$\boxed{W_q \geq \frac{\lambda^2\sigma_B^2 + \rho(\rho-2)}{2\lambda(1-\rho)}.} \tag{7.14}$$

Because of the term -2ρ in the numerator, this lower bound is positive if and only if $\sigma_B^2 > (2-\rho)/\lambda\mu$, and is thus not always of value. Nevertheless, it can be quite useful in a variety of important situations.

Figure 7.1 shows the bounds applied to an $M/G/1$ queue, with $\lambda = 1$ and $\mu = 2$. For this queue, the exact value of W_q is known; see Table 5.1. The lower bound, upper bound, and exact value are all quadratic in the standard deviation of service σ_B.

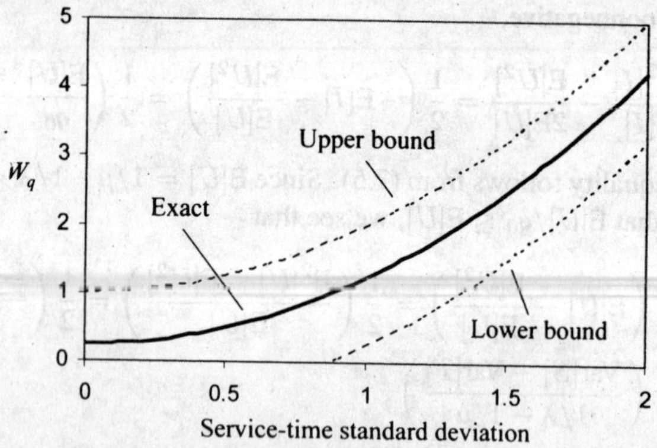

Figure 7.1 Bounds (7.13) and (7.14) applied to an $M/G/1$ queue.

Suppose it is now assumed that the interarrival and service distributions are both known and given by $A(t)$ and $B(t)$, respectively. Then another lower bound on the stationary line wait may be found to be $W_q \geq r_0$, where r_0 is the unique nonnegative root when $\rho < 1$ of

$$f(z) = z - \int_{-z}^{\infty} [1 - U(t)]\, dt = 0,$$

where $U(t)$ is the CDF of $U = S - T$.

To prove this assertion, we begin by observing that $f(z)$ does indeed have a unique nonnegative root. It is easily seen that $f(z)$ is monotonically increasing for $z \geq 0$, since

$$f'(z) = 1 - [1 - U(-z)] = U(-z) \geq 0.$$

When $z = 0$,

$$f(0) = -\int_{0}^{\infty} [1 - U(t)]\, dt < 0,$$

since the integrand is always positive; when z is large, say, $z = M$,

$$\begin{aligned} f(M) &= M - \int_{-M}^{\infty} [1 - U(t)]\, dt = M - \int_{-M}^{\infty} \int_{t}^{\infty} dU(x)\, dt \\ &= M - \int_{-M}^{\infty} \int_{-M}^{x} dt\, dU(x) = M - \int_{-M}^{\infty} (x + M)\, dU(x) \\ &\geq M - \int_{-\infty}^{\infty} (x + M)\, dU(x) = M - (\mathrm{E}[U] + M) \\ &= -\left(\frac{1}{\mu} - \frac{1}{\lambda}\right) = \frac{1 - \rho}{\lambda} > 0. \end{aligned}$$

Since $f(z)$ is monotonic and goes from $f(0) < 0$ to $f(M) > 0$, we have thus shown there is a unique nonnegative root when $\rho < 1$, which we call r_0. It therefore remains for us to show that $W_q \geq r_0$. Now rewrite $f(z)$ as $f(z) = z - f_1(z)$, where

$f_1(z) = \int_{-z}^{\infty}[1 - U(t)]dt$. Then we have that

$$f_1(z) = \int_{-z}^{\infty} [1 - U(t)]\, dt \quad \begin{cases} > z & (z < r_0), \\ \leq z & (z \geq r_0). \end{cases} \tag{7.15}$$

The function $f(z)$ is, in fact, continuous and convex, and we are going to apply Jensen's inequality for the expected value of a convex function of a nonnegative random variable (e.g., see Parzen, 1960) to get a relationship between W_q and f_1. Given that $W_q^{(n)}$ is (say) x, we see that

$$\begin{aligned}
\mathrm{E}[W_q^{(n+1)} | W_q^{(n)} = x] &= \mathrm{E}[\max(0, x + U^{(n)})] \\
&= \int_{-x}^{\infty} (x + t)\, dU^{(n)}(t) = \int_{-x}^{\infty} t\, dU^{(n)}(t) + x[1 - U^{(n)}(-x)] \\
&= \int_0^{\infty} \int_0^{t} dv\, dU^{(n)}(t) - \int_{-x}^{0} \int_t^{0} dv\, dU^{(n)} + x[1 - U^{(n)}(-x)] \\
&= \int_0^{\infty} \int_v^{\infty} dU^{(n)}(t)\, dv - \int_{-x}^{0} \int_{-x}^{v} dU^{(n)}(t)\, dv + x[1 - U^{(n)}(-x)].
\end{aligned}$$

Integrating over v gives

$$\begin{aligned}
&\mathrm{E}[W_q^{(n+1)} | W_q^{(n)} = x] \\
&= \int_0^{\infty} [1 - U^{(n)}(v)]\, dv - \int_{-x}^{0} [U^{(n)}(v) - U^{(n)}(-x)]\, dv + x[1 - U^{(n)}(-x)] \\
&= \int_0^{\infty} [1 - U^{(n)}(v)]\, dv \\
&\quad - \int_{-x}^{0} \{[1 - U^{(n)}(-x)] - [1 - U^{(n)}(v)]\}\, dv + x[1 - U^{(n)}(-x)] \\
&= \int_0^{\infty} [1 - U^{(n)}(v)]\, dv - x[1 - U^{(n)}(-x)] \\
&\quad - \int_{-x}^{0} [1 - U^{(n)}(v)]\, dv + x[1 - U^{(n)}(-x)] \\
&= \int_{-x}^{\infty} [1 - U^{(n)}(v)]\, dv = f_1(x).
\end{aligned}$$

Hence by the law of total probability, $\mathrm{E}[W_q^{(n+1)}] = \int_0^{\infty} f_1(x)\, dW_q^{(n)}(x)$. But Jensen's inequality then tells us that for a convex function f, $\mathrm{E}[f(x)] \geq f(\mathrm{E}[x])$, so that $\int_0^{\infty} f_1(x)\, dW_q^{(n)}(x) \geq f_1(\mathrm{E}[W_q^{(n)}])$, and hence $E[W_q^{(n+1)}] \geq f_1(\mathrm{E}[W_q^{(n)}])$, or, in the steady state,

$$W_q \geq f_1(W_q) = \int_{-W_q}^{\infty} [1 - U(t)]\, dt. \tag{7.16}$$

We now assume that $W_q < r_0$ and proceed to prove the result by contradiction. Equation (7.15) says that

$$\int_{-W_q}^{\infty} [1 - U(t)]\, dt > W_q,$$

for $W_q < r_0$. But this is a contradiction of (7.16); hence the result is shown and $r_0 \le W_q$. Putting the upper and lower bounds together gives

$$\boxed{\max\left(0, r_0, \frac{\lambda^2\sigma_B^2 + \rho(\rho - 2)}{2\lambda(1 - \rho)}\right) \le W_q \le \frac{\lambda(\sigma_A^2 + \sigma_B^2)}{2(1 - \rho)}.} \tag{7.17}$$

To illustrate, we apply the bounds to the $M/M/1$ queue. From (6.13),

$$U(t) = \begin{cases} \dfrac{\mu e^{\lambda t}}{\lambda + \mu} & (t < 0), \\ 1 - \dfrac{\lambda e^{-\mu t}}{\lambda + \mu} & (t \ge 0). \end{cases}$$

The lower bound r_0 is then found by solving $f(z) = 0$; that is,

$$0 = r_0 - \int_{-r_0}^{\infty} [1 - U(t)]\, dt = r_0 - \int_{r_0}^{0} \left(1 - \frac{\mu e^{\lambda t}}{\lambda + \mu}\right) dt - \int_{0}^{\infty} \frac{\lambda e^{-\mu t}}{\lambda + \mu}\, dt$$

$$= r_0 - r_0 + \frac{\mu(1 - e^{-\lambda r_0})}{\lambda(\lambda + \mu)} - \frac{\lambda}{\mu(\lambda + \mu)} = \frac{\mu^2 - \lambda^2 - \mu^2 e^{-\lambda r_0}}{\lambda\mu(\lambda + \mu)}.$$

So $\mu^2 - \lambda^2 = \mu^2 e^{-\lambda r_0}$, or $1 - \rho^2 = e^{-\lambda r_0}$. Hence finally,

$$r_0 = -\frac{1}{\lambda} \ln(1 - \rho^2).$$

The upper bound (7.13) for the $M/M/1$ queue is

$$\frac{\lambda(1/\lambda^2 + 1/\mu^2)}{2(1 - \rho)} = \frac{1 + \rho^2}{2\lambda(1 - \rho)}.$$

Both r_0 and the upper bound go to ∞ as ρ goes to 1, as expected. Figure 7.2 shows the bounds graphically. Indeed, this asymptotic behavior is true of the upper bound for all $G/G/1$ queues in the sense that the bound always gets asymptotically sharper. In turn, the lower bound gets sharper as ρ goes to zero.

In his paper, Marshall goes on to get some more results for special classes of possible arrival distributions, but it is not necessary to go into them here. The reader can pursue them in the cited reference. Suffice it to say that the bounds become tighter as more information is used on the interarrival and service times.

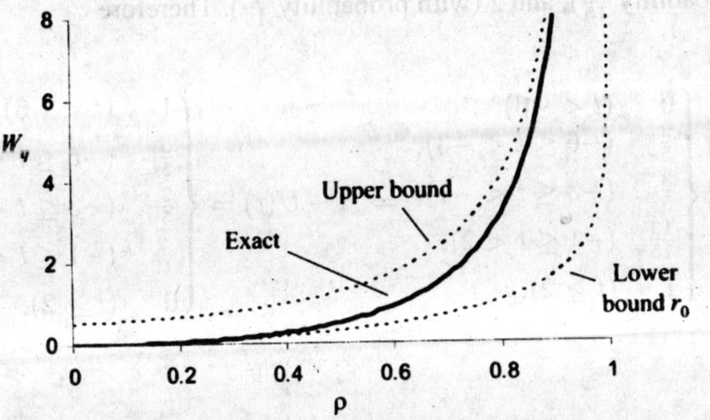

Figure 7.2 Bounds (7.17) applied to an $M/M/1$ queue.

■ EXAMPLE 7.1

To see how these results might apply to a real problem, let us examine their application to a $G/G/1$ queue in which both the interarrival times and service times are empirical. Assume that the service times are the same as in Example 5.3. That is, each service time is 9 minutes with probability $\frac{2}{3}$ and 12 minutes with probability $\frac{1}{3}$. Assume that each interarrival time is 10 minutes with probability $\frac{2}{5}$ and 15 minutes with probability $\frac{3}{5}$.

A little bit of calculation gives $\sigma_B^2 = 2\,\text{min}^2$ and $\sigma_A^2 = 6\,\text{min}^2$, with $\mu = \frac{1}{10}, \lambda = \frac{1}{13}$, and $\rho = \frac{10}{13}$. So we can immediately calculate the upper bound from (7.13) as

$$W_q \leq \frac{\frac{1}{13}(8)}{2(\frac{3}{13})} = \frac{4}{3}\,\text{min}.$$

Obtaining the lower bound is slightly more difficult, since we must find the nonnegative root of the nonlinear equation

$$f(r_0) = 0 = r_0 - \int_{-r_0}^{\infty} [1 - U(t)]\,dt.$$

(We use the second type of lower bound in view of the availability of so much information.) We begin by first calculating $U(t)$ directly from the empirical forms of the interarrival and service distributions. It is thus found that the only possible values of the random variable $U = S - T$ are $-6 = 9 - 15$ (with probability $(\frac{2}{3})(\frac{3}{5}) = \frac{2}{5}$), $-3 = 12 - 15$ (with probability $\frac{1}{5}$), $-1 = 9 - 10$

(with probability $\frac{4}{15}$), and 2 (with probability $\frac{2}{15}$). Therefore

$$U(t) = \begin{cases} 0 & (t < -6), \\ \frac{2}{5} & (-6 \le t < -3), \\ \frac{3}{5} & (-3 \le t < -1), \\ \frac{13}{15} & (-1 \le t < 2), \\ 1 & (t \ge 2), \end{cases} \Rightarrow 1 - U(t) = \begin{cases} 1 & (t < -6), \\ \frac{3}{5} & (-6 \le t < -3), \\ \frac{2}{5} & (-3 \le t < -1), \\ \frac{2}{15} & (-1 \le t < 2), \\ 0 & (t \ge 2). \end{cases}$$

So

$$\int_{-r_0}^{\infty} [1 - U(t)]\, dt = \int_{-r_0}^{2} [1 - U(t)]\, dt$$
$$= \begin{cases} \frac{2}{15} r_0 + \frac{4}{15} & (0 \le r_0 \le 1), \\ \frac{2}{5} r_0 & (1 < r_0 \le 3), \\ \frac{3}{5} r_0 - \frac{9}{15} & (3 < r_0 \le 6), \\ r_0 - 3 & (r_0 > 6). \end{cases}$$

Since the upper bound is just over one, we surmise that the lower bound should probably be less than one; that is to say, it is the solution to $r_0 = \frac{2}{15} r_0 + \frac{4}{15} \Rightarrow r_0 = \frac{4}{13}$, which must be correct, since r_0 is the unique nonnegative solution. Therefore

$$\tfrac{4}{13} \text{ min} \le W_q \le \tfrac{4}{3} \text{ min}.$$

But, of course, we should realize that an exact answer is indeed available for this problem. Recall that we did such a discrete example in Section 6.2.2, where we went directly to a discrete version of the $G/G/1$'s stationary delay equation given in (6.7) and solved for the CDF $W_q(t)$ as a regular Markov chain. The same is possible here and the outline of the solution is provided in the following.

The feasible values of the random variable U here are $(-6, -3, -1, 2)$, with respective probabilities $(\frac{2}{5}, \frac{1}{5}, \frac{4}{15}, \frac{2}{15})$. The stationary equations for the steady-state waiting probabilities $\{w_j, j \ge 0\}$ are found from (6.7) as follows:

$$\begin{aligned} w_0 &= w_0 p_{00} + w_1 p_{10} + w_2 p_{20} + w_3 p_{30} + w_4 p_{40} + w_5 p_{50} + w_6 p_{60}, \\ w_1 &= w_2 p_{21} + w_4 p_{41} + w_7 p_{71}, \\ w_j &= \sum_i w_i p_{ij} \quad (j \ge 2). \end{aligned}$$

The nonzero transition probabilities $\{p_{ij}\}$ are

$$
\begin{aligned}
p_{00} &= \Pr\{U \le 0\} = \tfrac{13}{15},\ p_{02} = \Pr\{U = 2\} = \tfrac{2}{15},\\
p_{10} &= \Pr\{U \le -1\} = \tfrac{13}{15},\ p_{13} = \Pr\{U = 2\} = \tfrac{2}{15},\\
p_{20} &= \Pr\{U \le -2\} = \tfrac{3}{5},\ p_{21} = \Pr\{U = -1\} = \tfrac{4}{15},\\
p_{24} &= \tfrac{2}{15},\ p_{30} = \tfrac{3}{5},\ p_{32} = \tfrac{4}{15},\ p_{35} = \tfrac{2}{15},\ p_{40} = \tfrac{2}{5},\ p_{41} = \tfrac{1}{5},\\
p_{43} &= \tfrac{4}{15},\ p_{46} = \tfrac{2}{15},\ p_{50} = \tfrac{2}{5},\ p_{52} = \tfrac{1}{5},\ p_{64} = \tfrac{4}{15},\ p_{57} = \tfrac{2}{15},\\
p_{i,i-6} &= \tfrac{2}{5},\ p_{i,i-3} = \tfrac{1}{5},\ p_{i,i-1} = \tfrac{4}{15},\ p_{i,i+2} = \tfrac{2}{15} \quad (i \ge 6).
\end{aligned}
$$

Thus the $\{w_j\}$ are found as the solution to

$$
\begin{aligned}
w_j &= \tfrac{2}{15} w_{j-2} + \tfrac{4}{15} w_{j+1} + \tfrac{1}{5} w_{j+3} + \tfrac{2}{5} w_{j+6} \quad (j \ge 2),\\
w_1 &= \tfrac{4}{15} w_2 + \tfrac{1}{5} w_4 + \tfrac{2}{5} w_7,\\
w_0 &= \tfrac{13}{15}(w_0 + w_1) + \tfrac{3}{5}(w_2 + w_3) + \tfrac{2}{5}(w_4 + w_5 + w_6).
\end{aligned}
\tag{7.18}
$$

The next step then is to obtain the roots of the eighth-degree operator polynomial equation formed from the top line of (7.18), namely,

$$6D^8 + 3D^5 + 4D^3 - 15D^2 + 2 = 0.$$

When a root-finding algorithm is applied, we find that there are four complex roots and three real roots in addition to the usual 1. They are (approximately)

$$0.3885, \quad -0.3481, \quad -1.2681, \quad 0.6353 \pm 1.0274i, \quad -0.5215 \pm 1.0296i.$$

Only the roots 0.3885 and -0.3481 have absolute values less than one, and hence the general solution is

$$w_j = C_1(0.3885)^j + C_2(-0.3481)^j. \tag{7.19}$$

To determine C_1 and C_2, we set up a pair of simultaneous linear equations derived from the second and third lines of (7.18) and the fact that the $\{w_j\}$ must sum to one. As a result, we find that $C_1 = 0.4349$ and $C_2 = 0.3894$.

There is an important theory that connects the solution of arbitrary $G/G/1$ queues to the sort of discrete forms just discussed. This is the concept of the *continuity of queues* (e.g., see Kennedy, 1972, and Whitt, 1974). The key idea is that if the interarrival and service-time distributions can be expressed, respectively, as the limits of sequences of distributions (say, $\{A_n\} \to A$ and $\{B_n\} \to B$) then the measures of effectiveness for the $G/G/1$ formed from A and B can be found as the limits of those obtained from the sequence of queues formed from A_n and B_n. If we now permit the sequences $\{A_n\}$ and $\{B_n\}$ to be constructed as increasingly accurate discrete approximations, then the previously discussed method can be used to measure the individual queues formed from A_n and B_n, and the limits estimated accordingly. This works theoretically, but unfortunately the method gets computationally overpowering

Table 7.1 Delay Probabilities for the First Three Customers

	Probability		
Delay	Customer 1	2	3
0	$\frac{13}{15} = .8667$	$\frac{187}{225} = .8311$	$\frac{2803}{3375} = .8305$
1	0	$\frac{8}{225} = .0356$	$\frac{104}{3375} = .0308$
2	$\frac{2}{15} = .1333$	$\frac{26}{225} = .1156$	$\frac{374}{3375} = .1108$
3	0	0	$\frac{32}{3375} = .0095$
4	0	$\frac{4}{225} = .0178$	$\frac{52}{3375} = .0154$
5	0	0	0
6	0	0	$\frac{8}{3375} = .0024$
Mean	$\frac{4}{15} = .2667$	$\frac{76}{225} = .3378$	$\frac{1204}{3375} = .3567$

rather quickly if convergence does not occur early. If that happens, we can simply use a small number of interarrival and service-time values as approximations to keep the computations manageable and then proceed as in Example 7.1.

The concept of stochastic dominance can also be used to bound and/or approximate. The key idea is that if the interarrival random variable of one queue (say, T_1) is stochastically smaller than that of a second queue (T_2) and the service-time variable of the second (S_2) is stochastically smaller than that of the first (S_1), then it follows that the waiting times of the first queue will be stochastically larger than those of the second. It will then result that W and W_q are larger for queue 1 than for queue 2. Thus we have a potential tool for bounding complicated $G/G/1$ systems by finding a solvable combination of interarrival and service-time CDFs which obey the ordering.

It is quite valuable to apply the concept of stochastic ordering directly to the calculations of Example 7.1. The point is that each successive probability of a wait j, w_j, gets closer to its stationary limit, with the kth customer's waiting time stochastically larger than its predecessor. Thus we may take the calculations as far as seems necessary for convergence. In a good many cases, there is rapid convergence to a good lower bound. For our example, the first three customers (after the initial customer) turn out to have the following delay probabilities in Table 7.1.

Using (7.19), we can calculate the true $\{w_j\}$, and calculate W_q as $\sum jw_j$. This gives

$$w_0 = 0.8244, \quad w_1 = 0.0334, \quad w_2 = 0.1128, \quad w_3 = 0.0091,$$
$$w_4 = 0.0156, \quad w_5 = 0.0019, \quad w_6 = 0.0022,$$

and using 50 terms in $\sum jw_j$ yields a W_q of 0.37724. Even with only three customers, the values in the table for customer 3 are not too far off.

7.1.3 Bounds for Multiserver Queues

It should be pretty clear, from the limited number of results we have presented thus far for queues with more than one server, that bounds can be particularly useful for analyzing the $G/G/c$ queue. To get some bounds for the $G/G/c$ queue, we recall some previous results in Chapters 2–5 on the relative merit of single-server queues as opposed to comparable multichannel queues (e.g., Problems 2.26 and 2.27). It is in this spirit that we proceed here.

To begin, assume that the $G/G/c$ queue we wish to bound has mean arrival rate λ and mean service rate μ at each channel. Let W_q denote the mean wait in queue for this system.

Upper Bound. To derive an upper bound for W_q, we consider the following modification to the original $G/G/c$ queue. Suppose that customers are assigned in *cyclic order* to the c servers. That is, the first customer is assigned to server 1, the second to server 2, . . ., the $(c+1)$st to server 1, and so forth. No jockeying is allowed. This modification creates inefficiencies since a customer is required to wait for his or her assigned server, even when other servers may be available.

Based on this modification, each server now becomes a separate single-server $G/G/1$ queue in which the interarrival distribution is the c-fold convolution of the original interarrival distribution. There is no change in the service distribution. For example, for an $M/G/4$ queue, cyclical assignment results in 4 separate $E_4/G/1$ queues, where the arrival rate to each queue is $\lambda/4$. The arrival streams to the separate queues are not independent.

Let W_{q1} denote the mean queue wait for one of the single-server queues. W_{q1} is an upper bound for W_q, since the cyclical assignment applies an added restriction on the original system (see Brumelle, 1971). In fact, we can make the stronger observation that the waiting times are stochastically ordered. Now, if the single-server queue is solvable itself, we can use the exact value for W_{q1} as an upper bound for W_q. Otherwise, we can use the previously derived upper bound for the $G/G/1$ queue in (7.13), but with the following modifications. Here, the arrival rate to each $G/G/1$ queue is λ/c and the arrival-time variance is $c\sigma_A^2$. Thus (7.13) is altered to give the following result:

$$W_q \le W_{q1} \le \frac{(\lambda/c)(c\sigma_A^2 + \sigma_B^2)}{2(1-\lambda/c\mu)} = \frac{\lambda(c\sigma_A^2 + \sigma_B^2)}{2c(1-\rho)} \quad (\rho = \lambda/c\mu).$$

A tighter upper bound was proposed by Kingman and proved by De Smit (1973), where the variance of the service time is divided by c, yielding

$$W_{q1} \le \frac{\lambda(c\sigma_A^2 + \sigma_B^2/c)}{2c(1-\rho)} \quad (\rho = \lambda/c\mu).$$

Lower Bound. To find a lower bound for W_q in the $G/G/c$ queue, we consider a similar $G/G/1$ queue in which the interarrival distribution is identical to the original system, but the server works c times faster. That is, the mean service time is $1/c\mu$, the service-time variance is σ_B^2/c^2, and the queue utilization is $\rho = \lambda/c\mu$.

To compare these two queues, we use an argument based on the remaining work in the system. The intuition is that both queues have the same arrival process; however, the single-server queue services customers at the rate $c\mu$ (when customers are present), while the multiserver queue services customers at a rate that is at most $c\mu$. So the remaining work in the single-server queue is less than in the multiserver queue.

Let ω be the average remaining work for the original $G/G/c$ queue. Then,

$$\omega = \frac{L_q}{\mu} + r \cdot \text{E[remaining service time per server]},$$

where $r = \lambda/\mu$ is the expected number of busy servers. Therefore

$$\omega = \frac{\lambda W_q}{\mu} + \left(\frac{\lambda}{\mu}\right)\frac{\text{E}[S^2]}{2/\mu} = \frac{\lambda W_q}{\mu} + \frac{\lambda(\sigma_B^2 + 1/\mu^2)}{2}.$$

Let ω_2 be the average remaining work for the modified $G/G/1$ queue. The convention is that each customer brings, on average, $1/\mu$ units of work to the system, just as in the multiserver case. But in the single-server case, the server processes work at the rate of c units per unit time (versus the conventional rate of one unit per unit time). With this convention,

$$\omega_2 = \frac{L_{q2}}{\mu} + \rho \cdot \text{E}[c \cdot \text{remaining service time}],$$

where $\rho = \lambda/c\mu$ is the expected number of busy servers. If we let S_2 denote a random service time in this modified queue, then

$$\omega_2 = \frac{\lambda W_{q2}}{\mu} + \left(\frac{\lambda}{c\mu}\right)\frac{c \cdot \text{E}[S_2^2]}{2/c\mu} = \frac{\lambda W_{q2}}{\mu} + \frac{\lambda(\sigma_B^2 + 1/\mu^2)}{2c}.$$

Brumelle (1971) has shown that $\omega \geq \omega_2$. This implies that

$$\frac{\lambda W_q}{\mu} + \frac{\lambda(\sigma_B^2 + 1/\mu^2)}{2} \geq \frac{\lambda W_{q2}}{\mu} + \frac{\lambda(\sigma_B^2 + 1/\mu^2)}{2c},$$

so

$$W_q \geq W_{q2} - \frac{\mu(c-1)(\sigma_B^2 + 1/\mu^2)}{2c}.$$

Again, we can use the exact value for W_{q2} (if known) or use a lower bound for W_{q2} based on (7.14). For the $G/G/1$ queue considered here, the mean service time is $1/c\mu$ and the service-time variance is σ_B/c^2. Thus (7.14) is altered to give the following result:

$$W_{q2} \geq \frac{\lambda^2\sigma_B^2/c^2 + \rho(\rho - 2)}{2\lambda(1-\rho)} \quad (\rho = \lambda/c\mu).$$

Combining the upper and lower bounds together gives

$$\left(\frac{\lambda^2\sigma_B^2 + c^2\rho(\rho-2)}{2\lambda c^2(1-\rho)} - \frac{\mu(c-1)(\sigma_B^2 + 1/\mu^2)}{2c}\right)^+ \le W_q \le \frac{\lambda(c\sigma_A^2 + \sigma_B^2/c)}{2c(1-\rho)},$$

where $(x)^+ \equiv \max(0, x)$.

7.2 Approximations

This section gives several approximations for evaluating the performance of $G/G/1$ and $G/G/c$ queues. In the previous section, we derived *bounds* for these queues. Bounds are provably valid all of the time. Approximations can be somewhat looser. While there is usually rigorous mathematical analysis that motivates an approximation, mathematical proof is not absolutely required to specify the accuracy of an approximation – whether the approximation is higher or lower than the exact value or how close the approximation is to the exact value.

This section considers three types of approximations, which we categorize using a scheme based on Bhat et al. (1979).

1. The first makes use of *bounds*. For example, one such scheme is a weighted average of the upper and lower bounds for a $G/G/1$ queue with the weighting factor depending on the traffic intensity.

2. The second type of approximation deals with using a known queueing system to approximate one for which results are not readily obtainable. We refer to this type of approximation as a *system approximation*. An example is the use of $M/E_k/c$ or $M/H_k/c$ models to approximate an $M/G/c$.

3. The third type of approximation involves approximating the queueing process itself by a process that is easier to deal with. Examples of such *process approximations* are replacement of the discrete queueing process by a continuous diffusion or fluid process, and using asymptotic or limiting results.

The following subsections present these ideas in more detail.

7.2.1 Using Bounds to Approximate

Based on the fact that the upper bound of (7.13) gets better as $\rho \to 1$, it might make sense to multiply the bound by a fractional function in ρ which itself approaches one as $\rho \to 1$. Marchal (1978) proposed the quotient

$$\frac{1 + \mu^2\sigma_B^2}{1/\rho^2 + \mu^2\sigma_B^2} = \frac{\rho^2 + \lambda^2\sigma_B^2}{1 + \lambda^2\sigma_B^2}.$$

This was chosen to make the approximation exact for the $M/G/1$ queue. The resulting approximation for W_q, based on multiplying the upper bound of (7.13) by the previous quotient, is

$$\hat{W}_q = \frac{\lambda(\sigma_A^2 + \sigma_B^2)}{2(1-\rho)}\left(\frac{\rho^2 + \lambda^2\sigma_B^2}{1+\lambda^2\sigma_B^2}\right).$$

To see that the approximation is exact for the $M/G/1$ queue, we let $\sigma_A^2 = 1/\lambda^2$. Then the approximation simplifies to

$$\frac{\lambda(1/\lambda^2 + \sigma_B^2)}{2(1-\rho)}\left(\frac{\rho^2 + \lambda^2\sigma_B^2}{1+\lambda^2\sigma_B^2}\right) = \frac{\rho^2 + \lambda^2\sigma_B^2}{2\lambda(1-\rho)},$$

which is precisely the PK formula. Marchal has shown that this formula also works well for $G/M/1$ queues. The performance of the approximation for arbitrary $G/G/1$ queues deteriorates as the service times or interarrival times deviate further from exponential. The accuracy of the approximation does, however, improve with increasing values of the traffic intensity, in light of the asymptotic sharpness of the upper bound.

Marchal also suggested the possibility of using a different weighting factor, namely,

$$\frac{\rho^2\sigma_A^2 + \sigma_B^2}{\sigma_A^2 + \sigma_B^2}.$$

Then a new approximation can be derived as

$$\hat{W}_q = \frac{\rho(\lambda^2\sigma_A^2 + \mu^2\sigma_B^2)}{2\mu(1-\rho)}.$$

This approximation is also exact for the $M/G/1$ queue. Furthermore, it is exact for the $D/D/1$ queue. The formula has a nice "product form" of three terms: (1) a variability factor, (2) a traffic-intensity factor, and (3) a time-scale factor. It can be written as

$$\hat{W}_q = \left(\frac{C_A^2 + C_B^2}{2}\right)\left(\frac{\rho}{1-\rho}\right)\left(\frac{1}{\mu}\right), \tag{7.20}$$

where C represents the coefficient of variation (standard deviation/mean). We can now observe the following progression in the expressions for W_q for the $M/M/1$ (2.27), $M/G/1$ (5.2), and $G/G/1$ queues:

$$W_q(M/M/1) = \left(\frac{1+1}{2}\right)\left(\frac{\rho}{1-\rho}\right)\left(\frac{1}{\mu}\right),$$

$$W_q(M/G/1) = \left(\frac{1+C_B^2}{2}\right)\left(\frac{\rho}{1-\rho}\right)\left(\frac{1}{\mu}\right),$$

$$W_q(G/G/1) \approx \left(\frac{C_A^2 + C_B^2}{2}\right)\left(\frac{\rho}{1-\rho}\right)\left(\frac{1}{\mu}\right).$$

In going from $M/M/1$ to $M/G/1$, we include the SCV of the service distribution. In going from $M/G/1$ to $G/G/1$, we include the SCV of the interarrival distribution. The first two equations are exact, the last equation is an approximation. In summary, the approximation for the $G/G/1$ queue can be seen as the wait in an analogous $M/M/1$ queue multiplied by the variability term $(C_A^2 + C_B^2)/2$.

Based on this observation, we can create a similar approximation for the $G/G/c$ queue. That is, we multiply the wait in an analogous $M/M/c$ queue by the variability term – specifically,

$$\begin{aligned}\hat{W}_q &= \left(\frac{C_A^2 + C_B^2}{2}\right) W_q(M/M/c) \\ &= \left(\frac{C_A^2 + C_B^2}{2}\right)\left(\frac{r^c p_0}{c \cdot c!(1-\rho)^2}\right)\left(\frac{1}{\mu}\right),\end{aligned} \tag{7.21}$$

where p_0 is the system-empty probability for the $M/M/c$ queue; see (2.32) and (2.34). This approximation is called the Allen–Cunneen (AC) approximation (Allen, 1990). Alternatively, let $C(c, r)$ denote the probability that all servers are busy in an $M/M/c$ queue. That is, from (2.38),

$$C(c, r) = \frac{r^c}{c!(1-\rho)} p_0.$$

Then the approximation can be written as

$$\hat{W}_q = \left(\frac{C_A^2 + C_B^2}{2}\right)\left(\frac{C(c, r)}{c(1-\rho)}\right)\left(\frac{1}{\mu}\right).$$

When $c = 1$, $C(c, r) = \rho$ and the AC approximation reduces to the one given previously by Marchal in (7.20).

7.2.2 System Approximations

As we noted earlier, examples of system approximations are the use of either an $M/E_k/c$ or an $M/H_k/c$ model to approximate an $M/G/c$. Recall our earlier comments that the Erlang family provides great modeling flexibility, particularly when it is generalized to the more global phase-type distributions first mentioned in Section 3.3.2. Recall also that both the usual Erlang and mixture of exponentials are easily expressed in phase form and that their use as service distributions leads to totally solvable systems.

Also in this spirit, we note some results of Section 6.2.1, where we showed that the $G/G/1$ problem can be greatly simplified if it may be assumed that both interarrival and service distributions can be expressed as convolutions of independent and not necessarily identical exponential random variables (usually called generalized Erlangs, *GE*). When the means of such exponentials are allowed to come in conjugate pairs (so that their Stieltjes transforms are inverse polynomials), Smith (1953) calls the family K_n (with n the degree of the defining polynomial). Other authors (e.g., Cohen, 1982) define K_n as the class of distributions whose transforms are rational functions (clearly including the inverse polynomials); but we shall instead call these R_n (with n the degree of the denominator's polynomial) or Coxian, after the early work of Cox

(1955). The class K_n includes all regular Erlangs, but not mixed exponentials and mixed Erlangs, which are, however, members of R_n. When the denominator polynomial of a K_n transform has real and distinct roots, the associated distribution is called a *GH*, for generalized hyperexponential (e.g., Botta and Harris, 1980). We also mention that the phase-type distributions (*PH*) have rational transforms as well, although not necessarily of the inverse polynomial form. Thus we may symbolically represent the relationship of those respective families as (see Harris, 1985)

$$GE \subset K_n \subset R_n, \quad GH \subset K_n \subset R_n, \quad \text{and } PH \subset R_n.$$

Now, under a double K_n assumption (i.e., the queue is $K_m/K_n/1$) it turns out (just as for *GE*/*GE*/1) that

$$W_q^*(s) = \frac{\prod_{i=1}^{n}(-z_i/\mu_i)(s+\mu_i)}{s\prod_{i=1}^{n}(s-z_i)},$$

where the $\{\mu_i\}$ are the individual parameters of the exponential decomposition of the service-time CDF and the $\{z_i\}$ are roots of the polynomial equation $A^*(-s)B^*(s) - 1 = 0$ with negative real parts. This result for the waiting-time transform is a nice simplification, since it now puts $W_q^*(S)$ into an invertible form. A partial-fraction expansion is then performed to give

$$W_q^*(s) = \frac{1}{s} + \sum_{i=1}^{n} \frac{k_i}{s - z_i} \quad \text{(all } z_i \text{ assumed distinct).}$$

Thus

$$W_q(t) = 1 + \sum_{i=1}^{n} k_i e^{z_i t},$$

where the $\{k_i\}$ would be determined in the usual way and are arbitrary in sign, with complex conjugates of the $\{k_i\}$ paired with the complex conjugates of the $\{z_i\}$. A completely parallel result exists for distributions in R_n.

Now, most importantly, Smith (1953) has also shown that a comparable result follows even for arbitrary interarrival times. That is, the $G/K_n/1$ and $G/R_n/1$ queues have mixed exponential waits independent of the form of G, where now the $\{z_i\}$ are the roots with negative real parts to the possibly transcendental equation $A^*(-s)B^*(s) - 1 = 0$. Thus we are now in an excellent position to well approximate the most awkward $G/G/1$ queues.

In closing here, we point out that the matching of any approximation to an original queue structure is likely to involve the use of procedures for parameter estimation. For this, the interested reader is referred to any basic statistical text and our later discussion in Section 8.3.2.

7.2.3 Process Approximations

As noted earlier, we define a process approximation as one where the actual problem is replaced by a nonqueueing one that is simpler to work with. The primary examples of interest are the use of creative probability arguments on random walks, stochastic convergence, and the like to solve heavy-traffic and nonstationary queueing problems, and the use of continuous-time diffusion models to solve queueing problems in heavy traffic.

7.2.3.1 Heavy-Traffic and Nonstationary Queues. The goal of this subsection is to give some interesting limit results and approximations for $G/G/1$ models in which the traffic intensity is just barely less than one $(1-\varepsilon<\rho<1)$ or is equal to or greater than one $(\rho \geq 1)$. The former will be said to be saturated or in heavy traffic, while the latter will be described as nonstationary, divergent, or unstable. Of course, any transient results derived earlier in the text are valid for heavy traffic and divergent queues, since no assumptions are made on the size of ρ.

For limit results, the virtual wait $V(t)$ (the wait a customer would undergo if that customer arrives at time t) and the actual wait $W^{(n)}$ (the actual waiting time of the nth customer) will be used to construct random sequences, which are then shown to stochastically converge. These convergence theorems will then be applied to obtain approximate distributions of the actual and virtual waits in queue of a customer and their averages. In addition, bounds on the average waiting times will be obtained, some of these directly from the limit results. While we shall present some nice theoretical results, the practicality of these is subject to question, since the results depend on allowing time to go to infinity. In reality, systems with congestion are often not allowed to run very long before corrective actions are taken.

Heavy Traffic. We begin by first discussing the heavy-traffic problem in the context of the $G/G/1$ queue. For this queue, the system is completely specified by the sequences of interarrival and service times. If each of these sequences contains IID random variables (where the sequences are also independent of each other), then the main result is that the queue waiting times in heavy traffic are approximately exponentially distributed with mean

$$W_q^{(H)} = \frac{1}{2}\frac{\text{Var}[T]+\text{Var}[S]}{1/\lambda - 1/\mu} = \frac{\lambda(\sigma_A^2+\sigma_B^2)}{2(1-\rho)}, \tag{7.22}$$

where T is a random interarrival time with variance σ_A^2, and S is a random service time with variance σ_B^2. Note that (7.22) is also the upper bound for W_q given by (7.13).

We now sketch a heuristic argument to motivate the result (see also Kingman, 1962a). The argument is based on random walks. Recall that we defined $U^{(n)} \equiv S^{(n)} - T^{(n)}$ to be the difference between the nth service time and the nth interarrival time [the time between the arrivals of the nth and $(n+1)$st customers]. We can derive $W_q^{(n)}$, the waiting time in queue of the nth customer, as an explicit function of $U^{(1)}, \ldots, U^{(n-1)}$ by recursively applying Lindley's equation:

$$\begin{aligned} W_q^{(n)} &= \max(W_q^{(n-1)} + U^{(n-1)}, 0) \\ &= \max(\max\{W_q^{(n-2)} + U^{(n-2)}, 0\} + U^{(n-1)},\ 0) \\ &= \max(\max\{W_q^{(n-2)} + U^{(n-2)} + U^{(n-1)},\ U^{(n-1)}\},\ 0) \\ &= \max(W_q^{(n-2)} + U^{(n-2)} + U^{(n-1)},\ U^{(n-1)},\ 0). \end{aligned}$$

Continuing in a recursive manner:

$$W_q^{(n)} = \max(U^{(1)} + \cdots + U^{(n-1)}, U^{(2)} + \cdots + U^{(n-1)}, \ldots, U^{(n-1)}, 0). \quad (7.23)$$

We have assumed that the queue begins in an empty state, so that $W_q^{(1)} = 0$ in the last recursion step. Thus $W_q^{(n)}$ is the maximum of partial sums of $U^{(n-1)}, U^{(n-2)}, \ldots, U^{(1)}$, where the summation is applied in reverse order. If we define the partial sum $P^{(k)}$ to be

$$P^{(k)} \equiv \sum_{i=1}^{k} U^{(n-i)},$$

then (7.23) can be rewritten as

$$W_q^{(n)} = \max_{n-1 \geq k \geq 0} P^{(k)}.$$

$P^{(k)}$ is assumed to be zero when $k = 0$. Also, $P^{(k)}$ is implicitly a function of n. Now, $P^{(k)}$ is a random walk, since it is the cumulative sum of IID jumps, where each jump follows the distribution of $U^{(i)}$ (the random walk terminates at $k = n-1$). Let the mean and variance of the jumps be denoted by the parameters α and β, defined as follows:

$$\alpha \equiv -\mathrm{E}[U^{(i)}] = -\left(\frac{1}{\mu} - \frac{1}{\lambda}\right) = \frac{1-\rho}{\lambda},$$

$$\beta^2 \equiv \mathrm{Var}[U^{(i)}].$$

If $\rho < 1$, then α is positive. We also have

$$\mathrm{E}[P^{(k)}] = -k\alpha, \quad \text{and} \quad \mathrm{Var}[P^{(k)}] = k\beta^2.$$

Figure 7.3 shows a sample path of $P^{(k)}$, where $n = 300$. Each jump corresponds to one observation of $U^{(i)}$. $P^{(k)}$ is the cumulative sum of the individual jumps, where k ranges from 0 to 299. $W_q^{(300)}$ is the maximum value of $P^{(k)}$ taken over this range. This particular example is an $E_2/E_4/1$ queue. Interarrival times follow an Erlang-2 distribution with rate $\lambda = 1$ (i.e., $\mathrm{E}[T^{(n)}] = 1/\lambda = 1$ and $\mathrm{Var}[T^{(n)}] = 1/(2\lambda^2) = 1/2$, see Section 3.3.1). Service times follow an Erlang-4 distribution with expected value $1/\mu = .95$ (i.e., $\mathrm{E}[S^{(n)}] = 1/\mu = .95$ and $\mathrm{Var}[S^{(n)}] = 1/(4\mu^2) = .95^2/4$). Thus $\rho = .95$, $\alpha = .05$, and $\beta^2 \doteq 0.73$. The expected downward drift, $\mathrm{E}[P^{(k)}] = -0.05k$, is also shown in the figure.

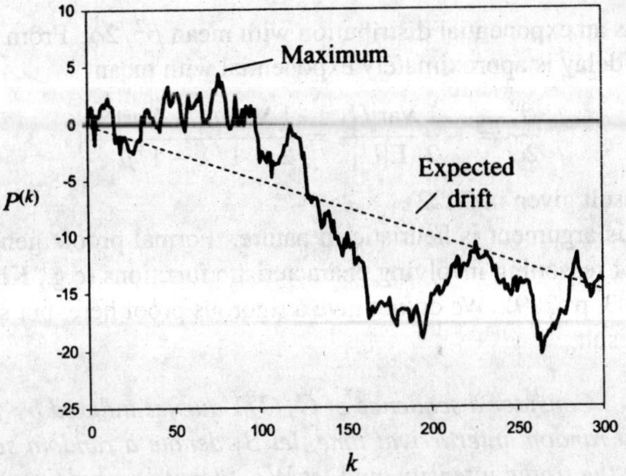

Figure 7.3 Example random walk for an $E_2/E_4/1$ queue.

For a finite n, $W_q^{(n)}$ is the maximum value of a random walk with n terms. In steady state, we let $n \to \infty$. Then, the waiting time distribution is the maximum (or supremum) of a random walk with negative drift:

$$W_q^{(\infty)} = \sup_{k \geq 0} P^{(k)}.$$

The previous discussion holds for any $G/G/1$ queue with $\rho < 1$. Now, we make use of the heavy-traffic assumption. The basic idea is that if ρ is very close to 1, then $P^{(k)}$ can be approximated by *Brownian motion*. For an introductory treatment of Brownian motion, see Ross (2007). Intuitively, Brownian motion can be thought of as the limit of a sequence of random walks where the jump sizes approach 0 and the times between jumps approach 0. While a random walk occurs in discrete time, Brownian motion occurs in continuous time. More specifically, a stochastic process $\{X(t), t \geq 0\}$ is a Brownian motion process with drift coefficient $-\alpha$ and variance parameter β^2 if $X(0) = 0$, $X(t)$ has stationary and independent increments, and $X(t) \sim N(-\alpha t, \beta^2 t)$ (a normal distribution with mean $-\alpha t$ and variance $\beta^2 t$).

Observe that $P^{(k)}$ has the first two properties: $P^{(0)} = 0$ and $P^{(k)}$ has independent and stationary increments (since the $U^{(i)}$ are IID). The third property is approximated when ρ is slightly less than 1. In this case, α is small and positive (assuming λ is not too small), so the jump sizes in the random walk are small. For large values of k, $P^{(k)}$ is the sum of a large number of independent random variables. The central limit theorem implies that $P^{(k)}$ has approximately a normal distribution with mean $-k\alpha$ and variance $k\beta^2$. This is the third property, with the discrete value k replacing the continuous value t.

In summary, the steady-state queue-wait distribution is the supremum of a random walk with negative drift. For ρ slightly less than 1, the random walk can be approximated by Brownian motion. Now, we use a result about the supremum of Brownian motion (e.g., Baxter and Donsker, 1957, Example 1). If $X(t)$ is Brownian motion with negative drift $-\alpha < 0$ and variance parameter β^2, then the supremum of $X(t)$,

$0 \le t < \infty$ has an exponential distribution with mean $\beta^2/2\alpha$. From this we deduce that the queue delay is approximately exponential with mean

$$\frac{\beta^2}{2\alpha} = -\frac{1}{2}\frac{\text{Var}[U]}{\text{E}[U]} = \frac{1}{2}\frac{\text{Var}[T] + \text{Var}[S]}{1/\lambda - 1/\mu}, \tag{7.24}$$

which is the result given in (7.22).

The previous argument is heuristic in nature. Formal proofs generally follow a different line of reasoning involving characteristic functions (e.g., Kingman, 1962a; Asmussen, 2003, p. 289). We do not give a rigorous proof here, but simply state the heavy-traffic result.

Theorem 7.1. *Consider a sequence of $G/G/1$ queues indexed by j. For queue j, let T_j denote a random interarrival time, let S_j denote a random service time, let $\rho_j < 1$ denote the traffic intensity, and let $W_{q,j}$ denote a random queue-wait for a customer in steady state. Let $\alpha_j = -E[S_j - T_j]$ and $\beta_j^2 = Var[S_j - T_j]$. Suppose that $T_j \xrightarrow{df} T$, $S_j \xrightarrow{df} S$, and $\rho_j \to 1$, where $\xrightarrow{df}$ denotes convergence in distribution. Then,*

$$\frac{2\alpha_j}{\beta_j^2} W_{q,j} \xrightarrow{df} Exp(1),$$

provided that (a) $Var[S - T] > 0$, and (b) for some $\delta > 0$, $E[S_j^{2+\delta}]$ and $E[T_j^{2+\delta}]$ are both less than some constant C for all j.

The last two conditions preclude special exceptions to the heavy-traffic situation. The first implies that the limiting distributions are not both deterministic. The last condition implies that T_j^2 and S_j^2 do not get arbitrarily large. In summary, it is not always true that if ρ is close to 1 then the queue is in heavy traffic. In particular, it is possible to construct queues where ρ is arbitrarily close to 1, but the queues are not in heavy traffic, as shown by the next two examples.

■ EXAMPLE 7.2

Consider a deterministic queue with interarrival times $T^{(n)} = 1$ and service times $S^{(n)} = 1-\epsilon$, where ϵ is a small, positive number. For this queue, $\rho = 1-\epsilon$ can be made arbitrarily close to 1. However, this is not a heavy-traffic situation because the queue wait for every customer is zero. This example violates the condition that $\text{Var}[S - T] > 0$.

■ EXAMPLE 7.3

(Kingman, 1962a) Consider a $G/G/1$ queue with $\rho < 1$. Add a fixed quantity B to every interarrival time $T^{(n)}$ and to every service time $S^{(n)}$. Then $U^{(n)} = S^{(n)} - T^{(n)}$ is unchanged, since B is added and subtracted from each term. Thus, even though $\rho \to 1$ as $B \to \infty$, this does not create a heavy-traffic queue, since the sequence of waiting times $W_q^{(n)}$ is unchanged. This example violates condition (b) in the theorem.

■ **EXAMPLE 7.4**

In this example, the heavy traffic theorem applies, and we can compare exact results to the approximation. Consider an $M/G/1$ queue with arrival rate λ, $E[S] = 1/\mu$, and $\text{Var}[S] = \sigma_B^2$. For this queue, the PK formula (see Section 5.1.1) gives the exact result:

$$W_q = \frac{\rho^2 + \lambda^2\sigma_B^2}{2\lambda(1-\rho)} = \frac{\lambda(1/\mu^2 + \sigma_B^2)}{2(1-\rho)}.$$

The heavy-traffic approximation from (7.22) is

$$W_q^{(H)} = \frac{\lambda(1/\lambda^2 + \sigma_B^2)}{2(1-\rho)}.$$

The relative error of the heavy-traffic approximation is

$$\frac{W_q^{(H)} - W_q}{W_q} = \frac{\lambda(1/\lambda^2 - 1/\mu^2)}{2(1-\rho)} \cdot \frac{2(1-\rho)}{\lambda(1/\mu^2 + \sigma_B^2)} = \frac{1-\rho^2}{\rho^2 + \lambda^2\sigma_B^2}.$$

Thus W_q goes to $W_q^{(H)}$ roughly at the rate that ρ^2 goes to 1.

We can also extend the $G/G/1$ bound to the $G/G/c$ queue by noting that the system is essentially always operating as a single-server queue with service rate $c\mu$. Hence it would be fair to expect exponential delays in heavy traffic with

$$W_q^{(H)} \approx \frac{\text{Var}[T] + (1/c^2)\text{Var}[S]}{2(1/\lambda - 1/c\mu)}.$$

Saturated Systems ($\rho \geq 1$). If we were to go back to Lindley's approach to waiting times for the $G/G/1$, or even to the $M/M/1$ case, and try to get results for $\rho \geq 1$, we would fail. The convergence theory that exists for stationary queues cannot be directly extended in either case, generally because the relevant quantities get excessively large as time increases, so that a completely new approach is required. This new approach is to appropriately scale and shift the random variable to permit some form of convergence. For example, when we study $W^{(n)}$ (the waiting time of the nth customer), the appropriate quantity to consider is $(W^{(n)} - an)/(b\sqrt{n})$, where a and b are suitably chosen constants.

It can be shown that the distribution of this new random variable converges to a normal distribution for $\rho > 1$. We again make use of

$$W_q^{(n+1)} = \max(0, W_q^{(n)} + U^{(n)}), \tag{7.25}$$

where $U^{(n)} = S^{(n)} - T^{(n)}$, but we shall now operate under the assumption that $E[U] > 0$. First, let $1/\mu - 1/\lambda = E[U] > 0$, that is, $\rho > 1$. We would then expect the difference $\Delta W_q^{(n)} \equiv W_q^{(n+1)} - W_q^{(n)}$ to behave like $U^{(n)}$ when n gets large, since by observation of (7.25) we see that it is unlikely that $W_q^{(n+1)}$ will ever be zero again when $\rho > 1$. But we have already stated earlier in this section that $\sum U^{(i)}$ properly normalized as

$$Y_n = \frac{\sum U^{(i)} + n\alpha}{\sqrt{n}\beta} \quad (\alpha < 0)$$

converges in distribution function to the unit normal. Since $W_q^{(0)} = 0$,

$$\sum_{i=0}^{n-1} \Delta W_q^{(i)} = W_q^{(n)} = \sum_{i=0}^{n-1} U^{(i)},$$

and hence

$$\Pr\left\{\frac{W_q^{(n)} + n\alpha}{\sqrt{n}\beta} \le x\right\} \to \int_{-\infty}^{x} \frac{e^{-t^2}/2}{\sqrt{2\pi}}\, dt. \tag{7.26}$$

One immediate result of (7.26) is the determination of an estimate for the probability of no wait for n large. This is given by

$$\Pr\{W_q^{(n)} = 0\} = \Pr\left\{\frac{W_q^{(n)} + n\alpha}{\sqrt{n}\beta} \le \frac{n\alpha}{\sqrt{n}\beta}\right\}$$
$$= \int_{-\infty}^{n\alpha/(\sqrt{n}\beta)} \frac{e^{-t^2/2}}{\sqrt{2\pi}}\, dt = \Phi(n\alpha/\sqrt{n}\beta),$$

where $\Phi(\cdot)$ is the CDF of the standard normal.

We can get an interesting alternative approximation to this probability which does not use the normal distribution by using Chebyshev's inequality, which says that

$$\Pr\{|X - \mu| \ge k\sigma\} \le \frac{1}{k^2}$$

for any variable X possessing two moments. Applying this result with $W_q^{(n)}$ in place of X, it is found that

$$\Pr\{W_q^{(n)} = 0\} = \Pr\{X \le 0\} = \tfrac{1}{2}\left(\Pr\{X \le 0\} + \Pr\{X \ge -2n\alpha\}\right)$$
$$= \tfrac{1}{2}\left(\Pr\{|X + n\alpha| \ge -n\alpha\}\right) \le \frac{\beta^2}{2n\alpha^2},$$

which provides a fairly reasonable upper bound for the probability that an arbitrary arrival encounters an idle system. This number does go to zero as n approaches ∞, as required.

7.2.3.2 Diffusion Approximations. Since we have already seen that heavy-traffic $G/G/1$ queues have exponential waiting times, it might therefore seem quite logical that these queues should be birth–death and of the $M/M/1$ type when $\rho = 1 - \epsilon$, independent of the form of the arrival and service distributions (remember that $M/G/1$ queues have exponential waits if $G = M$, and that $G/M/1$ queues also have exponential waits if $G = M$—see Problem 5.40). However, this is not correct, since we know from (7.24) that the interarrival- and service-time variances must be involved. But we can obtain a fairly simple approximate result for the transient system-size distribution of the $M/M/1$ in heavy traffic by using a continuous approximation of the diffusion type. After that we shall use another, but similar, approach to get a diffusion approximation for the transient distribution of the line delay for the $M/G/1$ in heavy traffic.

The $M/M/1$ diffusion result for the system-size distribution is found by first considering the elementary random-walk approximation of a birth–death process which changes every Δt units and has transition probabilities given by

$$\begin{aligned}
&\Pr\{\text{state goes up by one over unit interval}\} &&= \lambda\,\Delta t,\\
&\Pr\{\text{state goes down by one over unit interval}\} &&= \mu\,\Delta t,\\
&\Pr\{\text{no change}\} &&= 1-(\lambda+\mu)\Delta t,
\end{aligned}$$

assuming for the moment that since the queue is in heavy traffic, the walk is unrestricted with no impenetrable barrier at 0.

Now the elementary random walk is clearly a Markov chain, since its future behavior is only a probabilistic function of its current position and is independent of past history. More specifically, we may write that the probability $p_k(n)$ of finding the walk in position k after n steps is given by the CK equation

$$p_k(n+1) = p_k(n)[1-(\lambda+\mu)\Delta t] + p_{k+1}(n)\mu\,\Delta t + p_{k-1}(n)\lambda\,\Delta t,$$

which may be rewritten after a bit of algebra as

$$\begin{aligned}
\frac{p_k(n+1)-p_k(n)}{\Delta t} &= \frac{\mu+\lambda}{2}[p_{k+1}(n)-2p_k(n)+p_{k-1}(n)]\\
&\quad+\frac{\mu-\lambda}{2}[p_{k+1}(n)-p_k(n)]\\
&\quad+\frac{\mu-\lambda}{2}[p_k(n)-p_{k-1}(n)].
\end{aligned}\tag{7.27}$$

We then observe that (7.27) has been written in terms of the discrete version of derivatives: the left-hand side with respect to step (time) and the right-hand side with respect to the state variable. If we then appropriately take limits of (7.27) so that the time between transitions shrinks to zero, while simultaneously the size of the state steps goes to zero, we find ourselves ending up with a partial differential equation of the diffusion type.

To be more exact, let the length of a unit state change be denoted by θ and the step time by Δt; then (7.27) leads to

$$\begin{aligned}
\frac{p_{k\theta}(t+\Delta t)-p_{k\theta}(t)}{\Delta t} &= \frac{\mu+\lambda}{2}[p_{(k+1)\theta}(t)-2p_{k\theta}(t)+p_{(k-1)\theta}(t)]\\
&\quad+\frac{\mu-\lambda}{2}[p_{(k+1)\theta}(t)-p_{k\theta}(t)]\\
&\quad+\frac{\mu-\lambda}{2}[p_{k\theta}(t)-p_{(k-1)\theta}(t)].
\end{aligned}\tag{7.28}$$

Now let both Δt and θ go to 0, preserving the relationship that $\Delta t = \theta^2$ (which guarantees that the state variance is meaningful) and at the same time letting k increase to ∞ in such a way that $k\theta \to x$. Then $p_{k\theta} \to p(x, t|X_0 = x_0)$, which is now the probability density for the system state X, given that the queueing system began operation at a size of x_0. Utilizing the definitions of first and second derivatives, (7.28) becomes

$$\frac{\partial p(x,t|x_0)}{\partial t} = \left(\frac{\mu+\lambda}{2}\right)\frac{\partial^2 p(x,t|x_0)}{\partial x^2} + (\mu-\lambda)\frac{\partial p(x,t|x_0)}{\partial x}, \tag{7.29}$$

one form of the well-known diffusion equation, which among other things describes the movement of a particle under Brownian motion (e.g., see Prabhu, 1965b, or Heyman and Sobel, 1982). This particular form of the diffusion equation often goes under the name of Fokker–Planck and, in addition, turns out to be the version of the forward Kolmogorov equation that is found for the continuous-state Markov process.

So now we would like to solve (7.29) under the boundary conditions that

$$p(x,t|x_0) \geq 0,$$

$$\int_{-\infty}^{\infty} p(x,t|x_0)\,dx = 1,$$

$$\lim_{t\to 0} p(x,t|x_0) = 0 \qquad (x \neq x_0),$$

where the first two are usual properties of densities, while the third is essentially a continuity requirement at time 0. It can then be shown (Prabhu, 1965b) that the solution to (7.29) is given by

$$p(x,t|x_0) = \frac{e^{-[x-x_0+(\mu-\lambda)t]^2/2(\mu+\lambda)t}}{\sqrt{2\pi t(\mu+\lambda)}},$$

which is a Wiener process starting from x_0 with drift $-(\mu-\lambda)t$ and variance $(\mu+\lambda)t$. That is to say, the random process $X(t)$ is normal with mean $x_0-(\mu-\lambda)t$ and variance $(\mu+\lambda)t$ so that

$$\Pr\{n - \tfrac{1}{2} < X(t) < n + \tfrac{1}{2}|X_0 = x_0\} \approx \int_{n-1/2}^{n+1/2} N(x_0 - [\mu-\lambda]t, [\mu+\lambda]t)\,dt.$$

But we observe that this result is not very meaningful, since $\lambda < \mu$ and the drift is therefore negative, thus bringing the process eventually to one with a negative mean (this is due to neglecting the impenetrable barrier at 0; however, the approximation would be valid if x_0 were large, since it would then be unlikely for the queue to empty). In order to counter this possibility we must impose an impenetrable barrier upon the walk at $x = 0$. This is added to the problem in the form of the additional boundary condition that

$$\lim_{x\to 0} \frac{\partial p(x,t|x_0)}{\partial t} = 0 \qquad \text{(for all } t).$$

This is used because the process cannot move beyond zero to the negatives and therefore $p(x,t|x_0) = 0$. Thus $\Delta p(x,t|x_0) = 0$ for $x < 0$ and $\lim_{x\to 0} \Delta p(x,t|x_0) = 0$. The new solution is then given by

$$p(x,t|x_0) = \frac{1}{\sqrt{2\pi(\lambda+\mu)t}} \left[e^{-[x-x_0-(\lambda-\mu)t]^2/2(\mu+\lambda)t} \right.$$
$$+ e^{-2x(\mu-\lambda)/(\mu+\lambda)} \Big(e^{-[x+x_0-(\lambda-\mu)t]^2/2(\mu+\lambda)t}$$
$$\left. + \frac{2(\mu-\lambda)}{\mu+\lambda} \int_x^\infty e^{-[y+x_0-(\lambda-\mu)t]^2/2(\mu+\lambda)t}\, dy \Big) \right]. \tag{7.30}$$

This solution is then valid as an approximation for any $M/M/1$ provided $\rho = 1 - \epsilon$.

It is particularly interesting to make two additional computations. The first is to allow $\lambda = \mu$ throughout the derivation, and the second is to look at the limiting behavior of $p(x,t|x_0)$ as t goes to ∞. When $\lambda = \mu$, (7.29) is quite simplified and becomes

$$\frac{\partial p(x,t|x_0)}{\partial t} = \lambda \frac{\partial^2 p(x,t|x_0)}{\partial x^2}.$$

Under the same augmented boundary conditions as lead to (7.30), this differential equation has the solution

$$p(x,t|x_0) = \frac{1}{\sqrt{4\pi\lambda t}} (e^{-(x-x_0)^2/4\lambda t} + e^{-(x+x_0)^2/4\lambda t}),$$

again a Wiener process, but one with no drift. This one may be used to approximate the transient solution for any $M/M/1$ with $\rho = 1$.

On the other hand, when we let $t \to \infty$ in (7.29), it is found that

$$0 = \left(\frac{\mu+\lambda}{2}\right) \frac{d^2 p(x)}{dx^2} + (\mu - \lambda) \frac{dp(x)}{dx},$$

which is a homogeneous, second-order linear differential equation with solution

$$p(x) = C_1 + C_2 e^{-2[(\mu-\lambda)/(\mu+\lambda)]x}.$$

Since $p(x)$ must integrate to one, we see that $C_1 = 0$ and $C_2 = (\mu+\lambda)/[2(\mu-\lambda)]$. Thus $p(x)$ is an exponential density, as might have been expected, since $M/M/1$ lengths are geometric in distribution, which is just the discrete analog of the exponential. That this mean of $(\mu+\lambda)/[2(\mu-\lambda)] = (1+\rho)/[2(1-\rho)]$ makes sense can be seen by noting that for ρ nearly 1, $(1+\rho)/2 \approx \rho$; hence $(1+\rho)/[2(1-\rho)] \approx \rho/(1-\rho)$, the usual $M/M/1$ result.

Another approach to this approximation which leads to the same result is the use of the central limit theorem on the IID random variables making up the random walk. One then uses the same kind of limiting argument to get the results in terms of the same variables as before.

The $M/G/1$ heavy-traffic diffusion approximation for line delay is due to Gaver (1968). In order to approximate the conditional "density function," say, $w_q(x,t|x_0)$

(recall from Chapter 2 that this is not a true density, since there is a nonzero probability of a zero wait), of the virtual line delay $V(t)$(which is easier to use than the actual for the derivation here—keep in mind that both are the same for $M/G/1$), its mean μ and variance σ^2 are approximated by

$$\Delta t = \mathrm{E}[V(t+\Delta t) - V(t)|V(t)] = (\lambda \mathrm{E}[S] - 1)\Delta t + 0(\Delta t),$$
$$\sigma^2 \Delta t = \mathrm{Var}[V(t+\Delta t) - V(t)|V(t)] = \lambda \mathrm{E}[S^2]\Delta t + 0(\Delta t),$$

since the change in the virtual wait over the time increment Δt assuming a loaded system is the total service time needed to serve all arrivals over Δt, minus Δt. But it is known that $V(t)$ is a continuous-parameter continuous-state Markov process and hence will satisfy the Fokker–Planck equation for its conditional "density" $w_q(x, t|x_0)$ given by (see Newell, 1972)

$$\frac{\partial w_q(x,t|x_0)}{\partial t} = -\mu \frac{\partial w_q(x,t|x_0)}{\partial x} + \frac{\sigma^2}{2} \frac{\partial^2 w_q(x,t|x_0)}{\partial x^2},$$

subject to the boundary conditions

$$w_q(x,t|x_0) \geq 0,$$
$$\int_0^\infty w_q(x,t|x_0)\,dx = \frac{\lambda}{\mu},$$
$$\lim_{t\to 0} w_q(x,t|x_0) = 0 \qquad (x \neq 0).$$

From the earlier discussion of heavy traffic, it is to be expected that if

$$\frac{-\mu}{\sigma^2} = \frac{1 - \lambda \mathrm{E}[S]}{\lambda \mathrm{E}[S^2]}$$

is positive and small, then the diffusion solution should provide a good approximation. The final expression for $w_q(x,t|x_0)$ is found to be

$$w_q(x,t|x_0) = \frac{1}{\sqrt{2\pi t \sigma^2}} \left[e^{-(x-x_0-\mu t)^2/\sigma^2 t} + e^{2x\mu/\sigma^2} \left(e^{-(x-x_0-\mu t)^2/\sigma^2 t} + \frac{2\mu}{\sigma^2} \int_x^\infty e^{-(y-x_0-\mu t)^2/\sigma^2 t}\, dy \right) \right].$$

This section is not meant to exhaust the subject of diffusion approximations in queueing, but rather to give the reader a brief introduction; see also Feller (1971) and Karlin and Taylor (1975).

7.3 Network Approximations

Sections 4.2 and 4.3 analyzed *Jackson networks* in which service times and external interarrival times were exponentially distributed. For these networks, it was possible

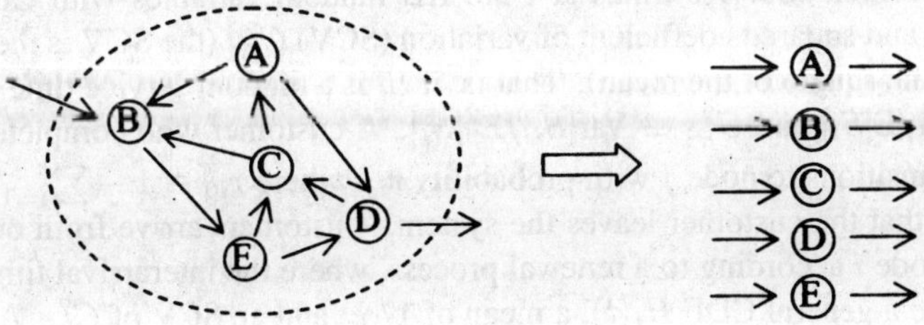

Figure 7.4 Basic approach of single-queue decomposition methods.

to derive exact analytical results. This section considers networks in which the service distributions are not exponential. In such networks, analytical results are not generally achievable and so approximation methods are needed.

There are many methods for approximating queueing performance in networks. This section describes one such technique. The technique is part of a class of methods called *decomposition methods*. These methods consist of decomposing a network into smaller subnetworks and then analyzing the subnetworks as separate and independent entities. Typically, the subnetworks are individual queues, as shown in Figure 7.4.

In a single-queue decomposition, the basic idea is to construct an arrival stream to each individual queue that matches as closely as possible the arrival stream in the original network. For example, the arrival stream to node B in the network decomposition should match the arrival stream to node B in the original network. In this example, the arrival stream to node B in the original network consists of customers coming from node A, node C, and the outside. Once an approximate arrival stream is constructed for node B, the queue is analyzed as a single queue in isolation. This can be done using various approximation techniques discussed earlier in this chapter.

In this section, we describe a *parametric decomposition method* to approximate the arrival stream into each node. The discussion is largely derived from Whitt (1983). We specifically discuss a two-parameter method. The method constructs an approximating renewal process where the first and second moments of the interarrival times approximately match those of the arrival stream to a given node in the original network. The fundamental assumption of this approach is that the arrival process to each node can be approximated well by a renewal process. In general, the arrival process to a node in a network is *not* a renewal process, because the times between successive arrivals are not necessarily independent. (The exception is that if the upstream nodes have Poisson arrivals and exponentially distributed service times, then the arrival process to the downstream node is renewal. See the discussion in Section 4.2.) For a discussion of methods for approximating a point process with a renewal process, see Whitt (1982) and Albin (1984).

7.3.1 Preliminaries and Notation

Before discussing the approximation method, we first state the fundamental assumptions and notation. Consider a network consisting of k nodes. Each node i consists of a single server. Service times at i are IID random variables with CDF $B_i(t)$, mean $1/\mu_i$, and squared coefficient of variation (SCV) C_{Bi}^2 (the SCV is the variance divided by the square of the mean). That is, if S_i is a random service time at node i, then $\mu_i \equiv 1/\mathrm{E}[S_i]$ and $C_{Bi}^2 \equiv \mathrm{Var}[S_i]/\mathrm{E}^2[S_i]$. A customer who completes service at node i transitions to node j with probability r_{ij}, where $r_{i0} \equiv 1 - \sum_{j=1}^{k} r_{ij}$ is the probability that the customer leaves the system. Customers arrive from outside the system to node i according to a renewal process, where the interarrival times of this process have a general CDF $H_i(t)$, a mean of $1/\gamma_i$, and an SCV of C_{0i}^2. All service times, external interarrival times, and routing transitions are independent of all else. Each node is assumed to be operating below capacity.

In summary, the network is like an open Jackson network (Section 4.2), but where the service times and external interarrival times follow general distributions, rather than being restricted to exponential. Table 7.2 summarizes the parameters associated with the network. In this discussion, we regard the first set of parameters as inputs, defining the network to be analyzed. The second set are outputs obtained through the analysis methodology that we describe in the next section.

Table 7.2 Network Notation

	Inputs
k	Number of nodes in network
r_{ij}	Transition probability from i to j
$B_i(t)$	CDF of service distribution at i
$H_i(t)$	CDF of external interarrival distribution to i
μ_i	Service rate at i
γ_i	External arrival rate to i
C_{Bi}^2	SCV of service distribution at i
C_{0i}^2	SCV of external interarrival distribution to i
	Outputs
λ_i	Throughput at i
ρ_i	Utilization at i ($\rho_i = \lambda_i/\mu_i$)
C_{Aj}^2	SCV of times between arrivals to j (approximate)
C_{Di}^2	SCV of times between departures from i (approximate)
C_{ij}^2	SCV of times between departures from i to j (approximate)
W_{qi}	Mean waiting time in queue at i (approximate)

7.3.2 Parametric Decomposition

We now describe a process for obtaining an approximate two-parameter characterization of the arrival process to each node. We have already seen a method for obtaining an approximate one-parameter characterization, via mean-flow rates. Specifically, for open Jackson networks, we obtained a set of equations (4.10a) balancing the mean rate of customers entering a node with the mean rate of customers leaving a node. Solving these equations provided the mean-flow rates λ_i of customers entering each node.

Now, the equations in (4.10a) apply equally well to the networks in this section, since the principle of balancing the mean-flow rates into and out of a node does not depend on the assumption of exponential service. Thus we have the following equations, identical to (4.10a):

$$\boxed{\lambda_i = \gamma_i + \sum_{j=1}^{k} \lambda_j r_{ji}, \qquad i = 1, \ldots, k.} \tag{7.31}$$

The equations are valid provided that all nodes are operating below capacity. Since the equations are linear, they can easily be solved to obtain λ_i. This provides the first parameter in the two-parameter decomposition method.

The second parameter we use in the characterization of the arrival process is C_{Ai}^2, the squared coefficient of variation of the interarrival times to node i. While the parameter λ_i is related to the first moment of the interarrival times, C_{Ai}^2 is related to the second moment. As before, the approach is to develop a set of equations involving C_{Ai}^2 based on relating the stream of customers entering a node to the stream of customers exiting a node. However, while the equations involving the mean-flow rates (7.31) are exact, the equations for C_{Ai}^2 are only approximate.

To obtain these balance equations, we break the flow of customers entering and exiting a node into three phases, as shown in Figure 7.5: superposition, queueing, and splitting. In the first phase, individual arrival streams originating from different nodes are merged into a single arrival stream to the given node. In the second phase, the combined arrival stream is processed by the queue. In the last phase, the departure stream is split into separate departure streams, identified by the destination of each customer.

The basic idea is to obtain equations relating the input and output streams of each phase. The equations for all three phases are then combined to yield a single set of equations that can be solved for C_{Ai}^2. Together, the exact values of λ_i and the approximated values of C_{Ai}^2 provide a two-parameter approximation for the arrival stream into node i. From this, performance metrics can be estimated using approximation formulas given earlier in this chapter.

We now describe this process in more detail, beginning with the derivation of balance equations for each phase in Figure 7.5.

Superposition. This section is based on material in Whitt (1982), Section 4.2. The arrival stream of customers into node i is the superposition of customers arriving to i coming from distinct sources indexed by j. This includes the stream of customers arriving from outside the network ($j = 0$). As we will argue, the following equation approximately relates the combined arrival stream to the component streams via the

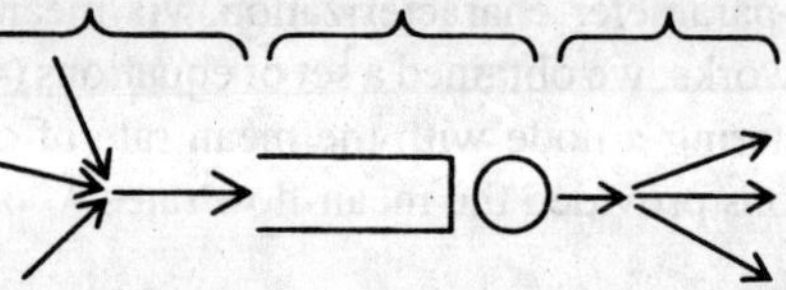

Figure 7.5 Customer flows at a node in the network.

parameters C_{Ai}^2 and C_{ji}^2:

$$C_{Ai}^2 = \frac{\gamma_i}{\lambda_i} C_{0i}^2 + \frac{1}{\lambda_i} \sum_{j=1}^{k} \lambda_j r_{ji} C_{ji}^2, \qquad i = 1, \ldots, k. \tag{7.32}$$

This equation is motivated by (a) approximating the component streams as independent renewal processes and (b) approximating the combined stream also as a renewal process. These assumptions are not generally true. Even when the first assumption is true, the second assumption does not generally follow from the first. In fact, the superposition of independent renewal processes is itself a renewal process if and only if all of the component processes are Poisson, in which case the combined process is also Poisson (Çinlar, 1972). As discussed in Section 4.2, the arrival process to a node is Poisson if and only if all upstream nodes have Poisson arrivals and exponentially distributed service times. Thus these assumptions, which we use to derive (7.32), typically are only approximations to the actual network behavior.

To derive (7.32), consider a node i in the network. Let $N_{ji}(t)$ be the number of arrivals from j to i that have occurred by time t [where $j = 0, 1, \ldots, k$ and $N_{0i}(t)$ represents the arrival process to i from the outside]. We assume that $N_{ji}(t)$ is a renewal process and that these processes are independent of each other. Let X_{ji} be a random interarrival time of this process. Let $N_i(t) \equiv N_{0i}(t) + \cdots + N_{ki}(t)$ be the combined arrival process to i. We assume that $N_i(t)$ is also a renewal process. Let X_i denote a random interarrival time of this process. Define the following limits, which we assume to exist:

$$\lambda_{ji} \equiv \lim_{t\to\infty} \frac{\mathrm{E}[N_{ji}(t)]}{t} \qquad \text{and} \qquad v_{ji} \equiv \lim_{t\to\infty} \frac{\mathrm{Var}[N_{ji}(t)]}{t}. \tag{7.33}$$

That is, λ_{ji} is the average rate of customers departing from j arriving at i. Since we have assumed the component processes are independent (an approximation),

$$\begin{aligned} \mathrm{E}[N_i(t)] &= \mathrm{E}[N_{0i}(t)] + \cdots + \mathrm{E}[N_{ki}(t)], \\ \mathrm{Var}[N_i(t)] &= \mathrm{Var}[N_{0i}(t)] + \cdots + \mathrm{Var}[N_{ki}(t)]. \end{aligned} \tag{7.34}$$

Then, (7.33) and (7.34) imply that

$$\lambda_i = \lambda_{0i} + \cdots + \lambda_{ki} \qquad \text{and} \qquad v_i = v_{0i} + \cdots + v_{ki},$$

where $\lambda_i \equiv \lim_{t\to\infty} \mathrm{E}[N_i(t)]/t$ and $v_i \equiv \lim_{t\to\infty} \mathrm{Var}[N_i(t)]/t$.

For a renewal process $N_i(t)$, there is a one-to-one relation between the first two moments of the interarrival distribution, $\mathrm{E}[X_i]$ and $\mathrm{E}[X_i^2]$, and the limiting values of the renewal process, λ_i and v_i. Specifically, these relationship are (Smith, 1959)

$$\lambda_i = \frac{1}{\mathrm{E}[X_i]} \qquad \text{and} \qquad v_i = \frac{\mathrm{E}[X_i^2] - \mathrm{E}^2[X_i]}{\mathrm{E}^3[X_i]}.$$

The first relationship simply states that the average arrival rate is the inverse of the average interarrival time. From these relationships, we have the following:

$$C_{Ai}^2 \equiv \frac{\mathrm{Var}[X_i]}{\mathrm{E}^2[X_i]} = \frac{\mathrm{E}[X_i^2] - \mathrm{E}^2[X_i]}{\mathrm{E}^2[X_i]} = v_i \mathrm{E}[X_i] = \frac{v_i}{\lambda_i}.$$

Similarly, the analogous relationship applies to each component renewal process – namely, $C_{ji}^2 = v_{ji}/\lambda_{ji}$. Combining these results gives

$$C_{Ai}^2 = \frac{v_i}{\lambda_i} = \frac{1}{\lambda_i}\sum_{j=0}^{k} v_{ji} = \frac{1}{\lambda_i}\sum_{j=0}^{k} \lambda_{ji} C_{ji}^2 = \frac{\gamma_i}{\lambda_i} C_{0i}^2 + \frac{1}{\lambda_i}\sum_{j=1}^{k} \lambda_j r_{ji} C_{ji}^2,$$

where we have used that $\gamma_i = \lambda_{0i}$ is the arrival rate from the outside to i and $\lambda_j r_{ji} = \lambda_{ji}$ is the rate from j to i.

This method of approximating the arrival stream is based on an *asymptotic method* (Whitt, 1982) in which the approximating renewal process is chosen so that the asymptotic values (e.g., $\lim_{t\to\infty} \mathrm{E}[N_i(t)]/t$ and $\lim_{t\to\infty} \mathrm{Var}[N_i(t)]/t$) of the approximating process match those of the original process. Many other approaches have also been suggested to approximate flows in a queue. For example, the *stationary-interval method* (Whitt, 1982) chooses an approximating renewal process so that the moments of the renewal interval match the moments of the stationary distribution of an interval from the original process. Albin (1984) and Whitt (1983) give hybrid approaches that combine results from the asymptotic and stationary-interval methods.

Queueing. Customers arriving at a node i in the network wait in the queue, receive service, and then depart from the node. The following equation relates (approximately) the arrival stream of customers to the departure stream of customers:

$$C_{Di}^2 = \rho_i^2 C_{Bi}^2 + (1-\rho_i^2) C_{Ai}^2, \qquad i = 1, \ldots, k. \tag{7.35}$$

The equation is motivated in part by considering the boundary conditions. When $\rho_i = 1$, the queue is saturated, so the interdeparture times match the service times, so $C_{Di}^2 = C_{Bi}^2$. When $\rho_i \approx 0$, arrivals are rare, so interarrival times are large. Thus the time between two departures is roughly equal to the time between two arrivals since the service time is relatively small, so $C_{Di}^2 \approx C_{Ai}^2$.

The specific form of the equation is motivated by approximating the arrival process to the queue as a renewal process. As mentioned previously, the arrival process to a queue in a network is not generally a renewal process. However, if we approximate the arrival process with a renewal process, then the queue behaves like a $G/G/1$ queue (i.e., the interarrival times are IID, so there is no correlation between successive interarrival times). Then we can make use of several previously developed results. In particular, in Section 7.1.1, we derived the following result (7.12) for a stationary $G/G/1$ with $\rho < 1$:

$$\mathrm{Var}[D] = 2\sigma_B^2 + \sigma_A^2 - 2W_q \left(\frac{1}{\lambda} - \frac{1}{\mu} \right).$$

Here, D is a random interdeparture time from the queue in steady state, σ_A^2 is the variance of the interarrival times, and σ_B^2 is the variance of the service times. Dividing both sides by $\mathrm{E}^2[D]$ gives

$$\frac{\mathrm{Var}[D]}{\mathrm{E}^2[D]} = 2\frac{\sigma_B^2}{\mathrm{E}^2[D]} + \frac{\sigma_A^2}{\mathrm{E}^2[D]} - 2\frac{W_q}{\mathrm{E}^2[D]} \left(\frac{1}{\lambda} - \frac{1}{\mu} \right).$$

Let S be a random service time and let T be a random interarrival time in steady state. Then

$$\frac{\mathrm{Var}[D]}{\mathrm{E}^2[D]} = 2\frac{\mathrm{E}^2[S]}{\mathrm{E}^2[D]} \left(\frac{\sigma_B^2}{\mathrm{E}^2[S]} \right) + \frac{\mathrm{E}^2[T]}{\mathrm{E}^2[D]} \left(\frac{\sigma_A^2}{\mathrm{E}^2[T]} \right) - 2\frac{W_q}{\mathrm{E}^2[D]} \left(\frac{1}{\lambda} - \frac{1}{\mu} \right).$$

Since $\rho < 1$ and the queue is in steady state, the average departure rate equals the average arrival rate – that is, $1/\mathrm{E}[D] = \lambda$. In particular, $\mathrm{E}[D] = \mathrm{E}[T]$ and $\mathrm{E}[S]/\mathrm{E}[D] = \lambda/\mu$. Thus the above equation can be written as

$$C_D^2 = 2\rho^2 C_B^2 + C_A^2 - 2\lambda W_q(1 - \rho).$$

Now, we approximate W_q using (7.20), a previously derived approximation for the $G/G/1$ queue (which is exact for Poisson arrivals):

$$W_q \approx \left(\frac{\rho}{1 - \rho} \right) \left(\frac{C_A^2 + C_B^2}{2} \right) \left(\frac{1}{\mu} \right).$$

Substituting this into the previous equation gives

$$\begin{aligned} C_D^2 &= 2\rho^2 C_B^2 + C_A^2 - \rho^2 (C_A^2 + C_B^2) \\ &= \rho^2 C_B^2 + (1 - \rho^2) C_A^2. \end{aligned}$$

Applying this result to each individual queue i yields the result (7.35). We note that there are better approximations than (7.35) (e.g., Whitt, 1983, 1984, 1995). The intent here is to provide an introductory treatment to these types of approximations.

Splitting. The departure stream of customers leaving node i is split into separate streams based on the destination j of each customer. The following equation relates (approximately) the split processes to the overall departure process, via the parameters C_{ij}^2 and C_{Di}^2:

$$C_{ij}^2 = r_{ij} C_{Di}^2 + (1 - r_{ij}), \qquad i = 1, \ldots, k. \qquad (7.36)$$

This equation is motivated by approximating the departure process from node i as a renewal process. In general, the departure process from a node is not a renewal process (e.g., Berman and Westcott, 1983).

To derive (7.36), consider a node i, and suppose that a customer has just departed from i to j. Let Y be the time until the next departure from i to j. Now, for the overall departure process (customers departing from i to *any* destination), let $Y_1, Y_2, \ldots$ be the times between successive departures. Since this is a renewal process by approximation, $Y_1, Y_2, \ldots$ are IID random variables. Also, a customer departing from i goes to j with probability r_{ij}, independent of all else. Thus

$$Y = \sum_{n=1}^{N} Y_n,$$

where N is a geometric random variable with mean $1/r_{ij}$. By conditioning on N,

$$\mathrm{E}[Y] = \mathrm{E}[\mathrm{E}[Y|N]] = \mathrm{E}[N\,\mathrm{E}[Y_n]] = \mathrm{E}[N]\,\mathrm{E}[Y_n] = \frac{\mathrm{E}[Y_n]}{r_{ij}}.$$

Similarly,

$$\begin{aligned}\mathrm{Var}[Y] &= \mathrm{E}[\mathrm{Var}[Y|N]] + \mathrm{Var}[\mathrm{E}[Y|N]] \\ &= \mathrm{E}[N\,\mathrm{Var}[Y_n]] + \mathrm{Var}[N\,\mathrm{E}[Y_n]] \\ &= \mathrm{E}[N]\,\mathrm{Var}[Y_n] + \mathrm{Var}[N]\,\mathrm{E}^2[Y_n] \\ &= \frac{1}{r_{ij}}\mathrm{Var}[Y_n] + \frac{1-r_{ij}}{r_{ij}^2}\mathrm{E}^2[Y_n].\end{aligned}$$

Then,

$$C_{ij}^2 \equiv \frac{\mathrm{Var}[Y]}{\mathrm{E}^2[Y]} = \mathrm{Var}[Y]\frac{r_{ij}^2}{\mathrm{E}^2[Y_n]} = r_{ij}\frac{\mathrm{Var}[Y_n]}{\mathrm{E}^2[Y_n]} + 1 - r_{ij} = r_{ij}C_{Di}^2 + 1 - r_{ij}.$$

In summary, (7.36) is exact under Markovian routing of a renewal process. Here, we have Markovian routing, but the departure process from i is not generally a renewal process.

Synthesis of Results. Equations (7.32), (7.35), and (7.36) can be combined to yield a single equation in C_{Aj}^2.

$$C_{Ai}^2 = \frac{\gamma_i}{\lambda_i}C_{0i}^2 + \sum_{j=1}^{k}\frac{\lambda_j}{\lambda_i}r_{ji}\left(r_{ji}[\rho_j^2 C_{Bj}^2 + (1-\rho_j^2)C_{Aj}^2] + 1 - r_{ji}\right), \qquad (7.37)$$

for $i = 1, \ldots, k$. These equations are linear in C_{Ai}^2, so they can be solved relatively easily on a computer, given values for r_{ij}, C_{0i}^2, and C_{Bi}^2 [also, $\rho_i = \lambda_i/\mu_i$ is determined by first solving (7.31) for λ_i].

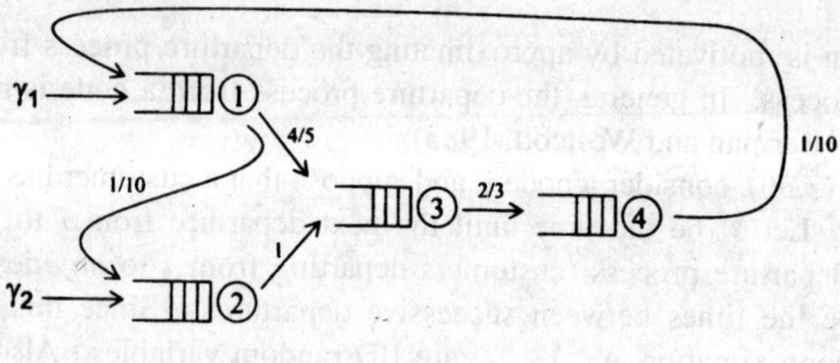

Figure 7.6 Example network.

■ **EXAMPLE 7.5**

Consider a queueing network with routing probabilities as shown in Figure 7.6. Routing probabilities that sum to less than one leaving a queue indicate the fraction of customers leaving the system from that queue. Each queue has a single server. Service times at node 1 follow an E_5 distribution with mean $1/20$. Similarly, service times at nodes 2, 3, and 4 also follow an E_5 distribution, but with means $1/10$, $1/20$, and $1/15$, respectively. Customers arrive from outside the network to nodes 1 and 2. The times between external arrivals to node 1 are E_2-distributed with mean $1/9$. The times between external arrivals to node 2 are E_2-distributed with mean $1/6$. All other assumptions discussed in this section apply. Approximate the mean and SCV of interarrival times to each node in the network in steady state.

The SCV of an E_k distribution is $1/k$; see (3.18) and (3.19). Thus the routing matrix and network parameters (defined in Table 7.2) are

$$\mathbf{R} = \begin{pmatrix} - & 1/10 & 4/5 & - \\ - & - & 1 & - \\ - & - & - & 2/3 \\ 1/10 & - & - & - \end{pmatrix} \quad \text{and}$$

i	γ_i	μ_i	C_{0i}^2	C_{Bi}^2
1	9	20	1/2	1/5
2	6	10	1/2	1/5
3	–	20	–	1/5
4	–	15	–	1/5

The flow balance equations (7.31) are

$$\begin{aligned}
\lambda_1 &= 9 + \tfrac{1}{10}\lambda_4, \\
\lambda_2 &= 6 + \tfrac{1}{10}\lambda_1, \\
\lambda_3 &= \tfrac{4}{5}\lambda_1 + \lambda_2, \\
\lambda_4 &= \tfrac{2}{3}\lambda_3.
\end{aligned}$$

Solving this set of equations gives the average net arrival rate λ_i to each queue, and from this the utilization ρ_i can be obtained:

$$\lambda_1 = 10, \quad \lambda_2 = 7, \quad \lambda_3 = 15, \quad \lambda_4 = 10,$$
$$\rho_1 = \tfrac{1}{2}, \quad \rho_2 = \tfrac{7}{10}, \quad \rho_3 = \tfrac{3}{4}, \quad \rho_4 = \tfrac{2}{3}.$$

Then, the SCV equations (7.37) for this network are

$$C_{A1}^2 = \frac{9}{10}\frac{1}{2} + \frac{10}{10}\frac{1}{10}\left(\frac{1}{10}\left[\left(\frac{2}{3}\right)^2\frac{1}{5} + \left(1-\left(\frac{2}{3}\right)^2\right)C_{A4}^2\right] + \frac{9}{10}\right),$$
$$C_{A2}^2 = \frac{6}{7}\frac{1}{2} + \frac{10}{7}\frac{1}{10}\left(\frac{1}{10}\left[\left(\frac{1}{2}\right)^2\frac{1}{5} + \left(1-\left(\frac{1}{2}\right)^2\right)C_{A1}^2\right] + \frac{9}{10}\right),$$
$$C_{A3}^2 = \frac{10}{15}\frac{4}{5}\left(\frac{4}{5}\left[\left(\frac{1}{2}\right)^2\frac{1}{5} + \left(1-\left(\frac{1}{2}\right)^2\right)C_{A1}^2\right] + \frac{1}{5}\right)$$
$$+\frac{7}{15}\left(\left[\left(\frac{7}{10}\right)^2\frac{1}{5} + \left(1-\left(\frac{7}{10}\right)^2\right)C_{A2}^2\right]\right),$$
$$C_{A4}^2 = \frac{15}{10}\frac{2}{3}\left(\frac{2}{3}\left[\left(\frac{3}{4}\right)^2\frac{1}{5} + \left(1-\left(\frac{3}{4}\right)^2\right)C_{A3}^2\right] + \frac{1}{3}\right).$$

These equations simplify to

$$C_{A1}^2 = \frac{1217}{2250} + \frac{1}{180}C_{A4}^2,$$
$$C_{A2}^2 = \frac{781}{1400} + \frac{3}{280}C_{A1}^2,$$
$$C_{A3}^2 = \frac{1303}{7500} + \frac{8}{25}C_{A1}^2 + \frac{119}{500}C_{A2}^2,$$
$$C_{A4}^2 = \frac{49}{120} + \frac{7}{24}C_{A3}^2.$$

These equations are linear in C_{Ai}^2. The solution is

$$C_{A1}^2 \doteq .5439, \quad C_{A2}^2 \doteq .5637, \quad C_{A3}^2 \doteq .4820, \quad C_{A4}^2 \doteq .5489.$$

The final step in analyzing a queueing network is to use the two-parameter characterization of each arrival process (i.e., λ_i and C_{Ai}^2) to estimate the performance of each queue. Since we are approximating the arrival processes to all queues as independent renewal processes, each queue can be treated as a separate $G/G/1$ queue. Thus the approximations given in Section 7.2.1 apply. Here, we use a variation of (7.20) proposed by Kraemer and Langenbach-Belz (1976). Specifically, the average wait in queue for a $G/G/1$ queue is approximated by

$$W_q \approx \left(\frac{\rho}{1-\rho}\right)\left(\frac{C_A^2 + C_B^2}{2}\right)\left(\frac{1}{\mu}\right) g(\rho, C_A^2, C_B^2), \tag{7.38}$$

where

$$g(\rho, C_A^2, C_B^2) = \begin{cases} \exp\left[-\dfrac{2(1-\rho)}{3\rho}\dfrac{(1-C_A^2)^2}{C_A^2 + C_B^2}\right] & (C_A^2 < 1), \\ 1 & (C_A^2 \geq 1). \end{cases} \tag{7.39}$$

When $C_A^2 \geq 1$, (7.38) and (7.39) reduce to (7.20).

In summary, the parametric decomposition method described in this section consists of the following steps:

1. Obtain the mean-flow rates λ_i using (7.31).

2. Obtain the SCVs C^2_{Ai} of the interarrival times to each node using (7.37).

3. Analyze the nodes individually as separate $G/G/1$ queues using approximation formulas, such as (7.38) and (7.39).

Aside from the first step to compute λ_i, which is exact, all other steps are approximate. In particular, (7.37) is the synthesis of three equations, each of which requires at least one approximating assumption. The final estimate of W_q is also an approximation. Although the method works very well in many cases, researchers have identified examples in which the methods do not work well – for example, see Suresh and Whitt (1990), Kim (2004). Thus care should be used when using these methods.

EXAMPLE 7.6

Continuing with the previous example, the expected wait in queue for node 1, using (7.38), is

$$W_{q1} \approx \left(\frac{1/2}{1-1/2}\right)\left(\frac{.5439+.2}{2}\right)\left(\frac{1}{20}\right) g(.5, .5439, 0.2) \doteq .0154.$$

Similarly, for the other nodes

$$W_{q2} \approx .0830, \quad W_{q3} \approx .0469, \quad W_{q4} \approx .0456.$$

7.3.2.1 Multiple Servers. The parametric decomposition method described previously can easily be extended to queues with multiple servers. There are two main changes: (a) the queueing equation (7.35) is generalized to account for multiple servers, and (b) the approximation equation for the $G/G/1$ queue is replaced by an approximation equation for the $G/G/c$ queue. The superposition and splitting equations, (7.32) and (7.36), remain unchanged since these processes do not involve any queueing.

To generalize the queueing equation (7.35), Whitt (1983) suggested the following:

$$\begin{aligned} C^2_{Di} &= 1 + (1-\rho_i^2)(C^2_{Ai}-1) + \frac{\rho_i^2}{\sqrt{c}}(C^2_{Bi}-1) \\ &= \frac{\rho_i^2}{\sqrt{c}}C^2_{Bi} + (1-\rho_i^2)C^2_{Ai} + \rho_i^2\left(1-\frac{1}{\sqrt{c}}\right). \end{aligned} \tag{7.40}$$

This approximation has the following properties. First, it reduces to (7.35) when $c = 1$. Second, it gives exact results for $M/M/c$ and $M/G/\infty$ queues, both of which have Poisson departure processes in steady state. In particular, for the $M/M/c$ queue ($C^2_{Ai} = 1$ and $C^2_{Bi} = 1$), (7.40) gives $C^2_{Di} = 1$ which is consistent with Poisson departures. Similarly, for an $M/G/\infty$ queue ($C^2_{Ai} = 1$ and $c = \infty$), (7.40) gives $C^2_{Di} = 1$. Combining the new queueing equation with the existing splitting and superposition equations (7.32) and (7.36) yields a modified version of (7.37):

$$C_{Ai}^2 = \frac{\gamma_i}{\lambda_i} C_{0i}^2 + \sum_{j=1}^{k} \frac{\lambda_j}{\lambda_i} r_{ji} \left(r_{ji} \left[\frac{\rho_j^2}{\sqrt{c}} C_{Bj}^2 + (1 - \rho_j^2) C_{Aj}^2 + \rho_j^2 \left(1 - \frac{1}{\sqrt{c}} \right) \right] + 1 - r_{ji} \right), \quad i = 1, \ldots, k. \tag{7.41}$$

Again, these equations are linear in C_{Ai}^2. For the $G/G/c$ queue, Section 7.2.1 gave an approximation for W_q, the average steady-state wait in queue; see (7.21). In summary, the multiserver extension of the approximation method consists of the following steps:

1. Obtain the mean-flow rates λ_i using (7.31).
2. Obtain the SCVs C_{Ai}^2 of the interarrival times to each node using (7.41).
3. Analyze the nodes individually as separate $G/G/c$ queues using (7.21).

PROBLEMS

7.1. The Bearing Straight Corporation (of Example 5.3 and Problem 6.4) is now in bad straits because it has found that an estimated machine-breakdown rate as low as 5/h was optimistic and that a more realistic estimate would be 6/h. It is then observed that the service rate is also 6/h, based on a time of 9 min two out of three times and 12 min one-third of the time. Use the heavy-traffic approximation of Section 7.2.3 to determine what decrease below the 9-min figure would guarantee Bearing machines an average wait twice what it was before, namely, now equal to 72 min.

7.2. Verify that (7.8) reduces to the PK formula for W_q, in the case of Poisson arrivals.

7.3. Give an example of a $G/G/1$ queue where the upper bound in (7.13) is exact.

7.4. Assemblies come to an inspection station with a single inspector. The interarrival times of assemblies to the station are found to be independent and having a hyperexponential distribution with density function

$$a(t) = 0.1e^{-t/3} + 0.07e^{-t/10}.$$

Approximately one-fourth of the assemblies are widgets, and approximately three-fourths of the assemblies are gadgets. The inspection time for a widget is exponential with a mean time to inspect of 9 min. The inspection time for a gadget is exponential with a mean time to inspect of 5 min. Assemblies are inspected in a FCFS manner, and the sequence of assembly types (widgets or gadgets) arriving to be inspected is independent. The production supervisor is interested to know what the worst-case average wait (total average time at the inspection station) might be.

7.5. Employ the bounds on W_q derived in Section 7.1 for $D/M/1$.

7.6. Do a Markov-chain exact analysis to find the stationary distribution for the delay in queue of a $D/G/1$ queue with interarrival times of 2 min and two equiprobable service times of 0 and 3 min.

7.7. Find upper and lower bounds for the W_q of the $D/G/1$ queue of Problem 7.6, and compare with the exact answer.

7.8. Considering Problem 7.1, what decrease below 9 min would guarantee Bearing-machine waits of more than 400 min less than 5% of the time?

7.9. The graduate assistant of Problem 2.8 now finds in the new semester that the arrival rate to the counter has increased to 20/h. If service times remain exponential with mean 4 min, use the result of Section 7.2 to find the approximate probability that the nth customer (n large) will have to wait. Then compare this result with the Chebyshev approximation.

7.10. Show that the density function $p(x, t|x_0)$ given by (7.30) satisfies the diffusion partial differential equation (7.29).

7.11. Apply the central limit theorem to the one-dimensional random walk which moves left with probability q, moves right with probability p, and stands still with probability r. Then use the limiting procedures of Section 8.1 which led to (7.30) to show that the continuous problem is a Wiener process, and that this Wiener density satisfies an equation of the same form as (7.29).

CHAPTER 8

NUMERICAL TECHNIQUES AND SIMULATION

Once we leave the arena of steady-state Markovian queues, nice closed-form analytical results are quite elusive. In order to obtain useful solutions in these cases, one must resort to simulation or numerical techniques. Typical examples of problems where concise closed-form solutions are particularly difficult to find are the $G/G/c$, non-phase-type $G/G/1$, transient problems, and non-Markovian networks.

In this chapter we examine three numerical-based techniques. First, we study some useful numerical techniques that cover both steady-state and transient results. Next, we study the issue of finding the inverse of a Laplace or Laplace-Stieltjes transform. This is important because a common solution technique in analyzing queueing situations is the use of transform methods. Sometimes finding the analytic inverse of the transform is very difficult. Hence we look at a method to numerically compute the inverse. Finally, we conclude this chapter with a brief overview of discrete event simulation methodology for queueing modeling.

8.1 Numerical Techniques

In contrast to the approximation techniques presented in the previous chapter, we now return our concentration to exact analysis of the situation of interest. In many

Fundamentals of Queueing Theory, Fourth Edition.
By D. Gross, J. F. Shortle, J. M. Thompson, and C. M. Harris
Copyright © 2008 John Wiley & Sons, Inc.

cases where we are unable to find neat, closed-solution analytical formulas, we can still analyze the system and obtain numerical answers, often to a desired, prespecified error tolerance.

For many steady-state situations, the analysis involves solving simultaneous linear-algebraic equations, perhaps large numbers of them; of interest are efficient procedures for solving such systems. In the case of transient solutions, we are often faced with solving sets of linear first-order differential equations, and a variety of numerical techniques for doing so exist.

The disadvantage of numerical solution procedures is that all parameters must be specified numerically before answers can be obtained, so that when parameters are changed, the calculations must be redone. However, where no other methods can suffice, this is a small price to pay for obtaining actual answers.

8.1.1 Steady-State Solutions

For queues in which a Markov analysis is possible, the steady-state solution is found by solving the stationary equations for a discrete-parameter Markov chain (DPMC)

$$\begin{aligned}\boldsymbol{\pi} &= \boldsymbol{\pi}\boldsymbol{P},\\ \boldsymbol{\pi}\boldsymbol{e} &= 1,\end{aligned}$$

or for a continuous-parameter Markov chain (CPMC)

$$\begin{aligned}\mathbf{0} &= \boldsymbol{p}\boldsymbol{Q},\\ 1 &= \boldsymbol{p}\boldsymbol{e},\end{aligned}$$

where $\boldsymbol{\pi}$ (or $\boldsymbol{p}$) is the steady-state probability vector, $\boldsymbol{P}$ the discrete-parameter Markov-chain transition probability matrix, $\boldsymbol{Q}$ the infinitesimal generator of the continuous-time Markov chain, and $\boldsymbol{e}$ a vector of ones (see Section 1.9.5). For situations such as finite-source queues, queues with limited waiting capacity, and so on, $\boldsymbol{P}$ and $\boldsymbol{Q}$ are finite-dimensional matrices, and numerically, the problem reduces to solving a system of simultaneous linear equations. The equation $\boldsymbol{\pi} = \boldsymbol{\pi}\boldsymbol{P}$ can always be written as $0 = \boldsymbol{\pi}\boldsymbol{Q}$, where $\boldsymbol{Q} = \boldsymbol{P} - \boldsymbol{I}$ for the DPMC case.

As an example, consider an $M/G/1/3$ queue. We desire to find the departure-point steady-state system-size probabilities. For this case, there are three possible states of the system at a departure point, namely, a departing customer can see an empty system, a system with one remaining customer, or a system with two remaining customers. Thus from Section 5.1.7, the $\boldsymbol{P}$ matrix for the imbedded Markov chain is

$$\boldsymbol{P} = \begin{pmatrix} k_0 & k_1 & 1-k_0-k_1 \\ k_0 & k_1 & 1-k_0-k_1 \\ 0 & k_0 & 1-k_0 \end{pmatrix},$$

where, as usual,

$$k_n = \Pr\{n \text{ arrivals during a service period}\} = \frac{1}{n!}\int_0^\infty (\lambda t)^n e^{-\lambda t}\, dB(t).$$

Thus if $B(t)$ and λ are specified, the values k_0 and k_1 can be calculated and the problem reduces to solving a 3×3 set of linear equations (one equation of $\boldsymbol{\pi} = \boldsymbol{\pi P}$ is always redundant):

$$(\pi_0, \pi_1, \pi_2) = (\pi_0, \pi_1, \pi_2) \begin{pmatrix} k_0 & k_1 & 1 - k_0 - k_1 \\ k_0 & k_1 & 1 - k_0 - k_1 \\ 0 & k_0 & 1 - k_0 \end{pmatrix},$$

$$\pi_0 + \pi_1 + \pi_2 = 1.$$

Let us now reconsider Example 5.3. Suppose that whenever three or more machines are down (a situation that is intolerable for production purposes), an outside repairman is called in, so that as far as the internal repair process is concerned, we have an $M/G/1/3$ system. For the two-point service distribution,

$$k_n = \frac{2}{3n!} e^{-3/4} (\tfrac{3}{4})^n + \frac{1}{3n!} e^{-1},$$

and thus

$$k_0 = \tfrac{2}{3} e^{-3/4} + \tfrac{1}{3} e^{-1} \doteq .43, \qquad k_1 = \tfrac{1}{2} e^{-3/4} + \tfrac{1}{3} e^{-1} \doteq .36.$$

To find $\boldsymbol{\pi}$ we must thus solve

$$(\pi_0, \pi_1, \pi_2) = (\pi_0, \pi_1, \pi_2) \begin{pmatrix} .43 & .36 & .21 \\ .43 & .36 & .21 \\ 0 & .43 & .57 \end{pmatrix},$$

$$\pi_0 + \pi_1 + \pi_2 = 1.$$

Rewriting in the $\mathbf{0} = \boldsymbol{\pi Q}$ form, we have

$$(0, 0, 0) = (\pi_0, \pi_1, \pi_2) \begin{pmatrix} -.57 & .36 & .21 \\ .43 & -.64 & .21 \\ 0 & .43 & -.43 \end{pmatrix},$$

$$1 = \pi_0 + \pi_1 + \pi_2.$$

Since one of the equations in $\mathbf{0} = \boldsymbol{\pi Q}$ is always redundant, we can replace the last column of the $\boldsymbol{Q}$ matrix by a column of ones and the last 0 element of the left-hand-side $\mathbf{0}$ vector by a one, thereby incorporating the summability to one condition. Thus we must solve

$$(0, 0, 1) = (\pi_0, \pi_1, \pi_2) \begin{pmatrix} -.57 & .36 & 1 \\ .43 & -.64 & 1 \\ 0 & .43 & 1 \end{pmatrix}, \tag{8.1}$$

a system of the form

$$\boldsymbol{b} = \boldsymbol{\pi A},$$

where $\boldsymbol{b}$ is the "modified" zero vector and $\boldsymbol{A}$ is the "modified" $\boldsymbol{Q}$ matrix. It suffices then to find $\boldsymbol{A}^{-1}$, since the solution is $\boldsymbol{b}\boldsymbol{A}^{-1}$. In fact, since $\boldsymbol{b}$ is a vector of all zeros except the last element, we need only find the last row $\boldsymbol{A}^{-1}$. This last row then contains the $\{\pi_i\}$. It is easy to find the last row of $\boldsymbol{A}^{-1}$ to be (.29, .38, .33), so that $\pi_0 = .29, \pi_1 = .38$, and $\pi_2 = .33$. Hence, solving these types of problems for large state spaces often boils down to obtaining or generating an efficient method for matrix inversion.

There is an alternative way of solving these problems for handling redundancy in $\mathbf{0} = \boldsymbol{\pi}\boldsymbol{Q}$. Instead of replacing one of the equations with the sum of the probabilities equal to 1, we can arbitrarily set one of the π_i equal to one, solve an $(n-1) \times (n-1)$ nonsingular system (i.e., solve for $n-1$ variables in terms of the remaining one), and then renormalize the resulting π_i. In the above example, suppose we set π_2 equal to one. Then the resulting $(n-1) \times (n-1)$ system of equations becomes

$$(0, -.43) = (\pi_0, \pi_1)\begin{pmatrix} -.57 & .36 \\ .43 & -.64 \end{pmatrix},$$

which still has the form $\boldsymbol{b} = \boldsymbol{\pi}\boldsymbol{A}$, but now $\boldsymbol{b}$ and $\boldsymbol{\pi}$ are two-component vectors and $\boldsymbol{A}$ is a 2×2 matrix. The solution is then $\boldsymbol{\pi} = \boldsymbol{b}\boldsymbol{A}^{-1}$, which turns out to be (0.88, 1.17). Now we must include $\pi_2 = 1$, so that renormalizing on the sum $0.88+1.17+1 = 3.05$ yields $\pi_0 = 0.88/3.05 = .29$, $\pi_1 = 1.17/3.05 = .38$, and $\pi_2 = 1/3.05 = .33$, the same answers as before.

In real systems, we may well end up with matrices with thousands or tens of thousands of rows and columns; one can easily conceive of queueing network problems with a million states. Thus efficient procedures for solving *large* systems of equations are crucial.

In many queueing applications, the matrix to be inverted is sparse; that is, most of the elements are zero. For example, in birth–death processes, only the elements on the main, super-, and subdiagonals can be nonzero; the other elements are zero [see Section 1.9.3, just prior to (1.30)]. Good sparse-matrix computer packages are available that can handle fairly large systems. Also, one can utilize a large-scale linear programming package, since the Markov queueing equations are linear. By formulating a linear program with equality constraints and a suitable objective function, such a package will yield solutions to queueing types of equations.

Iterative solution techniques have also been shown to be efficient for many large-scale Markovian queueing systems. For example, we could use the basic Markov-chain recursive equation (1.22), $\boldsymbol{\pi}^{(m)} = \boldsymbol{\pi}^{(m-1)}\boldsymbol{P}$, stopping at a suitably large m. Using this procedure for the above example with

$$\boldsymbol{\pi}^{(0)} = (1, 0, 0), \quad \boldsymbol{P} = \begin{pmatrix} .43 & .36 & .21 \\ .43 & .36 & .21 \\ 0 & .43 & .57 \end{pmatrix}$$

yields

$$\boldsymbol{\pi}^{(1)} = \boldsymbol{\pi}^{(0)}\boldsymbol{P} = (.43, .36, .21), \quad \boldsymbol{\pi}^{(2)} = \boldsymbol{\pi}^{(1)}\boldsymbol{P} = (.34, .37, .29),$$
$$\boldsymbol{\pi}^{(3)} = \boldsymbol{\pi}^{(2)}\boldsymbol{P} = (.31, .38, .31), \quad \boldsymbol{\pi}^{(4)} = \boldsymbol{\pi}^{(3)}\boldsymbol{P} = (.30, .38, .33),$$
$$\boldsymbol{\pi}^{(5)} = \boldsymbol{\pi}^{(4)}\boldsymbol{P} = (.29, .38, .33), \quad \boldsymbol{\pi}^{(6)} = \boldsymbol{\pi}^{(5)}\boldsymbol{P} = (.29, .38, .33),$$

and we see that after only five iterations, the vector $\boldsymbol{\pi}$ is, to two decimal places, equal to the steady-state solution previously obtained. This type of procedure, as we show a little later, can be used for general systems of linear equations [e.g., those given by (7.3)], and as applied above is called Jacobi stepping.

A variation of this procedure, which is called Gauss–Seidel stepping, uses each new $\pi_j^{(m)}$ as it is calculated for calculating $\pi_{j+1}^{(m)}$, rather than using only the $\pi^{(m-1)}$ elements. For example, designating $\boldsymbol{P}_i$ as the ith column of the $\boldsymbol{P}$ matrix,

$$\pi_0^{(1)} = \boldsymbol{\pi}^{(0)}\boldsymbol{P}_1 = (1, 0, 0)\begin{pmatrix} .43 \\ .43 \\ 0 \end{pmatrix} = .43.$$

Now adjust $\boldsymbol{\pi}^{(0)}$ from (1, 0, 0) to (.43, 0, 0), so that we now calculate

$$\pi_1^{(1)} = (.43, 0, 0)\boldsymbol{P}_2 = (.43, 0, 0)\begin{pmatrix} .36 \\ .36 \\ .43 \end{pmatrix} = .15.$$

Continuing,

$$\pi_2^{(1)} = (.43, .15, 0)\boldsymbol{P}_3 = (.43, .15, 0)\begin{pmatrix} .21 \\ .21 \\ .57 \end{pmatrix} = .12.$$

Now

$$\pi_0^{(2)} = (.43, .15, .12)\boldsymbol{P}_1 = .25, \quad \pi_1^{(2)} = (.25, .15, .12)\boldsymbol{P}_2 = .20,$$
$$\pi_2^{(2)} = (.25, .20, .12)\boldsymbol{P}_3 = .16.$$

Iterating again yields

$$\pi_0^{(3)} = (.25, .20, .16)\boldsymbol{P}_1 = .19, \quad \pi_1^{(3)} = (.19, .20, .16)\boldsymbol{P}_2 = .21,$$
$$\pi_2^{(3)} = (.19, .21, .16)\boldsymbol{P}_3 = .18,$$

and twice more gives

$$\pi_0^{(4)} = .17, \quad \pi_1^{(4)} = .21, \quad \pi_2^{(4)} = .18,$$
$$\pi_0^{(5)} = .16, \quad \pi_1^{(5)} = .21, \quad \pi_2^{(5)} = .18.$$

Normalizing after five iterations, we get an estimate of the steady-state π of

$$(.16/.56, .21/.56, .18/.56) = (.29, .38, .32),$$

which agrees, quite well, to two decimal places with what we obtained earlier by matrix inversion and Jacobi stepping.

These same types of techniques (Jacobi and Gauss–Seidel) can be used to solve any finite system of linear equations, such as the stationary equations $\mathbf{0} = \boldsymbol{pQ}$ for continuous-parameter Markov chains (e.g., Maron, 1982, Cooper, 1981). In fact, the Jacobi and Gauss–Seidel procedures really refer to solving sets of equations and not to stepping through finite discrete-parameter Markov chains (we took the liberty of using these names to describe the aforementioned Markov-chain stepping procedure for convenience).

Consider, for example, the finite system of equations for a continuous-parameter Markov chain,

$$\mathbf{0} = \boldsymbol{pQ},$$
$$\mathbf{1} = \boldsymbol{pe}.$$

Since, as mentioned earlier, one equation of the set $\mathbf{0} = \boldsymbol{pQ}$ is redundant, we can replace the last column of the $\boldsymbol{Q}$ matrix by ones and the last element of the $\mathbf{0}$ vector by a one. This incorporates the equation $\mathbf{1} = \boldsymbol{pe}$ so that we have the linear system of equations

$$\boldsymbol{b} = \boldsymbol{pA},$$

where $\boldsymbol{b} = (\mathbf{0}, 1)$ and

$$\boldsymbol{A} = \begin{pmatrix} -q_0 & q_{0,1} & \cdots & q_{0,N-2} & 1 \\ q_{10} & -q_1 & \cdots & q_{1,N-2} & 1 \\ \vdots & \vdots & & \vdots & \vdots \\ q_{N-1,0} & q_{N-1,1} & \cdots & q_{N-1,N-2} & 1 \end{pmatrix}. \tag{8.2}$$

Using straightforward solution techniques, as illustrated in the beginning of this section (and, in fact, as we mentioned before in the $M/G/1/3$ example), all we need is the last row of $\boldsymbol{A}^{-1}$, since only the last element of $\boldsymbol{b}$ is nonzero. Furthermore, the last element of $\boldsymbol{b}$ is a one, so we have

$$\boldsymbol{p} = (a_{N-1,0}, a_{N-1,1}, \ldots, a_{N-1,N}),$$

where $a_{N-1,j}$ is the jth element of the last row of $\boldsymbol{A}^{-1}$. We could, of course, have used the alternative formulation of setting one p_i equal to one, solving a nonsingular $(N-1)\times(N-1)$ system $\boldsymbol{b} = \boldsymbol{pA}$, and then renormalizing the $\{p_i\}$. We will mention more about this alternative problem setup a little later.

Standard inversion techniques such as Gauss–Jordan pivoting are quite adequate for moderately sized systems; however, to model large-state-space systems (for networks, N's of 10,000, 50,000, or even 100,000 are not unrealistic), Jacobi and Gauss–Seidel iterative techniques appear to be more efficient.

The Jacobi technique for solving sets of simultaneous equations is "mechanically" similar to stepping through a discrete-parameter Markov chain. If we break up the matrix $\boldsymbol{A}$ into a lower-triangular matrix $\boldsymbol{L}$, a diagonal matrix $\boldsymbol{D}$, and an upper-triangular $\boldsymbol{U}$ matrix, so that $\boldsymbol{L}+\boldsymbol{D}+\boldsymbol{U}=\boldsymbol{A}$, then the equations $\boldsymbol{b}=\boldsymbol{p}\boldsymbol{A}$ can be written $\boldsymbol{b}=\boldsymbol{p}(\boldsymbol{L}+\boldsymbol{D}+\boldsymbol{U})$ or $\boldsymbol{p}\boldsymbol{D}=\boldsymbol{b}-\boldsymbol{p}(\boldsymbol{L}+\boldsymbol{U})$. We start with some arbitrary $\boldsymbol{p}^{(0)}$ and iterate using the equation

$$\boldsymbol{p}^{(n+1)}\boldsymbol{D}=\boldsymbol{b}-\boldsymbol{p}^{(n)}(\boldsymbol{L}+\boldsymbol{U})$$

until we satisfy some stopping criterion. Convergence is not, in general, guaranteed; it often depends on the order in which the equations are written and on the particular problem generating the equations. For equations emanating from continuous-time Markov chains as illustrated previously, Cooper and Gross (1991) prove that convergence is guaranteed when using the nonsingular $(N-1)\times(N-1)$ set generated by setting one of the p_i to one, although in our example, we shall see that we also obtain convergence for the $N\times N$ system given by (8.2).

The Gauss–Seidel procedure is a modification of the Jacobi technique in that as each new element of $\boldsymbol{p}^{(n+1)}$ is calculated, it is used to replace the old element of $\boldsymbol{p}^{(n)}$ in calculating the next element of $\boldsymbol{p}^{(n+1)}$, as was done when stepping through a Markov chain.

To illustrate the Jacobi and Gauss–Seidel methods for solving equations, we use the previous *discrete-time* Markov-chain example for the $N\times N$ system of equations given in (8.1). The matrix of that example

$$\boldsymbol{A}=\begin{pmatrix}-.57, & .36 & 1\\ .43 & -.64 & 1\\ 0 & .43 & 1\end{pmatrix}$$

can be written as $\boldsymbol{A}=\boldsymbol{L}+\boldsymbol{D}+\boldsymbol{U}$, where

$$\boldsymbol{L}=\begin{pmatrix}0 & 0 & 0\\ .43 & 0 & 0\\ 0 & .43 & 0\end{pmatrix},\quad \boldsymbol{D}=\begin{pmatrix}-.57, & 0 & 0\\ 0 & -.64 & 0\\ 0 & 0 & 1\end{pmatrix},\quad \boldsymbol{U}=\begin{pmatrix}0 & .36 & 1\\ 0 & 0 & 1\\ 0 & 0 & 0\end{pmatrix},$$

so that $\boldsymbol{\pi}^{(n+1)}\boldsymbol{D}=\boldsymbol{b}-\boldsymbol{\pi}^{(n)}(\boldsymbol{L}+\boldsymbol{U})$ yields

$$\begin{aligned}(\pi_0^{(n+1)},\pi_1^{(n+1)},\pi_2^{(n+1)})&\begin{pmatrix}-.57, & 0 & 0\\ 0 & -.64 & 0\\ 0 & 0 & 1\end{pmatrix}\\ &=(0,0,1)-(\pi_0^{(n)},\pi_1^{(n)},\pi_2^{(n)})\begin{pmatrix}0 & .36 & 1\\ .43 & 0 & 1\\ 0 & .43 & 0\end{pmatrix}.\end{aligned}$$

Hence

$$\begin{aligned}-.57\pi_0^{(n+1)}&=0-.43\pi_1^{(n)},\\ -.64\pi_1^{(n+1)}&=0-.36\pi_0^{(n)}-.43\pi_2^{(n)},\\ \pi_2^{(n+1)}&=1-\pi_0^{(n)}-\pi_1^{(n)},\end{aligned}$$

or equivalently,

$$\begin{aligned} \pi_0^{(n+1)} &= .75\pi_1^{(n)}, \\ \pi_1^{(n+1)} &= .56\pi_0^{(n)} + .67\pi_2^{(n)}, \\ \pi_2^{(n+1)} &= 1 - \pi_0^{(n)} - \pi_1^{(n)}. \end{aligned} \tag{8.3}$$

Thus for the Jacobi method, we pick a starting vector $\boldsymbol{\pi}^{(0)}$ [say, (1, 0, 0)] and we find

$$\pi_0^{(1)} = .75(0) = 0, \quad \pi_1^{(1)} = .56(1) + .67(0) = .56, \quad \pi_2^{(1)} = 1 - 1 - 0 = 0,$$
$$\pi_0^{(2)} = .75(.56) = .42, \quad \pi_1^{(2)} = .56(0) + .67(0) = 0, \quad \pi_2^{(2)} = 1 - 0 - .56 = .44,$$

and so on.

For the Gauss–Seidel procedure (8.3) is modified as follows:

$$\begin{aligned} \pi_0^{(n+1)} &= 0.75\pi_1^{(n)}, \\ \pi_1^{(n+1)} &= 0.56\pi_0^{(n+1)} + 0.67\pi_2^{(n)}, \\ \pi_2^{(n+1)} &= 1 - \pi_0^{(n+1)} - \pi_1^{(n+1)}. \end{aligned}$$

This modification, in matrix notation, is $(\boldsymbol{U}^T + \boldsymbol{D})\boldsymbol{\pi}^{(n+1)} = \boldsymbol{b} - \boldsymbol{L}^T\boldsymbol{\pi}^{(n)}$, $\boldsymbol{\pi}^{(n+1)}$ and $\boldsymbol{\pi}^{(n)}$ now being column vectors, and T indicating transpose. The calculations proceed as

$$\pi_0^{(1)} = .75(0) = 0, \quad \pi_1^{(1)} = .56(0) + .67(0) = 0, \quad \pi_2^{(1)} = 1 - 0 - 0 = 1,$$
$$\pi_0^{(2)} = .75(0) = 0, \quad \pi_1^{(2)} = .56(0) + .67(0) = .67, \quad \pi_2^{(2)} = 1 - 0 - .67 = .33,$$

and so on. Table 8.1 shows 21 iterations, rounded to two decimal places.

It appears that, to two decimal places, Gauss–Seidel converged in about 14 iterations, while it is not clear that Jacobi has as yet converged after 21 iterations. As mentioned previously, convergence is often a problem with these techniques. There are many examples wherein these iterative techniques do not converge for some ordering of the equations but do for others. Also, the starting guess can affect convergence. However, experience has shown that if a limiting distribution exists for a set of equations generated from a Markov process, convergence appears to result (see Cooper, 1981). Cooper and Gross (1991), as stated earlier, have proved that convergence is guaranteed for a *continuous-time* Markov process if one uses the $(N-1) \times (N-1)$ formulation of the stationary equations (setting one of the $p_i = 1$, solving for the reduced set of $N - 1$ $\{p_i\}$, and then renormalizing), but not necessarily for the $N \times N$ formulation where one of the equations is replaced by the $\sum p_i = 1$ as shown in (8.2).

In comparing the Markov-chain iterative procedure $\boldsymbol{\pi}^{(n+1)} = \boldsymbol{\pi}^{(n)}\boldsymbol{P}$, which we know from Markov theory *will* converge for appropriate $\boldsymbol{P}$, we see that after the five iterations presented previously, we are closer to the solution than with either the Jacobi or the Gauss–Seidel method applied to the stationary equations. [No general

Table 8.1 Jacobi and Gauss–Seidel Calculations

Iteration Number	Jacobi π_0	Jacobi π_1	Jacobi π_2	Gauss–Seidel π_0	Gauss–Seidel π_1	Gauss–Seidel π_2
0	1.0	.00	.00	1.0	.00	.00
1	.00	.56	.00	.00	.00	1.0
2	.42	.00	.44	.00	.67	.33
3	.00	.53	.58	.50	.50	.00
4	.40	.39	.47	.38	.21	.42
5	.29	.54	.21	.16	.37	.48
6	.40	.31	.17	.27	.47	.25
7	.23	.34	.29	.36	.37	.28
8	.25	.32	.42	.28	.34	.38
9	.24	.43	.42	.25	.40	.34
10	.32	.42	.32	.30	.40	.30
11	.31	.40	.26	.30	.37	.33
12	.30	.35	.29	.28	.38	.35
13	.26	.36	.35	.28	.39	.33
14	.27	.38	.38	.29	.38	.32
15	.29	.40	.35	.29	.38	.33
16	.30	.39	.31	.28	.38	.33
17	.30	.38	.30	.29	.38	.33
18	.29	.37	.33	.29	.38	.33
19	.28	.38	.35	.28	.38	.33
20	.28	.39	.35	.29	.38	.33
21	.29	.39	.32	.29	.38	.33

conclusions should be drawn as to the relative merits of these procedures, however; the interested reader is referred to Gross et al. (1984).]

For continuous-parameter Markov chains, if we wish to utilize the Markov-chain stepping theory instead of solving the stationary equations, an appropriate imbedded discrete-parameter chain must be found which yields the same steady-state probabilities. For example, in birth–death processes one might think of using the imbedded jump (transition point) chain, but since this is periodic, no steady state exists. We will, however, present in the next section on transient solution techniques an imbedded chain that does have a steady-state solution and for which the imbedded discrete-parameter process probabilities are identical to the general-time (continuous-parameter process) steady-state probabilities.

Another major problem with these iterative procedures is choosing a stopping criterion. Generally, the Cauchy criterion is used, namely, the calculations stop when

$\max_i |p_i^{(n+1)} - p_i^{(n)}| < \epsilon$. This is not always a good choice, depending on the type and rate of convergence, and in certain cases can lead to significant errors (see Gross et al. 1984).

There are ways of speeding up Gauss–Seidel convergence by using a weighting scheme (called overrelaxation; again, see Maron, 1982). Rather than dwell any further on solving steady-state equations at this time, we turn our attention toward numerical techniques for obtaining transient solutions, and we will mention a way to utilize these also for steady state.

8.1.2 Transient Solutions

For Markovian queues, transient solutions, conceptually, can be obtained by solving the Kolmogorov differential equations,

$$\boldsymbol{p}'(t) = \boldsymbol{p}(t)\boldsymbol{Q}.$$

Again, as long as the $\boldsymbol{Q}$ matrix is of finite dimension, numerical techniques for solving these linear, first-order differential equations can be employed. Numerical integration methods such as the Euler, Taylor, Runge–Kutta (RK), or predictor–corrector methods have long been employed in solving systems of differential equations.

Another method that is particularly well suited for queueing models is referred to as the *randomization technique*. It also has the advantage of having a probabilistic interpretation, and we shall derive it by a direct probabilistic analysis. We illustrate some of these techniques in the following sections.

8.1.2.1 Numerical Integration Methods.

Numerical integration methods can be employed to solve a general system of ordinary differential equations described by

$$\boldsymbol{p}'(t) \equiv \begin{pmatrix} p_1'(t) \\ p_2'(t) \\ \vdots \\ p_k'(t) \end{pmatrix} = \begin{pmatrix} f_1(p_1,\ldots,p_k;t) \\ f_2(p_1,\ldots,p_k;t) \\ \vdots \\ f_k(p_1,\ldots,p_k;t) \end{pmatrix} \equiv \boldsymbol{f}(\boldsymbol{p},t)$$

with known initial value $\boldsymbol{p}(t_0)$. The standard techniques are generally variations of Euler, Taylor, RK, or predictor–corrector methods.

Taylor and RK methods are based on formulas that approximate the Taylor series solutions

$$p_i(t+h) = p_i(t) + hp'(t) + \frac{h^2}{2}p''(t) + \cdots + \frac{h^n}{n!}p_i^{(n)}(t) + R_n,$$

$i = 1,\ldots,k$. RK methods use approximations for the second and higher order derivatives, rather than doing the exact differentiation as prescribed for the Taylor methods. Euler's method is a special RK method, with $k = 1$. These methods have been used by several authors (e.g., Bookbinder and Martell, 1979; Grassmann, 1977; Liitschwager and Ames, 1975; and Neuts, 1973) to find transient solutions in queueing systems.

Predictor–corrector methods require information about several previous points in order to evaluate the next point. These methods involve using one formula to predict the next $\boldsymbol{p}(t)$ value, followed by the application of a more accurate corrector formula. Unlike the Taylor and RK methods, predictor–corrector methods are not self-starting; hence they must use the RK or Taylor methods to obtain the first $\boldsymbol{p}(t)$ value. Predictor–corrector methods can provide an estimate of the local truncation error at each step in the calculations, in contrast to the Taylor and RK methods, which cannot obtain such an estimate. Predictor–corrector methods have been used by Ashour and Jha (1973) for queueing problems.

We illustrate numerical integration methodology by considering some simple cases. For further details we refer the interested reader to Maron (1982). Consider an $M/M/1/1$ queue for which direct analytical methods do yield the transient solution (see Section 2.11.1). The differential equation to be solved (2.70) is

$$\frac{dp_1(t)}{dt} = -\mu p_1(t) + \lambda p_0(t) = -\mu p_1(t) + \lambda[1 - p_1(t)] = -(\mu + \lambda)p_1(t) + \lambda. \quad (8.4)$$

Letting $\lambda = 1, \mu = 2$, and assuming $p_1(0) = 0$, we have the solution from (2.71) as $p_1(t) = (1 - e^{-3t})/3$.

Let us consider Euler's method of solving (8.4) numerically. For small Δt,

$$\frac{dp_1(t)}{dt} \approx \frac{p_1(t + \Delta t) - p_1(t)}{\Delta t},$$

so that (8.4) is approximately

$$p_1(t + \Delta t) \approx p_1(t) - \Delta t(\mu + \lambda)p_1(t) + \lambda\, \Delta t \approx p_1(t) - 3\,\Delta t\, p_1(t) + \Delta t. \quad (8.5)$$

We can solve this recursively for $t = 0, \Delta t, 2\Delta t$, and so on, and obtain the recursive relationship

$$\begin{aligned} p_1([n + 1]\Delta t) &= (1 - 3\,\Delta t)p_1(n\,\Delta t) + \Delta t \quad (n \geq 1), \\ p_1(\Delta t) &= (1 - 3\,\Delta t)p_1(0) + \Delta t = \Delta t. \end{aligned}$$

Table 8.2 shows the solutions for $t = 0, \Delta t, 2\Delta t$, and $3\Delta t$, where $\Delta t = 0.01$, and is compared to the analytical solution. For greater accuracy, Δt can be made smaller.

Euler's method is called a first-order method, for the following reason. Consider a Taylor series expansion of $p_1(t)$ about Δt, namely,

$$p_1(t + \Delta t) = p_1(t) + \Delta t\, p_1'(t) + \frac{\Delta t^2}{2!} p_1''(t) + \cdots + \frac{(\Delta t)^{n-1}}{(n-1)!} p_1^{(n-1)}(t) + R_n.$$

A first-order expansion, or approximation, is merely $p_1(t + \Delta t) \approx p_1(t) + \Delta t\, p_1'(t)$, which, substituting for $p_1'(t)$ from (8.4), gives (8.5). We could get greater accuracy in a second-order approximation, say,

$$p_1(t + \Delta t) \approx p_1(t) + \Delta t\, p_1'(t) + \frac{\Delta t^2}{2} p_1''(t).$$

Table 8.2 Euler's Method versus Analytical Solution

n	$t = n\,\Delta t$	$p_1(n\,\Delta t)$	Exact $p_1(t)$	Error
0	.00	.0000	.0000	.0000
1	.01	.0100	.0099	.0001
2	.02	.0197	.0194	.0003
3	.03	.0291	.0287	.0004

Table 8.3 Comparison of First- and Second-Order Approximations

t	Exact	First Order	Error	Second Order	Error
.00	.0000	.0000	.0000	.0000	.0000
.01	.0099	.0100	.0001	.0099	.0000
.02	.0194	.0197	.0003	.0195	.0001
.03	.0287	.0291	.0004	.0288	.0001

We have $p_1'(t)$ from our original equation, (8.4). To get $p_1''(t)$, we differentiate the right-hand side of (8.4), which yields $p_1''(t) = -(\mu+\lambda)p_1'(t)$. Now the second-order approximation becomes

$$\begin{aligned} p_1(t+\Delta t) &\approx p_1(t) + \Delta t[-(\mu+\lambda)p_1(t)+\lambda] + \frac{\Delta t^2}{2}[-(\mu+\lambda)p_1'(t)] \\ &= [1-(\mu+\lambda)\Delta t]p_1(t) + \lambda\,\Delta t - \frac{\Delta t^2}{2}(\mu+\lambda)[-(\mu+\lambda)p_1(t)+\lambda] \\ &= \left(1-(\mu+\lambda)\Delta t + (\mu+\lambda)^2\frac{\Delta t^2}{2}\right)p_1(t) + \lambda\,\Delta t - (\mu+\lambda)\lambda\frac{\Delta t^2}{2} \\ &= \left(1 - 3\,\Delta t + \frac{9\,\Delta t^2}{2}\right)p_1(t) + \Delta t - \frac{3\,\Delta t^2}{2}. \end{aligned}$$

Table 8.3 shows the second-order approximation versus the first-order and exact solutions. We can see, of course, the increased accuracy of the higher-order approximation at the expense of more effort in setting up the approximating formula.

The numerical integration procedures for more complex problems are similar, except one must work with sets of equations, instead of just one as in the preceding example (see Problem 8.1). Nevertheless, if the set of equations is fully determined at some time t, it can be determined at time $t+\Delta t$, since $p_0(t+\Delta t), p_1(t+\Delta t), \ldots, p_N(t+\Delta t)$ become functions of $p_0(t), p_1(t), \ldots, p_N(t)$, that is,

$$p_n(t+\Delta t) = f(p_0(t), p_1(t), \ldots, p_N(t)) \quad (n = 0, 1, \ldots, N).$$

Numerical integration procedures then calculate the set of equations recursively in steps of Δt, starting with the known values $p_0(0), p_1(0), \ldots, p_N(0)$, as was done for the special case of $N = 1$ above.

A final comment on numerical integration methods: Why are they called *numerical integration* methods? The answer lies in the fact that the formulas can be obtained from the fundamental theorem of calculus, namely,

$$p(t + \Delta t) = p(t) + \int_t^{t+\Delta t} p'(t)\,dt.$$

Since we do not know $p(t)$, we cannot use the above equation directly. What these methods do, then, is equivalent to computing the integral numerically; for example, a first-order approximation to the integral would be $\Delta t\, p'(t)$.

8.1.2.2 Randomization Technique. Although the randomization procedure is a computational technique for solving the set of differential equations, $\boldsymbol{p}'(t) = \boldsymbol{p}(t)\boldsymbol{Q}$, arising from Markovian queueing systems, we develop the procedure by analyzing the stochastic process, rather than looking at the numerical solution of differential equations. Consider a finite birth–death process with a $\boldsymbol{Q}$ matrix given as follows:

$$\boldsymbol{Q} = \begin{pmatrix} -q_{00} & \lambda_0 & 0 & 0 & \cdots & & & 0 \\ \mu_1 & -q_{11} & \lambda_1 & 0 & \cdots & & & 0 \\ 0 & \mu_2 & -q_{22} & \lambda_2 & & & & 0 \\ \vdots & & \ddots & \ddots & \ddots & & & \vdots \\ 0 & 0 & 0 & 0 & \cdots & \mu_{N-1} & -q_{N-1,N-1} & \lambda_{N-1} \\ 0 & 0 & 0 & 0 & \cdots & 0 & \mu_N & -q_{NN} \end{pmatrix}, \tag{8.6}$$

where q_{ii} is the sum of the nondiagonal elements in row i. Suppose the process is in state n (n customers in the system). It remains in state n until either a birth or a death occurs. The time until a birth (arrival) is exponential with mean $1/\lambda_n$, and the time until a death (service completion) is exponential with mean $1/\mu_n$. The process leaves state n when the first of these two possible events occurs. Thus the time it spends in state n is the minimum of these two exponential random variables, which is itself an exponential random variable with mean $1/(\lambda_n + \mu_n)$.

When a transition occurs, the probability that it is a birth (see Section 1.9.4) is $\lambda_n/(\lambda_n + \mu_n)$, and the probability that it is a death is $\mu_n/(\lambda_n + \mu_n)$. Thus we can view the system as a Markov process with exponential holding times [mean of $(\lambda_n + \mu_n)^{-1}$ for holding in state n] and transition probabilities of $\lambda_n/(\lambda_n + \mu_n)$ and $\mu_n/(\lambda_n + \mu_n)$ for going up one or down one, respectively, from state $n (\mu_0 = \lambda_N = 0)$. Furthermore, it can be shown that the occurrence of birth or death is independent of the holding time.

Suppose we wish to generate this process, for example, by simulating it in a Monte Carlo way. We could first create the transition-time occurrence and then generate, using simple Bernoulli probabilities, the change of state—that is, whether the process increases or decreases its state by one. The generation of the transition times is somewhat complicated in that we would have to sample from an exponential distribution with a state-dependent parameter $\lambda_n + \mu_n$.

To avoid this, we can reproduce this process in the following way. Consider the *minimum* mean holding time (time until the next transition occurs) of this process. This will correspond to the *maximum* value of $\lambda_n + \mu_n$, or equivalently, the minimum diagonal element of $\boldsymbol{Q}$, the generator of the process. Call this value Λ. Let us denote the diagonal elements of the $\boldsymbol{Q}$ matrix by $-q_n$, that is,

$$q_n = \begin{cases} \lambda_0 & (n = 0), \\ \lambda_n + \mu_n & (n = 1, 2, \ldots, N-1), \\ \mu_N & (n = N). \end{cases}$$

To reproduce our desired transition occurrence process, we could generate a true Poisson process with rate Λ (exponential holding times with constant mean $1/\Lambda$) and then *thin* the process to get the desired state-dependent transition rate. By thinning, we mean that whenever an occurrence is generated by the Poisson (Λ) process, if we are in state n, we draw from a Bernoulli probability distribution with success probability q_n/Λ, where a *success* indicates counting the occurrence as a transition and a *failure* (probability $1 - q_n/\Lambda$) indicates ignoring the occurrence. This thinning procedure on the Poisson (Λ) process generates our desired underlying state-dependent transition process.

Thus to reproduce our process we can:

1. generate a Poisson process with rate Λ;

2. thin the Poisson (Λ) process by a Bernoulli *switch* with acceptance probability q_n/Λ and rejection probability $1 - q_n/\Lambda$;

3. generate the state change (up or down one unit) by another Bernoulli switch with an up probability of $\lambda_n/(\lambda_n + \mu_n)$ and a down probability of $\mu_n/(\lambda_n + \mu_n)$.

Denote the Poisson process as $\{N(t), t \geq 0\}$, that is, $N(t)$ equals the number of occurrences in [0, t], and let Y_k equal the state of the system after the kth occurrence of the Poisson process, that is, Y_k equals $X(T_k)$, where $T_k = \min\{t : N(t) \geq k\}$. Viewing the process in this manner will allow us to derive the randomization computing algorithm that will yield $\boldsymbol{p}(t)$.

Consider an element of the transient state probability vector $\boldsymbol{p}(t)$, say, $p_n(t) = \Pr\{X(t) = n\}$. Letting $p_{in}(t) = \Pr\{X(t) = n | X(0) = i\}$ gives

$$p_n(t) = \sum_{i=0}^{N} p_i(0) p_{in}(t). \tag{8.7}$$

Now considering our process as described above, we have, using the law of total probability,

$$p_{in}(t) = \sum_{k=0}^{\infty} \Pr\{Y_k = n | Y_0 = i\} \Pr\{N(t) = k\} = \sum_{k=0}^{\infty} \tilde{p}_{in}^{(k)} \frac{e^{-\Lambda t}(\Lambda t)^k}{k!}, \tag{8.8}$$

where $\tilde{p}_{in}^{(k)}$ represents the probability of going from state i to state n in k occurrences of the Poisson process. But

$$\begin{aligned}\tilde{p}_{in} &= \Pr\{\text{going from } i \text{ to } n \text{ in one occurrence}\}\\ &= \Pr\{\text{accepting the occurrence as a transition}\}\\ &\quad\times \Pr\{\text{transiting from } i \text{ to } n \mid \text{a transition takes place}\},\end{aligned}$$

which is

$$\tilde{p}_{in}^{(1)} = \begin{cases} \dfrac{q_i}{\Lambda}\dfrac{\lambda_i}{\lambda_i+\mu_i} = \dfrac{\lambda_i}{\Lambda} & (n = i+1),\\ \dfrac{q_i}{\Lambda}\dfrac{\mu_i}{\lambda_i+\mu_i} = \dfrac{\mu_i}{\Lambda} & (n = i-1),\\ 1 - \dfrac{\lambda_i+\mu_i}{\Lambda} & (n = i),\\ 0 & (\text{elsewhere}).\end{cases}$$

Denote the matrix with elements $\tilde{p}_{in}^{(1)}$ by $\tilde{\boldsymbol{P}}$; note that $\tilde{\boldsymbol{P}} = \boldsymbol{Q}/\Lambda + \boldsymbol{I}$. We can obtain $\tilde{p}_{in}^{(k)}$ as elements of a matrix formed by multiplying $\tilde{\boldsymbol{P}}$ by itself k times, that is,

$$\tilde{\boldsymbol{P}}^{(k)} \equiv \{\tilde{p}_{in}^{(k)}\} = \tilde{\boldsymbol{P}}^{k}.$$

Substituting this into (8.7) and (8.8) yields

$$\boldsymbol{p}(t) = \sum_{k=0}^{\infty} \boldsymbol{p}(0)\tilde{\boldsymbol{P}}^{(k)} \frac{e^{-\lambda t}(\Lambda t)^k}{k!}. \tag{8.9}$$

For computing purposes, one problem remains, namely, the infinite summation in (8.9). Thus we must truncate the sum at some value, say, $T(t,\epsilon)$. We can set this value to guarantee a truncation error of less than a prespecified amount, say, ϵ, since we are throwing away the tail of a Poisson distribution. From (8.9) we have

$$\begin{aligned} p_n(t) &= \sum_{k=0}^{\infty}\sum_{i=0}^{N} p_i(0)\tilde{p}_{in}^{(k)} \frac{e^{-\Lambda t}(\Lambda t)^k}{k!}\\ &= \sum_{k=0}^{T(t,\epsilon)}\sum_{i=0}^{N} p_i(0)\tilde{p}_{in}^{(k)} \frac{e^{-\Lambda t}(\Lambda t)^k}{k!} + \sum_{T(t,\epsilon)+1}^{\infty}\sum_{i=0}^{N} p_i(0)\tilde{p}_{in}^{(k)} \frac{e^{-\Lambda t}(\Lambda t)^k}{k!}.\end{aligned}$$

We desire

$$\sum_{T(t,\epsilon)+1}^{\infty}\sum_{i=0}^{N} p_i(0)\tilde{p}_{in}^{(k)} \frac{e^{-\Lambda t}(\Lambda t)^k}{k!} \equiv R_T < \epsilon.$$

But

$$R_T < \sum_{T(t,\epsilon)+1}^{\infty} \frac{e^{-\Lambda t}(\Lambda t)^k}{k!},$$

so that by finding $T(t,\epsilon)$ such that

$$\sum_{T(t,\epsilon)+1}^{\infty} < \epsilon \quad \Rightarrow \quad \sum_{k=0}^{T(t,\epsilon)} \frac{e^{-\Lambda t}(\Lambda t)^k}{k!} > 1-\epsilon, \tag{8.10}$$

we have an error bound on $p_n(t)$ of ϵ.

While we have developed this procedure for a birth–death model, the computing formula (in vector-matrix form)

$$\boldsymbol{p}(t) = \sum_{k=0}^{T(t,\epsilon)} \boldsymbol{p}(0)\tilde{\boldsymbol{P}}^{(k)} \frac{e^{-\Lambda t}(\Lambda t)^k}{k!} \tag{8.11}$$

holds for any Markov process with infinitesimal generator

$$\boldsymbol{Q} = \begin{pmatrix} -q_0 & q_{01} & q_{02} & \cdots & q_{0N} \\ q_{10} & -q_1 & q_{12} & \cdots & q_{1N} \\ \vdots & \vdots & \vdots & & \vdots \\ q_{N0} & q_{N1} & q_{N2} & \cdots & -q_N \end{pmatrix},$$

where $q_i = \sum_{j \neq i} q_{ij}, i = 0, 1, 2, \ldots, N$. The general expression for $\tilde{p}_{in}^{(1)}$ is

$$\tilde{p}_{in}^{(1)} = \begin{cases} q_{ij}/\Lambda & (i \neq n), \\ 1 - q_i/\Lambda & (i = n). \end{cases} \tag{8.12}$$

With respect to computational difficulties in problems with large state spaces, the major computing effort is raising the matrix $\tilde{\boldsymbol{P}}$ to the kth power, since $T(t,\epsilon)$ can be sizable. This can be avoided by a recursive computing scheme as follows.

In (8.11), consider the product $\boldsymbol{p}(0)\tilde{\boldsymbol{P}}^{(k)}$, and call this $\boldsymbol{\phi}^{(k)}$. This vector is the system state probability vector after k occurrences of the underlying Poisson (Λ) process, that is, the probability distribution of the state of the system of a discrete-parameter Markov chain after k transitions, namely, the Markov chain Y_k with transition probability matrix $\tilde{\boldsymbol{P}}$. From Markov-chain theory we know that

$$\boldsymbol{\phi}^{(k)} = \boldsymbol{\phi}^{(k-1)}\tilde{\boldsymbol{P}} \quad (\tilde{\boldsymbol{P}} = \boldsymbol{Q}/\Lambda + \boldsymbol{I}), \tag{8.13}$$

so that we can write (8.11) as

$$\boldsymbol{p}(t) = \sum_{k=0}^{T(t,\epsilon)} \boldsymbol{\phi}^{(k)} \frac{e^{-\Lambda t}(\Lambda t)^k}{k!} \tag{8.14}$$

and calculate $\boldsymbol{\phi}^{(k)}$ recursively using (8.13).

What we have essentially done is to reduce the complex calculations of the continuous-parameter Markov-chain $X(t)$ with infinitesimal generator $\boldsymbol{Q}$ to calculations on the discrete-parameter Markov chain Y_k with transition probability matrix

$\tilde{\boldsymbol{P}}$ relating k to t by the Poisson (Λ) process. That is, for the Y_k process (often referred to as the uniformized, imbedded Markov chain), we measure time in number of occurrences of the Poisson (Λ) process and relate it to clock time through the Poisson probabilities.

It is very often the case for queueing problems that $\tilde{\boldsymbol{P}}$ is a *sparse* matrix; that is, most of the elements are zero. For example, consider the $\boldsymbol{Q}$ matrix of (8.6), keeping in mind that $\tilde{\boldsymbol{P}} = \boldsymbol{Q}/\Lambda + \boldsymbol{I}$. Of the $(N+1)^2$ elements, fewer than $3(N+1)$ are nonzero. Thus in doing the matrix multiplication of (8.13), many zero multiplications transpire. There are ways to avoid this, and we refer the reader to Gross and Miller (1984) for one such procedure.

The randomization procedure can also be used as a vehicle for obtaining steady-state solutions. One way to do this is simply to make t large, or keep trying successive t values until $\boldsymbol{p}(t)$ does not change appreciably with t. Another route is to consider only the uniformized imbedded discrete-parameter Markov chain Y_k with transition probability matrix $\tilde{\boldsymbol{P}}$ and use (8.13) recursively until $\boldsymbol{\phi}^{(k)}$ appears independent of time. It can easily be shown that this uniformized imbedded discrete-parameter Markov chain with transition probability matrix $\tilde{\boldsymbol{P}}$ has the same steady-state probability distribution as the original continuous-parameter Markov chain, that is, $\lim_{k\to\infty} \boldsymbol{\phi}^{(k)} = \lim_{t\to\infty} \boldsymbol{p}(t)$, because

$$\boldsymbol{\phi} = \boldsymbol{\phi}\tilde{\boldsymbol{P}} \Rightarrow \boldsymbol{\phi} = \boldsymbol{\phi}\left(\frac{\boldsymbol{Q}}{\Lambda} - \boldsymbol{I}\right) \Rightarrow \boldsymbol{0} = \boldsymbol{\phi}\frac{\boldsymbol{Q}}{\Lambda} \Rightarrow \boldsymbol{0} = \boldsymbol{\phi}\boldsymbol{Q}.$$

It would seem that using (8.13) should be a more efficient means of obtaining a steady-state distribution than using (8.14), although the mixing with Poisson probabilities done in (8.14) might tend to act as a smoothing procedure and in some cases might conceivably converge faster. One could also, of course, use Gauss–Seidel stepping on the discrete-parameter Markov chain of (8.13). Which of these alternatives is the best is an open question and obviously will depend, at least in part, on the specific problem being considered.

Our interest in this section is not to present a treatise on numerical analysis, but to point out that this can be a very important contribution in applying queueing theory—an area that unfortunately has not gotten the attention in the past that is its due. It is certainly nice to obtain closed-form solutions where that is possible, but in those many cases where it is not, numerical methods can provide a way to obtain meaningful answers.

8.2 Numerical Inversion of Transforms

Analysis of queues often involves manipulation of Laplace transforms, or Laplace–Stieltjes transforms. For an introduction to these transforms see Appendix C. This text has given several examples in which the distribution of interest is expressed as a transform. For example, the steady-state system wait of an $M/G/1$ queue was given as an LST in (5.33):

$$W^*(s) = \frac{(1-\rho)sB^*(s)}{s - \lambda[1 - B^*(s)]}.$$

In this equation, $W^*(s)$ is the LST of the system wait and $B^*(s)$ is the LST of the service distribution. Other examples of transform solutions in this text include the busy-period distribution of the $M/G/1$ queue (5.37) and the steady-state queue-wait of the $G/G/1$ queue (6.12).

An important final step in using transforms is to invert the desired transform back into a probability distribution. In some cases, transform inversion can be done analytically. For example, in the previous equation, if G is an exponential distribution, then $W^*(s)$ can be inverted analytically as the next example shows.

■ **EXAMPLE 8.1**

Find $W(t)$ for the $M/M/1$ queue, with arrival rate λ and service rate μ, using (5.33). Solution: The LST of the service distribution is $B^*(s) = \mu/(s+\mu)$. Thus

$$W^*(s) = \frac{(1-\rho)s \cdot \mu/(s+\mu)}{s - \lambda[1-(\mu/(s+\mu))]} = \frac{(1-\rho)s \cdot \mu/(s+\mu)}{s - \lambda[s/(s+\mu)]}$$
$$= \frac{(1-\rho)\mu/(s+\mu)}{1-\lambda/(s+\mu)} = \frac{(1-\rho)\mu}{s+\mu-\lambda} = \frac{\mu-\lambda}{s+\mu-\lambda}.$$

This is the LST of an exponential distribution with mean $1/(\mu-\lambda)$. Thus $W(t) = 1 - e^{-(\mu-\lambda)t}$, in agreement with (2.29).

As another example, Problem 5.18 asks the reader to find $W(t)$ for an $M/E_2/1$ queue via inversion of the transform. However, depending on the specific form of the transform, analytical inversion may be difficult or impossible. In some cases, it is not even possible to explicitly write out the transform itself in closed form. One example is the busy-period transform of the $M/G/1$ queue. This transform is expressed *implicitly* as the solution to the equation given in (5.37).

Thus a general *numerical* procedure is useful to invert transforms, regardless of the specific form of the transform or how it is computed. The basic approach in developing a numerical inversion procedure is to start with an *exact* inversion formula. One example of an exact formula is the Bromwich contour integral, as given in the following theorem.

First, to define notation, let $f(t)$ be a real-valued function defined on the interval $0 \le t < \infty$ and let $\bar{f}(s)$ be its Laplace transform:

$$\mathcal{L}\{f(t)\} \equiv \bar{f}(s) \equiv \int_0^\infty e^{-st} f(t)\,dt, \tag{8.15}$$

where s is a complex variable.

Theorem 8.1. *The function $f(t)$ can be determined from its Laplace transform $\bar{f}(s)$ using the following contour integral in the complex plane:*

$$f(t) = \frac{1}{2\pi i}\int_{b-i\infty}^{b+i\infty} e^{st}\bar{f}(s)\,ds, \qquad t > 0, \tag{8.16}$$

where b is a real number to the right of all singularities of $\bar{f}$. Furthermore, when $t < 0$, the integral equals zero.

Equation (8.16) is not the only exact formula to invert (8.15). There are many others. Furthermore, given an exact inversion formula such as (8.16), there are many ways to numerically approximate the exact formula. For example, there are many ways to numerically approximate the integral in (8.16). In summary, many different methods exist in the literature to numerically invert transforms.

This section gives a brief introduction to numerical transform inversion. We discuss one method in detail – the Fourier-series method. Other methods are discussed briefly in Section 8.2.4. The material in this section is based largely on material from the overview paper by Abate et al. (1999). For further details, see this paper and references therein.

8.2.1 The Fourier-Series Method

The Fourier-series method is based on the Bromwich inversion integral, given in (8.16). In developing this method, we first convert the complex-valued integral in (8.16) to a real-valued integral, given in (8.21). To do this, we first apply a change of variables, $s = b + iu$:

$$\begin{aligned} f(t) &= \frac{1}{2\pi}\int_{-\infty}^{\infty} e^{(b+iu)t}\bar{f}(b+iu)\,du \\ &= \frac{1}{2\pi}\int_{-\infty}^{\infty} e^{bt}(\cos ut + i\sin ut)\bar{f}(b+iu)\,du \\ &= \text{Re}\left[\frac{1}{2\pi}\int_{-\infty}^{\infty} e^{bt}(\cos ut + i\sin ut)\bar{f}(b+iu)\,du\right]. \end{aligned} \tag{8.17}$$

The last equality follows since $f(t)$ is real. (Also, it can be checked directly that the imaginary part of the integral is zero; see Problem 8.23.) We split the integral into two parts, $\int_{-\infty}^{0}(\cdot)\,du + \int_{0}^{\infty}(\cdot)\,du$, and work with the first integral.

$$\begin{aligned} &\text{Re}\left[\frac{1}{2\pi}\int_{-\infty}^{0} e^{bt}(\cos ut + i\sin ut)\bar{f}(b+iu)\,du\right] \\ &= \text{Re}\left[\frac{e^{bt}}{2\pi}\int_{0}^{\infty} [\cos(-ut) + i\sin(-ut)]\bar{f}(b-iu)\,du\right] \\ &= \text{Re}\left[\frac{e^{bt}}{2\pi}\int_{0}^{\infty} [\cos ut - i\sin ut]\bar{f}(b-iu)\,du\right] \\ &= \frac{e^{bt}}{2\pi}\int_{0}^{\infty} \left[\text{Re}(\bar{f}(b-iu))\cos ut + \text{Im}(\bar{f}(b-iu))\sin ut\right] du \\ &= \frac{e^{bt}}{2\pi}\int_{0}^{\infty} \left[\text{Re}(\bar{f}(b+iu))\cos ut - \text{Im}(\bar{f}(b+iu))\sin ut\right] du. \end{aligned} \tag{8.18}$$

The first equality follows from a change of variables (replacing u with $-u$). The second equality follows since $\cos(-x) = \cos(x)$ and $\sin(-x) = -\sin(x)$. The third equality follows since $\mathrm{Re}\,[(a-bi)(c+di)] = ca+db$. The last equality follows since $\mathrm{Re}[\bar{f}(b-iu)] = \mathrm{Re}[\bar{f}(b+iu)]$ and $\mathrm{Im}[\bar{f}(b-iu)] = -\mathrm{Im}[\bar{f}(b+iu)]$; see Problem 8.22. Similarly, working with the second integral $\int_0^\infty (\cdot)\,du$ gives

$$\begin{aligned}\mathrm{Re}&\left[\frac{1}{2\pi}\int_0^\infty e^{bt}(\cos ut + i\sin ut)\bar{f}(b+iu)\,du\right]\\ &= \frac{e^{bt}}{2\pi}\int_0^\infty \left[\mathrm{Re}(\bar{f}(b+iu))\cos ut - \mathrm{Im}(\bar{f}(b+iu))\sin ut\right] du. \qquad (8.19)\end{aligned}$$

Combining (8.17), (8.18), and (8.19),

$$f(t) = \frac{e^{bt}}{\pi}\int_0^\infty \left[\mathrm{Re}(\bar{f}(b+iu))\cos ut - \mathrm{Im}(\bar{f}(b+iu))\sin ut\right] du. \qquad (8.20)$$

Now suppose $t > 0$. By Theorem 8.1, $f(-t) = 0$, as defined by the original contour integral in (8.16). Since $f(-t) = 0$, (8.20) implies that

$$\int_0^\infty \mathrm{Re}(\bar{f}(b+iu))\cos ut\,du = -\int_0^\infty \mathrm{Im}(\bar{f}(b+iu))\sin ut\,du.$$

Thus (8.20) becomes

$$\boxed{f(t) = \frac{2e^{bt}}{\pi}\int_0^\infty \mathrm{Re}(\bar{f}(b+iu))\cos(ut)\,du.} \qquad (8.21)$$

Also,

$$f(t) = -\frac{2e^{-bt}}{\pi}\int_0^\infty \mathrm{Im}(\bar{f}(b+iu))\sin(ut)\,du.$$

We will use (8.21) in the development of the inversion procedure.

■ **EXAMPLE 8.2**

We demonstrate (8.21) on the PDF of an exponential random variable, $f(t) = \lambda e^{-\lambda t}$. The corresponding Laplace transform is $\bar{f}(s) = \lambda/(s+\lambda)$ (e.g., see Appendix C). Then,

$$\mathrm{Re}(\bar{f}(b+iu)) = \mathrm{Re}\left[\frac{\lambda}{b+iu+\lambda}\right] = \mathrm{Re}\left[\frac{\lambda(b+\lambda-iu)}{(b+\lambda)^2+u^2}\right] = \frac{\lambda(b+\lambda)}{(b+\lambda)^2+u^2}.$$

Thus (8.21) becomes

$$f(t) = \frac{2e^{bt}}{\pi}\int_0^\infty \frac{\lambda(b+\lambda)}{(b+\lambda)^2+u^2}\cos(ut)\,du.$$

This integral can be evaluated using the theory of complex variables (or using a table of integrals) to obtain $f(t) = \lambda e^{-\lambda t}$.

In summary, so far we have converted an exact complex-valued integral (8.16) into an exact real-valued integral (8.21). It still remains to develop a numerical procedure to evaluate (8.21), since this integral cannot, in general, be evaluated analytically.

One of the simplest techniques to numerically approximate an integral is the trapezoidal rule:

$$\int_a^b g(x)\,dx \approx (b-a)\frac{g(a)+g(b)}{2}.$$

The formula is exact when $g(x)$ is linear, in which case the integrated area is a trapezoid. On a large interval, we can apply the rule multiple times by breaking the interval into smaller subintervals. For example, breaking the interval $[0,\infty)$ into subintervals of equal length h gives

$$\int_0^\infty g(u)\,du \approx h\frac{g(0)}{2} + h\sum_{k=1}^{\infty} g(kh).$$

Applying this to (8.21) gives

$$f(t) \approx f_h(t) \equiv \frac{he^{bt}}{\pi}\mathrm{Re}(\bar{f}(b)) + \frac{2he^{bt}}{\pi}\sum_{k=1}^{\infty}\mathrm{Re}(\bar{f}(b+ikh))\cos(kht). \qquad (8.22)$$

Although more sophisticated numerical integration procedures are available, it turns out that the trapezoidal rule is quite effective here. The reason will be discussed in the next section on error bounds. Now, the infinite sum in (8.22) must also be approximated. A natural way to do this is by truncation:

$$f_h(t) \approx \frac{he^{bt}}{\pi}\mathrm{Re}(\bar{f}(b)) + \frac{2he^{bt}}{\pi}\sum_{k=1}^{K}\mathrm{Re}(\bar{f}(b+ikh))\cos(kht). \qquad (8.23)$$

Roughly speaking, the accuracy of the approximation improves as h gets smaller and K gets larger. More specifically, as $h \to 0$ and $hK \to \infty$, then (8.23) converges to $f(t)$, provided that f is continuous at t. [If f is discontinuous at t and f is a CDF, then (8.23) converges to $(f(t^-)+f(t))/2$.]

Instead of simple truncation, a more efficient numerical approach can be developed using a summation acceleration technique. There are many such techniques (e.g., Wimp, 1981). One that works well in this context is *Euler summation*. Consider the series $a = \sum_{k=1}^{\infty} a_k$ and let s_n be the partial sum $s_n \equiv \sum_{k=1}^{n} a_k$. Euler summation, rather than approximating a with a single partial sum s_n, approximates a with a *weighted average* of the partial sums, $s_n, s_{n+1}, \ldots, s_{n+m}$:

$$a \approx \sum_{k=0}^{m}\binom{m}{k}2^{-m}s_{n+k}.$$

The weights in Euler summation correspond to a binomial probability distribution. Euler summation is particularly effective when the terms a_k alternate in sign.

In order to effectively apply Euler summation to (8.22), we choose the parameters b and h to achieve an eventually alternating series. To do this, let $h = \pi/(2t)$ and $b = A/(2t)$, where A is a new constant. In this case, $\cos(kht)$ takes on the values -1, 0, or 1. Then, (8.22) becomes

$$f_h(t) = f_A(t) \equiv \frac{e^{A/2}}{2t}\left[\bar{f}\left(\frac{A}{2t}\right) + 2\sum_{k=1}^{\infty}(-1)^k \mathrm{Re}\left[\bar{f}\left(\frac{A+2k\pi i}{2t}\right)\right]\right]. \quad (8.24)$$

(Note: $\mathrm{Re}[\bar{f}(A/2t)] = \bar{f}(A/2t)$, since $A/2t$ is real.) A truncated version of this expression is

$$f_{A,n}(t) \equiv \frac{e^{A/2}}{2t}\left[\bar{f}\left(\frac{A}{2t}\right) + 2\sum_{k=1}^{n}(-1)^k \mathrm{Re}\left[\bar{f}\left(\frac{A+2k\pi i}{2t}\right)\right]\right]. \quad (8.25)$$

The terms in the sum are eventually alternating in sign – provided that the real part of $\bar{f}((A+2k\pi i)/(2t))$ is eventually of fixed sign for large k. Applying Euler summation to this series gives

$$f(t) \approx f_{A,m,n}(t) \equiv \sum_{k=0}^{m}\binom{m}{k} 2^{-m} f_{A,n+k}(t). \quad (8.26)$$

In summary, a function $f(t)$ can be *exactly* derived from its Laplace transform $\bar{f}(s)$ using (8.21). The function $f(t)$ can be *approximately* computed by discretizing the integral and truncating the infinite sum, as in (8.23). The function $f(t)$ can be approximated *more efficiently* using Euler summation, rather than direct truncation. This requires choosing specific values for h and b to achieve the (eventually) alternating series in (8.24). Note that (8.24) requires a single input parameter A, rather than the two input parameters h and b in (8.22). Euler summation introduces two additional input parameters n and m, which specify the starting index and number of terms in the sum. The complete algorithm is summarized below.

Algorithm 8.1 *Fourier-series method with Euler summation. This algorithm estimates $f(t)$ (for a given t), given the Laplace transform $\bar{f}(s)$ of the function. Parameters that must be specified in the algorithm are A, m, n.*

1. Compute $f_{A,j}(t)$ using (8.25), for $j = n, n+1, \ldots, n+m$.

2. Approximate $f(t)$ using (8.26).

Note that the input to this algorithm is the *ordinary* Laplace transform $\bar{f}(s)$ of the function $f(t)$, not the Laplace–Stieltjes transform $F^*(s)$ of a distribution. Appendix C discusses the relation between the two transforms. In addition, Section 8.2.3 gives several examples applying the algorithm on different transforms.

If multiple values of t are required for evaluation, the binomial coefficients in (8.26) can be precomputed. The parameters A, m, and n are specified by the user. As a default, Abate et al. (1999) use the values $A = 19$, $m = 11$, and $n = 38$. The next section looks more closely at numerical errors in using this method.

8.2.2 Error Analysis

There are three sources of error in using the Fourier-series method to numerically invert a transform:

1. Discretization error, associated with approximating the integral in (8.21) with the infinite sum in (8.22).

2. Truncation error, associated with approximating the infinite sum in (8.22) with a finite sum.

3. Round-off error, associated with the finite number of digits used in calculations by the computer.

We discuss these sources of error and how they can be controlled using the input parameters of the algorithm.

Discretization Error. The discretization error is defined to be $f_A(t) - f(t)$, where $f_A(t)$ is given in (8.24). It can be shown that this error is (e.g., Abate et al., 1999, p. 266)

$$f_A(t) - f(t) = \sum_{k=1}^{\infty} e^{-kA} f((2k+1)t), \tag{8.27}$$

provided that t is a continuity point of $f(\cdot)$. An upper bound for the discretization error can be given if $|f(x)| \leq C$ for $x > 3t$:

$$|f_A(t) - f(t)| \leq \sum_{k=1}^{\infty} Ce^{-kA} = C\frac{e^{-A}}{1-e^{-A}} \approx Ce^{-A}. \tag{8.28}$$

The equality follows since the infinite sum is a geometric series. The approximation assumes that A is not small, so the denominator is approximately 1. (The assumption that $|f(x)|$ is bounded for $x > 3t$ is not restrictive – for example, if $f(x)$ is a CDF, then $|f(x)| \leq 1$ for all x. Also, most common PDF's $f(x)$ satisfy this property.) In summary, the discretization error is approximately Ce^{-A} and can be made arbitrarily small by letting A be appropriately large. However, increasing A also increases round-off error, so there is a trade-off, as we will discuss.

For details on the proof of (8.27), see, for example, Abate et al. (1999). We comment briefly that the derivation involves constructing a periodic function based on $f(t)$ and then writing the Fourier series of this periodic function. Multiplying the result by $e^{A/2}$ yields the approximation $f_A(t)$ in (8.24). In other words, the approximation $f_A(t)$ can be motivated using an argument involving Fourier series – without using the Bromwich inversion integral. This motivates the name of the method.

Truncation Error. There is no exact formula for the truncation error, like there is for the discretization error (8.27). However, the error can be estimated by comparing one estimate of $f(t)$ with a better estimate – for example, the difference of successive terms in Euler summation:

$$\begin{aligned} f_{A,m,n}(t) - f(t) &\approx f_{A,m,n}(t) - f_{A,m,n+1}(t) \\ &= [f_{A,m,n}(t) - f(t)] - [f_{A,m,n+1}(t) - f(t)]. \end{aligned} \tag{8.29}$$

The idea is that $f_{A,m,n+1}(t)$ is a much better estimate of $f(t)$ than $f_{A,m,n}(t)$. Thus the right-hand term in brackets is much smaller than the left-hand term.

Round-off Error. The discretization error in (8.27) can be made arbitrarily small by letting A be arbitrarily large. However, increasing A comes at the cost of increasing round-off error. Note that (8.25) is premultiplied by the factor $e^{A/2}$. This term grows rapidly in A. The terms inside the brackets in (8.25) can generally be computed with a round-off error on the order of machine precision, say, 10^{-14}. Thus the overall round-off error in computing $f_{A,K}(t)$ is about $e^{A/2}$ times the machine precision. In other words, as the discretization error is getting better, the round-off error is getting worse, as a function of A. Abate et al. (1999) give an extension to the Fourier-series method that can be used to control round-off error. We do not give the details here. The basic idea is to choose parameters to balance the discretization and round-off errors.

In summary, to use the Fourier-series method with Euler summation, the parameter values A, m, and n are specified. The value A can be chosen to achieve a desired discretization error, via the approximate upper bound Ce^{-A} in (8.28). The overall error [i.e., $f_{A,m,n}(t) - f(t)$] including the truncation error is estimated as $f_{A,m,n}(t) - f_{A,m,n+1}(t)$. If the estimated overall error is too large, higher values of m and n can be chosen to reduce the error. To simultaneously reduce round-off error, see the method in Abate et al. (1999). As a default, they use $A = 19$, $m = 11$, and $n = 38$.

8.2.3 Examples

This section gives several examples related to the $M/G/1$ queue. Our goal is to calculate the CDF of the system wait $W(t)$ and the CDF of the queue wait $W_q(t)$. The transforms of these distributions, $W^*(s)$ and $W_q^*(s)$, are given by (5.33) and (5.34), respectively. However, it is not always possible to invert these transforms analytically, so we use numerical methods here.

To use the Fourier-series method and (8.25), we need the *ordinary* Laplace transforms of $W(t)$ and $W_q(t)$ rather than Laplace–Stieltjes transforms. The ordinary transforms, $\bar{W}(s)$ and $\bar{W}_q(s)$, are related to the Laplace–Stieltjes transforms, $W^*(s)$ and $W_q^*(s)$, as follows:

$$\bar{W}(s) = \frac{W^*(s)}{s} \quad \text{and} \quad \bar{W}_q(s) = \frac{W_q^*(s)}{s}.$$

This relationship is discussed in more detail in Appendix C. Thus from (5.33),

$$\bar{W}(s) = \frac{W^*(s)}{s} = \frac{(1-\rho)B^*(s)}{s - \lambda[1 - B^*(s)]}.$$

To perform computations with this equation, we now write complex numbers in their component form, say, $s = a + bi$:

$$\bar{W}(a+bi) = \frac{(1-\rho)B^*(a+bi)}{a+bi-\lambda[1-B^*(a+bi)]}.$$

Furthermore, let us write $B^*(a+bi) \equiv c + di$. That is, $c \equiv \text{Re}[B^*(a+bi)]$ and $d \equiv \text{Im}[B^*(a+bi)]$. (The values c and d are implicitly functions of a and b.) Then,

$$\bar{W}(a+bi) = \frac{(1-\rho)(c+di)}{a+bi-\lambda[1-(c+di)]} = \frac{(1-\rho)(c+di)}{(a-\lambda+\lambda c)+(b+\lambda d)i}.$$

So,

$$\text{Re}\left[\bar{W}(s)\right] = (1-\rho)\frac{c(a-\lambda+\lambda c)+d(b+\lambda d)}{(a-\lambda+\lambda c)^2+(b+\lambda d)^2}. \tag{8.30}$$

EXAMPLE 8.3

For an $M/M/1$ queue, $B^*(s) = \mu/(s+\mu)$. Thus

$$\begin{aligned} B^*(a+bi) &= \frac{\mu}{a+bi+\mu} = \frac{\mu(a+\mu-bi)}{(a+\mu)^2+b^2} \\ &= \left(\frac{\mu(a+\mu)}{(a+\mu)^2+b^2}\right) + \left(\frac{-\mu b}{(a+\mu)^2+b^2}\right) i. \end{aligned}$$

The expressions for c and d that are substituted into (8.30) are

$$c = \frac{\mu(a+\mu)}{(a+\mu)^2+b^2} \quad \text{and} \quad d = \frac{-\mu b}{(a+\mu)^2+b^2}. \tag{8.31}$$

To use Algorithm 8.1, we need to make repeated evaluations of the transform in (8.25), namely,

$$\text{Re}\left[\bar{W}\left(\frac{A+2k\pi i}{2t}\right)\right]. \tag{8.32}$$

For the $M/G/1$ queue, this is given by (8.30), where $a = A/(2t)$, $b = k\pi/t$, and c and d depend on the service distribution. For example, if the service distribution is exponential, then c and d are given by (8.31). The estimate for $W(t)$ is then obtained by evaluating the finite sums in (8.25) and (8.26). This can be repeated for multiple values of t to obtain the complete CDF $W(t)$.[†] Figure 8.1 shows sample results for the $M/M/1$ queue, with $\lambda = 0.2$, $\mu = 1.0$, and the default parameters values in Algorithm 8.1. Evaluation of $W_q(t)$ is discussed in a moment.

[†] Equation (8.25) cannot be evaluated directly for $t = 0$. However, the function of interest is often known at $t = 0$, in which case a numerical approximation is not needed at $t = 0$.

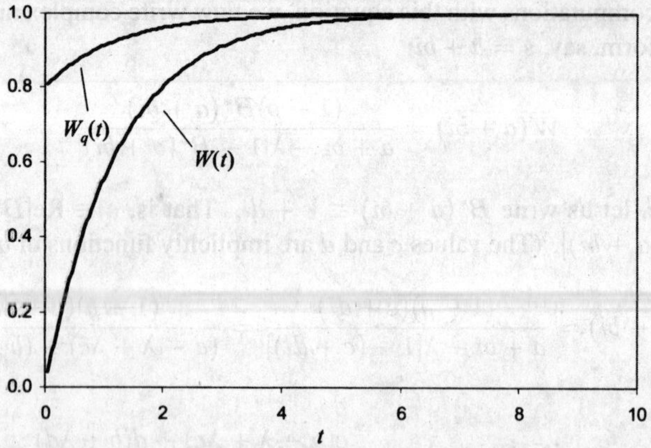

Figure 8.1 Numerical results for $M/M/1$ queue with $\mu = 1$ and $\lambda = 0.8$.

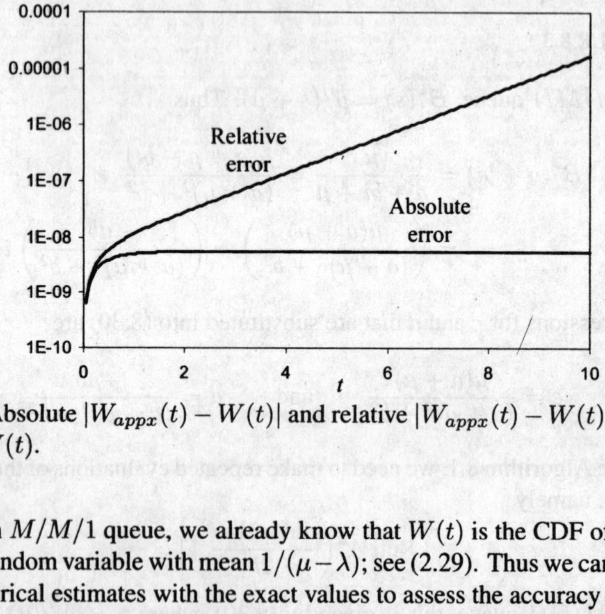

Figure 8.2 Absolute $|W_{appx}(t) - W(t)|$ and relative $|W_{appx}(t) - W(t)|/[1 - W(t)]$ errors for $W(t)$.

For an $M/M/1$ queue, we already know that $W(t)$ is the CDF of an exponential random variable with mean $1/(\mu - \lambda)$; see (2.29). Thus we can compare the numerical estimates with the exact values to assess the accuracy of the numerical method. Figure 8.2 shows the numerical errors in estimating $W(t)$. Let $W_{appx}(t)$ be the numerical estimate. The absolute error is $|W_{appx}(t) - W(t)|$ and the relative error is $|W_{appx}(t) - W(t)|/[1 - W(t)]$. The relative error is the relative error in estimating the tail probability $W^c(t) = 1 - W(t)$, which decreases rapidly in t. For this example, the absolute error is less than 10^{-8} for all values of t. The relative error increases in t, since the tail probability $W^c(t)$ decreases in t.

EXAMPLE 8.4

Numerically estimate $W_q(t)$ for an $M/M/1$ queue. From (5.34), the Laplace transform of $W_q(t)$ is

$$\bar{W}_q(s) = \frac{W_q^*(s)}{s} = \frac{1-\rho}{s - \lambda[1 - B^*(s)]}.$$

As before, if we write $s = a + bi$ and $B^*(s) = c + di$, then this becomes

$$\bar{W}_q(a+bi) = \frac{1-\rho}{a + bi - \lambda[1-(c+di)]} = \frac{1-\rho}{(a + \lambda c - \lambda) + (b + \lambda d)i},$$

and

$$\text{Re}\left[\bar{W}_q(a+bi)\right] = \frac{(1-\rho)(a+\lambda c - \lambda)}{(a+\lambda c-\lambda)^2 + (b+\lambda d)^2}. \tag{8.33}$$

Again, c and d are implicit functions of a and b. For exponential service, c and d are given by (8.31).

We can proceed as before to obtain the CDF $W_q(t)$ using Algorithm 8.1 (with the default parameters). The results are plotted in Figure 8.1 (where $\lambda = 0.2$ and $\mu = 1.0$ for this queue). The relative errors in estimating $W_q(t)$ are similar to those for $W(t)$ shown in Figure 8.2.

EXAMPLE 8.5

Numerically estimate $W_q(t)$ for an $M/D/1$ queue (i.e., where service times are deterministic and equal to a value D). For deterministic service, $B^*(s) = e^{-sD}$. Thus

$$\bar{W}_q(s) = \frac{W_q^*(s)}{s} = \frac{1-\rho}{s - \lambda[1 - B^*(s)]} = \frac{1-\rho}{s - \lambda[1 - e^{-sD}]}.$$

To evaluate $\text{Re}[\bar{W}_q(a+bi)]$, we use (8.33) (which is valid for any $M/G/1$ queue) and substitute appropriate expressions for c and d based on the service distribution:

$$c \equiv \text{Re}\left[B^*(a+bi)\right] = \text{Re}\left[e^{-(a+bi)D}\right] = e^{-aD}\cos(bD),$$

$$d \equiv \text{Im}\left[B^*(a+bi)\right] = \text{Im}\left[e^{-(a+bi)D}\right] = -e^{-aD}\sin(bD).$$

As before, the estimate for $W_q(t)$ is then obtained by evaluating the finite sums in (8.25) and (8.26), as in Algorithm 8.1.

Now, $W_q(t)$ is also known analytically for this queue (e.g., Erlang, 1909) and is given by the following formula:

$$W_q(t) = (1-\rho)\sum_{i=0}^{\lfloor t/D \rfloor} e^{-\lambda(iD-t)}\frac{(iD-t)^i}{i!}\lambda^i,$$

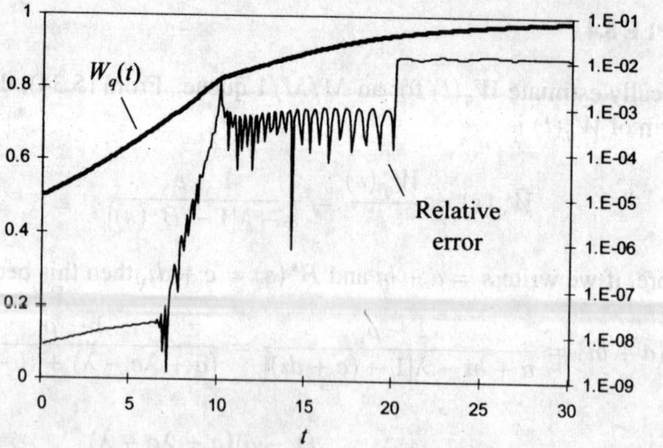

Figure 8.3 Numerical estimates for $W_q(t)$ for $M/D/1$ queue.

where $\lfloor x \rfloor$ is the greatest integer less than or equal to x. Figure 8.3 shows sample results for this queue with $\lambda = 0.05$ and $D = 10$ ($\rho = \lambda D = 0.5$). The figure shows the numerical estimate for $W_q(t)$ and the relative error $|W_{q,appx}(t) - W_q(t)|/[1 - W_q(t)]$. Although the CDF $W_q(t)$ is continuous for $t > 0$, its *derivative* is discontinuous at $t = D$ ($D = 10$ for this example). The relative error gets significantly worse at $t = D$. This illustrates a general phenomenon that numerical transform inversion can have difficulties at points where the function of interest is discontinuous, or in this case, at a point where the derivative of the function is discontinuous. See Sakurai (2004) for techniques to better handle these issues.

■ EXAMPLE 8.6

Numerically compute $W_q(t)$ for an $M/E_2/1$ queue. To evaluate $\text{Re}[\bar{W}_q(a + bi)]$, we use (8.33) where c and d are determined from the service distribution. Since an Erlang-2 distribution is the convolution of two exponential random variables, its LST is the product of two exponential transforms:

$$B^*(s) = \frac{\mu}{s+\mu} \cdot \frac{\mu}{s+\mu}.$$

Thus

$$\begin{aligned} B^*(a+bi) &= \frac{\mu^2}{(a+bi+\mu)^2} = \frac{\mu^2}{a^2 - b^2 + \mu^2 + 2a\mu + 2abi + 2\mu bi} \\ &= \frac{[\mu^2(a^2 - b^2 + \mu^2 + 2a\mu)] - [\mu^2(2ab + 2\mu b)]i}{(a^2 - b^2 + \mu^2 + 2a\mu)^2 + (2ab + 2\mu b)^2}. \end{aligned}$$

The real and imaginary parts of this expression are substituted for c and d, respectively, in (8.33). The numerical inversion proceeds as in the previous examples.

8.2.4 Other Numerical Inversion Methods

So far, we have presented one numerical inversion technique, the Fourier-series method. (In addition, we discussed an acceleration method, Euler summation, to improve the convergence of the method.) The Fourier-series method can be derived from the Bromwich inversion integral (8.16). Now, this is not the only exact formula that can be used to invert a transform. In this section, we briefly discuss a few other inversion formulas. In addition, we discuss numerical methods based on these formulas. We do not discuss these methods in detail. The purpose is to provide selected references for the interested reader. For further references, an extensive bibliography is given in Piessens (1975) and Piessens and Dang (1976); see also the literature review in Abate and Whitt (1992).

Another exact inversion formula that can be used as a basis for a numerical method is the Post–Widder inversion formula (e.g., Feller, 1971, Eq. 6.6, p. 233):

$$f(t) = \lim_{n\to\infty} \frac{(-1)^n}{n!}\left(\frac{n+1}{t}\right)^{n+1} \bar{f}^{(n)}\left(\frac{n+1}{t}\right),$$

where $\bar{f}^{(n)}$ is the nth derivative of $\bar{f}$. The discrete analog of this formula is (Gaver, 1966)

$$f(t) = \lim_{n\to\infty} f_n(t) \equiv \lim_{n\to\infty} (-1)^n \frac{\ln 2}{t} \frac{(2n)!}{n!(n-1)!} \Delta^n \bar{f}\left(\frac{n \ln 2}{t}\right), \tag{8.34}$$

where $\Delta \bar{f}(n\alpha) = \bar{f}(n\alpha + \alpha) - \bar{f}(n\alpha)$ and $\Delta^k = \Delta(\Delta^{k-1})$. In other words, the latter formula uses finite differences rather than derivatives. A sufficient condition for both inversion formulas to hold is that $f(\cdot)$ is a bounded real-valued function that is continuous at t. In contrast to the Bromwich inversion integral, $f(t)$ is expressed here via derivatives (or differences) of the transform $\bar{f}$, rather than via integration. The first formula is the basis for the Jagerman–Stehfest procedure (Jagerman, 1978, 1982). The second formula is the basis for the Gaver–Stehfest procedure (Gaver, 1966). Both methods are summarized in (Abate and Whitt, 1992, Section 8). The methods have been adapted using an acceleration technique given by Stehfest (1970).

The next formula is the basis for the *Laguerre method* of transform inversion. The exact inversion formula expands $f(t)$ as an infinite sum of Laguerre polynomials:

$$f(t) = e^{-t/2} \sum_{n=0}^{\infty} q_n L_n(t), \qquad \text{where} \qquad L_n(t) = \sum_{k=0}^{n} \binom{n}{k} \frac{(-t)^k}{k!}$$

are the Laguerre polynomials. The coefficients q_n are defined via the series expansion:

$$\sum_{n=0}^{\infty} q_n z^n \equiv \frac{1}{1-z} \bar{f}\left(\frac{1+z}{2(1-z)}\right).$$

Thus $f(t)$ is obtained via the coefficients q_n, which are determined from the Laplace transform $\bar{f}(\cdot)$. The basic idea in the numerical implementation is to compute a finite number of the coefficients q_n and then to approximate $f(t)$ using a finite sum. For further details on the Laguerre method, see Weeks (1966) and Abate et al. (1996, 1997). One advantage of the Laguerre method is that it provides the whole function $f(t)$. In other words, to evaluate $f(t)$ at multiple values of t, the coefficients q_n only need to be determined once. As noted in Abate et al. (1996), the method "is more likely to yield one function that is a good approximation for a large set of t. However, ... if $f(t)$ is badly behaved at just one value of t, then the Laguerre method has difficulties at all values of t."

Abate and Whitt (2006) provide a framework that unifies the treatment of many numerical transform inversion techniques. In this framework, $f(t)$ is approximated by a finite linear combination of the transform $\bar{f}$ evaluated at different points:

$$f(t) \approx \frac{1}{t} \sum_{k=0}^{n} \omega_k \bar{f}\left(\frac{\alpha_k}{t}\right),$$

where ω_k and α_k are constants. This framework was also suggested in earlier papers by Zakian (1969, 1970, 1973). Many inversion methods, such as the Fourier-series method with Euler summation, are specific instances of this general framework.

8.3 Discrete-Event Stochastic Simulation

It often turns out that the models and analysis techniques discussed thus far cannot adequately represent a particular queueing system. This can be due to the characteristics of the input or service mechanisms, the complexity of the system design, the nature of the queue discipline, or combinations of all of the these. For example, a multistation multiserver system with some recycling, where service times are (truncated) normally distributed and a complex priority system is in effect, is impossible to model analytically. Furthermore, even some of the models treated previously in the text provided only steady-state results, and if one were interested in transient effects or if the probability distributions were to change with time, it might not be possible to develop analytical solutions or efficient numerical schemes in these cases. For such problems as these, it may be necessary to resort to analyses by simulation. It should be emphasized, however, that if analytical models are achievable, they should be used, and that simulation should be resorted to only in cases where either analytical models are not achievable and approximations not acceptable or they are so complex that the solution time is prohibitive.

While simulation may offer a way out for many analytically intractable models, it is not in itself a panacea. There are a considerable number of pitfalls one may encounter in using simulation. Since simulation is comparable to analysis by experimentation, one has all the usual problems associated with running experiments in order to make inferences concerning the real world, and must be concerned with such things as run length, number of replications, and statistical significance. However, the theory of statistics (including, of course, experimental design) can be of help here.

Another drawback to simulation analyses occurs if one is interested in optimal design of a queueing system. Suppose that it is desired to determine the optimal number of channels or the optimal service rate for a particular system where conflicting system costs are known. If an analytical model can be developed, the mathematics of optimization (differential calculus, mathematical programming, etc.) can be utilized. However, if it is necessary to study the system using simulation, then one must rely on techniques for searching experimental output. These search techniques are often not as neat as the mathematics of optimization for analytical functions. Frequently the experimenter will merely try a few alternatives and simply choose the best among them. It might well be that none of the alternatives tried is optimal nor even near optimal. How close one gets to optimality in a simulation study often depends on how clever the analyst is in considering the alternatives to be investigated. Because of these potential drawbacks, simulation analysis has often been referred to as an "art." Nevertheless, with the advances in simulation methodology in these areas (see, e.g., Rubinstein, 1986), it is becoming increasingly competitive with analytical modeling in some situations; and in many situations, simulation is the only way to proceed. Simulation has found major uses in modeling transportation, manufacturing, and communication systems. Such systems are usually stochastic in nature, with a variety of random processes interacting in complex ways. Without simplifying assumptions about the nature of the randomness, routing probabilities, and so on, analytical modeling is usually not an option.

The following presents an overview of what we believe to be the key points in discrete-event stochastic simulation; for more details, the interested reader is referred to basic simulation texts such as Banks et al. (2005), Fishman (2001), Law and Kelton (2000), and Leemis and Park (2006).

8.3.1 Elements of a Simulation Model

One can look at simulation modeling as being composed of three major elements: (1) input distribution selection (sometimes called input modeling) and generation, (2) bookkeeping, and (3) output analysis. Since we are interested in modeling stochastic systems, it is necessary to select and then generate the appropriate stochastic phenomena in the computer. For example, a telecom center may consist of a network of queues with a variety of different interarrival-time and service-time distributions. We must decide on which probability distributions we wish to use to represent these arrival and service mechanisms (sometimes we may use an empirical distribution made up from actual collected data). Then, random variates from these different distributions must be generated so that the system can be observed in action. Once these distributions are chosen and random variates generated, the bookkeeping phase keeps track of transactions moving around the system and keeps counters on ongoing processes in order to calculate appropriate performance measures. Output analysis has to do with computing measures of system effectiveness and employing the appropriate statistical techniques required to make valid statements concerning system performance.

The following very simple hypothetical example illustrates the three basic elements described above.

Table 8.4 Input Data

Time between orders:	–, 7, 2, 6, 7, 6, 7, 2, 5, 4, 5, 3, 2, 6, 2, 4, 2, 6, 5, 5
Service times:	1, 3, 2, 3, 6, 5, 4, 5, 1, 1, 3, 1, 3, 2, 2, 6, 5, 1, 3, 5

■ **EXAMPLE 8.7**

A small manufacturer of specialty items has signed a contract with a prestigious customer for 20 orders of its premiere product. Management is concerned with current capacity and wishes to analyze the situation using discrete-event simulation. The customer will place orders at random times and of course would like them filled as soon as possible. Orders are placed only at the beginning of a month and could come as frequently as 2 months apart or as infrequently as 7 months apart, or anything in between, all with equal probability. Currently, the production capability for this product is such that orders are shipped only at the end of a month and the order filling time is equally likely between 1 and 6 months, inclusive. Only one order at a time can be processed, so that if a second order comes in while one is being prepared, it must wait until the order ahead of it is completed. For this capability, management would like to get an idea of the average number of orders in the system, the average time an order spends in the system, the maximum time an order spends in the system, and the percentage of time the system is idle. The date of the first order is known, and the production line will be set up just in time to receive the first order. The production line will be taken down after the last (20th) order is completed.

Input modeling has been simplified here, since we have decided that the probability distributions are discrete uniform distributions. This also makes generating the random input data easy, since it can be done by (the equivalent of a) roll of a fair die. Times between placement of orders are discrete-uniform (2,7) and service times are discrete-uniform (1,6). Thus for generating the interarrival times, we simply roll the die 19 times and add one to each value to get the times between successive orders after the first one.

For the service times, the value of the roll itself suffices and we simply need to roll the die 20 more times. Table 8.4 gives the results of using a fair die to generate the input data.

Using the input data in Table 8.4, we can construct an abbreviated bookkeeping table as shown in Table 8.5 (note that Tables 8.4 and 8.5 are in the same spirit as Tables 1.3 and 1.4, respectively, of Section 1.6). At clock 0, the first order comes into the system, has a service time of 1 month, and is due to depart at clock 1. At clock 1, the next arrival, order 2, is due in at clock $0 + 7 = 7$, and since no order is in the system, will depart at its arrival time plus service time, that is, $7 + 3 = 10$. The clock is advanced to time 7, and the next arrival (order 3) scheduled at $7 + 2 = 9$. Since order 3 arrives before order

Table 8.5 Bookkeeping[a]

Master Clock Time	Next Events Arrival	Next Events Departure	Transaction in Queue	Transaction in Service
0	[2], 7	[1], 1		→[1]
1	[2], 7	[2], 10		[1]→
7	[3], 9	[2], 10		→[2]
9	[4], 15	[2], 10	→[3]	[2]
10	[4], 15	[3], 12	[3]→	→[3] [2]→
⋮	⋮	⋮	⋮	⋮
81	[20], 86	[19], 84		→[19]
84	[20], 86			[19]→
86		[20], 91		→[20]
91				[20]→

[a]Key: [n], t = [transaction number], time of occurrence.

2 leaves, the clock is advanced to time 9, the arriving order 3 enters the queue, and order 2 is still in service, but due to depart at time 10. Order 4 is due in at $9 + 6 = 15$. The clock is then advanced to time 10, where order 2 leaves the system, and order 3 enters service and is scheduled to depart at $10 + 2 = 12$. Order 4 is next to arrive, and it is due in at 15, so the clock advances to 12. The bookkeeping continues on in this fashion until the 20th order is processed.

Such a bookkeeping table (which can get very complicated for realistic, complex systems) allows one to make the performance-measure calculations. It is straightforward to use Table 8.5 and obtain the queue wait and total time in system for each order. For example, order 1 came into the system at time 0, went right into service, and left at time 1, spending zero time in queue and 1 month in the system. Order 2 arrived at time 7, also went directly into processing, and left at time 10, spending 3 months in the system. Order 3, however, arriving at time 9, had to enter the queue, since 2 was still in process. It left the queue for processing at time 10 and exited the system at time 12 (not shown in the abbreviated table), spending 1 month in queue waiting for processing and 3 months total time in the system. Average waiting times and maximum waiting times can then be easily calculated. Queue-size and system-size values, as well as idle periods, can also be obtained from Table 8.5, although it is a little more work getting average figures, since the sizes must be weighted by the amount of time the queue and system stayed at their various sizes (see Section 1.6).

In this example, the maximum number of orders in the queue was 1, the maximum number of orders in the system was 2, the maximum time an order spent in the system waiting to be processed was 4 months (order 17), and the maximum time an order spent in the system in toto was 9 months (also order 17). The average queue size was 0.13, the average system size was 0.81, the average percentage of the time the system was empty and idle was 32%, and the average waiting times in queue and system, respectively, were 0.6 and 3.7 months.

8.3.2 Input Modeling and Random-Variate Generation

In this subsection, we treat the topics of choosing the appropriate distributions to represent the stochastic mechanisms of the system (input modeling) and the generation of random variates from the chosen distributions, and also include a brief treatment of pseudorandom-number generators.

8.3.2.1 Input Modeling. Input modeling is not only appropriate for simulation but is necessary for any probabilistic modeling, including analytical and numerical treatments as well. It is most important, since the output of any model can only be as good as its input. The two major problems in input modeling are the selection of a family of distributions (e.g., exponential, Erlang normal), and, once the family is selected, estimating its parameters. We start with the easier of the two problems first, namely, parameter estimation once a distribution family is selected. Because of the importance of the exponential and its relatives (e.g., Erlang) in analytical modeling, we use these as our primary illustrations. However, the reader should keep in mind that with modern simulation packages, almost any known statistical distribution can easily be utilized, and this is one of the major advantages of going to simulation modeling.

8.3.2.2 Parameter Estimation. We assume that we have chosen the distribution family and have data available on actual transactions (interarrival times or service times). Let us assume we have a random sample of size n, say, $t_1, t_2, \ldots, t_n$. The two classical methods of parameter estimation are the method of maximum likelihood, with its resulting estimators referred to as MLEs (we have encountered these before in Section 6.7) and the method of moments (MOM) estimators.

Maximum-likelihood estimators have some nice statistical properties and can be obtained as follows. First consider the case where we believe our underlying distribution is exponential with parameter θ and we wish to estimate θ from the sample data. We form the likelihood function (joint density function of the sample), assuming the observations are *independent*, as

$$L(\theta) = \prod_{i=1}^{n} \theta e^{-\theta t_i} = \theta^n e^{-\theta \sum_{i=1}^{n} t_i},$$

where t_i is the ith sample observation (say, interarrival or service time). The MLE of θ is the value that maximizes L. It is often more convenient to find the maximum of $\ln L = \mathcal{L}$. Thus $\hat{\theta}$, the MLE of θ, is the value for which

$$\max_{\theta} \mathcal{L}(\theta) = \max_{\theta} \left(n \ln \theta - \theta \sum_{i=1}^{n} t_i \right)$$

is attained. Taking $d\mathcal{L}/d\theta$ and setting it equal to zero yields the maximizing value as

$$\frac{n}{\hat{\theta}} - \sum_{i=1}^{n} t_i = 0 \quad \Rightarrow \quad \hat{\theta} = \frac{n}{\sum_{i=1}^{n} t_i} = \frac{1}{\bar{t}},$$

where $\bar{t} \equiv \left(\sum_{i=1}^{n} t_i\right)/n$ is the sample mean.

This is the same estimator that results when the empirical and theoretical means are equated and then solved for θ. This procedure (equating theoretical and sample moments) is called the MOM and is usually a very fast way to get estimates, though often giving quite different answers from the MLE. However, the MOM does not in general lead to statistics with all the good properties we can ascribe to the MLE. In our context, one particularly important concern is whether an estimator is consistent in the sense that $\hat{\theta} \to \theta$ in probability as the sample size goes to ∞. It does indeed turn out that the likelihood equation is guaranteed to have a consistent solution under fairly general conditions; but the MOM is not. On the other hand, we know how important moments are in queueing theory. For example, the PK formula for the $M/G/1$ depends only on the mean and the variance of the service-time distribution. The Kingman–Marshall upper bound and the heavy-traffic approximation depend only on the first two moments (mean and variance) of the interarrival and service-time distributions. Thus for estimating parameters for queueing models, even though MOM estimators are less desirable than MLEs in a statistical sense, one might well be better off using them.

As another example, let us consider finding MLEs and MOM estimates for the two parameters of the Erlang density. We will first derive the MLEs. For the Erlang density,

$$f(t) = \frac{\phi(\phi t)^{k-1} e^{-\phi t}}{(k-1)!},$$

so the likelihood function may be written as

$$L(\phi, k) = \prod_{i=1}^{n} \frac{\phi(\phi t_i)^{k-1} e^{-\phi t_i}}{(k-1)!} = \frac{\phi^{nk} e^{-\phi \sum_{i=1}^{n} t_i} \left(\prod_{i=1}^{n} t_i\right)^{k-1}}{[(k-1)!]^n}.$$

Thus the log likelihood is

$$\mathcal{L}(\phi, k) = nk \ln \phi - \phi \sum_{i=1}^{n} t_i + (k-1) \sum_{i=1}^{n} \ln t_i - n \ln(k-1)!.$$

Therefore

$$\frac{\partial \mathcal{L}}{\partial \phi} = \frac{nk}{\phi} - \sum_{i=1}^{n} t_i \quad \Rightarrow \quad \hat{\phi} = \frac{k}{\bar{t}},$$

where, again, $\bar{t}$ is the sample mean of the data. Now to get the complete pair $(\hat{\phi}, \hat{k})$, consider k to be a continuous variable (say, x) and proceed in the usual way to obtain the MLE $\hat{x}$ of x as the numerical solution to

$$\frac{\partial \mathcal{L}}{\partial x} = 0 = n \ln \hat{\phi} + \sum_{i=1}^{n} \ln t_i - n\psi(\hat{x})$$

$$= n(\ln \hat{x} - \ln \bar{t}) + \sum_{i=1}^{n} \ln t_i - n\psi(\hat{x}),$$

where $\psi(x)$ is the logarithmic derivative of the Γ function, that is, $\psi(x) \equiv d \ln \Gamma(x)/dx$. The function $\psi(x)$ is tabulated in Abramowitz and Stegun (1964), and a good approximation to it when x is not too small (say, $\geq$3) is

$$\psi(x) \approx \ln(x - \tfrac{1}{2}) + \frac{1}{24(x - \frac{1}{2})^2}.$$

Hence the MLE $\hat{k}$ of k is either $[\hat{x}]$ or $[\hat{x}] + 1$ (where $[x]$ is the greatest integer in x), depending on which pair, $([\hat{x}]/\bar{t}, [\hat{x}])$ or $(([\hat{x}] + 1)/\bar{t}, [\hat{x}] + 1)$, gives a higher value to the log likelihood.

In the case of the Erlang, the MOM answer follows much more easily. Here, since there are two parameters, two equations are needed, namely,

$$\bar{t} = \frac{\tilde{k}}{\tilde{\phi}} \quad \text{and} \quad s^2 = \frac{\tilde{k}}{\tilde{\phi}^2},$$

where s^2 is the sample variance. Thus from the simultaneous solution of the above two equations, the moment estimates are found as

$$\tilde{\phi} = \frac{\bar{t}}{s^2} \quad \text{and} \quad \tilde{k} = \left[\frac{\bar{t}^2}{s^2}\right] \quad \text{or} \quad \left[\frac{\bar{t}^2}{s^2}\right] + 1.$$

For the Erlang case (and for most distributions), MOM and ML give different estimators for the parameters. If the mean and variance using the MLEs gave very different mean and variance values from those of the sample data, we would be hesitant in using the MLEs and would go with the MOM.

8.3.2.3 Distribution Selection. We now turn to the much more difficult but very important topic of how to decide on which distribution family to choose to represent the input distributions (interarrival and service times). The choice of appropriate candidate probability distributions hinges upon knowing as much as possible about the characteristics of the potential distributions and the "physics" of the situation to be modeled. Generally, we have first to decide which probability functions are appropriate to use for the arrival and service processes. For example, we know the exponential distribution has the Markovian (memoryless) property. Is this a reasonable condition for the actual situation under study? Let us say we are looking to describe a service mechanism consisting of a single server. If the service for all customers is fairly repetitive, we might feel that the longer the customer is in service, the greater the probability of a completion in a given interval of time (nonmemoryless). In this case, the exponential distribution would not be a reasonable candidate for consideration. On the other hand, if the service is mostly diagnostic in nature (we must find the trouble in order to fix it), or there is a wide variation of service required from customer to customer, the exponential might indeed suffice.

The actual shape of the density function also gives quite a bit of information, as do its moments (see Law and Kelton, 2000, for pictures of most standard distribution families). One particularly useful measure is the ratio of the standard deviation to the mean, called the coefficient of variation ($C = \sigma/\mu$). The exponential distribution has $C = 1$, while E_k (Erlang type k), $k > 1$, has $C < 1$, and H_k (hyperexponential), $k > 1$, has $C > 1$. Hence choosing the appropriate distribution is a combination of knowing as much as possible about distribution characteristics, the "physics" of the situation to be modeled, and statistical analyses when data are available.

To help in characterizing probability distributions for consideration as candidates in describing interarrival or service times, we present a concept that emanates from the area of *reliability theory*, namely, the hazard rate (or as it is also called, the failure rate) function. We will relate this to the Markov property for the exponential distribution and point out its use as a way to gain general insight into probability distributions.

Suppose we desire to choose a probability distribution to describe a continuous random variable T (say, interarrival or service time) with CDF $F(t)$. The density function, $f(t) = dF(t)/dt$, can be interpreted as

$$f(t)\,dt \approx \Pr\{t \le T \le t + dt\},$$

that is, as the approximate probability that the random time will be in a neighborhood about a value t. The CDF $F(t)$ is, of course, the probability that the time will be less than or equal to the value t. We define a conditional type of probability as follows:

$$h(t)\,dt \approx \Pr\{t \le T \le t + dt | T \ge t\},$$

which is the approximate probability that the time will be in a neighborhood about a value t, given that the time is already t. For example, if we are dealing with interarrival times, it is the approximate probability that an arrival occurs in an interval dt, given that it has been t since the last arrival. If we are dealing with service times, $h(t)$ is the approximate probability that a customer is completed in dt, given that the customer has already been in service for a time t.

From the law of conditional probability we have

$$\begin{aligned} h(t)\,dt &\approx \Pr\{t \le T \le t + dt \,|\, T \ge t\} \\ &= \frac{\Pr\{t \le T \le t + dt \text{ and } T \ge t\}}{\Pr\{T > t\}} = \frac{f(t)\,dt}{1 - F(t)}. \end{aligned}$$

Therefore

$$h(t) = \frac{f(t)}{1 - F(t)}. \tag{8.35}$$

This hazard or failure rate function, $h(t)$, can be increasing in t (called an increasing failure rate, or IFR), decreasing in t (DFR), constant (considered to be *both* IFR and DFR), or a combination. The constant case implies the memoryless or *ageless* property, and we shall shortly show this for the exponential distribution.

If we believe that service is consistent enough that the longer a customer has been in service, the more likely it is that the service is completed in the next dt, then we desire an $f(t)$ for which $h(t)$ is increasing in t, that is, an IFR distribution.

From (8.35) we can obtain $h(t)$ from $f(t)$. Thus the hazard rate is another important source [as is the shape of $f(t)$ itself] for obtaining knowledge concerning candidate $f(t)$'s that may be considered for modeling arrival and service patterns.

Consider the exponential distribution, $f(t) = \theta e^{-\theta t}$. We desire to find $h(t)$. From (8.35),

$$h(t) = \frac{\theta e^{-\theta t}}{e^{-\theta t}} = \theta.$$

Thus the exponential distribution has a constant failure (hazard) rate and is memoryless. Suppose we feel in a particular queueing situation that we need an IFR distribution for describing service times. It turns out that the Erlang ($k > 1$) has this property. The density function, from Chapter 3, Section 3.3.1, where we now let $\theta = k\mu$, is

$$f(t) = \frac{\theta^k t^{k-1} e^{-\theta t}}{(k-1)!},$$

and (see Problem 8.3)

$$F(t) = \frac{\theta^k}{(k-1)!} \int_0^t e^{-\theta x} x^{k-1}\, dx = 1 - \sum_{i=0}^{k-1} \frac{(\theta t)^i e^{-\theta t}}{i!}. \tag{8.36}$$

Thus

$$h(t) = \frac{\theta^k t^{k-1} e^{-\theta t}}{(k-1)! \sum_{i=0}^{k-1} (\theta t)^i e^{-\theta t}/i!} = \frac{\theta(\theta t)^{k-1}}{(k-1)! \sum_{i=0}^{k-1} (\theta t)^i/i!}.$$

Without doing numerical work, it is difficult to ascertain the direction of change of $h(t)$ with t. It can be shown, however (see Problem 8.4), that $h(t)$ can be written as

$$h(t) = \frac{1}{\int_0^\infty (1 + u/t)^{k-1} e^{-\theta u}\, du}. \tag{8.37}$$

Now it is fairly easy to see that for $k > 1$, as t increases, the integrand in the denominator decreases, so that the integral decreases, and hence $h(t)$ is increasing with t (IFR). Furthermore, it has an asymptote of θ as t goes to infinity, and $h(0) = 0$. Since $h(t)$ has an asymptote, even though $h(t)$ increases with t, it does so at an ever slower rate, and eventually approaches the constant θ.

Now suppose we were to desire the opposite IFR condition, that is, an accelerating rate of increase with t. There is a distribution called the Weibull [with $1 - F(t) = e^{-\theta t^\alpha}$] for which we can obtain this condition. In fact, depending on how we pick the shape parameter α of the Weibull, we can obtain an IFR with decreasing acceleration, constant acceleration (linear in t), or increasing acceleration, or even obtain a DFR or the constant-failure-rate exponential. Even though the Weibull does not lend itself to analytical treatment (except when it reduces to the exponential) in queueing situations, it still can be a candidate distribution for a simulation analysis. In many situations, more than one particular distribution may be a reasonable candidate. For example, if we decide we want an IFR with a deceleration, we could consider the Weibull or Erlang family of distributions.

We present one further example in the process of choosing an appropriate candidate distribution for modeling. Let us say we are satisfied with an IFR that has a deceleration effect, such as the Erlang, but we believe that the C might be greater than one. This latter condition eliminates the Erlang from consideration. But we know that a mixture of exponentials does have $C > 1$. It is also known (see Barlow and Proschan, 1975) that any mixture of exponentials is DFR. In fact, Barlow and Proschan prove that all IFR distributions have $C < 1$, while all DFR distributions have $C > 1$. However, the converse is not true, that is, $C < 1$ does not imply IFR, $C > 1$ does not imply DFR, and $C = 1$ does not imply a constant failure rate (CFR). An example of this is the lognormal family of distributions, which, depending on the value of its parameters, can yield C's less than, equal to, or greater than one. Its hazard function is both IFR and DFR; that is, over a certain range it is IFR and over a certain range it is DFR. So, if we are convinced that we have an IFR situation, we must accept $C < 1$. Intuitively, this can be explained as follows. Situations that have $C > 1$ generally are ones where, say, service times are mixtures (say, of exponentials). Thus if a customer has been in service a long time, chances are it is of a type requiring a lot of service, so the chance of its being completed in the next *dt* diminishes. Situations for which we have an IFR condition indicate a more consistent service pattern among customers, thus yielding a $C < 1$.

In summary, we again make the point that choosing an appropriate probability model is a combination of knowing as much as possible about the characteristics of the probability distribution family being considered, and as much as possible about the actual situation being modeled. In most cases, data on the processes we are trying to model are available, but not always. Later we treat the case where data are available or can be collected and how these data can help us in choosing the appropriate family of distributions. But first we briefly comment that, in those cases where we have no data or cannot gather any (e.g., a new system we are modeling), the considerations we have mentioned previously become paramount. Sometimes in the absence of data, the triangular distribution is chosen, where the modeler is asked to set the minimum, maximum, and most likely values, which will then yield the three parameters of this three-parameter distribution. A more flexible distribution family is the beta, which can yield a large variety of shapes, and is employed along with "expert opinion" in deciding on the appropriate parameters (the beta is a two-parameter distribution, with both parameters affecting the distribution shape). Further discussions of choosing distributions with no data available can be found in Law and Kelton (2000) and Banks et al. (2005).

We now turn to the case where we have observations available on interarrival and service times. We will also assume the data come from a homogeneous time period (i.e., the process from which the data were gathered was not changing in time—say, a system during the peak traffic hours). It is very important to check the data to see that this is so. We also assume that all observations are independent. It is important to check the data for these conditions (IID), and there are some tests one can perform to do so (we refer the interested reader to Leemis, 1996, or Law and Kelton, 2000).

We now assume we have data that form an IID sample. One of the first things we can do in trying to decide which family of distributions might be appropriate is to calculate the sample mean, sample standard deviation, and sample C. A histogram plot of the data can also be very useful, although the shape of the resultant histogram depends on the length of intervals used to accumulate the frequencies (the number of observations falling in each interval). A rule of thumb is to choose interval lengths such that there are a minimum of five observations in each interval and that there are at least five intervals, but it is a good idea to try a variety of interval lengths. Suppose now, after looking at histogram plots and considering the physical characteristics of the system we are modeling, as well as the characteristics of distribution families, we choose a potential candidate distribution. There are a variety of statistical tests we can perform to see if our candidate distribution is a reasonable choice. We mention three general tests applicable for any family of distributions and one specific test if we believe our situation calls for choosing an exponential distribution.

The most commonly used (but not necessarily the best) test is the χ^2 goodness-of-fit tests. It assumes we have our data in histogram form, with each block of the histogram (referred to as a frequency class) providing the number of observations in the interval that is covered by the block. Generally, the intervals are of equal length, spanning the range of the data, and the number of observations in each interval varies to yield a picture approximating a density function. Figure 8.4 shows a histogram of the 25 service-time observations given below:

$$\begin{gathered}27.6, 28.9, 3.8, 16.6, 13.3, 3.3, 7.8, 55.3, 12.6, 1.8, 12.9, 4.8, 12.6, 8.8,\\ 3.3, 2.7, 0.6, 1.3, 1.1, 21.3, 11.3, 14.9, 15.7, 8.6, 9.6.\end{gathered}$$

The mean, standard deviation, and C are, respectively, 12.02, 11.91, and 0.991. In reality, we should have considerably more than 25 observations, but we use this small data set in order to make our discussion easier to follow.

Looking at the C, which is near one, we might consider the exponential family as a candidate distribution family. Both the ML and MOM estimators of the mean are 12.02. The histogram of Figure 8.4 has a roughly exponential distribution shape, and we first try the χ^2 test on these data to see if the exponential distribution is a reasonable choice. The χ^2 test statistic is

$$\chi_k^2 = \sum_{i=1}^{n} \frac{(o_i - c_i)^2}{e_i},$$

where o_i is the number observed in the ith frequency class, e_i the number expected in the ith frequency class if the hypothesized distribution were correct, and k the number of degrees of freedom, always equal to the total number of classes minus one

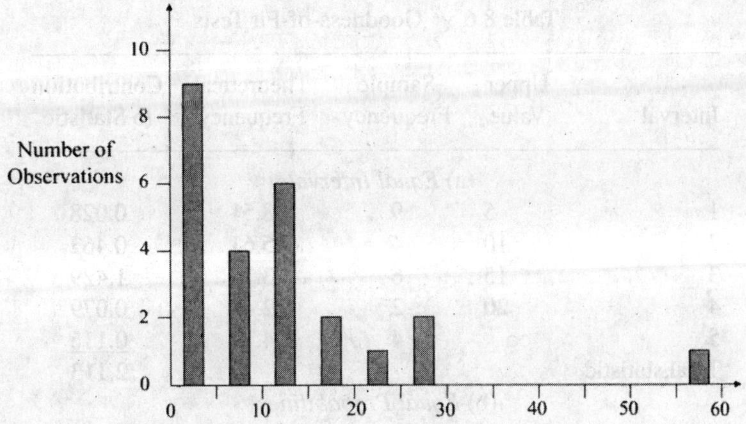

Figure 8.4 Histogram of sample service-time data.

and then minus one for each parameter estimated. Of course, the usual precautions must be taken to keep the number of observations in a class from being too small (a rule of thumb being five or more).

To get e_i, we integrate the theoretical distribution over the interval; for example, if i^- is the lower point of the ith interval and i^+ is the upper point, then for the exponential distribution,

$$e_i = n \int_{i^-}^{i^+} \theta e^{-\theta t}\, dt = n(e^{-\theta i^-} - e^{-\theta i^+}),$$

where θ is replaced by the MLE $\hat{\theta}$ and n is the sample size. For our case, $\hat{\theta}$ is 1/12.02.

There are two variations of the χ^2 test: equal intervals or equal probabilities. The first has equal values for $i^+ - i^-$: in our case above, 5.0. The second has equal values for e_i, so that the $i^+ - i^-$ values vary. We perform both versions, and the results are given in Table 8.6. For the tail frequency, we combined the last eight intervals into one interval, >20 (20, ∞), for the equal-interval test. For the equal-probability version, we set the intervals so that the theoretical frequency in each would be 5.

Since there are three degrees of freedom, the critical value at the 5% level is 7.815 (see any statistical textbook), and neither test rejects the hypothesis that the data come from an exponential distribution.

Great care should always be exercised in doing χ^2 goodness-of-fit tests, and the analyst would, of course, be well advised to study for a definitive treatise on the subject in the statistical literature. The basic weaknesses of the χ^2 test are its requirement for large samples (our case of 25 is much too small), its heavy dependence on the choice of the number and position of the time-axis intervals, and its possibly very high Type 2 error (i.e., the probability of accepting a false hypothesis) for some feasible alternative distributions.

Table 8.6 χ^2 Goodness-of-Fit Tests

Interval	Upper Value	Sample Frequency	Theoretical Frequency	Contribution to Statistic
		(a) *Equal Intervals*		
1	5	9	8.51	0.028
2	10	4	5.61	0.462
3	15	6	3.70	1.429
4	20	2	2.44	0.079
5	∞	4	4.74	0.115
Total statistic				2.113
		(b) *Equal Probabilities*		
1	2.682	4	5	0.2
2	6.140	5	5	0
3	11.014	4	5	0.2
4	19.345	8	5	1.8
5	∞	4	5	0.2
Total statistic				2.4

Another popular goodness-of-fit test is the Kolmogorov–Smirnov (KS) test. The KS test compares deviations of the empirical CDF from the theoretical CDF, and uses as its test statistic a modified maximum absolute deviation, namely,

$$K = \max_j \max \left\{ \left| \frac{j}{n} - F(t_j) \right|, \left| \frac{j-1}{n} - F(t_j) \right| \right\}, \tag{8.38}$$

where t_j is the jth ordered (ascending) observation, and $F(t_j)$ is the value of the hypothesized distribution function at the jth observation. Unfortunately, as we see, the test presupposes that the CDF F is completely known. In case there is interest in performing a KS test, tables for critical values of the test statistic are widely available in the published literature. If, however, the parameters of the hypothesized CDF are unknown and are to be estimated from the data, then a special KS table must be established, or the test statistic must be modified for the particular family from which F came, or both. For example, see the work of Lilliefors (1967, 1969) on the normal and exponential, respectively. Use of the general KS table for distributions with estimated means will give very conservative results in the sense that the actual significance level achieved will be much lower than that indicated. For our example (again we point out that 25 observations is too small a sample size to do any meaningful goodness-of-fit testing, but it serves our purpose for illustration of the technique), we find that the maximum deviation K obtained from applying (8.38) is 0.11739. Modifying this according to Stephens (1974), who suggests using as the test statistic

$$\tilde{K} = \left(K - \frac{0.2}{n}\right)\left(\sqrt{n} + 0.26 + \frac{0.5}{\sqrt{n}}\right)$$

and using the modified tables (see Law and Kelton, 2000), the 5%-level critical value is 1.094, and since $\tilde{K}$ turns out to be $(0.117 - 0.2/25)(5 + 0.26 + 0.5/5) = 0.415$, we again have no reason to reject the hypothesis.

A variation on the KS test is the Anderson–Darling (AD) test, which instead of using only the maximum deviation as KS does, uses all the deviations (actually, a weighted average of the squared deviation, with the weights being the largest at the tails of the distribution). Again, special tables are available for certain distributions, the exponential being one of them. For our example, the test statistic turns out to be 0.30045, and the critical value at the 5% level is 1.29, so that again we cannot reject the hypothesis.

Although goodness-of-fit tests require large sample sizes to really discriminate, and many are limited if we have to estimate parameters from the sample, there are specific tests for the exponential distribution. Since the exponential distribution is so important in analytical queueing modeling, we would like to point out a specific test for exponentiality which is quite powerful (power meaning ability to discern false hypotheses) against almost any alternative hypothesis and will usually outperform these other tests. It is the F test.

To perform the F test, $r(\approx n/2)$ and $n - r$ of a set of n interoccurrence times t_i are randomly grouped. It follows that the quantity

$$F = \frac{\sum_{i=1}^{r} t_i/r}{\sum_{i=r+1}^{n} t_i/(n-r)} \tag{8.39}$$

is the ratio of two Erlangs and is distributed as an F distribution with $2r$ and $2(n-r)$ degrees of freedom when the hypothesis of exponentiality is true. Therefore a two-tailed F test will be performed on the F calculated from a set of data in order to determine whether the stream is indeed truly exponential. This argument for the F test can be extended easily to the case in which there are randomly occurring incomplete interoccurrence periods, as is common in repairlike problems. Tables of critical points for the F distribution are available in most standard statistics books. We again illustrate this test on our earlier sample data.

Since the data are in random order, we sum the first 13 observations to get 201.3 and the last 12 to get 99.2. Calculating the F statistic according to (8.39) gives $F = (201.3/13)/(99.2/12) = 1.873$. The 95% critical values for $F_{26,24}$ are $1/2.23 \doteq 0.45$ and 2.26, respectively, so we accept the hypothesis that the data are exponential.

To close this discussion on input modeling, we mention that there are software packages that run data and recommend the distribution that "best" represents the data (e.g., ExpertFit). One note of caution is that most of these packages use ML to estimate parameters and have multiple criteria (besides goodness-of-fit testing) to select their recommendations. Often the model selected as best has considerable differences in the second and higher moments from those of the sample. We know from certain analytical theory (e.g., PK formula and heavy-traffic approximations) that the first and second moments are very important, and for these cases they are the *only* thing that matters—the actual distribution does not even enter into the formulas. Juttijudata (1996) and Gross and Juttijudata (1997) have shown how important moments are, and we caution the reader against choosing a distribution for which the theoretical moments (especially the first three or four) differ significantly from those of the sample.

Many simulation modelers recommend that, instead of trying to choose a theoretical probability distribution, one simply use the empirical distribution, that is, the sample histogram (this is very closely akin to what is referred to in statistical circles as *bootstrapping*). For a discussion on the pros and cons of this debate, see Fox (1981) and Kelton (1984).

8.3.2.4 Generation of Random Variates. Once the appropriate probability distributions are selected for representing the input processes (interarrival and service times), it is necessary to generate typical observations from them for "running" the simulated system. Determining how many to generate (i.e., how long to observe the simulated system) will be treated in a later section.

The procedure for generating IID random variates from a given specified probability distribution, say, $f(x)$, generally consists of two phases: (1) generation of pseudorandom numbers distributed uniformly on (0, 1) and (2) using the pseudorandom numbers to obtain variates (observations) from $f(x)$. We first discuss generation of uniform (0, 1) pseudorandom numbers and then how to use these to generate random variates from a specified $f(x)$.

Most computer programming languages contain a pseudorandom-number generator. These are "pseudo" in that they are completely reproducible by a mathematical algorithm, but "random" in the sense that they have passed statistical tests, which basically test for equal probability of all values and statistical independence. Most computer routines are based on linear-congruential methods that involve modulo arithmetic. It is a recursive algorithm of the form

$$r_{n+1} = (kr_n + a) \bmod m, \tag{8.40}$$

where k, a, and m are positive integers ($k < m, a < m$). That is, r_{n+1} is the remainder when $kr_n + a$ is divided by m. We must choose an initial value, r_0, which is called the *seed*, and this should be less than m.

For example, if $k = 4$, $a = 0$, and $m = 9$, and we choose r_0 initially as 1, we generate the numbers

$$1, 4, 7, 1, 4, 7, 1, 4, 7, \ldots.$$

Since the smallest number (remainder on division by 9) could be 0 and the largest number could be 8, the range is [0–8]. To normalize to (0, 1), all numbers are divided by $m = 9$—note that since 0 is a possibility, we are really normalizing to [0, 1). The results are then

$$0.111, 0.444, 0.778, 0.111, 0.444, 0.778, \ldots.$$

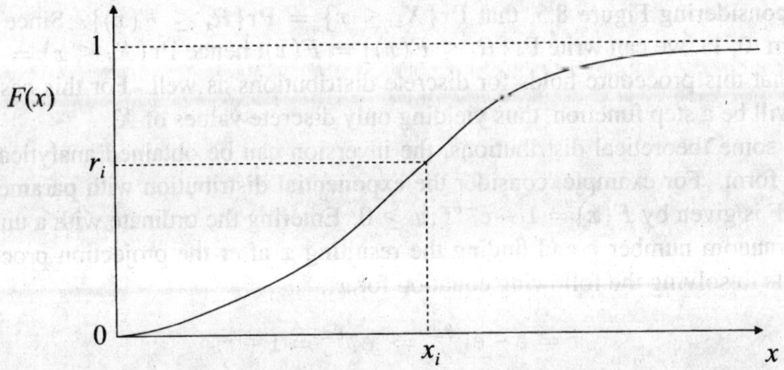

Figure 8.5 Inversion technique for generating random variates.

It is clear that this sequence is not acceptable. First, only three of the possible nine numbers appear. Second, we see that the sequence is cyclic with a cycle length of three. If one desired more than three random numbers, this sequence would be unusable. So let us instead change k to 5, a to 3, and m to 16 with an r_0 of 7 (an example from Law and Kelton, 2000), which yields

$$7, 6, 1, 8, 11, 10, 5, 12, 15, 14, 9, 0, 3, 2, 13, 4, 7, 6, 1, 8, \ldots$$

Here all possible numbers [0–15] are generated, so that we have a *full* cycle generator, but the cycle length is only 16 [the maximum cycle length of any stream using (8.40) is $m - 1$]. To normalize on [0, 1), we divide by 16 [if we truly desire uniform random numbers on (0, 1), then we can discard the 0's generated].

Thus we see that very careful consideration must be given to selecting k, a, and m (and to some extent r_0 also). Some values of m that seem to work well and appear in simulation software packages are $2^{31} - 1$ and 2^{48}, which work well with 32- and 64-bit machines in accomplishing the modulo arithmetic. For a more detailed discussion of random-number generation, the interested reader is referred to L'Ecuyer (2006).

We desire now to generate representative observations from any specified probability distribution, with CDF (say) $F(x)$. Again, there are several methods of achieving this. We present the most popular and refer the reader to the aforementioned references for further detail, if desired.

The main method we offer is sometimes referred to as the inverse or probability transformation method, or generation by inversion. It can best be described graphically by considering a plot of the CDF from which we desire to generate random variates. Such a plot is shown in Figure 8.5. The procedure is to first generate uniform (0, 1) random variates, say, $r_1, r_2, \ldots$. To obtain x_1, the first random variate corresponding to $F(x)$, we simply enter the ordinate with r_1 and project over and down, as shown in Figure 8.5; the resulting value from the abscissa is x_1. Repeating the procedure with $r_2, r_3 \ldots$ will yield $x_2, x_3, \ldots$.

To prove that this procedure works, we would like to show that a random variate (say, X_i) generated by this procedure obeys the relation $\Pr\{X_i \le x\} = F(x)$. We have, considering Figure 8.5, that $\Pr\{X_i \le x\} = \Pr\{R_i \le F(x)\}$. Since R_i is uniform (0, 1), we can write $\Pr\{R_i \le F(x)\} = F(x)$; hence $\Pr\{X_i \le x\} = F(x)$. Note that this procedure holds for discrete distributions as well. For this case, the CDF will be a step function, thus yielding only discrete values of X.

For some theoretical distributions, the inversion can be obtained analytically in closed form. For example, consider the exponential distribution with parameter θ. Its CDF is given by $F(x) = 1 - e^{-\theta x}, x \ge 0$. Entering the ordinate with a uniform (0, 1) random number r and finding the resulting x after the projection procedure amounts to solving the following equation for x:

$$r = 1 - e^{-\theta x} \Rightarrow e^{-\theta x} = 1 - r.$$

Since r is uniform (0, 1), it is immaterial whether we use r or $1 - r$ as our random number, and hence we can write $e^{-\theta x} = r$. Taking natural logarithms of both sides finally gives

$$x = \frac{-\ln r}{\theta}. \tag{8.41}$$

Unfortunately, analytical inversion is not possible for all probability distributions. In the case of some continuous distributions, we can find alternative ways to generate the variates, sometimes using numerical integration techniques, or more often making use of the fundamentals of probability theory to aid in the random-variate generation procedure. As an example of how we may take advantage of statistical theory, we examine the Erlang type k. Instead of attempting inversion on the Erlang CDF, we can merely take sums of k exponential random variates, which are quite easy to generate by inversion, as we have shown above. This was actually done in generating the data for Problem 3.24 of Chapter 3. Thus if we desired to generate random variates from an Erlang type k with mean $1/\mu$, we could obtain this type of random variate, say, x, from the uniform (0, 1) random variates $r_1, r_2, \ldots, r_k$ by

$$x = \sum_{i=1}^{k} \left(-\frac{\ln r_i}{k\mu} \right) = -\frac{\ln \Pi_{i=1}^{k} r_i}{k\mu}.$$

Another useful continuous distribution in the modeling of queues is the mixed exponential. To generate mixed-exponential variates according to the density

$$f(t) = \sum_{i=1}^{n} p_i \lambda_i e^{-\lambda_i t},$$

we recognize that each observation is essentially the result of two independent probabilistic events. That is, we first select the relevant exponential subpopulation using the discrete mixing probabilities $\{p_i\}$, and then, given that the jth population is indeed selected, an exponential variate is generated from (8.41) using mean $1/\lambda_j$. For procedures for generating other random variates (e.g., normal, gamma, lognormal), we refer the reader to Law and Kelton (2000).

It is possible, therefore, to generate random variates from any probability distribution using procedures such as those given earlier, although in some cases it may be time consuming and/or approximate. Nevertheless, in most cases it can be done without too much difficulty, and most modern simulation software has procedures for doing so.

We next turn our attention to consideration of the bookkeeping phase of a simulation analysis.

8.3.3 Bookkeeping Aspects of Simulation Analysis

As mentioned earlier, the bookkeeping phase of a simulation model must keep track of the transactions moving around the system, and set up counters on ongoing processes in order to calculate various measures of system performance. The simulation modeler has a large variety of languages and packages from which to choose. These can be general-purpose languages such as C++, Java, and Visual Basic, which allow the most flexibility in modeling, but require the most effort to program, or a variety of simulation language packages. One can get a feel from the very simple example given in Section 8.3.1 of the programming effort involved in generating variates from the input probability distributions, keeping track of which transactions are at what places at what times, and performing the statistical calculations needed for obtaining output measures of performance.

Requirements for bookkeeping, random-variate generation, and data collection necessary for statistical analyses of output are similar for large classes of simulation models, and this situation has given rise to the development of a variety of simulation language packages. These simulation language packages (e.g., Arena, ProModel, and AutoMod, to mention a few) make programming a simulation model much, much easier. However, some flexibility in modeling is sacrificed, since the model must fit into the package language environment. For most cases, this is not a problem. As a general rule, one can expect that the easier the programming becomes, the less flexibility there is in deviating from the language environment.

Most of the popular simulation languages use a next-event approach to bookkeeping (as opposed to fixed time increments), in that the master clock is advanced, as in Example 8.7, to the next event scheduled to occur, rather than the clock being advanced in fixed increments of time, where, for many of the increments, nothing may have happened (see Tables 1.4 and 8.5). Furthermore, most of the event-oriented routines employ a transaction-process technique that keeps track of the entire experience of a transaction as it proceeds along its way in the system. For more detailed discussion on programming languages, the reader is referred to Banks et al. (2005) or Law and Kelton (2000). For a survey on simulation software packages, see Swain (2005). This survey is generally updated every two years or so.

For small, demo simulations, the QtsPlus package has a simulation module and will simulate single-server and multiserver queues, queues with priorities, queues with multiple customer classes, and small networks of queues. It is somewhat limited in the choices for input distributions and has no automatic statistical output analysis. It is not meant to compete with commercial simulation packages, but allows the reader to gain some familiarity with a discrete event simulation program, for which the only package needed is Excel.

8.3.4 Output Analysis

Reaching reliable conclusions from simulation output requires a great deal of thought and care. When simulating stochastic systems, a single run yields output values that are statistical in nature, so that sound experimental design and sound statistical analyses are required for valid conclusions. Unlike sampling from a population in the classical sense, where great effort is made to have random samples with independent observations, we often purposely induce correlation in simulation modeling as a variance reduction technique, so that classical, off-the-shelf statistical techniques for analyzing sample data are often not appropriate. We present next some basic procedures for analyzing simulation output.

There are two major types of simulation models: terminating and continuing (nonterminating). A terminating model has a natural start and stop time, for example, a bank opens its doors at 9:00 A.M. and closes its doors at 3:00 P.M. (Example 7.2 is also a terminating model.) On the other hand, a continuing model does not have a start and stop time, for example, a manufacturing process where, at the beginning of a shift, things are picked up exactly as they were left at the end of the previous shift, so that, in a sense, the process runs continuously. In these latter cases, steady-state results are usually of interest, and in simulating such a system, a determination must be made as to when the initial transients are damped out and the simulation is in steady state.

Considering first a terminating simulation such as Example 7.2, from a single run we cannot make any statistical statements. For example, the maximum waiting time is a single observation, that is, a sample of one. What can be done is to *replicate* the experiment (repeated runs), using different random number streams for the order arrival and processing times for each run, and thus generate a sample of independent observations to which classical statistics can be applied. Assuming we replicate n times, we then will have n values for the maximum waiting time, say, $w_1, w_2, \ldots, w_n$. Assuming n is large enough to employ the central limit theorem, we can get a $100(1-\alpha)\%$ confidence interval (CI) by first calculating the mean and sample standard deviation of the maximum waiting time by

$$\bar{w} = \frac{\sum_{i=1}^{n} w_i}{n}$$

and

$$s_w = \sqrt{\frac{\sum_{i=1}^{n} (w_i - \bar{w})^2}{n-1}},$$

and then obtaining the CI as

$$\left[\bar{w} - \frac{t(n-1, 1-\alpha/2)s_w}{\sqrt{n}}, \bar{w} + \frac{t(n-1, 1-\alpha/2)s_w}{\sqrt{n}}\right],$$

where $t(n-1, 1-\alpha/2)$ is the upper $1-\alpha/2$ critical value for the t distribution with $n-1$ degrees of freedom.

For continuing simulations for which we are interested in steady-state results, we know from the ergodic theory of stochastic processes that

$$\lim_{T\to\infty} \frac{1}{T}\int_0^T X^n(t)\,dt = \mathrm{E}[X^n],$$

so that if we run long enough, we shall get close to the limiting average value. But it is not clear how long is long enough, and we often wish to be able to obtain a CI statement. Thus we have two additional problems: determining when we reach steady state and deciding when to terminate the simulation run. Assuming for the moment that these problems are solved and we decide to run the simulation for n transactions after reaching steady state and measure the time a customer spends waiting in a particular queue for service, we can obtain n queue wait values, which we shall again denote by w_i (now these are actual waits, not maximum waits). It might be tempting to calculate the average and standard deviation of these n values and proceed as above to form a CI. However, these w_i are correlated, and using the above formula for s_w greatly underestimates the true variance. There are procedures for estimating the required correlations and obtaining an estimate of the standard deviation from these correlated data, but this requires a great deal of estimation from this single data set, which has its drawbacks in statistical precision.

To get around the correlation problem, we may again replicate the run m times, using a different random number seed each time, as we did in the terminating case. For each run, we still calculate the mean of the w_i, denoting the mean for the jth replication by $\bar{w}_j$, that is,

$$\bar{w}_j = \frac{\sum_{i=1}^{n} w_{ij}}{n},$$

where w_{ij} is the waiting time for transaction i on replication j, $i = 1, 2, \ldots, n$ and $j = 1, 2, \ldots, m$. The $\bar{w}_j$ are now independent and in an analogous fashion to that of the terminating simulation, we can form a $100(1-\alpha)\%$ CI by calculating

$$\bar{w} = \frac{\sum_{j=1}^{m} \bar{w}_j}{m}$$

and

$$s_{\bar{w}_j} = \sqrt{\frac{\sum_{j=1}^{m} (\bar{w}_j - \bar{w})^2}{m-1}},$$

and then the CI becomes

$$\left[\bar{w} - \frac{t(m-1, 1-\alpha/2)s_{\bar{w}_j}}{\sqrt{m}}, \bar{w} + \frac{t(m-1, 1-\alpha/2)s_{\bar{w}_j}}{\sqrt{m}}\right].$$

Returning to the previous two problems mentioned for continuing simulations (viz., when steady state is reached and when to stop each run), we first discuss the latter and then the former. The run length (n) and the number of replications (m) will both influence the size of the standard error ($s_{\bar{w}_j}/\sqrt{m}$) required for making the CI above. The smaller the standard error, the more precise is the CI (narrower limits) for a given confidence $(1-\alpha)$. We know that the standard error goes down by the square root of m, so that the more replications made, the more precise the CI. Also, we might expect that as the run length n is increased, the computed value of $s_{\bar{w}_j}$ itself will be smaller for a given number of replications, so that longer run lengths will also increase the precision of the CI. Thus for a fixed amount of computer running time, we can trade off size of n versus size of m. Setting n and m is still something of an art, and trial runs can be made to evaluate the trade-offs.

The warmup period (the initial amount of time required to bring the process near steady-state conditions) is also not an easy thing to determine. Often in practice, it is essentially ignored in that it is hoped that there is enough data points in the run so that the transient effects are simply *swamped* by the portion of the data taken after near steady-state conditions are reached. A more common and somewhat more scientific approach is to divide the run into two portions: a transient period and a steady-state period. The basic idea is that if we could compute the time or number of transactions required so that the process is near steady state, we could simply not start recording data for calculating output measures until the master clock passed that point (i.e., discard the observatons during the transient period). Finding where that point is is one of the more difficult tasks in output analysis. A variety of procedures have been developed, and the reader is referred to the basic simulation texts referenced previously for more detailed discussion. We mention in a little detail a few procedures.

One particular method that seems to work fairly well (again, the warmup period analysis is still more of an art than a science) is that due to Welch (1983) and also presented in Law and Kelton (2000). The procedure involves choosing one of the output performance measures (e.g., average waiting time for service), calculating means of this measure over the replications for *each* transaction (i.e., if we have m replications of run length n, we average the m values we obtain for transaction i from each replication, for $i = 1, 2, \ldots, n$), taking moving averages of neighboring values of these transaction averages, plotting these, and visually determining when the graph appears to be stabilizing. It is recommended to try various moving average windows (the number of adjacent points in the moving average). The size of the moving average window and the point at which the graph settles down are again judgment calls.

Another approach using the transaction averages calculated as described above is a regression approach suggested by Kelton and Law (1983). The transaction average data stream of n values is segmented into b batches. A regression line is fitted for the last batch, and if the slope of the line is not significantly different from zero, steady state is assumed. Then the next to last batch of data is included in the regression, the test made again, and if not significantly different from zero, the last $2b$ values are said to be in steady state. This procedure is continued, and when the test for slope becomes significantly different from zero, that batch and all preceeding ones are considered to be in the transient region. This approach is predicated on the assumption that the performance measure is monotonic in time, which should be the case when the initial conditions assume the system starts empty and idle. Again, decisions on the values for n, m, and b must be made subjectively.

A third approach differs from the two previous approaches in that, rather than attempting to find a point at which the process enters steady state, the bias of the transient effects is estimated in order to determine if the data do show an initial conditions bias (Schruben, 1982). This procedure could be used in conjunction with one of the two above to check whether the warmup period chosen was adequate to remove the initial-conditions bias.

Assuming we have decided on what the warmup period should be, in order to avoid throwing away the initial warmup period observations for each of the m replications in a nonterminating simulation experiment, the procedure of *batch means* has been suggested (Law, 1977; Schmeiser, 1982). Rather than replicating, a single long run (say, mn transactions) is made and then broken up into m segments (batches) of n each. The performance measures for the segments are assumed to be approximately independent (if the segments are long enough, the correlation between segments should be small), so that the classical estimate of the standard deviation can be employed; that is, the segments act as if they were independent replications. The methodology for determining the CIs is identical to that for m independent replications. But now, one only has to discard a single warmup period instead of m as before.

Other methods have been suggested, such as the regenerative method (Crane and Iglehart, 1975; Fishman, 1973a,b) and time series analyses (Fishman, 1971; Schruben, 1982).

In comparing two alternative system designs, the technique most commonly used is a paired t CI on the difference of a given performance measure for each design. For example, if we are simulating a queueing system, and one design has two servers serving at a particular rate at a service station in the system and a competing design replaces the two servers with automatic machines, we may be interested in the average holding time of a transaction. We make a run for design 1, calculate the average holding time, make a run for design 2, calculate the average holding time, and then compute the difference between the two average holding times. Then we replicate the pair of runs m times, obtaining m differences, $d_i, i = 1, 2, \ldots, m$. The mean and standard deviation of the d_i are calculated and the t distribution is used to form a $100(1-\alpha)\%$ CI on the mean difference in a manner analogous to that described earlier, yielding

$$\left[\bar{d} - \frac{t(m-1, 1-\alpha/2)s_{d_i}}{\sqrt{m}},\ \bar{d} + \frac{t(m-1, 1-\alpha/2)s_{d_i}}{\sqrt{m}}\right].$$

Whenever possible, the same random number stream(s) should be used for each design *within* a replication, so that the difference observed depends only on the design-parameter change and not the variation due to the randomness of the random variates generated. Of course, different random-number streams are used *between* the replications. This is a *variance reduction technique* (VRT) called *common random numbers* (discussed in more detail later) and is quite effective in narrowing the CI limits.

Often, we wish to compare more than two designs, which necessitates using multiple comparison techniques. There are a variety of procedures that can be of help. All systems could be compared in a pairwise fashion using the methodology for comparing only two systems. However, if the confidence level of a CI for single pair is $1 - \alpha$, and we have k pairs, the confidence associated with a statement concerning all the pairs simultaneously drops to $1 - k\alpha$ (Bonferroni inequality). Therefore if we desire the overall confidence to be $1 - \alpha$, then it is necessary to have the confidence for each pair be $1 - \alpha/k$.

Also available are a variety of ranking and selection procedures, such as selecting the best of k systems, selecting a subset of size r containing the best of the k systems, or selecting the r best of k systems. These are treated in some detail in Law and Kelton (2000).

We now turn our attention to some variance reduction techniques (VRTs). Unlike sampling from the real world, the simulation modeler has control over the randomness generated in the system. Often, purposely introducing correlation among certain of the random variates in a simulation run can reduce variance and provide narrower CIs. One example of this was mentioned above in forming a paired t CI by using *common random numbers* (CRNs) within a replication, which introduces positive correlation between the two performance measures within a replication, yielding a smaller variance of the mean difference over the replications.

Another technique, called *antithetic variates* (AVs), introduces negative correlation between two successive replications of a given design with the idea that a large random value in one of the pairs will be offset by a small random value in the other. The performance measures for the pairs are averaged to give a single "observed" performance measure. Hence if m replications are run, only $m/2$ independent values end up being averaged for the CI calculation, but the variance of these values should be considerably lower than for m independent observations.

Among other variance reduction techniques are *indirect estimation, conditioning, importance sampling*, and *control variates*, and again we refer the reader to one of the basic simulation texts.

One drawback mentioned at the beginning of this simulation section was the difficulty in finding an optimal system design. Multiple comparison procedures will help us in finding the best of those tried, but finding the true optimal design is quite difficult. There has been quite a bit of attention paid lately to sensitivity analyses and optimization, and we refer the reader to Rubinstein (1986), Rubinstein and Melamed (1998), and Andradottir (1998). In spite of many excellent researchers working in this area, it is still a very complex and difficult task, much more so than for analytical modeling.

We end this discussion with the very important topic of *model validation.*

8.3.5 Model Validation

Model validation is a very important step in a simulation study, which often is glossed over by modelers. Prior to embarking on developing a simulation model, it behooves the simulation analyst to become very familiar with the system being studied, to involve the managers and operating personnel of the system, and thus to agree on the level of detail required to achieve the goal of the study. The appropriate level of detail is always the coarsest that can still provide the answers required. One problem with simulation modeling is that since any level of detail can be modeled, models are often developed in more detail than necessary and this can be very inefficient and counterproductive.

Validity is closely associated with *verification* and *credibility*. Verification has to do with program debugging to make sure the computer program does what is intended. This is generally the most straightforward of the three goals to accomplish, as there are well-known and established methods for debugging computer programs.

Validation deals with how accurate a representation of reality the model provides, and credibility deals with how believable the model is to the users. To establish validity and credibility, users must be involved in the study early and often. Goals of the study, appropriate system performance measures, and level of detail must be agreed upon and kept as simple as possible. A log book of assumptions should be kept, updated frequently, and signed off periodically by the model builders and users.

When possible, simulation model output should be checked against actual system performance, if the system being modeled is in operation. If the model can duplicate (in a statistical sense) *actual* data, both validity and credibility are advanced. If no system currently exists, then if the model can be run under conditions where theoretical results are known (e.g., in studying a queueing system, one can compare the simulation results with known queueing-theoretic results), and if the simulation results duplicate theoretical results, then verification is confirmed. The model can be run under a variety of conditions, and results examined by the users for plausibility, thus providing some validity and credibility checks. Most simulation texts have at least one chapter devoted to this important topic. Other references include Carson (1986), Law (2005), Gass and Thompson (1980), Sargent (2003), and Schruben (1980).

PROBLEMS

8.1. For a machine-repair problem, $M = Y = c = 1$, $\lambda = 1$, and $\mu = 1.5$, find $p_2(t)$, $p_1(t)$, and $p_0(t)$ for $t = \frac{1}{2}$ and 1, where $p_0(0) = 1$ and $p_2(0) = p_1(0) = 0$:

(a) Exactly by using the Laplace transform.

(b) By Euler's method using $\Delta t = 0.10$.

(c) By the randomization procedure using $\epsilon = 0.01$.

8.2. For the previous problem, find the approximate steady-state probability distribution by:

(a) Jacobi stepping on the uniformized imbedded discrete-parameter Markov chain with transition probability matrix $\tilde{\boldsymbol{P}}$.

(b) Gauss–Seidel stepping using $\tilde{\boldsymbol{P}}$.

(c) In view of the results of the randomization technique used in Problem 8.1(c), comment on the merit, in this case, of using it to obtain $\boldsymbol{p}(t)$ for t suitably large.

8.3. Show that the CDF $F(t)$ for the Erlang distribution can be written as (8.36).

8.4. Show that the hazard rate function $h(t)$ for the Erlang distribution can be written as (8.37).

8.5. Use a pseudorandom-number generator from your spreadsheet software to create five observations from the following distributions:
(a) Uniform between 5 and 15.
(b) Exponential, mean 5.
(c) Erlang type 3, mean 5.

8.6. Use a pseudorandom number generator to create ten observations from the following distributions:
(a) Mixed exponential, with mixing probabilities $(\frac{1}{3}, \frac{2}{3})$ and subpopulation means of 5 and 10.
(b) Gamma, with mean 5 and variance 10.

8.7. Use a pseudorandom-number generator to create five observations from the following distributions:
(a) The triangular distribution, where

$$f(x) = \begin{cases} 2x/3 & (0 \le x \le 1), \\ 1 - x/3 & (1 \le x \le 3). \end{cases}$$

(b) Poisson, with mean 2.
(c) The discrete distribution given by

Value:	1	3	4	5
Probability:	.1	.3	.2	.4

8.8. Use the $G/G/1$ simulation model in QtsPlus to simulate the following single-server queue. Estimate L, L_q, W, and W_q.

Interarrival time, min :	4	5	6
Probability of occurrence :	.1	.3	.6
Service time, min :	4	5	6
Probability of occurrence :	.5	.3	.2

8.9. Write an event-oriented simulation program, in a computer language of your choice, giving the expected system size and waiting time in queue for a single-channel queueing model where interarrival times and service times are generated from subroutines to be specified by the user. Check the program by using exponential interarrival and service times and comparing results with known results for the $M/M/1$ queue.

8.10. Write a $G/G/1$ simulator in a computer language of your choice. The simulator should be able to simulate an arbitrary number of customers. Determine waiting times for an $M/G/1$ queue. Use any service-time distribution (besides exponential) with ρ less than one. Start with a run size of

10,000 customers. Compare your answer with that expected from the PK formula. Then increase your run size and continue comparisons. What is your conclusion?

8.11. Use a simulation language (e.g., ProModel, Arena) to program the model of Problem 8.9. Compare programming effort.

8.12. Write a simulator in a computer language of your choice for a single-server machine-repair problem, with exponential lifetimes of mean 2, exponential service times with mean 2, and four machines.

8.13. Suppose you have programmed a general single-channel queueing simulator that allows for any input and service patterns. To validate the model, you decide to make runs with exponential interarrival times (mean 10) and exponential service times (mean 8). The following are results of average system-size calculations for 20 replications:

$$5.21, 3.63, 4.18, 2.10, 4.05, 3.17, 4.42, 4.91, 3.79, 3.01,$$
$$3.71, 2.98, 4.31, 3.27, 3.82, 3.41, 5.00, 3.26, 3.19, 3.63.$$

Based on these values, what can you conclude?

8.14. Write a simulation program for estimating the mean stationary waiting time of a $G/G/1$ queue with interarrival and service times distributed as follows. The interarrival CDF is the mixed exponential

$$A(t) = 1 - \frac{e^{-t}}{2} - \frac{e^{-2t}}{2},$$

while service times are distributed as

$$B(t) = 1 - e^{-(4n/3)t},$$

where n is the number in the system at the instant service begins.

8.15. Table 8.7 shows a comparison of two alternative designs for a queueing system. The table gives the mean waiting times observed for each design based on simulation of the system. Fifteen replications were conducted, and the two designs were compared on the same random number stream for each replication. Does it appear that one design is preferable?

8.16. A third design is proposed for Problem 8.15. Results for 15 replications (using the same random number streams as before) are presented in Table 8.8. Is design 3 any better than design 2? Comment from both a statistical and a practical point of view.

8.17. Calculate the LST of a hyperexponential distribution, with PDF

$$f(t) = p_1\lambda_1 e^{-\lambda_1 t} + (1 - p_1)\lambda_2 e^{-\lambda_2 t}.$$

Table 8.7 Data for Problem 8.15

Replication Number	Mean Waiting Time Design 1	Mean Waiting Time Design 2
1	23.02	23.97
2	25.16	24.98
3	19.47	21.63
4	19.06	20.41
5	22.19	21.93
6	18.47	20.38
7	19.00	21.97
8	20.57	21.31
9	24.63	23.17
10	23.91	23.09
11	27.19	26.93
12	24.61	24.82
13	21.22	22.18
14	21.37	21.99
15	18.78	20.61

Table 8.8 Data for Problem 8.16

Replications 1–5:	23.91, 24.95, 21.52, 20.37, 21.90,
Replications 6–10:	20.17, 21.90, 21.26, 23.10, 23.02,
Replications 11–15:	26.90, 24.67, 22.09, 21.91, 20.60.

8.18. Consider a two-phase Erlang distribution, with PDF $f(t) = 4\mu^2 te^{-2\mu t}$. This is also the convolution of two exponential distributions, each with mean $1/2\mu$. Calculate the Laplace–Stieltjes transform (LST) of this Erlang distribution two different ways:

(a) Using the definition of the LST and direct integration; see (C.3) in Appendix C.

(b) Using the convolution property of transforms.

8.19. From (8.17), show that

$$\text{Im}\left[\frac{1}{2\pi}\int_{-\infty}^{\infty} e^{bt}(\cos ut + i\sin ut)\bar{f}(b+iu)\,du\right] = 0.$$

[*Hint*: Follow the derivation of (8.18) and (8.19).]

8.20. Show the following:

(a) $\text{Re}(\bar{f}(a-bi)) = \text{Re}(\bar{f}(a+bi))$.

(b) $\text{Im}(\bar{f}(a-bi)) = -\text{Im}(\bar{f}(a+bi))$.

8.21. From Example 8.2, show that

$$f(t) = \frac{2e^{bt}}{\pi}\int_0^\infty \frac{\lambda(b+\lambda)}{(b+\lambda)^2+u^2}\cos(ut)\,du = \lambda e^{-\lambda t}.$$

(a) Show that the integral can be written using a complex variable z:

$$f(t) = \text{Re}\left[\frac{e^{bt}}{\pi}\int_{-\infty}^{\infty}\frac{\lambda(b+\lambda)}{(b+\lambda)^2+z^2}e^{izt}\,dz\right].$$

(b) Argue that this integral is approximately

$$f(t) \approx \text{Re}\left[\frac{e^{bt}}{\pi}\int_{\gamma_R}\frac{\lambda(b+\lambda)}{(b+\lambda)^2+z^2}e^{izt}\,dz\right],$$

where γ_R is a large semicircular contour with radius R and a base extending along the real axis from $-R$ to R (covering the positive imaginary plane), for large R.

(c) Find the poles and residues of the integrand.

(d) The residue theorem from complex variables states that the contour integral $\int_{\gamma_R}$ equals $2\pi i$ times the sum of the residues of the poles within the contour. Using this fact, derive the integral.

8.22. In the development of the Fourier-series method for numerical transform inversion, show that $\text{Re}[\bar{f}(b-iu)] = \text{Re}[\bar{f}(b+iu)]$ and $\text{Re}[\bar{f}(b-iu)] = \text{Re}[\bar{f}(b+iu)]$.

8.23. In the development of the Fourier-series method for numerical transform inversion, show that the imaginary part of (8.17) is zero. That is, show that

$$\text{Im}\left[\frac{1}{2\pi}\int_{-\infty}^{\infty} e^{bt}(\cos ut + i\sin ut)\bar{f}(b+iu)\,du\right] = 0.$$

(This requires the result from Problem 8.22.)

REFERENCES

ABATE, J., CHOUDHURY, G., AND WHITT, W. 1999. An introduction to numerical transform inversion and its application to probability models. In *Computational Probability*, W. Grassman, Ed. Kluwer, Boston, 257–323.

ABATE, J., CHOUDHURY, G. L., AND WHITT, W. 1996. On the Laguerre method for numerically inverting Laplace transforms. *INFORMS Journal on Computing 8*, 413–427.

ABATE, J., CHOUDHURY, G. L., AND WHITT, W. 1997. Numerical inversion of multidimensional Laplace transforms by the Laguerre method. *Performance Evaluation 31*, 229–243.

ABATE, J. AND WHITT, W. 1989. Calculating time-dependent performance measures for the $M/M/1$ queue. *IEEE Transactions on Communications 37*, 10, 1102–1104.

ABATE, J. AND WHITT, W. 1992. The Fourier-series method for inverting transforms of probability distributions. *Queueing Systems 10*, 5–88.

ABATE, J. AND WHITT, W. 2006. A unified framework for numerically inverting Laplace transforms. *INFORMS Journal on Computing 18*, 4, 408–421.

ABRAMOWITZ, M. AND STEGUN, I. A. 1964. Handbook of mathematical functions. In *Applied Mathematics Series 55*. National Bureau of Standards, Washington, DC.

ALBIN, S. L. 1984. Approximating a point process by a renewal process, II: Superposition arrival processes to queues. *Operations Research 32*, 5, 1133–1162.

ALLEN, A. O. 1990. *Probability, Statistics, and Queueing Theory with Computer Science Applications*, 2nd ed. Academic Press, New York.

By D. Gross, J. F. Shortle, J. M. Thompson, and C. M. Harris
Copyright © 2008 John Wiley & Sons, Inc.

ANDRADOTTIR, S. 1998. A review of simulation optimization techniques. In *Proceedings of the 1998 Winter Simulation Conference*. Washington, DC, 151–158.

ARTALEJO, J. R. 1999. Retrial queues. *Mathematical and Computer Modeling 30*, 1–6.

ASHOUR, S. AND JHA, R. D. 1973. Numerical transient-state solutions of queueing systems. *Simulation 21*, 117–122.

ASMUSSEN, S. 2003. *Applied Probability and Queues*, 2nd ed. Springer, New York.

AVI-ITZHAK, B. AND NAOR, P. 1963. Some queuing problems with the service station subject to breakdown. *Operations Research 11*, 3, 303–320.

BAILEY, N. T. J. 1954. A continuous time treatment of a single queue using generating functions. *Journal of the Royal Statistical Society: Series B 16*, 288–291.

BANKS, J., CARSON, J. S., AND NELSON, B. L. 2005. *Discrete-Event System Simulation*, 4 ed. Prentice-Hall, Upper Saddle River, NJ.

BARBOUR, A. D. 1976. Networks of queues and the methods of stages. *Advances in Applied Probability 8*, 584–591.

BARLOW, R. E. AND PROSCHAN, F. 1975. *Statistical Theory of Reliability and Life Testing*. Holt, Rinehart and Winston, New York.

BASKETT, F., CHANDY, K. M., MUNTZ, R. R., AND PALACIOS, F. G. 1975. Open, closed and mixed networks of queues with different classes of customers. *Journal of the Association for Computing Machinery 22*, 248–260.

BAXTER, G. AND DONSKER, M. D. 1957. On the distribution of the supremum functional for processes with stationary independent increments. *Transactions of the American Mathematical Society 85*, 1, 73–87.

BENEŠ, V. E. 1957. A sufficient set of statistics for a simple telephone exchange model. *Bell System Technical Journal 36*, 939–964.

BENGTSSON, B. 1983. On some control problems for queues. *Linköping Studies in Science and Technology, Dissertation No. 87*.

BERMAN, M. AND WESTCOTT, M. 1983. On queueing systems with renewal departure processes. *Advances in Applied Probability 15*, 657–673.

BHAT, U. N., SHALABY, M., AND FISCHER, M. J. 1979. Approximation techniques in the solution of queueing problems. *Naval Research Logistics Quarterly 26*, 311–326.

BILLINGSLEY, P. 1961. *Statistical Inference for Markov Processes*. University of Chicago Press, Chicago.

BILLINGSLEY, P. 1995. *Probability and Measure*, 3rd ed. Wiley, New York.

BODILY, S. E. 1986. Spreadsheet modeling as a stepping stone. *Interfaces 16*, 5 (September-October), 34–52.

BOLCH, G., GREINER, S., DE MEER, H., AND TRIVEDI, K. S. 2006. *Queueing Networks and Markov chains*, 2nd ed. Wiley, Hoboken, NJ.

BOOKBINDER, J. H. AND MARTELL, P. L. 1979. Time-dependent queueing approach to helicopter allocation for forest fire initial attack. *Infor. 17*, 58–70.

BOTTA, R. F. AND HARRIS, C. M. 1980. Approximation with generalized hyperexponential distribution: Weak convergence results. *Queueing Systems 1*, 169–190.

BRIGHAM, G. 1955. On a congestion problem in an aircraft factory. *Journal of the Operations Research Society of America 3*, 412–428.

BRILL, P. H. 2008. *Level Crossing Methods in Stochastic Models*. Springer, New York.

BRUELL, S. C. AND BALBO, G. 1980. *Computational Algorithm for Closed Queueing Networks*. North Holland, Operating and Programming Systems Series, P. J. Denning (Ed.). New York, Oxford.

BRUMELLE, S. L. 1971. Some inequalities for parallel-server queues. *Operations Research 19*, 402–413.

BUNDAY, B. D. AND SCRATON, R. E. 1980. The *G/M/r* machine interference model. *European Journal of Operational Research 4*, 399–402.

BURKE, P. J. 1956. The output of a queueing system. *Operations Research 4*, 699–714.

BURKE, P. J. 1969. The dependence of service in tandem *M/M/s* queues. *Operations Research 17*, 754–755.

BUZEN, J. P. 1973. Computational algorithms for closed queueing networks with exponential servers. *Communications of the ACM 16*, 527–531.

CARSON, J. S. 1986. Convincing users of model's validity is challenging aspect of modeler's job. *Industrial Engineering 18*, 74–85.

ÇINLAR, E. 1972. Superposition of point processes. In *Stochastic Point Processes: Statistical Analysis, Theory, and Applications*, P. A. W. Lewis, Ed. Wiley, Hoboken, NJ, 549–606.

ÇINLAR, E. 1975. *Introduction to Stochastic Processes*. Prentice-Hall, Englewood Cliffs, NJ.

CHAMPERNOWNE, D. G. 1956. An elementary method of solution of the queueing problem with a single server and a constant parameter. *Journal of the Royal Statistical Society: Series B 18*, 125–128.

CHAUDHRY, M. L., HARRIS, C. M., AND MARCHAL, W. G. 1990. Robustness of rootfinding in single-server queueing models. *ORSA Journal on Computing 2*, 273–286.

CHAUDHRY, M. L. AND TEMPLETON, J. G. C. 1983. *A First Course in Bulk Queues*. Wiley, Hoboken, NJ.

CLARKE, A. B. 1957. Maximum likelihood estimates in a simple queue. *The Annals of Mathematical Statistics 28*, 1036–1040.

COBHAM, A. 1954. Priority assignment in waiting line problems. *Operations Research 2*, 70–76; correction, **3**, 547.

COHEN, J. W. 1982. *The Single Server Queue*, 2nd ed. North Holland, New York.

COOPER, R. B. 1981. *Introduction to Queueing Theory*, 2nd ed. North Holland, New York.

COOPER, R. B. AND GROSS, D. 1991. On the convergence of Jacobi and Gauss–Seidel iteration for steady-state probabilities of finite-state continuous-time Markov chains. *Stochastic Models 7*, 185–189.

COX, D. R. 1955. A use of complex probabilities in the theory of stochastic processes. *Proceedings of the Cambridge Philosophical Society 51*, 313–319.

COX, D. R. 1965. Some problems of statistical analysis connected with congestion. In *Proceedings of the Symposium on Congestion Theory*, W. L. Smith and W. E. Wilkinson, Eds. University of North Carolina Press, Chapel Hill, NC.

CRABILL, T. B. 1968. Sufficient conditions for positive recurrence of specially structured Markov chains. *Operations Research 16*, 858–867.

CRABILL, T. B., GROSS, D., AND MAGAZINE, M. 1977. A classified bibliography of research on optimal design and control of queues. *Operations Research 28*, 219–232.

CRANE, M. A. AND IGLEHART, D. L. 1975. Simulating stable stochastic systems, III: Regenerative processes and discrete-event simulations. *Operations Research 23*, 33–45.

CROMMELIN, C. D. 1932. Delay probability formulae when the holding times are constant. *P. O. Electrical Engineering Journal 25*, 41–50.

DE SMIT, J. H. A. 1973. Some general results for many server queues. *Advances in Applied Probability 5*, 153–169.

DISNEY, R. L. 1981. Queueing networks. *American Mathematical Society Proceedings of the Symposium on Applied Mathematics 25*, 53–83.

DISNEY, R. L. 1996. Networks of queue. In *Encyclopedia of Operations Research & Management Science*, S. I. Gass and C. M. Harris, Eds. Kluwer Academic, Boston.

DISNEY, R. L., MCNICKLE, D. C., AND SIMON, B. 1980. The *M/G/*1 queue with instantaneous Bernoulli feedback. *Naval Research Logistics Quarterly 27*, 635–644.

ERLANG, A. K. 1909. The theory of probabilities and telephone conversations. *Nyt Tidsskrift Mat. B 20*, 33–39.

ERLANG, A. K. 1917. Solution of some problems in the theory of probabilities of significance in automatic telephone exchanges. *Electroteknikeren 13*, 5–13. English translation, *P. O. Elec. Eng. J.* **10**, 189-197 (1917-1918).

FABENS, A. T. 1961. The solution of queueing and inventory models by semi-Markov processes. *Journal of the Royal Statistical Society: Series B 23*, 113–127.

FALIN, G. I. AND TEMPLETON, J. G. C. 1997. *Retrial Queues*. Chapman & Hall, New York.

FELLER, W. 1957. *An Introduction to Probability Theory and Its Applications*. Vol. 1. Wiley, Hoboken, NJ.

FELLER, W. 1971. *An Introduction to Probability Theory and Its Applications*, 2nd ed. Vol. 2. Wiley, Hoboken, NJ.

FISHMAN, G. S. 1971. Estimating sample size in computer simulation experiments. *Management Science 18*, 21–38.

FISHMAN, G. S. 1973a. *Concepts and Methods in Discrete Event Digital Simulation*. Wiley, New York.

FISHMAN, G. S. 1973b. Statistical analysis for queueing simulations. *Management Science 20*, 363–369.

FISHMAN, G. S. 2001. *Discrete-Event Simulation Modeling, Programming and Analysis*. Springer-Verlag, New York.

FOSTER, F. G. 1953. On stochastic matrices associated with certain queuing processes. *The Annals of Mathematical Statistics 24*, 355–360.

FOX, B. L. 1981. Fitting "standard" distributions to data is necessarily good: Dogma or myth. *Proceedings of the 1981 Winter Simulation Conference Institute of Electrical and Electronic Engineers, Piscataway, NJ*, 305–307.

FRY, T. C. 1928. *Probability and Its Engineering Uses*. Van Nostrand, Princeton, NJ.

GANS, N., KOOLE, G., AND MANDELBAUM, A. 2003. Telephone call centers: Tutorial, review, and research prospects. *Manufacturing and Service Operations Management 5*, 79–141.

GASS, S. I. AND THOMPSON, B. W. 1980. Guidelines for model evaluation. *Operations Research 28*, 431–439.

GAVER, D. P. 1966. Observing stochastic processes and approximate transform inversion. *Operations Research 14*, 444–459.

GAVER, D. P., J. 1968. Diffusion approximations and models for certain congestion problems. *Journal of Applied Probability 5*, 607–623.

GEBHARD, R. F. 1967. A queueing process with bilevel hysteretic service-rate control. *Naval Research Logistics Quarterly 14*, 55–68.

GELENBE, E. AND PUJOLLE, G. 1998. *Introduction to Queueing Networks*, 2nd ed. Wiley, New York.

GORDON, W. J. AND NEWELL, G. F. 1967. Closed queuing systems with exponential servers. *Operations Research 15*, 254–265.

GRADSHTEYN, I. S. AND RYZHIK, I. M. 2000. *Table of Integrals, Series, and Products*, 6th ed. Academic Press, New York.

GRASSMANN, W. 1977. Transient solutions in Markovian queueing systems. *Computers and Operations Research 4*, 47–56.

GREENBERG, I. 1973. Distribution-free analysis of $M/G/1$ and $G/M/1$ queues. *Operations Research 21*, 629–635.

GROSS, D. AND HARRIS, C. M. 1985. *Fundamentals of Queueing Theory*, 2nd ed. Wiley, Hoboken, NJ.

GROSS, D. AND INCE, J. 1981. The machine repair problem with heterogeneous populations. *Operations Research 29*, 532–549.

GROSS, D. AND JUTTIJUDATA, M. 1997. Sensitivity of output measures to input distributions in queueing simulation modeling. In *Proceedings of the 1997 Winter Simulation Conference*. Institute of Electrical and Electronics Engineers, Piscataway, New Jersey.

GROSS, D., KIOUSSIS, L. C., MILLER, D. R., AND SOLAND, R. M. 1984. Computational aspects of determining steady-state availability for Markovian multi-echelon repairable item inventory models. Tech report, imse, The George Washington University, Washington, DC.

GROSS, D. AND MILLER, D. R. 1984. The randomization technique as a modeling tool and solution procedure for transient Markov processes. *Operations Research 32*, 343–361.

GROSS, D., MILLER, D. R., AND SOLAND, R. M. 1983. A closed queueing network model for multi-echelon repairable item provisioning. *IIE Transactions 15*, 344–352.

GROSSMAN, T. A. 1999. Teacher's forum: spreadsheet modeling and simulation improves understanding of queues. *Interfaces 29*, 3 (May–June), 88–103.

HALFIN, S. AND WHITT, W. 1981. Heavy-traffic limits for queues with many exponential servers. *Operations Research 29*, 3, 567–588.

HARRIS, C. M. 1974. Some new results in the statistical analysis of queues. In *Mathematical Methods in Queueing Theory*, A. B. Clarke, Ed. Springer-Verlag, Berlin.

HARRIS, C. M. 1985. A note on mixed exponential approximations for $GI/G/1$ queues. *Computers and Operations Resesarch 12*, 285–289.

HARRIS, C. M. AND MARCHAL, W. G. 1988. State dependence in $M/G/1$ server-vacation models. *Operations Research 36*, 560–565.

HEYMAN, D. P. 1968. Optimal operating policies for $M/G/1$ queuing systems. *Operations Research 16*, 362–382.

HEYMAN, D. P. AND SOBEL, M. J. 1982. *Stochastic Models in Operations Research.* Vol. I. McGraw-Hill, New York.

HEYMAN, D. P. AND SOBEL, M. J. 1984. *Stochastic Models in Operations Research.* Voi. II. McGraw-Hill, New York.

HILLIER, F. S. AND LIEBERMAN, G. J. 1995. *Introduction to Operations Research*, 6th ed. McGraw-Hill, New York.

HUNT, G. C. 1956. Sequential arrays of waiting lines. *Operations Research 4*, 674–683.

JACKSON, J. R. 1957. Networks of waiting lines. *Operations Research 5*, 518–521.

JACKSON, J. R. 1963. Jobshop-like queueing systems. *Management Science 10*, 131–142.

JAGERMAN, D. L. 1978. An inversion technique for the Laplace transform with applications. *Bell System Technical Journal 57*, 669–710.

JAGERMAN, D. L. 1982. An inversion technique for the Laplace transform. *Bell System Technical Journal 61*, 1995–2002.

JAISWAL, N. K. 1968. *Priority Queues.* Academic Press, New York.

JUTTIJUDATA, M. 1996. Sensitivity of output performance measures to input distributions in queueing simulation modeling. Ph.D. thesis, Department of Operations Research, The George Washington University, Washington, DC.

KAO, E. P. C. 1991. Using state reduction for computing steady state probabilities of $GI/PH/1$ types. *ORSA Journal on Computing 3*, 231–240.

KARLIN, S. AND TAYLOR, H. M. 1975. *A First Course on Stochastic Processes.* Academic Press, New York.

KEILSON, J., COZZOLINO, J., AND YOUNG, H. 1968. A service system with unfilled requests repeated. *Operations Research 16,* 6, 1126–1137.

KELLY, F. P. 1975. Networks of queues with customers of different types. *Journal of Applied Probability 12*, 542–55.

KELLY, F. P. 1976. Networks of queues. *Advances in Applied Probability 8*, 416–432.

KELLY, F. P. 1979. *Reversibility and Stochastic Networks.* Wiley, Hoboken, NJ.

KELTON, W. D. 1984. Input data collection and analysis. *Proceedings of the 1984 Winter Simulation Conference Institute of Electrical and Electronics Engineers, Piscataway, NJ*, 305–307.

KELTON, W. D. AND LAW, A. M. 1983. A new approach for dealing with the startup problem in discrete event simulation. *Naval Research Logistics Quarterly 30*, 641–658.

KENDALL, D. G. 1953. Stochastic processes occurring in the theory of queues and their analysis by the method of imbedded Markov chains. *The Annals of Mathematical Statistics 24*, 338–354.

KENNEDY, D. P. 1972. The continuity of the single-server queue. *Journal of Applied Probability 9*, 370–381.

KESTEN, H. AND RUNNENBURG, J. T. 1957. Priority waiting line problems I, II. *Koninkz. Ned. Akad. Wetenschap. Proc. Ser. A 60*, 312–336.

KIM, S. 2004. The heavy-traffic bottleneck phenomenon under splitting and superposition. *European Journal of Operations Research 157*, 736–745.

KINGMAN, J. F. C. 1962a. On queues in heavy traffic. *Journal of the Royal Statistical Society. Series B (Methodological) 24*, 2, 383–392.

KINGMAN, J. F. C. 1962b. Some inequalities for the queue $GI/G/1$. *Biometrika 49*, 315–324.

KOENIGSBERG, E. 1966. On jockeying in queues. *Management Science 12*, 412–436.

KOLESAR, P. AND GREEN, L. 1998. Insights on service system design from a normal approximation to Erlang's formula. *Production and Operations Management 7*, 3, 282–293.

KOSTEN, L. 1948. On the validity of the Erlang and Engset loss formulae. *Het P.T.T. Bedriff 2*, 22–45.

KRAEMER, W. AND LANGENBACH-BELZ, M. 1976. Approximate formulae for the delay in the queueing system $GI/G/1$. *Congressbook, Eighth International Teletraffic Congress*, 235.1–235.8.

KRAKOWSKI, M. 1973. Conservation methods in queueing theory. *RAIRO 7 V-1*, 63–84.

KRAKOWSKI, M. 1974. Arrival and departure processes in queues. Pollaczek-Khintchine formulas for bulk arrivals and bounded systems. *RAIRO 8 V-1*, 45–56.

LAVENBERG, S. S. AND REISER, M. 1979. Stationary state probabilities at arrival instants for closed queueing networks with multiple types of customers. Research Report RC 759, IBM T. J. Watson Research Center, Yorktown Heights, NY.

LAW, A. M. 1977. Confidence intervals in discrete event simulation: a comparison of replication and batch means. *Naval Research Logistics Quarterly 27*, 667–678.

LAW, A. M. 2005. How to build credible and valid simulation models. *Proceedings of the 2005 Winter Simulation Conference Orlando*, 27–32.

LAW, A. M. AND KELTON, W. D. 2000. *Simulation Modeling and Analysis*, 3rd ed. McGraw-Hill, New York.

L'ECUYER, P. 2006. Random number generation. In *Elsevier Handbooks in Operations Research and Management Science: Simulation*, S. G. Henderson and B. Nelson, Eds. Chap. 3.

LEDERMANN, W. AND REUTER, G. E. 1954. Spectral theory for the differential equations of simple birth and death process. *Philosophical Transactions of the Royal Society of London Series A 246*, 321–369.

LEEMIS, L. M. 1996. Discrete-event simulation input process modeling. *Proceedings 1996 Winter Simulation Conference San Diego, CA, Institute of Electrical and Electronics Engineers, Piscataway, NJ*, 39–46.

LEEMIS, L. M. AND PARK, S. K. 2006. *Discrete-Event Simulation, A First Course*. Prentice Hall, Upper Saddle River, NJ.

LEMOINE, A. J. 1977. Networks of queues—a survey of equilibrium analysis. *Management Science 24*, 464–481.

LEON, L., PRZASNYSKI, Z., AND SEAL, K. C. 1996. Spreadsheets and OR/MS models: an end-user perspective. *Interfaces 26,* 2 (March-April), 92–104.

LIITSCHWAGER, J. AND AMES, W. F. 1975. On transient queues—practice and pedagogy. 206.

LILLIEFORS, H. W. 1966. Some confidence intervals for queues. *Operations Research 14,* 723–727.

LILLIEFORS, H. W. 1967. On the Kolmogorov-Smirnov statistic for normality with mean and variance unknown. *Journal of the American Statistical Association 62,* 399–402.

LILLIEFORS, H. W. 1969. On the Kolmogorov-Smirnov statistic for the exponential distribution with mean unknown. *Journal of the American Statistical Association 64,* 387–389.

LINDLEY, D. V. 1952. The theory of queues with a single server. *Proceedings of the Cambridge Philosophical Society 48,* 277–289.

LITTLE, J. D. C. 1961. A proof for the queuing formula $L = \lambda W$. *Operations Research 9,* 383–387.

MARCHAL, W. G. 1978. Some simpler bounds on the mean queuing time. *Operations Research 26,* 1083–1088.

MARON, M. J. 1982. *Numerical Analysis, A Practical Approach.* Macmillan, New York.

MARSHALL, K. T. 1968. Some inequalities in queuing. *Operations Research 16,* 651–665.

MELAMED, B. 1979. Characterization of Poisson traffic streams in Jackson queueing networks. *Advances in Applied Probability 11,* 422–438.

MILLER, D. R. 1981. Computation of the steady-state probabilities for $M/M/1$ priority queues. *Operations Research 29,* 945–958.

MODER, J. J. AND PHILLIPS, C. R., J. 1962. Queuing with fixed and variable channels. *Operations Research 10,* 218–231.

MOLINA, E. C. 1927. Application of the theory of probability to telephone trunking problems. *Bell Syst. Tech. J. 6,* 461–494.

MORSE, P. M. 1958. *Queues, Inventories and Maintenance.* Wiley, New York.

NEUTS, M. F. 1973. The single server queue in discrete time—numerical analysis, I. *Naval Research Logistics Quarterly 20,* 297–304.

NEUTS, M. F. 1981. *Matrix-Geometric Solutions in Stochastic Models.* Johns Hopkins University Press, Baltimore.

NEWELL, G. F. 1972. *Applications of Queueing Theory.* Chapman & Hall, London.

PALM, C. 1938. Analysis of the Erlang traffic formulae for busy-signal arrangements. *Ericsson Tech. 6,* 39–58.

PAPOULIS, A. 1991. *Probability, Random Variables and Stochastic Processes,* 2nd ed. McGraw-Hill, New York.

PARZEN, E. 1960. *Modern Probability and Its Applications.* Wiley, Hoboken, NJ.

PARZEN, E. 1962. *Stochastic Processes.* Holden-Day, San Francisco.

PERROS, H. 1994. *Queueing Networks with Blocking.* Oxford University Press, New York.

PHIPPS, T. E., J. 1956. Machine repair as a priority waiting-line problem. *Operations Research 4,* 76–85. (Comments by W. R. Van Voorhis, *4,* 86).

PIESSENS, R. 1975. A bibliography on numerical inversion of the Laplace transform and applications. *Journal of Computational and Applied Mathematics 1*, 115–128.

PIESSENS, R. AND DANG, N. D. P. 1976. A bibliography on numerical inversion of the Laplace transform and applications: a supplement. *Journal of Computational and Applied Mathematics 2*, 225–228.

POLLACZEK, F. 1932. Lösung eines geometrischen wahrscheinlichkeits-problems. *Math. Z. 35*, 230–278.

POSNER, M. AND BERNHOLTZ, B. 1968. Closed finite queueing networks with time lags. *Operations Research 16*, 962–976.

PRABHU, N. U. 1965a. *Queues and Inventories.* Wiley, Hoboken, NJ.

PRABHU, N. U. 1965b. *Stochastic Processes.* Macmillan, New York.

PRABHU, N. U. 1974. Stochastic control of queueing systems. *Naval Research Logistics Quarterly 21*, 411–418.

PUTERMAN, M. L. 1991. *Markov Decision Processes.* Wiley, Hoboken, NJ.

RAINVILLE, E. D. AND BEDIENT, P. E. 1969. *A Short Course in Differential Equations.* Macmillan, New York.

RAO, S. S. 1968. Queueing with balking and reneging in $M/G/1$ systems. *Metrika 12*, 173–188.

REICH, E. 1957. Waiting times when queues are in tandem. *The Annals of Mathematical Statistics 28*, 768–773.

RESNICK, S. I. 1992. *Adventures in Stochastic Processes.* Birkhauser, Boston.

ROMANI, J. 1957. Un modelo de la teoria de colas con número variable de canales. *Trabajos Estadestica 8*, 175–189.

ROSS, S. M. 1996. *Stochastic Processes*, 2nd ed. Wiley, New York.

ROSS, S. M. 2007. *An Introduction to Probability Models*, 9th ed. Academic Press, New York.

RUBINSTEIN, R. Y. 1986. *Monte Carlo Optimization, Simulation and Sensitivity of Queueing Networks.* Wiley, Hoboken, NJ.

RUBINSTEIN, R. Y. AND MELAMED, B. 1998. *Modern Simulation and Modeling.* Wiley, Hoboken, NJ.

RUE, R. C. AND ROSENSHINE, M. 1981. Some properties of optimal control policies for entries to an $M/M/1$ queue. *Naval Research Logistics Quarterly 28*, 525–532.

SAATY, T. L. 1961. *Elements of Queueing Theory with Applications.* McGraw Hill, New York.

SAKURAI, T. 2004. Numerical inversion for Laplace transforms of functions with discontinuities. *Advances in Applied Probability 36*, 2, 616–642.

SARGENT, R. G. 2003. Verification and validation of simulation models. *Proceedings of the 2003 Winter Simulation Conference New Orleans*, 37–48.

SCHMEISER, B. W. 1982. Batch size effects in the analysis of simulation output. *Operations Research 30*, 556–568.

SCHRAGE, L. E. AND MILLER, L. W. 1966. The queue $M/G/1$ with the shortest remaining processing time discipline. *Operations Research 14*, 670–684.

SCHRUBEN, L. W. 1980. Establishing the credibility of simulations. *Simulation 34*, 101–105.

SCHRUBEN, L. W. 1982. Detecting initialization bias in simulation output. *Operations Research 30*, 569–590.

SERFOZO, R. F. 1981. Optimal control of random walks, birth and death processes, and queues. *Advances in Applied Probability 13*, 61–83.

SERFOZO, R. F. AND LU, F. V. 1984. *M/M/*1 queueing decision processes with monotone hysteretic optimal policies. *Operations Research 32*, 1116–1132.

SEVICK, K. C. AND MITRANI, I. 1979. The distribution of queueing network states at input and output instants. In *Proceedings of the 4th International Symposium on Modelling and Performance Evaluation of Computer Systems.* Vienna.

SIMON, B. AND FOLEY, R. D. 1979. Some results on sojourn times in cyclic Jackson networks. *Management Science 25*, 1027–1034.

SMITH, W. L. 1953. On the distribution of queueing times. *Proceedings of the Cambridge Philosophical Society 49*, 449–461.

SMITH, W. L. 1959. On the cumulants of renewal processes. *Biometrika 46*, 1–29.

SOBEL, M. J. 1969. Optimal average cost policy for a queue with start-up and shut-down costs. *Operations Research 17*, 145–162.

SOBEL, M. J. 1974. Optimal operation of queues. In *Mathematical Methods in Queueing Theory*, A. B. Clarke, Ed. Lecture Notes in Economics and Mathematical Systems 98. Springer-Verlag, Berlin, 231–236.

STEHFEST, H. 1970. Algorithm 368. Numerical inversion of Laplace transforms [D5]. *Communications of the ACM 13*, 1, 47–49.

STEPHENS, M. A. 1974. Edf statistics for goodness of fit and some comparisons. *Journal of the American Statistical Association 69*, 730–737.

STIDHAM, S., J. 1970. On the optimality of single-server queuing systems. *Operations Research 18*, 708–732.

STIDHAM, S., J. 1982. Optimal control of arrivals to queues and network of queues. In *21st IEEE Conference on Decision and Control.* IEEE. in Orlando, FL.

STIDHAM, S., J. AND PRABHU, N. U. 1974. Optimal control of queueing systems. In *Mathematical Methods in Queueing Theory*, A. B. Clarke, Ed. Lecture Notes in Economics and Mathematical Systems 98. Springer-Verlag, Berline, 263–294.

SURESH, S. AND WHITT, W. 1990. The heavy-traffic bottleneck phenomenon in open queueing networks. *Operations Research Letters 9*, 355–362.

SWAIN, J. J. 2005. "Gaming" reality: biennial survey of discrete-event simulation software tools. *ORMS Today 32*, 6, 44–55.

TAKÁCS, L. 1962. *Introduction to the Theory of Queues.* Oxford University Press, Oxford, England.

TAKÁCS, L. 1969. On Erlang's formula. *The Annals of Mathematical Statistics 40*, 71–78.

VAN DIJK, N. M. 1993. *Queueing Networks and Product Forms: A Systems Approach.* Wiley, New York.

VAULOT, A. E. 1927. Extension des formules d'erlang au cas où les durées des conversations suivent une loi quelconque. *Rev. Gíeln. Electricitíel* 22, 1164–1171.

WALRAND, J. 1988. *An Introduction to Queueing Networks.* Prentice Hall, Englewood Cliffs, NJ.

WEEKS, W. T. 1966. Numerical inversion of Laplace transforms using Laguerre functions. *Journal of ACM 13*, 419–426.

WELCH, P. D. 1983. The statistical analysis of simulation results. In *The Computer Performance Modeling Handbook*, S. S. Lavenberg, Ed. Academic Press, New York.

WHITE, H. AND CHRISTIE, L. S. 1958. Queuing with preemptive priorities or with breakdown. *Operations Research 6*, 1, 79–95.

WHITT, W. 1974. The continuity of queues. *Advances in Applied Probability 6*, 175–183.

WHITT, W. 1982. Approximating a point process by a renewal process, I: Two basic methods. *Operations Research 30*, 1, 125–147.

WHITT, W. 1983. The queueing network analyzer. *The Bell System Technical Journal 62*, 9, 2779–2815.

WHITT, W. 1984. Approximations for departure processes and queues in series. *Naval Research Logistics Quarterly 31*, 499–521.

WHITT, W. 1995. Variability functions for parametric-decomposition approximations of queueing networks. *Management Science 41*, 1704–1715.

WIMP, J. 1981. *Sequence Transformations and Their Applications.* Academic Press, New York.

WOLFF, R. W. 1965. Problems of statistical inference for birth and death queuing models. *Operations Research 13*, 343–357.

WOLFF, R. W. 1989. *Stochastic Modeling and the Theory of Queues.* Prentice Hall, Englewood Cliffs, NJ.

YADIN, M. AND NAOR, P. 1967. On queueing systems with variable service capacities. *Naval Research Logistics Quarterly 14*, 43–54.

ZAKIAN, V. 1969. Numerical inversion of Laplace transform. *Electronics Letters 5*, 120–121.

ZAKIAN, V. 1970. Optimisation of numerical inversion of Laplace transforms. *Electronics Letters 6*, 677–679.

ZAKIAN, V. 1973. Properties of i_{MN} approximants. In *Padé Approximants and their Applications*, P. R. Graves-Morris, Ed. Academic Press, New York, 141–144.

APPENDIX A
SYMBOLS AND ABBREVIATIONS

This appendix contains definitions of common symbols and abbreviations used frequently and consistently throughout the text. Symbols that are used only occasionally in isolated sections of the text are not always included here. The symbols are listed in alphabetical order. Greek symbols are filed according to their English names; for example, λ (lambda) is found under L. Within an alphabetical category, Latin symbols precede Greek. Listed at the end are nonliteral symbols such as primes and asterisks.

Fundamentals of Queueing Theory, Fourth Edition.
By D. Gross, J. F. Shortle, J. M. Thompson, and C. M. Harris
Copyright © 2008 John Wiley & Sons, Inc.

$A/B/X/Y/Z$	Notation for describing queueing models, where A indicates interarriv pattern, B indicates service pattern, X indicates number of channel Y indicates system capacity limit, Z indicates queue discipline
a.s.	Almost surely
$A(t)$	Cumulative distribution function of interarrival times
$a(t)$	Probability density of interarrival times
$B(t)$	Cumulative distribution function of service times
$b(t)$	Probability density of service times
b_n	(1) Probability of n services during an interarrival time; (2) discouragement function in queueing models with balking
C	(1) Arbitrary constant; (2) cost per unit time, often used in functional notation as $C(\cdot)$; (3) coefficient of variation ($\equiv \sigma/\mu$)
c	Number of parallel channels (servers)
CDF	Cumulative distribution function
CK	Chapman–Kolmogorov
CTMC	Continuous-time Markov chain
CV	Coefficient of variation ($\equiv \sigma/\mu$)
C_s	Marginal cost of a server per unit time
c_n	Probability that the batch size is n
C_W	Cost of customer wait per unit time
$C(t)$	CDF of the interdeparture process
$C(z)$	Probability generating function of $\{c_n\}$
$c(t)$	Probability density of the interdeparture process
D	(1) Deterministic interarrival or service times; (2) linear difference operator, $Dx_n = X_{n+1}$; (3) linear differential operator, $Dy(x) = dy/dx$
$\bar{d}$	Mean observed interdeparture time of a queueing system
DFR	Decreasing failure rate
DTMC	Discrete-time Markov chain
df	Distribution function
Δy_n	First finite difference; that is, $\Delta y_n = y_{n+1} - y_n$
E_k	Erlang type-k distributed interarrival or service times
$\mathrm{E}[\cdot]$	Expected value
η_{ij}	Mean time spent in state i before going to j
FCFS	First-come, first-served queue discipline
$F_n(t)$	Joint probability that n are in the system at time t after the last depar and t is less than the interdeparture time
$F_{ij}(t)$	Conditional probability that, given that a process begins in state i an next goes to state j, the transition time is $\leq t$ (a conditional CDF)
f_{ij}	Probability that state j of a process is ever reached from state i
$f_{ij}^{(n)}$	Probability that the first passage of a process from state i to state j occurs in exactly n steps
G	General distribution for service and/or interarrival times
GCD	Greatest common divisor

GD	General queue discipline
$G(N)$	Normalizing constant in a closed network
$G(t)$	Cumulative distribution function of the busy period for $M/G/1$ and $G/M/1$ models
$G(z)$	Generating function associated with Erlang-service steady-state probabilities $\{p_{n,i}\}$
$G_j(t)$	Conditional probability that, given that a process starts in state i, the time to the next transition is $\leq t$ (a conditional CDF)
γ_i	External flow rate to node i of a network
H_k	Mixture of k exponentials used as distribution for interarrival and/or service times
H	Hyperexponential (a balanced H_2) distribution for service and/or interarrival times
$H(z, y)$	(1) Probability generating function for $\{p_{n,i}\}$; (2) joint generating function for a two-priority queueing model
$H_{ij}(t)$	Cumulative distribution function of time until first transition of a process into state j beginning at state i
$H_r(y, z)$	Generating function associated with $P_{mr}(z)$ for a two-priority queueing model
$h(u)$	Failure or hazard rate of a probability distribution
I	Phase of service the customer is in for Erlang service models (a random variable)
IFR	Increasing failure rate
IID	Independent and identically distributed
I_u	Expected useful server idle time
$I_n(\cdot)$	Modified Bessel function of the first kind
$\tilde{I}(t)$	Probability of a server being idle for a time $> t$ (a complementary CDF)
$\tilde{I}_n(t)$	Conditional probability that one of the $c - n$ idle servers remains idle for a time $> t$ (a conditional complementary CDF)
$i(t)$	Probability density of idle time
$J_n(\cdot)$	Regular Bessel function
K	System capacity limit (truncation point of system size)
K_q	Greatest queue length at which an arrival would balk (a random variable)
$K(z)$	Probability generating function of $\{k_n\}$
$K_i(z)$	Probability generating function of $\{k_{n,i}\}$
k_n	Probability of n arrivals during a service time
$k_{n,i}$	Probability of n arrivals during a service time, given i in the system when service began
L	Expected system size
LCFS	Last-come, first-served queue discipline
LST	Laplace—Stielties transform
LT	Laplace transform
$L^{(D)}$	Expected system size at departure points

$L^{(P)}$	Expected number of phases in the system of an Erlang queueing model
$L^{(n)}$	(1) Expected number of customers of type n in system; (2) expected system size at station n in a series or cyclic queue
$L_{(k)}$	The kth factorial moment of system size
L_q	Expected queue size
L_q'	Expected queue size of nonempty queues
$L_q^{(D)}$	Expected queue size at departure points
$L_q^{(P)}$	Expected number of phases in the queue of an Erlang queueing model
$L_q^{(n)}$	(1) Expected queue size for customers of type n; (2) expected queue size in front of station n in a series or cyclic queue
$L_{q(k)}^{(D)}$	The kth factorial moment of the departure-point queue size
$L(\cdot)$	Likelihood function
$\mathcal{L}(\cdot)$	(1) Log-likelihood function; (2) Laplace transform
Λ	Minimum diagonal element of $\boldsymbol{Q}$
λ	Mean arrival rate (independent of system size)
λ_n	(1) Mean arrival rate when there are n in the system; (2) mean arrival rate of customers of type n
M	(1) Poisson arrival or service process (or equivalently exponential interarrival or service times); (2) finite population size
MC	Markov chain
MGF	Moment generating function
MLE	Maximum-likelihood estimator
MOM	Method of moments (estimator)
$M_x(t)$	Moment generating function of the random variable X
m_i	Mean time a process spends in state i during a visit
m_{ij}	Mean first passage time of a process from state i to state j
m_{jj}	Mean recurrence time of a process to state j
μ	Mean service rate (independent of system size)
$\mu^{(B)}$	Mean service rate for a bulk queueing model
μ_n	(1) Mean service rate when there are n in the system; (2) mean service rate of server n; (3) mean service rate for customers of type n
N	Steady-state number in the system (a random variable)
N_q	Steady-state number in the queue (a random variable)
$N(t)$	Number in the system at time t (a random variable)
$N_q(t)$	Number in the queue at time t (a random variable)
n_a, n_{ae}, n_b	Number of observed arrivals to a system, to an empty system, and to a busy system, respectively ($n_a = n_{ae} + n_b$)
$o(\Delta t)$	Order Δt; that is, $\lim_{\Delta t \to 0} o(\Delta t)/\Delta t = 0$
ω	Expected remaining work
$\boldsymbol{P}$	Single-step transition probability matrix of a DTMC
PDE	Partial differential equation
PK	Pollaczek–Khintchine formula

PR	Priority queue discipline
$P(z), P(z,t)$	Probability generating function of $\{p_n\}$ and $\{p_n(t)\}$, respectively
$P_{mr}(z)$	Probability generating function of priority steady-state probabilities $\{P_{mnr}\}$
$\boldsymbol{p}$	Steady-state probability vector of a CTMC
p_n	(1) Steady-state probability of n in the system; (2) steady-state probability that a CTMC is in state n
$p_n^{(B)}$	Steady-state probability of n in a bulk queueing system
$p_n^{(P)}$	Steady-state probability of n phases in an Erlang queueing system
$p_n(t)$	Probability of n in the system at time t
p_{ij}	Single-step transition probability of going from state i to state j
$p_{n,i}$	Steady-state probabilities for Erlang models of n in the system and the customer in service (if service is Erlangian) or next to arrive (if arrivals are Erlangian) in phase i
$p_{ij}^{(n)}$	Transition probability of going from state i to state j in n steps
$p_{n,i}(t)$	Probability that, in an Erlang queueing model at time t, n are in the system and the customer in service (if the service is Erlangian) or next to arrive (if arrivals are Erlangian) is in phase i
$p_{mnr}(t)$	Probability at time t of m units of priority 1, n units of priority 2 in the system, and a unit of priority r in service ($r = 1$ or 2)
$p_{i,j}(u,s)$	Transition probability of moving from state i to state j in time beginning at u and ending at s
$p_{n1,n2,\ldots,n_k}(t)$	Probability of n_1 customers at station 1, n_2 at station 2, ..., n_k at station k in a series queue at time t
$p_{n1,n2,\ldots,n_k}$	Steady-state probability of $p_{n1,n2,\ldots,n_k}(t)$
p–c	Predictor–corrector
$\Pi(z)$	Probability generating function of $\{\pi_n\}$
π	Steady-state probability vector of a DTMC
π_n	(1) Steady-state probability of n in the system at a departure point; (2) steady-state probability that a DTMC is in state n
$\boldsymbol{Q}$	Infinitesimal generator matrix of a CTMC
$Q_{ij}(t)$	Joint conditional probability that, given that a process begins in state i, the next transition will be to state j in an amount of time t (a conditional CDF)
q_n	Steady-state probability that an arriving customer finds n in the system
$\boldsymbol{R}$	Network routing probability matrix
$R(t)$	Distribution function of remaining service time
Re	Real portion of a complex number
RK	Runge–Kutta
RSS	Random selection for service
RV	Random variable
r	Defined as λ/μ for multichannel models; defined as $\lambda/k\mu$ for Erlang service models

r_i	The ith root of a polynominal equation (if there is only one root, r_0 is used)
r_n	The nth uniform (0, 1) random number
r_{ij}	Routing probability in a queueing network of a customer going to station j after being served at station i
$r(n)$	Reneging function
ρ	Traffic intensity ($=\lambda/\mu$ for single-channel and *all* network models, and $=\lambda/c\mu$ for other multichannel models)
S	Steady-state service time (a random variable)
SMP	Semi-Markov process
$S^{(n)}$	Service time of the nth arriving customer (a random variable)
$S_k^{(n)}$	Service time of the nth arriving customer of type k
$S_k(S'_k)$	Time it takes to serve $n_k(n'_k)$ waiting customers of type k (a random variable)
S_0	Time required to finish customer in service (remaining time of service; a random variable)
s_n	Probability that n servers are busy ($c-n$ idle) in a multichannel system
S_X	Sample standard deviation of the random variable X
σ^2, σ_B^2	Variance of the service-time distribution
σ_A^2	Variance of the interarrival-time distribution
σ_k	Sum of traffic intensities in a priority queueing model; that is, $\sigma_k = \Sigma\lambda_i/\mu_i$
T	(1) Time spent in system (a random variable), with expected value W; (2) steady-state interarrival time (a random variable); (3) steady-state interdeparture time (a random variable)
$T^{(n)}$	Interarrival time between the nth and $(n+1)$st customers (a random variable)
T_A	Instant of arrival
T_i	Length of time a stochastic process spends in state i (a random variable)
T_S	Instant of service completion
T_{busy}	Length of a busy period (a random variable)
T_q	Time spent in queue (a random variable), with expected value W_q
$T_{b,i}$	Length of i channel busy period for $M/M/c$ (a random variable)
t_b, t_e, t	Observed time a system is busy, observed time a system is empty, and total observed time, respectively ($t = t_b + t_e$)
τ_i	Time at which the ith arrival to a Poisson process occurred
U	Steady-state difference between service time and interarrival time, $U = S - T$ (a random variable)
$U^{(n)}$	Service time of nth customer minus interarrival time between customer $n+1$ and n; that is $U^{(n)} = S^{(n)} - T^{(n)}$ (a random variable)
$U(t)$	(1) Cumulative distribution function of $U = S - T$; (2) cumulative distribution function of the time back to the most recent transition

$U^{(n)}(t)$	Cumulative distribution function of $U^{(n)}$
$U_i(t)$	Cumulative distribution function of the time back to the most recent transition, given the process starts in state i
Var[·]	Variance
V	Expected virtual wait
$V(t)$	Virtual waiting-time function
v_j	(1) Steady-state probability of a semi-Markov process being in state j; (2) relative throughput in a closed network
W	Expected waiting time in system
$W^{(n)}$	Waiting time including service at station n of a series or cyclic queue
W_k	Ordinary kth moment of waiting time in system
W_q	Expected waiting time in queue
$W_{q,k}$	Regular kth moment of waiting time in queue
$W(t)$	Cumulative distribution function of waiting time in system
$W_q^{(H)}$	Expected time in queue for a system in heavy traffic
$W_q^{(n)}$	(1) Waiting time in queue for the nth arriving customer (a random variable); (2) expected wait in queue for customers of priority class n; (3) expected time in queue at station n of a network of queues
$W_q(t)$	Cumulative distribution function of waiting time in queue
$\tilde{W}_q(t\|j)$	Probability in $M/M/c$ model that the delay undergone by an arbitrary arrival who joined when $c+j$ were in the system is more than t
$X(t)$	Stochastic process with state space X and parameter t
x, x_i	Observed interval of a queueing system, and observed interval of type i (busy, empty, etc.) of a queueing system, respectively
$[x]$	Greatest integer value $\leq x$
$\doteq$	Approximately equal to
$\sim$	Asymptotic to
$\in$	Set membership
$*$	(1) LST; (2) used for various other purposes as specifically defined in text
-	Laplace transform
$\binom{n}{c}$	Binomial coefficient, $n!/[(n-c)!c!]$
[·]	Batch queueing model
(·)	(1) Order of convolution; (2) order of differentiation, (3) number of steps (transitions) in a discrete-parameter Markov chain.
	(1) Differentiation; (2) conditional; for example, p_n' is a conditional probability distribution of n in the system given system not empty; (3) used for various other purposes as specifically defined in text
	(1) Complementary CDF; (2) used for various other purposes as specifically defined in text

APPENDIX B

TABLES

This appendix provides three tables. The first table summarizes the models treated in the book and the types of results obtained. The second and third tables summarize key results from applied probability for continuous and discrete distributions, respectively.

Fundamentals of Queueing Theory, Fourth Edition.
By D. Gross, J. F. Shortle, J. M. Thompson, and C. M. Harris
Copyright © 2008 John Wiley & Sons, Inc.

Table B.1 Summary of models treated and types of results

Model (Notation Explained in Table 1.1, Section 1.3)	Types of Results[a] for: $\{p_n\}$	Expected-Value Measures (L, L_q, W, W_q)	Waiting-Time Distribution	Section	QTS Module
		(a) Steady State			
FCFS					
$M/M/1$	a	a	a	2.2	√
$M/M/1/K$	a	a	a	2.5	√
$M/M/c$	a	a	a	2.3	√
$M/M/c/K$	a	a	a^{b}	2.5	√
$M/M/c/c$	a	a	—	2.6	√
$M/M/\infty$	a	a	—	2.7	√
Finite-source $M/M/c$	a	a	a	2.8	√
State-dep. serv. $M/M/1$	a, n^{c}	a, n^{c}	0	2.9	√
Impatience $M/M/c$	a, n^{c}	a, n^{c}	a, n^{c}	2.10	—
$M^{[X]}/M/1$	g	a	0	3.1	√
$M/M^{[Y]}/1$	a^{d}	a^{d}	0	3.2	√
$M/M^{[Y]}/c$		Results indicated		3.2	—
$M/E_k/1(M/D/1)$	g, (a)	a	0	3.3.3	√
$M/D/1$	a	a	a	5.1, 8.2.3	√
$M/D/c$	g	a	0	6.3	√
$E_k/M/1(D/M/1)$	a^{d}	n	n	3.3.4	√
$E_j/E_k/1$		Some numerical results possible—reference given		3.3.5	√
$M/M/1$ retrial	a	a	0	3.5	√

M/G/1	n	a	n	5.1–5.1.6	√
M/G/1/K	a	a	0	5.1.7	—
Impatience *M/G/1*		Results indicated—reference given		5.1.8	—
Finite source *M/G/1*		Reference given		5.1.8	—
$M^{[X]}/G/1$	g, n^{c}	a, n^{c}	0	5.1.9	—
$M/G^{[K]}/1$		Reference given		5.1.8	—
State-dep. serv. *M/G/1*	n	n	0	5.1.10	—
M/G/c		Little's relation on higher moments		5.2.1	—
M/G/c/c	a	a	—	5.2.2	√
$M/G/\infty$	a	a	—	5.2.2	√
G/M/1	a^{d}	a^{d}	a^{d}	5.3.1, 6.4	√
G/M/c	a^{d}	n	a^{d}	5.3.2	√
G/M/1/K		Results indicated—reference given		5.3.2	—
G/M/c/K		Results indicated—reference given		5.3.2	—
Impatience *G/M/1*		Results indicated—reference given		5.3.2	—
Impatience *G/M/c*		Results indicated—reference given		5.3.2	—
$G/M^{[Y]}/1$		Results indicated—reference given		5.3.2	—
$G/M^{[Y]}/c$		Results indicated—reference given		5.3.2	—
$G/E_k/1$		Results indicated—reference given		6.1	—
$G^{[K]}/M/1$				6.1	—
$G/PH_k/1$		Results indicated—reference given		6.1	—
G/G/1	0	n, bounds, approx.	n	6.2, 7.1, 7.2	√
G/G/c	0	bounds, approx.	0	7.1.3, 7.2.1	—
Priority					
M/M/1, two priorities	g	a	0	3.4.1	√
M/M/1, many priorities	0	a	0	3.4.2	√

Table B.1 (continued)

Model (Notation Explained in Table 1.1, Section 1.3)	Types of Results[a] for: $\{p_n\}$	Expected-Value Measures (L, L_q, W, W_q)	Waiting-Time Distribution	Section	QTS Module
$M/M/1$ preemptive	g	a	0	3.4.3	√
$M/G/1$, many priorities	0	0	a	5.1.8	√
$M/M/c$, many priorities	0	a	0	3.4.2	√
Series					
$M/M/c$	a	a	a	4.1.1	√
$M/M/1$ with blocking		Partial results, depending on the model		4.1.2	—
Cyclic					
$M/M/1$, $M/M/c$	a	a	0	4.4	√
Networks					
$M/M/1$, $M/M/c$	a	a	0	4.2, 4.3	√
$G/G/c$	—	approx.	0	7.3	√
(b) Transient					
$M/M/1/1$	a	n^{b}	—	2.11.1	—
$M/M/1$	a, n	n^{b}	0	2.11.2	√
$M/M/\infty$	a, n	a, n^{b}	—	2.11.3, 8.1.2	√
$M/M/c$	n	n^{b}	0	8.1.2	—
$M/M/c/k$	n	n^{b}	0	8.1.2	—

Finite source $M/M/c$	n	n^{b}	0	8.1.2	—
General birth–death	n	n^{b}	0	8.1.2	—
$M/G/1$		Results indicated—reference given		5.1.8	—
$M/G/\infty$		Results indicated—reference given		5.2.2	—
$G/M/1$		Results indicated—reference given		5.3.2	—

[a] Notation: a, analytical results; n, numerical results; g, results in form of generating function or Laplace transform; 0, no results, —, not applicable.

[b] Indicated but not presented.

[c] Depends on particular model.

[d] Analytical results follow after a root to a nonlinear equation is found; finding the root may require numerical analysis.

Table B.2 Continuous probability distributions, moments and generating functions

Name	Probability Function $f(x)$, continuous	Parameters	Mean $\mathrm{E}[X]$	Variance $\mathrm{E}[X-\mathrm{E}[X]]^2$	Moment Generating Function $\mathrm{E}[e^{tx}]$
Uniform	$f(x) = \frac{1}{b-a}$	$-\infty < a < b < \infty$	$\frac{a+b}{2}$	$\frac{(b-a)^2}{12}$	$\frac{e^{tb}-e^{ta}}{t(b-a)}$
Normal	$f(x) = \frac{1}{\sigma\sqrt{2\pi}} e^{-(x-\mu)^2/(2\sigma^2)}$	$-\infty < \mu < \infty$ $\sigma > 0$	μ	σ^2	$e^{t\mu+(t^2\sigma^2)/2}$
Exponential	$f(x) = \theta e^{-\theta x}$	$\theta > 0$	$\frac{1}{\theta}$	$\frac{1}{\theta^2}$	$\frac{\theta}{\theta - t}$
Gamma	$f(x) = \frac{1}{\Gamma(\alpha)\beta^\alpha} x^{\alpha-1} e^{-x/\beta}$	$\alpha, \beta > 0$	$\alpha\beta$	$\alpha\beta^2$	$\left(\frac{1/\beta}{1/\beta - t}\right)^\alpha$
Erlang-k	$f(x) = \frac{(\theta k)^k}{(k-1)!} x^{k-1} e^{-k\theta x}$	$\theta > 0$ $k = 1, 2, \cdots$	$\frac{1}{\theta}$	$\frac{1}{k\theta^2}$	$\left(\frac{k\theta}{k\theta - t}\right)^k$
2-Term hyperexponential	$f(x) = p\theta_1 e^{-\theta_1 x}$ $+(1-p)\theta_2 e^{-\theta_2 x}$	$0 < p < 1$ $\theta_1, \theta_2 > 0$	$\frac{p}{\theta_1} + \frac{1-p}{\theta_2}$	$\frac{p}{\theta_1^2} + \frac{1-p}{\theta_2^2}$ $-\frac{2p(1-p)}{\theta_1\theta_2}$	$\frac{p\theta_1}{\theta_1 - t} + \frac{(1-p)\theta_2}{\theta_2 - t}$

Table B.2 (continued)

Name	Probability Function $f(x)$, continuous	Parameters	Mean E[X]	Variance $E[X - E[X]]^2$	Moment Generating Function $E[e^{tx}]$
Chi-square	$f(x) = \frac{1}{2^{n/2}\Gamma(n/2)} x^{(n/2)-1} e^{-x/2}$	$n = 1, 2, \cdots$	n	$2n$	$\left(\frac{1/2}{1/2 - t}\right)^{n/2}$
Beta	$f(x) = \frac{\Gamma(\alpha + \beta}{\Gamma(\alpha)\Gamma(\beta)} x^{\alpha-1}(1-x)^{\beta-1}$	$\alpha, \beta > 0$	$\frac{\alpha}{\alpha+\beta}$	$\frac{\alpha\beta}{(\alpha+\beta)^2(\alpha+\beta+1)}$	$\frac{\Gamma(\alpha+\beta)}{\Gamma(\alpha)} \cdot \sum_{j=0}^{\infty} \frac{\Gamma(\alpha+j)(t)^j}{\Gamma(\alpha+\beta+j)\Gamma(j+1)}$

The Characteristic Function, $E[e^{it}]$, can be obtained from the Moment Generating Function by replacing t with it.

Table B.3 Discrete probability distributions, moments and generating functions

Name	Probability Function $p(x)$, discrete	Parameters	Mean $\mathrm{E}[X]$	Variance $\mathrm{E}[X-\mathrm{E}[X]]^2$	Moment Generating Function $\mathrm{E}[e^{tx}]$	Probability Generating Function $\sum_{m=0}^{\infty} p(m) z^m$
Bernoulli	$p(x) = \begin{cases} p & (x=1) \\ 1-p & (x=0) \end{cases}$	$0 \le p \le 1$	p	$p(1-p)$	$pe^t + 1 - p$	$1 - p + pz$
Binomial	$p(x) = \binom{n}{x} p^x (1-p)^{n-x}$	$n = 1, 2, \cdots$ $0 \le p \le 1$	np	$np(1-p)$	$(pe^t + 1 - p)^n$	$(pz + 1 - p)^n$
Poisson	$p(x) = \frac{e^{-\lambda}\lambda^x}{x!}$	$\lambda > 0$	λ	λ	$e^{\lambda(e^t - 1)}$	$e^{-\lambda(1-z)}$
Geometric	$p(x) = p(1-p)^x$	$0 \le p \le 1$	$\frac{1-p}{p}$	$\frac{1-p}{p^2}$	$\frac{p}{1-(1-p)e^t}$	$\frac{p}{1-(1-p)z}$
Negative binomial	$p(x) = \binom{k+x-1}{x} p^k (1-p)^x$	$k > 0$ $0 \le p \le 1$	$\frac{k(1-p)}{p}$	$\frac{k(1-p)}{p^2}$	$\left(\frac{p}{1-(1-p)e^t}\right)^k$	$\left(\frac{p}{1-(1-p)z}\right)^k$

The Characteristic Function, $E[e^{it}]$, can be obtained from the Moment Generating Function by replacing t with it.

APPENDIX C

TRANSFORMS AND GENERATING FUNCTIONS

In queueing theory, it is sometimes difficult to explicitly obtain the probability distribution of interest — for example, the CDF of the waiting time in queue, or the probability distribution of the number of customers in the system. It is often easier to obtain these distributions in terms of related transforms or generating functions. This appendix briefly describes the concepts of transforms and generating functions and gives key properties that are useful in queueing analysis.

C.1 Laplace Transforms

A transform is a mapping of a function from one space to another. While it may be very difficult to solve certain equations directly for a particular function of interest, it is often easier to solve the equations in terms of a transform of the function. The resulting solution gives the transform of the function, which must then be inverted

Fundamentals of Queueing Theory, Fourth Edition.
By D. Gross, J. F. Shortle, J. M. Thompson, and C. M. Harris
Copyright © 2008 John Wiley & Sons, Inc.

back into the original space to give the function of interest. One particular transform that is useful in queueing analysis is the *Laplace transform*.

Let $f(t)$ be a real-valued function defined on the interval $0 \le t < \infty$. The Laplace transform (LT) of f is

$$\mathcal{L}\{f(t)\} \equiv \bar{f}(s) \equiv \int_0^\infty e^{-st} f(t)\,dt, \tag{C.1}$$

where s is a complex variable. Under broad conditions, it can be shown that $\bar{f}(s)$ is analytic in the half-plane where $\text{Re}(s) > \alpha$, for some constant α. Table C.1 gives the LT of several common functions.

■ **EXAMPLE C.1**

Calculate the LT of $f(t) = e^{-at}$. Using (C.1),

$$\bar{f}(s) = \int_0^\infty e^{-st} e^{-at}\,dt = \lambda \int_0^\infty e^{-(s+a)t}\,dt = \frac{1}{s+a}. \tag{C.2}$$

A related transform is the *Laplace-Stieltjes transform*. Let $F(t)$ be a real-valued function defined on the interval $0 \le t < \infty$. The Laplace-Stieltjes transform (LST) of $F(t)$ is

$$\mathcal{L}^*\{F(t)\} \equiv F^*(s) \equiv \int_0^\infty e^{-st}\,dF(t),$$

where the integral is the Lebesgue-Stieltjes integral. For our purposes, we consider $F(t)$ to be the CDF of a nonnegative random variable X, so

$$F^*(s) \equiv \mathrm{E}\left[e^{-sX}\right] = \int_0^\infty e^{-st}\,dF(t). \tag{C.3}$$

■ **EXAMPLE C.2**

Calculate the LST of an exponential random variable with mean $1/\lambda$.

$$F^*(s) = \int_0^\infty e^{-st}\,dF(t) = \int_0^\infty e^{-st} \lambda e^{-\lambda t}\,dt = \frac{\lambda}{s+\lambda}. \tag{C.4}$$

■ **EXAMPLE C.3**

Calculate the LST of a random variable uniformly distributed on the interval $[a, b]$ (where $a > 0$ and $b > a$).

$$F^*(s) = \int_0^\infty e^{-st}\,dF(t) = \int_a^b e^{-st} \frac{1}{b-a}\,dt = \left.\frac{e^{-st}}{b-a}\right|_{t=a}^{t=b} = \frac{e^{-sb} - e^{-sa}}{b-a}.$$

Table C.1 Table of Laplace Transforms

$f(t)$	$\bar{f}(s)$	
1	$\frac{1}{s}$	
t	$\frac{1}{s^2}$	
t^n	$\frac{n!}{s^{n+1}}$	$n = 0, 1, \ldots$
e^{-at}	$\frac{1}{s+a}$	
te^{-at}	$\frac{1}{(s+a)^2}$	
$t^n e^{-at}$	$\frac{n!}{(s+a)^{n+1}}$	$n = 0, 1, \ldots$
$\cos bt$	$\frac{s}{s^2+b^2}$	
$\sin bt$	$\frac{b}{s^2+b^2}$	
$e^{-at}\cos bt$	$\frac{s+a}{(s+a)^2+b^2}$	
$e^{-at}\sin bt$	$\frac{b}{(s+a)^2+b^2}$	

■ EXAMPLE C.4

Calculate the LST of a discrete random variable X, where $\Pr\{X = 2\} = 2/5$ and $\Pr\{X = 7\} = 3/5$.

$$F^*(s) = \int_0^\infty e^{-st}\, dF(t) = \frac{2}{5}e^{-2s} + \frac{3}{5}e^{-7s}.$$

The ordinary Laplace transform and the Laplace–Stieltjes transform can be related when the random variable X has a density function $f(t)$. In this case, (C.3) simplifies to

$$F^*(s) = \int_0^\infty e^{-st}\, dF(t) = \int_0^\infty e^{-st} f(t)\, dt = \bar{f}(s). \tag{C.5}$$

In other words, the LST of the CDF $F(t)$ equals the LT of the PDF $f(t)$. We can further manipulate the equation above using integration by parts.

$$F^*(s) = \int_0^\infty e^{-st}\, dF(t) = e^{-st}F(t)\Big|_{t=0}^{t=\infty} + \int_0^\infty se^{-st}F(t)\, dt.$$

Now, since $F(t)$ is a CDF (so $F(t) \leq 1$), $\lim_{t\to\infty} e^{-st}F(t) = 0$. Also, since we have assumed that X is a nonnegative random variable with a density function $f(t)$, then $F(0) = 0$. In summary, we have

$$F^*(s) = s\bar{F}(s). \tag{C.6}$$

More generally, (C.6) holds for any CDF $F(t)$ of a nonnegative random variable, regardless of whether or not the distribution has a density function (e.g., Billingsley, 1995, p. 236).

The following two examples illustrate the use of (C.6). In the first example, the random variable has a density function; in the second example, it does not.

■ **EXAMPLE C.5**

Use (C.6) to calculate the LST of an exponential random variable with mean $1/\lambda$.

$$\bar{F}(s) = \int_0^\infty e^{-st}F(t)\,dt = \int_0^\infty e^{-st}(1 - e^{-\lambda t})\,dt$$
$$= \int_0^\infty e^{-st}\,dt - \int_0^\infty e^{-(s+\lambda)t}\,dt = \frac{1}{s} - \frac{1}{s+\lambda} = \frac{\lambda}{s(s+\lambda)}.$$

Applying (C.6) gives $F^*(s) = s\bar{F}(s) = \lambda/(s+\lambda)$, which is the same as (C.4).

■ **EXAMPLE C.6**

Calculate the LST of the CDF $W_q(t)$, where

$$W_q(t) = 1 - \rho e^{-\mu(1-\rho)t}, \qquad t \geq 0.$$

This is the steady-state distribution of queue wait for the $M/M/1$ queue (2.28). Now, $W_q(t)$ does not have a PDF at $t = 0$, because $W_q(t)$ is discontinuous there. In particular, $W_q(0^-) = 0$, while $W_q(0) = 1 - \rho$, so there is a point mass of probability $(1 - \rho)$ at $t = 0$. In other words, with probability $(1 - \rho)$ the queue wait is *exactly* zero.

We calculate $W_q^*(s)$ two different ways. First, we directly apply (C.3). Aside from the point at $t = 0$, $W_q(t)$ is continuous and differentiable, so

$$dW_q(t) = \rho\mu(1-\rho)e^{-\mu(1-\rho)t}\,dt, \quad (t > 0).$$

Thus (C.3) gives

$$W_q^*(s) = \int_0^\infty e^{-st}\,dW_q(t) = (1-\rho) + \int_0^\infty e^{-st}\lambda(1-\rho)e^{-\mu(1-\rho)t}\,dt,$$

where $(1-\rho)$ is the LST of the point mass at $t=0$. Continuing the integration,

$$W_q^*(s) = (1-\rho) + \frac{\lambda(1-\rho)}{s+\mu(1-\rho)}$$
$$= \frac{s(1-\rho)+\mu(1-\rho)^2+\lambda(1-\rho)}{s+\mu(1-\rho)}$$
$$= \frac{(s+\mu)(1-\rho)}{s+\mu(1-\rho)}.$$

In the second approach, we take the ordinary Laplace transform of $W_q(t)$, and then apply (C.6).

$$\bar{W}_q(s) = \int_0^\infty e^{-st} W_q(t)\,dt = \int_0^\infty e^{-st}\left(1-\rho e^{-\mu(1-\rho)t}\right) dt$$
$$= \int_0^\infty e^{-st}\,dt - \int_0^\infty \rho e^{-(s+\mu(1-\rho))t}\,dt$$
$$= \frac{1}{s} - \frac{\rho}{s+\mu(1-\rho)}.$$

Then,

$$W_q^*(s) = s\bar{W}_q(s) = 1 - \frac{\rho s}{s+\mu(1-\rho)}$$
$$= \frac{s+\mu(1-\rho)-\rho s}{s+\mu(1-\rho)} = \frac{(s+\mu)(1-\rho)}{s+\mu(1-\rho)},$$

which is the same result.

Two useful properties of Laplace transforms are the following. First, there is a one-to-one correspondence between probability distributions and their transforms. Thus a probability distribution can be uniquely determined from its LST. Second, the LST of the convolution of independent random variables is the product of the LSTs of the individual random variables. That is, if $F^*(s)$ is the LST of X and $G^*(s)$ is the LST of Y, then $F^*(s)G^*(s)$ is the LST of $X+Y$, provided X and Y are independent. Because the inverse of the LST is unique, this allows one to determine the distribution of a sum of random variables by inverting the LST resulting from the product of individual LSTs.

■ **EXAMPLE C.7**

Let X_1 and X_2 be IID exponential random variables with mean $1/\lambda$. The LST of X_1 is $F^*(s) = \lambda/(\lambda+s)$, as given in (C.4). Similarly, the LST of X_2 is $F^*(s) = \lambda/(\lambda+s)$. Thus the LST $G^*(s)$ of X_1+X_2 is

$$G^*(s) = F^*(s)F^*(s) = \frac{\lambda^2}{(s+\lambda)^2}.$$

We see from Table B.2 that this is the LST of an Erlang-2 random variable with mean $2/\lambda$. (The LST of a random variable evaluated at s is the same as its moment generating function evaluated at $-s$; see the next section.)

C.1.1 Moment Generating Functions

Closely related to the Laplace-Stieltjes transform of a random variable X is the moment generating function. This is defined as

$$M_X(t) \equiv \mathrm{E}\left[e^{tX}\right]. \tag{C.7}$$

If X is a nonnegative random variable with CDF $F(x)$, then

$$M_X(t) = \int_0^\infty e^{tx}\, dF(x). \tag{C.8}$$

This is similar to the LST in (C.3), but with t replacing $-s$. In other words, $M_X(t) = F^*(-t)$.

■ **EXAMPLE C.8**

The moment generating function of an exponential random variable with mean $1/\lambda$ is

$$\begin{aligned}
M_X(t) = \mathrm{E}\left[e^{tX}\right] &= \int_0^\infty e^{tx}\lambda e^{-\lambda x}\, dx \\
&= \lambda \int_0^\infty e^{-(\lambda - t)x}\, dx \\
&= \frac{\lambda}{\lambda - t} \qquad (t < \lambda).
\end{aligned}$$

This is the same as the LST in (C.4) but with t replacing $-s$.

■ **EXAMPLE C.9**

The moment generating function of a Poisson random variable with mean λ is

$$\begin{aligned}
M_X(t) = \mathrm{E}\left[e^{tX}\right] &= \sum_{n=0}^{\infty} e^{tn}\frac{e^{-\lambda}\lambda^n}{n!} \\
&= e^{-\lambda}\sum_{n=0}^{\infty}\frac{(\lambda e^t)^n}{n!} \\
&= e^{-\lambda}e^{\lambda e^t} \\
&= \exp[\lambda(e^t - 1)].
\end{aligned}$$

The MGF can be used to find the moments of the random variable X. This can be seen by expanding (C.7) to get

$$M_X(t) = \mathrm{E}\left[1 + tX + \frac{t^2X^2}{2!} + \frac{t^3X^3}{3!} + \cdots\right].$$

Then, assuming we can switch the expectation and infinite sum*

$$M_X(t) = 1 + \mathrm{E}[X]t + \frac{\mathrm{E}[X^2]}{2!}t^2 + \frac{\mathrm{E}[X^3]}{3!}t^3 + \cdots.$$

The moments of X can be found from the derivatives of $M_X(t)$:

$$\begin{aligned} M_X(0) &= 1, \\ M_X'(0) &= \mathrm{E}[X], \\ M_X''(0) &= \mathrm{E}[X^2], \\ &\vdots \end{aligned}$$

More generally,

$$\mathrm{E}[X^n] = \left.\frac{d^{(n)}M_X(t)}{dt^n}\right|_{t=0}.$$

If $M_X(t)$ can be written in a power series, then $\mathrm{E}[X^n]$ is the coefficient in front of t^n multiplied by $n!$.

■ **EXAMPLE C.10**

For an exponential random variable

$$M_X(t) = \frac{\lambda}{\lambda - t} = \frac{1}{1 - t/\lambda} = \sum_{n=0}^{\infty}\left(\frac{t}{\lambda}\right)^n.$$

Thus, $\mathrm{E}[X^n] = n!/\lambda^n$.

■ **EXAMPLE C.11**

For a Poisson random variable with mean λ,

$$\begin{aligned} M_X(t) &= \exp[\lambda(e^t - 1)], \\ M_X'(t) &= \exp[\lambda(e^t - 1)] \cdot \lambda e^t, \\ M_X''(t) &= \exp[\lambda(e^t - 1)](\lambda e^t)^2 + \lambda e^t \exp[\lambda(e^t - 1)]. \end{aligned}$$

*If $M_X(t)$ is finite for t in some open interval $(-\epsilon, \epsilon)$ around zero, then X has finite moments of all orders (e.g., Billingsley, 1995, p. 278), and $M_X(t)$ is analytic on this interval. A situation where this breaks down is when X follows a heavy-tailed distribution [e.g., a Pareto distribution like $F^c(x) = 1/(1+x)^3$]. In this case, the integral in (C.8) is infinite for $t > 0$, and the moments cannot be recovered by expanding $M_X(t)$ in a power series.

Then, $\mathrm{E}[X] = M'_X(0) = \lambda$ and $\mathrm{E}[X^2] = M''_X(0) = \lambda^2 + \lambda$. Thus, $\mathrm{Var}[X] = \mathrm{E}[X^2] - \mathrm{E}^2[X] = \lambda$.

Moment generating functions have properties similar to Laplace transforms. In particular, there is a one-to-one correspondence between moment generating functions and probability distributions, and the MGF of the sum of independent random variables is the product of the MGFs of the individual random variables.

C.2 Generating Functions

Consider a function $G(z)$ that has a power series expansion

$$G(z) = \sum_{n=0}^{\infty} g_n z^n = g_0 + g_1 z + g_2 z^2 + g_3 z^3 + \cdots . \tag{C.9}$$

If the series converges for some range of z, $G(z)$ is called the generating function (GF) of the sequence $g_0, g_1, g_2, \ldots$. (The generating function as defined here is closely related to the z-transform often used in engineering work and defined as $\sum_{n=0}^{\infty} g_n z^{-n}$.) Just as Laplace transforms are useful in solving certain differential equations, generating functions are useful in solving certain *difference* equations.

■ **EXAMPLE C.12**

Find the solution to the following difference equation:

$$g_{n+2} - 2g_{n+1} + g_n = a, \qquad (n = 0, 1, 2, \ldots),$$

with boundary condition $g_0 = 0$ and $g_1 = 0$.

One way to find the solution is to use the characteristic equation, as described in Appendix D.2. Here, we give an alternate solution using generating functions. First multiply all terms by z^n and then sum from 0 to ∞:

$$\sum_{n=0}^{\infty} g_{n+2} z^n - 2\sum_{n=0}^{\infty} g_{n+1} z^n + \sum_{n=0}^{\infty} g_n z^n = a \sum_{n=0}^{\infty} z^n$$

$$z^{-2}\sum_{n=0}^{\infty} g_{n+2} z^{n+2} - 2z^{-1}\sum_{n=0}^{\infty} g_{n+1} z^{n+1} + \sum_{n=0}^{\infty} g_n z^n = \frac{a}{1-z}$$

$$z^{-2}\sum_{n=2}^{\infty} g_n z^n - 2z^{-1}\sum_{n=1}^{\infty} g_n z^n + \sum_{n=0}^{\infty} g_n z^n = \frac{a}{1-z}.$$

Substituting $\sum_{n=0}^{\infty} g_n z^n = G(z)$ gives

$$z^{-2}(G(z) - zg_1 - g_0) - 2z^{-1}(G(z) - g_0) + G(z) = \frac{a}{1-z}.$$

Using the boundary conditions, $g_0 = 0$ and $g_1 = 0$, gives

$$z^{-2}G(z) \quad 2z^{-1}G(z) + G(z) - \frac{a}{1-z}.$$

Solving for $G(z)$ gives

$$G(z) = \frac{az^2}{(1-z)^3}.$$

Thus we have the generating function $G(z)$ for the sequence $g_0, g_1, \ldots$. To obtain the actual values for this sequence, we expand $G(z)$ in a power series. A general way to do this is using a Maclaurin series

$$G(z) = G(0) + G'(0)z + \frac{G''(0)z^2}{2} + \cdots + \frac{G^{(n)}(0)z^n}{n!} + \cdots.$$

Finding this series requires successive differentiation of $G(z)$, and in particular a general expression for $G^{(n)}(0)$. This can be quite cumbersome to find.

Rather, for this problem, we take an approach that makes use of the known series expansion for $1/(1-z)$. That is,

$$\frac{1}{1-z} = \sum_{n=0}^{\infty} z^n, \qquad (|z| < 1). \tag{C.10}$$

Since $G(z)$ has $(1-z)^3$ in the denominator, we seek a way to manipulate (C.10) to match the expression for $G(z)$. This can be done by taking the derivative of (C.10) twice:

$$\frac{2}{(1-z)^3} = \sum_{n=2}^{\infty} n(n-1)z^{n-2}.$$

Hence

$$\begin{aligned} G(z) = \frac{az^2}{(1-z)^3} &= \frac{az^2}{2} \cdot \frac{2}{(1-z)^3} \\ &= \frac{az^2}{2} \sum_{n=2}^{\infty} n(n-1)z^{n-2} \\ &= \sum_{n=2}^{\infty} \frac{an(n-1)}{2} z^n. \end{aligned}$$

Then, g_n is the coefficient in front of z^n, so

$$g_n = \frac{an(n-1)}{2} \qquad (n \geq 2).$$

The formula is also valid for $n = 0$ and $n = 1$, since it yields the boundary conditions $g_0 = 0$ and $g_1 = 0$.

This preceding example involves a somewhat arbitrary difference equation, with no particular relation to queueing theory. In queueing theory, difference equations often arise as a relationship between successive values of a probability distribution $p_0, p_1, \ldots$. For example, for the $M/M/1$ queue, let p_n denote the long-run fraction of time that there are n customers in the system. As discussed in Section 2.2, the probabilities are related as follows (2.6):

$$(\lambda + \mu)p_n = \mu p_{n+1} + \lambda p_{n-1} \qquad (n \geq 1),$$

or

$$\mu p_{n+1} - (\lambda + \mu)p_n + \lambda p_{n-1} = 0 \qquad (n \geq 1).$$

This is a second-order difference equation. This type of application motivates a more particular type of generating function.

C.2.1 Probability Generating Functions

A *probability generating function* (PGF) is a particular type of generating function where the sequence $g_0, g_1, g_2, \ldots$ corresponds to the probabilities of a non-negative integer-valued random variable. Specifically, let X be a random variable with

$$\Pr\{X = n\} = p_n \qquad (n = 0, 1, 2, \ldots),$$

and $\sum_{n=0}^{\infty} p_n = 1$. Then

$$P(z) \equiv \mathrm{E}\left[z^X\right] = \sum_{n=0}^{\infty} p_n z^n \qquad \text{(C.11)}$$

is the probability generating function for the random variable X. The probability generating function for an integer-valued random variable is similar to the moment generating function defined in (C.7), but with z replacing e^t.

As discussed in the previous section on generating functions, the PGF can be used to determine the probabilities $p_0, p_1, \ldots$. This is done by writing $P(z)$ as a power series and observing that p_n is the coefficient in front of z^n in the series.

Probability generating functions can also be used to *directly* obtain moments of the random variable, without requiring a derivation of the values $\{p_n\}$. In some cases, it is not easy to explicitly write out $P(z)$ in a series expansion – even though $P(z)$ is known in closed form. Nevertheless, knowing $P(z)$ is still quite useful, since it can be used to obtain the moments of the distribution, as we now show.

The first two derivatives of $P(z)$ are

$$P'(z) = \sum_{n=1}^{\infty} n p_n z^{n-1},$$

$$P''(z) = \sum_{n=2}^{\infty} n(n-1) p_n z^{n-2}.$$

Evaluating $P(z)$, $P'(z)$, and $P''(z)$ at $z = 1$ gives

$$P(1) = \sum_{n=0}^{\infty} p_n = 1,$$

$$P'(1) = \sum_{n=1}^{\infty} np_n = \mathrm{E}[X],$$

$$P''(1) = \sum_{n=2}^{\infty} n(n-1)p_n = \mathrm{E}[X(X-1)].$$

More generally, we have

$$\boxed{P^{(n)}(1) = \mathrm{E}[X(X-1)(X-2)\cdots(X-n+1)].}$$

In other words, the nth derivative of $P(z)$ evaluated at $z = 1$ gives the nth factorial moment of X. Since it is possible to relate factorial moments to regular moments ($\mathrm{E}[X], \mathrm{E}[X^2], \mathrm{E}[X^3], \ldots$), $P(z)$ can be used to determine the moments of a distribution.

■ EXAMPLE C.13

Determine the expected value, variance, and probability distribution for a random variable whose PGF is

$$P(z) = e^{-\lambda(1-z)}.$$

The first two derivatives of $P(z)$ are

$$P'(z) = \lambda e^{-\lambda(1-z)},$$
$$P''(z) = \lambda^2 e^{-\lambda(1-z)}.$$

Then

$$\mathrm{E}[X] = P'(1) = \lambda,$$
$$\mathrm{E}[X(X-1)] = P''(1) = \lambda^2.$$

With a little manipulation, $\mathrm{Var}[X]$ can be obtained from $\mathrm{E}[X(X-1)]$ and $\mathrm{E}[X]$:

$$\begin{aligned}\mathrm{Var}[X] &= \mathrm{E}[X^2] - \mathrm{E}^2[X]\\ &= (\mathrm{E}[X^2] - \mathrm{E}[X]) + \mathrm{E}[X] - \mathrm{E}^2[X]\\ &= \lambda^2 + \lambda - \lambda^2\\ &= \lambda.\end{aligned}$$

To obtain the probability distribution $\{p_n\}$, we expand $P(z)$ in a power series:

$$P(z) = e^{-\lambda(1-z)} = e^{-\lambda}e^{\lambda z} = e^{-\lambda}\sum_{n=0}^{\infty}\frac{(\lambda z)^n}{n!} = \sum_{n=0}^{\infty}\frac{e^{-\lambda}\lambda^n}{n!}z^n.$$

The coefficient in front of z^n is p_n. Thus

$$p_n = \frac{e^{-\lambda}\lambda^n}{n!},$$

which is the distribution of a Poisson random variable (with mean λ and variance λ).

APPENDIX D
DIFFERENTIAL AND DIFFERENCE EQUATIONS

Differential and difference equations play a key role in the solution of most queueing models. In this appendix we review some of the fundamentals concerning these types of equations.

D.1 Ordinary Differential Equations

A differential equation is an equation involving a function and its derivatives. An example of such an equation might be

$$3\frac{d^2y}{dx^2} + 14x\frac{dy}{dx} - x^3y = 6e^x, \tag{D.1}$$

where y is a function of x, that is, $y = y(x)$. The problem is to determine the most general $y(x)$ that satisfies (D.1). Prior to discussing methods of solution to such equations, we first discuss the nomenclature involved with categorizing differential equations.

Fundamentals of Queueing Theory, Fourth Edition.
By D. Gross, J. F. Shortle, J. M. Thompson, and C. M. Harris
Copyright © 2008 John Wiley & Sons, Inc.

D.1.1 Classification

A differential equation is called *ordinary* if it involves only total (as opposed to partial) derivatives. Differential equations are further categorized by *order* and *degree*. Thus a differential equation of the form

$$a_0(x)\frac{d^n y}{dx^n} + a_1(x)\frac{d^{n-1}y}{dx^{n-1}} + \cdots + a_{n-1}(x)\frac{dy}{dx} + a_n(x)y = f(x) \qquad \text{(D.2)}$$

is called a *linear* ordinary differential equation of *order n*. The order refers to the highest derivative in the equation, while the degree (linear in this case) refers to the exponent on the dependent variable y and its derivatives. When the coefficients $a_n(x)$ are independent of x, the equation is said to be *constant coefficients*. If the right-hand side of (D.2) is zero, then the equation is called *homogeneous*. Thus the equation

$$a_0\frac{d^n y}{dx^n} + a_1\frac{d^{n-1}y}{dx^{n-1}} + \cdots + a_{n-1}\frac{dy}{dx} + a_n y = 0$$

is a linear, homogeneous differential equation of order n with constant coefficients. The descriptor "ordinary" is understood and generally omitted unless one is dealing simultaneously with ordinary and partial differential equations.

D.1.2 Solutions

Discussion in this appendix is restricted to solutions of linear ordinary differential equations. Solution techniques for nonlinear differential equations are extremely complex, and furthermore, the types of differential equations that arise from our interest in queueing analyses are usually linear.

Consider the following linear differential equation of second order with constant coefficients, namely,

$$y'' + 3y' + 2y = 6e^x, \qquad \text{(D.3)}$$

where the prime notation is now used to denote differentiation. One solution to (D.3) is

$$y = e^x, \qquad \text{(D.4)}$$

which can be verified by substitution. This is referred to as a *particular* solution to (D.3). Another solution to (D.4) is

$$y = C_1 e^{-x} + e^x, \qquad \text{(D.5)}$$

where C_1 is any constant. This solution can also be verified by substitution. It contains the particular solution of (D.4) and is a more general solution. We desire the most general solution to any differential equation, which we refer to simply as the *general solution*. It turns out that the general solution of (D.3) is given by

$$y = C_1 e^{-x} + C_2 e^{-2x} + e^x. \qquad \text{(D.6)}$$

Any particular solution can be obtained by specifying the arbitrary constants C_1 and C_2. For example, the particular solution given by (D.4) results from (D.6) when $C_1 = C_2 = 0$. The number of arbitrary constants appearing in a general solution of a linear ordinary differential equation can be shown to be equal to the order n. Since (D.3) is of order two, two constants appear in the general solution given by (D.6).

Another way of looking at the solution given by (D.6) is to first consider solutions to a homogeneous equation obtained from (D.3) by setting the right-hand side to zero. The homogeneous equation then becomes linear:

$$y'' + 3y' + 2y = 0. \tag{D.7}$$

We note that C_1e^{-x} and C_2e^{-2x} are both solutions to (D.7). Also, e^x is a solution to the original nonhomogeneous equation (D.3), so that the general solution consists of a linear combination of all solutions to the homogeneous equation (the general solution to the homogeneous equation) plus a particular solution to the nonhomogeneous equation. It can be proved that for a linear ordinary differential equation of order n there are n solutions to the homogeneous equation, so that the general solution is comprised of a linear combination of the n solutions (thus yielding n arbitrary constants) plus a particular solution to the nonhomogeneous equation. See, for example, Rainville and Bedient (1969).

To determine the constants of a general solution, that is, which particular solution is desired, one must utilize boundary conditions. A boundary condition is a condition on the function $y(x)$ for a specific x, and results from the model which the differential equation represents. For the equation given by (D.3), suppose one knows from the physical situation that generated (D.3) that both the function and its derivative must be zero when x is zero, that is,

$$y(0) = y'(0) = 0.$$

Using these conditions in (D.6) yields two equations in two unknowns, namely,

$$0 = C_1 + C_2 + 1,$$
$$0 = -C_1 - 2C_2 + 1,$$

which result in $C_1 = -3$ and $C_2 = 2$, giving the particular solution and the general solution as

$$y = -3e^{-x} + 2e^{-2x} + e^x.$$

We see then, for an nth order equation, n boundary conditions are required to obtain a particular solution from the general solution. Thus the fundamental approach presented here in solving differential equations is to first find the general solution and then, using the boundary conditions, find the particular solution desired. Emphasis in this appendix is on finding general solutions.

D.1.3 Separation of Variables

The easiest type of differential equation to solve is one for which *separation of variables* is possible. The general solution can then be obtained by integrating both sides. For example, consider the equation

$$y\frac{dy}{dx} = 3x^2 + 2e^x.$$

We can write

$$ydy = (3x^2 + 2e^x)dx.$$

Integrating both sides and combining the arbitrary constants arising from indefinite integration yields

$$\frac{y^2}{2} = x^3 + 2e^x + C.$$

If, in general, we have an equation of the form [even for $g(y)$, $u(y)$ nonlinear]

$$f(x)g(y)\frac{dy}{dx} = h(x)u(y),$$

we can separate variables to obtain

$$\frac{g(y)}{u(y)}dy = \frac{h(x)}{f(x)}dx,$$

and the general solution is

$$\int \frac{g(y)}{u(y)}dy = \int \frac{h(x)}{f(x)}dx + C. \tag{D.8}$$

Although the examples thus far have been linear differential equations of the first order, it may also be possible to separate variables in higher-order linear equations. For example, the solution for

$$\frac{d^2y}{dx^2} = f(x)$$

can be obtained by integrating twice to yield

$$y = \int\left[\int f(x)dx\right]dx + C_1x + C_2$$

since

$$\frac{d^2y}{dx^2} = \frac{d(dy/dx)}{dx},$$

and integrating the first time gives a solution

$$\frac{dy}{dx} = \int f(x)dx + C_1.$$

■ EXAMPLE D.1

Find the general solution of

$$y'' = 6x^2.$$

Integrating once gives

$$y' = 2x^3 + C_1$$

and integrating a second time yields

$$y = \tfrac{1}{2}x^4 + C_1 x + C_2.$$

D.1.4 Linear Differential Equations of First Order

The linear differential equation of the first order can be written in general terms as

$$\frac{dy}{dx} + a(x)y = f(x). \tag{D.9}$$

If we can determine a function $g(x)$ so that when both sides of (D.9) are multiplied by it, the equation can be put in the form

$$\frac{d(gy)}{dx} = gf, \tag{D.10}$$

then the solution can be determined by separating variables, that is, the solution becomes

$$gy = \int gf\,dx + C$$

or

$$y = \frac{1}{g}\int gf\,dx + \frac{C}{g}. \tag{D.11}$$

Such a function as g is referred to as an *integrating factor*. We can, for linear first-order differential equations, find g as follows. Using the product rule of differentiation, and dividing through by g, (D.10) can be written as

$$\frac{dy}{dx} + \frac{y}{g}\frac{dg}{dx} = f. \tag{D.12}$$

For (D.12) to be equivalent to (D.9) we must have

$$\frac{1}{g}\frac{dg}{dx} = a(x).$$

Integrating both sides yields

$$\ln g = \int a(x)dx + C_1$$

or

$$g = e^{C_1} e^{A(x)},$$

where $A(x) = \int a(x)dx$. Since we are seeking only a particular g that will yield equivalency for (D.9) and (D.12), we are free to set the constant C_1 to any value we desire. It is most convenient to set $C_1 = 0$. Hence a suitable integrating factor is

$$g = e^{A(x)}. \tag{D.13}$$

Using (D.12) in (D.11) yields the final solution for y, namely,

$$y = e^{-A(x)} \int e^{A(x)} f(x)dx + Ce^{-A(x)}. \tag{D.14}$$

Unfortunately, no such similar method is possible for obtaining solutions to higher-order linear differential equations. We will consider, however, some higher-order equations of specific types.

D.1.5 Linear Differential Equations with Constant Coefficients

The simplest linear equation of higher order is one where the coefficients are independent of x, namely,

$$a_0 \frac{d^n y}{dx^n} + a_1 \frac{d^{n-1} y}{dx^{n-1}} + \cdots + a_{n-1} \frac{dy}{dx} + a_n y = f(x). \tag{D.15}$$

The approach here is to first find the n solutions to the homogeneous equation

$$a_0 \frac{d^n y}{dx^n} + a_1 \frac{d^{n-1} y}{dx^{n-1}} + \cdots + a_{n-1} \frac{dy}{dx} + a_n y = 0, \tag{D.16}$$

and then find a particular solution for the nonhomogeneous equation.

The form of (D.16) suggests that the homogeneous solutions are of the form e^{rx}, since the nth derivative is a multiple of the function itself, that is,

$$\frac{d^n e^{rx}}{dx} = r^n e^{rx}. \tag{D.17}$$

Now if e^{rx} is a solution to (D.16), then we have

$$(a_0 r^n + a_1 r^{n-1} + \cdots + a_{n-1} + a_n)e^{rx} = 0,$$

which implies for a nontrivial solution ($y = e^{rx} \neq 0$) that

$$a_0 r^n + a_1 r^{n-1} + \cdots + a_{n-1} + a_n = 0. \tag{D.18}$$

Equation (D.18) is called the *characteristic* or *operator* equation. The characteristic equation can also be obtained directly by looking at the derivative as an operator, say,

D, so that

$$\begin{aligned} Dy &= \frac{dy}{dx}, \\ D^2y = D(Dy) &= \frac{d^2y}{dx^2}, \\ &\vdots \\ D^ny = D(D^{n-1}y) &= \frac{d^ny}{dx^n}. \end{aligned}$$

Hence (D.16) can be written as

$$(a_0D^n + a_1D^{n-1} + \cdots + a_{n-1}D + a_n)y = 0,$$

where the characteristic equation is in terms of D instead of r.

Denoting the n roots of the characteristic equation by $r_1, r_2, \ldots, r_n$, we can write

$$(r - r_1)(r - r_2)\cdots(r - r_n)y = 0,$$

and hence theoretically the roots can be found by factorization.* If the n roots are distinct, we then have n solutions $e^{r_i x}(i = 1, 2, \ldots, n)$ of the homogeneous equation (D.16). The most general solution of (D.16) is then

$$y = C_1e^{r_1x} + C_2e^{r_2x} + \cdots + C_ne^{r_nx}.$$

If the roots are not all distinct, we have less than n solutions. To find the missing solution we can proceed as follows. Suppose that r_1 is a double root of the characteristic equation. Then we have

$$(r - r_1)^2(r - r_2)\cdots(r - r_{n-1})e^{rx} = 0. \tag{D.19}$$

Observing that

$$\frac{\partial(r - r_1)^2}{\partial r} = 2(r - r_1),$$

we find that the partial derivative with respect to r evaluated at $r = r_1$ also vanishes, so that if e^{r_1x} is a solution, so too is $\partial e^{rx}/\partial r|_{r=r_1} = xe^{r_1x}$. To verify that xe^{r_1x} is a solution consider solutions of the form xe^{rx}. Putting this in for y in (D.16) yields

$$a_0\frac{d^nxe^{rx}}{dx^n} + a_1\frac{d^{n-1}xe^{rx}}{dx^{n-1}} + \cdots + a_{n-1}\frac{dxe^{rx}}{dx} + a_nxe^{rx} = 0.$$

Since

$$xe^{rx} = \frac{\partial e^{rx}}{\partial r},$$

we can write

$$a_0\frac{\partial^n(\partial e^{rx}/\partial r)}{\partial x^n} + \cdots + a_{n-1}\frac{\partial(\partial e^{rx}/\partial r)}{\partial x} + a_n\frac{\partial e^{rx}}{\partial r} = 0,$$

*Depending on the characteristic equation that results, factorization could be impossible and numerical methods might be required.

and changing the order of differentiation gives

$$a_0 \frac{\partial(\partial^n e^{rx}/\partial x^n)}{\partial r} + \cdots + a_{n-1} \frac{\partial(\partial e^{rx}/\partial x)}{\partial r} + a_n \frac{\partial e^{rx}}{\partial r} = 0.$$

Hence we can write

$$\frac{\partial}{\partial r}\left(a_0 \frac{\partial^n e^{rx}}{\partial x^n} + \cdots + a_{n-1}\frac{\partial e^{rx}}{\partial x} + a_n e^{rx} \right) = 0$$

or

$$\frac{\partial}{\partial r}\left[\left(a_0 r^n + a_1 r^{n-1} + \cdots + a_{n-1} r + a_n \right) e^{rx} \right] = 0.$$

But we have said that the characteristic equation factors into $n-1$ roots as given in (D.19) so that

$$\frac{\partial}{\partial r}\left\{ \left[(r-r_1)^2 (r-r_2)\cdots(r-r_{n-1}) \right] e^{rx} \right\} = 0.$$

This equation does hold for $r = r_1$, since the partial with respect to r vanishes at that point. Thus the two solutions for a double root r_1 are

$$C_1 e^{r_1 x} + C_2 x e^{r_1 x}.$$

This can be generalized to roots of multiplicity k; that is, if r_1 has multiplicity k, the solution associated with r_1 is

$$C_1 e^{r_1 x} + C_2 x e^{r_1 x} + C_3 x^2 e^{r_1 x} + \cdots + C_k x^{k-1} e^{r_1 x}.$$

When we have multiple roots, if factorization is not possible and we must resort to numerical methods, we might only be able to find (say) $n-k$ distinct roots to the characteristic equation. To find which root (or roots) have multiplicity we can simply take partial derivatives of the characteristic equation and check for which root (or roots) vanish. The roots for which only the first partial derivative of the characteristic equation vanishes have multiplicity two. If a root causes the first, second, ..., kth partial derivatives to vanish, it is of multiplicity $k+1$.

■ EXAMPLE D.2

Find the general solution for

$$\frac{d^3y}{dx^3} - 4\frac{dy}{dx} = 0.$$

The characteristic equation is

$$D^3 - 4D = 0,$$

which factors into

$$D(D+2)(D-2) = 0;$$

hence the roots are $r_1 = 0$, $r_2 = -2$, and $r_3 = +2$. The general solution is

$$y = C_1 + C_2 e^{-2x} + C_3 e^{2x}.$$

■ **EXAMPLE D.3**

Find the general solution for

$$\frac{d^3y}{dx^3} - 4\frac{d^2y}{dx^2} + 5\frac{dy}{dx} - 2y = 0.$$

The characteristic equation is

$$D^3 - 4D^2 + 5D - 2 = 0,$$

which factors into

$$(D-2)(D-1)^2 = 0.$$

Thus the roots are $r_1 = 2$ and $r_2 = r_3 = 1$, and we have

$$y = C_1e^{2x} + C_2e^x + C_3xe^x.$$

Had we not been able to factor the characteristic equation but had determined that 2 and 1 were all distinct roots, we know that since the characteristic equation is cubic, one root must be double. To find which root it is, we take the partial derivative of the characteristic equation, which gives

$$3D^2 - 8D + 5,$$

and evaluating $D = 2$ and $D = 1$ yields

$$3(2)^2 - 8(2) + 5 = 1$$

and

$$3(1)^2 + 8(1) + 5 = 0$$

so that root $D = 1$ is the double root.

It remains now to discuss the determination of a particular solution for the nonhomogeneous linear differential equation with constant coefficients. There are four methods for finding a particular solution to the nonhomogeneous equation: (1) undetermined coefficients, (2) variation of parameters, (3) differential operators, and (4) Laplace transforms. We briefly discuss the first and third methods here. Laplace transforms are also presented in Appendix C.

D.1.6 Undetermined Coefficients

If the right-hand side of the differential equation given in (D.15) is of the form x^m (m is an integer), $\sin(bx)$, $\cos(bx)$, e^{bx}, and/or products of two or more such functions, we can employ the method of *undetermined coefficients* to find a particular solution. We first define a family of a function $f(x)$ and its derivatives. The functions specified above are functions with a finite number of derivatives for which the function and its

Table D.1 Functions and Their Families

Function	Family
x^m	$x^m, x^{m-1}, x^{m-2}, \ldots, x^2, x, 1$
$\sin bx$	$\sin bx, \cos bx$
$\cos bx$	$\cos bx, \sin bx$
e^{bx}	e^{bx}

derivatives are linearly independent. Table D.1 lists the families of the aforementioned functions. The family of a function consisting of a product of n terms of this type consists of all possible products of the family membeis of each of the n terms. For example, the family of $x^2 \cos x$ is $x^2 \cos x$, $x \cos x$, $\cos x$, $x^2 \sin x$, $x \sin x$, and $\sin x$. The method works as follows in three steps:

1. Assuming $f(x)$ is a linear combination of functions or products of functions given in Table D.1, construct the family for each, eliminating families that are included in other families.

2. If any family has a member that is also a solution to the homogeneous equation, replace that family by a new one, obtained by multiplying the original family by x (or the lowest power of x necessary) so that the new family has no members that are also solutions to the homogeneous equation.

3. The particular solution is assumed to be a linear combination of all members of the constructed families. The constants of the linear combination are then found by substituting this particular solution into the differential equation.

EXAMPLE D.4

Find the general solution for

$$y''' - y' = 2x + 1 - 4\cos x + 2e^x$$

The general homogeneous solution can be found from previous methods to be

$$y = C_1 + C_2 e^x + C_3 e^{-x}.$$

The families for the right-hand side function are, respectively,

$$\{x, 1\}, \{1\}, \{\cos x, \sin x\}, \{e^x\}.$$

Since $\{1\}$ is included in $\{x, 1\}$, we omit this. Furthermore, since 1 and e^x are in the homogeneous solution, their families are replaced by $\{x^2, x\}$ and $\{xe^x\}$, respectively. Then the resulting terms to be used are

$$\{x^2, x, \cos x, \sin x, xe^x\}$$

and the particular solution is of the form

$$y_p = Ax^2 + Bx + C\cos x + D\sin x + Exe^x.$$

Substituting y_p into the differential equation yields

$$\begin{aligned} &C\sin x - D\cos x + E(xe^x + 3e^x) \\ &\qquad - [2Ax + B - C\sin x + D\cos x + E(xe^x + e^x)] \\ &\qquad\qquad = 2x + 1 - 4\cos x + 2e^x, \end{aligned}$$

or simplifying we get

$$-2Ax - B + 2C\sin x - 2D\cos x + 2Ee^x = 2x + 1 - 4\cos x + 2e^x.$$

Matching coefficients of like terms yields

$$A = -1, B = -1, C = 0, D = 2, E = 1.$$

Hence the particular solution is

$$y_p = -x^2 - x + 2\sin x + xe^x$$

and the general solution becomes

$$y = C_1 + C_2e^x + C_3e^{-x} - x^2 - x + 2\sin x + xe^x.$$

D.1.7 Differential Operators

We illustrate the use of differential operators on the same equations used in the previous example. The equation

$$y''' - y' = 2x + 1 - 4\cos x + 2e^x$$

can be written in operator notation as

$$D^3y - Dy = g,$$

where g, as before, is the right-hand side. This can be factored as

$$D(D+1)(D-1)y = g.$$

We let

$$y_1 = (D+1)(D-1)\,y; \tag{D.20}$$

hence we have

$$Dy_1 = g$$

or

$$\frac{dy_1}{dx} = g.$$

Solving directly by integration gives

$$\begin{aligned} y_1 &= \int g dx + C_1 \\ &= \int (2x + 1 - 4\cos x + 2e^x)\, dx + C_1 \\ &= x^2 + x - 4\sin x + 2e^x + C_1. \end{aligned}$$

Substituting y_1 into (D.20) yields the differential equation

$$(D+1)(D-1)\, y = x^2 + x - 4\sin x + 2e^x + C_1.$$

We next let

$$y_2 = (D-1)\, y \tag{D.21}$$

and get

$$(D+1)\, y_2 = x^2 + x - 4\sin x + +2e^x + C_1$$

or

$$\frac{dy_2}{dx} + y_2 = x^2 + x - 4\sin x + 2e^x + C_1. \tag{D.22}$$

Equation (D.22) is now a first-order equation that can be solved by the solution previously derived in Section D.1.4 and given by (D.14), which yields

$$\begin{aligned} y_2 &= e^{-x} \int e^x \left(x^2 + x - 4\sin x + 2e^x + C_1\right) dx + C_2 e^{-x} \\ &= x^2 - x + 1 - 2\sin x + 2\cos x + e^x + C_1 + C_2 e^{-x}. \end{aligned}$$

Now substituting y_2 into (D.21) yields another first-order equation

$$(D-1)\, y = x^2 - x + 1 - 2\sin x + 2\cos x + e^x + C_1 + C_2 e^{-x}.$$

Again using the solution for first-order equations we get

$$\begin{aligned} y &= e^x \int e^{-x} \left(x^2 - x + 1 - 2\sin x + 2\cos x + e^x + C_1 + C_2 e^{-x}\right) dx + C_3 e^x \\ &= -x^2 - x - 2 + 2\sin x + xe^x - C_1 - \frac{C_2}{2} e^{-x} + C_3 e^x, \end{aligned}$$

which agrees with our previous solution upon redefining the arbitrary constants. This method of using operators applies only for equations with constant coefficients. Essentially, any equation with constant coefficients can be written in operator notation as

$$f(D) y = g(x).$$

If a function $f^{-1}(D)$ can be found where

$$f(D) f^{-1}(D) = 1,$$

the solution to the equation is

$$y = f^{-1}(D)g(x).$$

The material referenced above deals with determining such inverse differential operations.

D.1.8 Reduction of Order

Leaving the topic of particular solutions and returning now to the topic of solutions in general, if one homogeneous solution of a linear differential equation of order n is known, the remainder of the solution can be determined by solving a new linear differential equation of order $n-1$ in much the same way one can reduce the degree of an algebraic equation when one root is known. Consider the following second-order equation:

$$y'' + a_1 y' + a_2 y = g, \tag{D.23}$$

where y and g are functions of x and the coefficients a_1 and a_2 may be also. Suppose one solution to the homogeneous equation can be found from inspection and we denote it by $y_1(x)$, that is,

$$y_1'' + a_1 y_1' + a_2 y = 0. \tag{D.24}$$

Then if we let

$$y = y_1 v,$$

then y is a solution to (D.23) if

$$(y_1 v)'' + a_1 (y_1 v)' + a_2 y = g$$

or

$$y_1 v'' + 2y_1' v' + y_1'' v + a_1 (y_1 v' + v y_1') + a_2 y_1 v = g.$$

Simplifying we obtain

$$y_1 v'' + (2y_1' + a_1 y_1) v' + (y_1'' + a_1 y_1' + a_2 y_1) v = g. \tag{D.25}$$

But since y_1 is a homogeneous solution, using (D.24) and (D.25) gives

$$y_1 v'' + (2y_1' + a_1 y_1) v' = g.$$

Letting $u = v'$ we get the following first-order equation involving u, namely,

$$y_1' u' + (2y_1' + a_1 y_1) u = g, \tag{D.26}$$

which can be solved by (D.14) of Section D.1.4. Finally, we can get v from $v' = u$ by integration and the general solution $y = y_1 v$ results.

■ **EXAMPLE D.5**

Consider the equation

$$y'' - y = x.$$

From inspection we can see that one solution to the homogeneous equation is

$$y_1 = e^x.$$

Thus using this in (D.26) we have the first-order equation

$$e^x u' + 2e^x u = x$$

or

$$u' + 2u = xe^{-x},$$

which by (D.14) yields

$$u = e^{-x}(x-1) + C_1 e^{-2x}.$$

Integrating u we obtain v as

$$v = -xe^{-x} - \frac{C_1}{2}e^{-2x} + C_2,$$

and finally (redefining the constant C_1)

$$y = y_1 v = -x + C_1 e^{-x} + C_2 e^x.$$

D.1.9 Systems of Linear Differential Equations

This section considers systems of simultaneous linear differential equations with constant coefficients. To begin, consider the following system of two equations in two unknowns:

$$\frac{d^2 y_1}{dx^2} - y_1 - 2y_2 = g_1(x)$$

$$\frac{d^2 y_2}{dx^2} - 2y_2 - 3y_1 = g_2(x) \tag{D.27}$$

These can be rewritten using operator notation as

$$\left(D^2 - 1\right) y_1 - 2y_2 = g_1,$$

$$-3y_1 + \left(D^2 - 2\right) y_2 = g_2.$$

Using Cramer's rule the solution yields the following two differential equations of a single variable:

$$\begin{vmatrix} (D^2-1) & -2 \\ -3 & (D^2-2) \end{vmatrix} y_1 = \begin{vmatrix} g_1 & -2 \\ g_2 & (D^2-2) \end{vmatrix},$$

$$\begin{vmatrix}(D^2-1) & -2 \\ -3 & (D^2-2)\end{vmatrix} y_2 = \begin{vmatrix}(D^2-1) & g_1 \\ -3 & g_2\end{vmatrix},$$

or rewriting

$$(D^4 - 3D^2 - 4)y_1 = (D^2 - 2)g_1 + 2g_2,$$
$$(D^4 - 3D^2 - 4)y_2 = (D^2 - 1)g_2 + 3g_1.$$

Since g_1 and g_2 are known functions of x, the differentiation implied by the operator can be performed and the right-hand side represented by two known function, $h_1(x)$ and $h_2(x)$, yielding

$$(D^4 - 3D^2 - 4)y_1 = h_1(x),$$
$$(D^4 - 3D^2 - 4)y_2 = h_2(x).$$

Both equations have identical characteristic equations (which can always be obtained from the determinant of the left-hand side of the system of equations) so the general solution to the homogeneous equations are of the same form. Denoting the four roots to the characteristic equation by r_1, r_2, r_3, and r_4, we have the homogeneous solutions

$$y_1 = C_1e^{r_1x} + C_2e^{r_2x} + C_3e^{r_3x} + C_4e^{r_4x}$$
$$y_2 = C_5e^{r_1x} + C_6e^{r_2x} + C_7e^{r_3x} + C_8e^{r_4x}. \quad \text{(D.28)}$$

While there are eight constants, they are not all independent and their relationships can be obtained by substitution of (D.28) in either equation of (D.27) with the right-hand side equal to zero, yielding

$$C_5 = \frac{r_1^2 - 1}{2}C_1, \quad C_6 = \frac{r_2^2 - 1}{2}C_2, \quad C_7 = \frac{r_3^2 - 1}{2}C_3, \quad C_8 = \frac{r_4^2 - 1}{2}C_4.$$

Equations (D.28) are the homogeneous solutions to (D.27). To obtain particular solutions, one can use the method of undetermined coefficients.

■ EXAMPLE D.6

Solve the following for y_1 and y_2:

$$y_1' - 2y_1 + 2y_2' = 2 - 4e^{2x},$$
$$2y_1' - 3y_1 + 3y_2' - y_2 = 0. \quad \text{(D.29)}$$

Considering first the homogeneous solutions, we rewrite the equations in operator notation as

$$(D - 2)y_1 + 2Dy_2 = 0,$$
$$(2D - 3)y_1 + (3D - 1)y_2 = 0, \quad \text{(D.30)}$$

and the characteristic equation is then

$$\begin{vmatrix} (D-2) & 2D \\ (2D-3) & (3D-1) \end{vmatrix} = 0,$$

which upon expanding yields

$$-D^2 - D + 2 = 0.$$

Factoring gives the two roots as 1 and −2 so that the homogeneous solutions are

$$y_1 = C_1 e^x + C_2 e^{-2x},$$
$$y_2 = C_3 e^x + C_4 e^{-2x}.$$

To determine the relationship among the constants, we substitute the above in either equation of (D.30) to get

$$C_1 = 2C_3, \quad C_2 = -C_4,$$

and thus

$$y_1 = C_1 e^x + C_2 e^{-2x},$$
$$y_2 = \frac{C_1}{2} e^x - C_2 e^{-2x}.$$

To obtain the particular solution, we use the method of undetermined coefficients. The family to be considered is $\{1, e^{2x}\}$ and we proceed as follows:

$$y_{1,p} = A + Be^{2x},$$
$$y_{2,p} = C + De^{2x}.$$

Substituting into (D.29) gives

$$2Be^{2x} - 2A - 2Be^{2x} + 4De^{2x} = 2 - 4e^{2x},$$
$$4Be^{2x} - 3A - 3Be^{2x} + 6De^{2x} - C - De^{2x} = 0,$$

or upon simplifying we have

$$-2A + 4De^{2x} = 2 - 4e^{2x},$$
$$-3A - C + (B + 5D)e^{2x} = 0.$$

Now equating coefficients of like terms yields

$$-2A = 2, \quad 4D = -4, \quad -3A - C = 0, \quad B + 5D = 0,$$

which finally gives

$$A = -1, \quad B = 5, \quad C = 3, \quad D = -1,$$

and the general solutions are

$$y_1 = C_1 e^x + C_2 e^{-2x} - 1 + 5e^{2x},$$

$$y_2 = \frac{C_1}{2} e^x - C_2 e^{-2x} + 3 - e^{2x}.$$

The procedure, of course, generalizes to systems of size greater than two. If we have n simultaneous equations, the characteristic equation is obtained from evaluating an $n \times n$ determinant.

D.1.10 Summary

In solving ordinary linear differential equations, the first approach should be to determine whether the variables are separable. If they are, the general solution can be obtained directly by integration as discussed in Section D.1.3. If separation of variables is not possible, but the equation is first order, the solution can be obtained from (D.14) as derived in Section D.1.4.

For higher-order equations with constant coefficients, the general solution to the homogeneous equation can be obtained by finding the roots of the characteristic equation (Section D.1.5) and then finding the particular solution via undetermined coefficients (Section D.1.6). Use of operators (Section D.1.7) can also be employed to determine general solutions for nonhomogeneous linear equations with constant coefficients. If one (or more) solutions to the homogeneous equation are known, the order of the equation can be reduced (Section D.1.8), thereby yielding equations of lower order, which may be solved more readily.

Finally, in Section D.1.9 solutions of systems of simultaneous linear differential equations with constant coefficients are discussed.

D.2 Difference Equations

Consider a function of an independent variable x, where x is now a discrete variable; that is, it can take only integer values. Then the function exists only at discrete points (integer values of x) and we denote this type of function by y_x instead of $y(x)$. The first finite difference of y_x is given as

$$\Delta y \equiv y_{x+1} - y_x,$$

the second finite difference as

$$\Delta^2 y = \Delta(\Delta y) = (y_{x+2} - y_{x+1}) - (y_{x+1} - y_x)$$
$$= y_{x+2} - 2y_{x+1} + y_x,$$

and the nth finite difference as

$$\Delta^n y = \Delta(\Delta^{n-1} y).$$

We define an operator D to be

$$\begin{aligned} Dy_x &= y_{x+1}, \\ D^2 y_x &= D(Dy_x) = y_{x+2}, \\ &\vdots \\ D^n y_x &= D(D^{n-1} y_x) = y_{x+n}. \end{aligned}$$

One can easily see the relationship between Δ and D as

$$D^n = (\Delta + 1)^n .$$

D.2.1 Linear Difference Equations with Constant Coefficients

An equation involving y_x of the type

$$y_{x+n} + a_1 y_{x+n-1} + \cdots + a_{n-1} y_{x+1} + a_n y_x = g_x \tag{D.31}$$

is called a linear difference equation of order n with constant coefficients. We shall not treat here the case where the coefficients are also dependent on x.

One can see many similarities between difference equations and differential equations and, indeed, the solution techniques are often quite similar. The technique for solving (D.31) is very much like that used for linear differential equations with constant coefficients. In fact, it can be shown that a general solution of (D.31) consists of a linear combination of all solutions to the homogeneous equation (g_x replaced by zero) plus a particular solution to (D.31). Also, for nth degree equation, there are n arbitrary constants associated with the homogeneous solution, which in any particular case can be found from n boundary conditions.

To find the solution to the homogeneous equation we proceed in a manner similar to Section D.1.5. We first rewrite (D.31) using operator notation to get

$$(D^n + a_1 D^{n-1} + \cdots + a_{n-1} D + a_n) y_x = 0.$$

The homogeneous solutions are of the form r^x (as opposed to e^{rx} for differential equations), where r is a root to the characteristic equation

$$D^n + a_1 D^{n-1} + \cdots + a_{n-1} D + a_n = 0.$$

To see this, we let $y_x = r^x$ in (D.31) and get

$$r^{x+n} + a_1 r^{x+n-1} + \cdots + a_{n-1} r^{x+1} + a_n r^x = 0,$$

whereupon factoring out r^x we have

$$r^x \left(r^n + a_1 r^{n-1} + \cdots + a_{n-1} r + a_n \right) = 0.$$

But since r is a root to the characteristic equation, the left-hand equals zero.

Since the characteristic equation has n roots, the general solution to the homogeneous equation is

$$y_x = C_1 r_1^x + C_2 r_2^x + \cdots + C_n r_n^x.$$

Multiple roots can be handled in a manner analogous to differential equations in that for a root of multiplicity k, the first $k-1$ derivatives of the characteristic equation with respect to D must vanish and the k solutions are of the form r^x, xr^x, $x(x-1)r^x$, ..., $x(x-1)\cdots(x-k+1)r^x$, since in taking the ith derivative or r^x one obtains $x(x-1)\cdots(x-i+1)r^x r^{-i}$ and the r^{-i} can be absorbed in the arbitrary constant.

To find a particular solution to (D.31), the method of undetermined coefficients can be employed. We illustrate the procedures on the following example.

■ EXAMPLE D.7

Consider the difference equation

$$y_{x+2} + 6y_{x+1} + 9y_x = 16x^2.$$

The homogeneous equation in operator notation is

$$\left(D^2 + 6D + 9\right) y_x = 0,$$

and the solution to the characteristic equation has two roots at -3. Hence the solution is

$$y_x = C_1(-3)^x + C_2 x(-3)^x.$$

To find the particular solution, the family of x^2 gives terms $\{x^2, x, 1\}$. Therefore

$$y_{x,p} = Ax^2 + Bx + C,$$

and substituting this into the original equation gives

$$A(x+2)^2 + B(x+2) + C + 6[A(x+1)^2 + B(x+1) + C] + 9[Ax^2 + Bx + C] = 16x^2,$$

or

$$16Ax^2 + (16A + 16B)x + 10A + 8B + 16C = 16x^2.$$

Equating like coefficients yields the conditions

$$16A = 16, \quad 16A + 16B = 0, \quad 10A + 8B + 16C = 0,$$

or finally

$$A = 1, \quad B = -1, \quad C = -\tfrac{1}{8}.$$

Thus the particular solution is

$$y_{x,p} = x^2 - x - \tfrac{1}{8},$$

and the general solution becomes

$$y_x = C_1(-3)^x + C_2 x(-3)^x + x^2 - x - \tfrac{1}{8}.$$

D.2.2 Systems of Linear Difference Equations

The solution to systems of difference equations is analogous to the procedure used in Section D.1.9 for differential equations. One first writes the equation in operator notation, finds the characteristic equation using the determinant of the left-hand side "coefficients," solves for the roots, and obtains the homogeneous solution as linear combinations of r_i^x (instead of $e^{r_i, x}$ as for differential equations). The number of constants are then reduced as before by substituting the homogeneous solutions into the homogeneous equations. Then a particular solution can be found (if the equations are nonhomogeneous) by the method of undetermined coefficients.

■ **EXAMPLE D.8**

Consider the following system of difference equations to be solved for y and z.

$$\begin{aligned} y_{x+1} - 3y_x + z_{x+1} - 3z_x &= 2, \\ 2y_{x+1} - 5y_x + 3z_{x+1} - 3z_x &= 6(4)^x. \end{aligned} \qquad \text{(D.32)}$$

We first obtain the homogeneous solutions by solving the characteristic equation obtained after writing in operator notation. The characteristic equation is

$$\begin{vmatrix} (D-3) & (D-3) \\ (2D-5) & (3D-3) \end{vmatrix} = 0,$$

which upon calculating the determinant yields

$$D^2 - D - 6 = 0.$$

The roots can be found by factoring to be 3 and -2. Thus the homogeneous solutions are

$$\begin{aligned} y_x &= C_1(3)^x + C_2(-2)^x, \\ z_x &= C_3(3)^x + C_4(-2)^x. \end{aligned}$$

To reduce the number of arbitrary constants we substitute the above in the original equations of (D.32) with the right-hand side set to zero. This yields the relations

$$C_3 = -\tfrac{1}{6}C_1, \qquad C_4 = -C_2,$$

and hence the homogeneous solutions become

$$\begin{aligned} y_x &= C_1(3)^x + C_2(-2)^x, \\ z_x &= -\frac{C_1}{6}(3)^x - C_2(-2)^x. \end{aligned}$$

To obtain the particular solution, we employ undetermined coefficients. The family of the first right-hand side of (D.32) is $\{1\}$ and the family of the second is $\{4^x\}$. Thus we have

$$y_{x,p} = A + B(4)^x,$$
$$z_{x,p} = C + D(4)^x.$$

Substituting into (D.32) we get

$$A + 4B(4)^x - 3A - 3B(4)^x + C + 4D(4)^x - 3C - 3D(4)^x = 2,$$
$$2A + 8B(4)^x - 5A - 5B(4)^x + 3C + 12D(4)^x - 3C - 3D(4)^x = 6(4)^x.$$

Upon simplification we obtain

$$-2A - 2C + (B + D)(4)^x = 2,$$
$$-3A + (3B + 9D)(4)^x = 6(4)^x.$$

Equating like coefficients yields

$$-2A - 2C = 2, \quad B + D = 0, \quad -3A = 0, \quad 3B + 9D = 6,$$

which gives

$$A = 0, \quad B = -1, \quad C = -1, \quad D = 1.$$

The general solution is then

$$y_x = C_1(3)^x + C_2(-2)^x - (4)^x,$$
$$z_x = -\frac{C_1}{6}(3)^x - C_2(-2)^x - 1 + (4)^x.$$

This method of finding particular solutions for systems of equations through the use of undetermined coefficients does not always work. For example, one can verify that undetermined coefficients do not yield a particular solution of the following set of equations:

$$y_{x+1} - 2y_x + 2z_x = 2,$$
$$2y_{x+1} - 3y_x + 3z_{x+1} - z_x = 6(4)^x.$$

For such cases, other methods are necessary: However, further detailed treatment of finding particular solutions is not necessary since differential and difference equations encountered in queueing theory are, for the most part, homogeneous.

APPENDIX E
QTSPLUS SOFTWARE

For the current edition, there are two versions of the QtsPlus software: Excel and OpenOffice. The Excel version is available for Excel 97 and above. The OpenOffice version is supported by OpenOffice 1.1.x and above. Instructions for downloading the software are found at the end of the appendix.

The software is organized into eight categories based on model type, for example, single-server models or multiserver models. When loaded properly, a menu comes up that allows the user to select a particular model in a category. Figure E.1 is an example of the menu in the Excel version (the OpenOffice version is virtually identical).

To run a model, first open the **QtsPlus.xls** workbook (OpenOffice: open **QtsPlus4Calc.sxc**). For example, to run the $M/G/c/c$ Pure Overflow Model, select the **Multi-Server** entry from the combobox. From the resulting list, select the $M/G/c/c$ model and press the **Run Model** button or double-click the model name in the list. A workbook will appear for the model with explanation and/or instructions at the top, if necessary. To solve a particular instance of a model, change the INPUT PARAMETERS as desired, and the spreadsheet automatically calculates new RESULTS (output measures of performance). As an illustration, suppose we desire to run a case

Fundamentals of Queueing Theory, Fourth Edition.
By D. Gross, J. F. Shortle, J. M. Thompson, and C. M. Harris
Copyright © 2008 John Wiley & Sons, Inc.

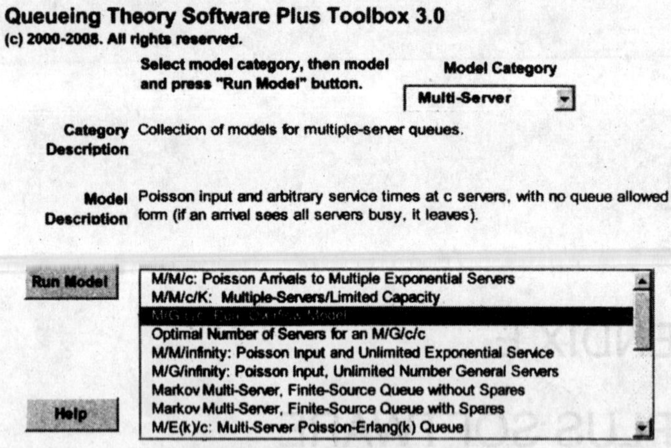

Figure E.1 QtsPlus main menu.

with $\lambda = 4$, $\mu = 1/2$ (or $1/\mu = 2$), and $c = 5$. We enter these new input values by writing over the current ones, and the output values are changed accordingly as shown in Figure E.2.

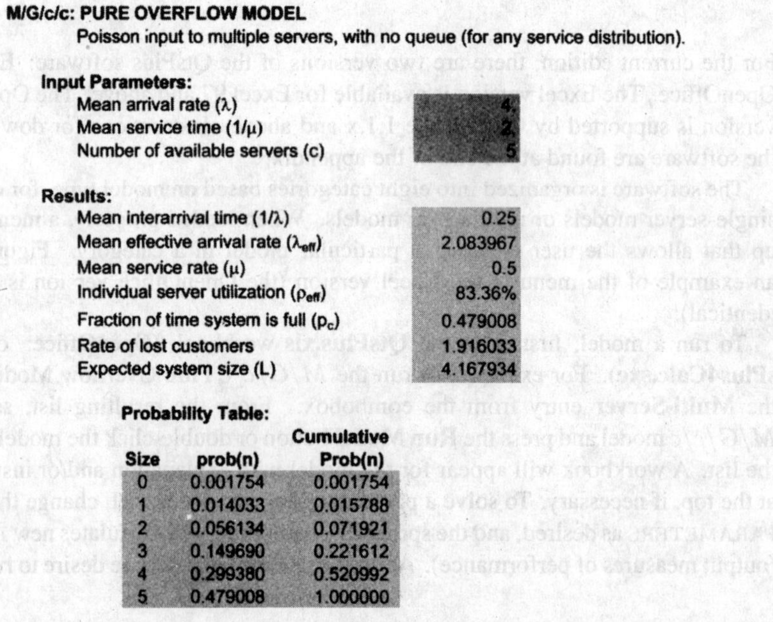
M/G/c/c: PURE OVERFLOW MODEL

Poisson input to multiple servers, with no queue (for any service distribution).

Input Parameters:

Mean arrival rate (λ)	4
Mean service time (1/μ)	2
Number of available servers (c)	5

Results:

Mean interarrival time (1/λ)	0.25
Mean effective arrival rate (λ_{eff})	2.083967
Mean service rate (μ)	0.5
Individual server utilization (ρ_{eff})	83.36%
Fraction of time system is full (p_c)	0.479008
Rate of lost customers	1.916033
Expected system size (L)	4.167934

Probability Table:

Size	prob(n)	Cumulative Prob(n)
0	0.001754	0.001754
1	0.014033	0.015788
2	0.056134	0.071921
3	0.149690	0.221612
4	0.299380	0.520992
5	0.479008	1.000000

Figure E.2 $M/G/c/c$ pure overflow model with new inputs.

In Figure E.3 we present one additional module, a Markov Single-Server, Finite-Source Queue without Spares from the Single-Server category. This model has self-contained instructions at the top. To solve a particular situation, enter the INPUT PARAMETERS and press the **Solve** button. A sample problem is illustrated in Figure E.3. In addition to system performance measures, such as W, W_q, L and L_q, the model computes stationary probabilities for system size and provides a plot of the probabilities as shown in Figure E.4.

MARKOV SINGLE-SERVER, FINITE-SOURCE QUEUE WITHOUT SPARES

(Machine Repair with Single Repairman)

After entering input parameters, press "Solve" button.

Input Parameters:

Mean interarrival time for a single customer ($1/\lambda$)	30	(i.e., Mean time between failures)
Mean time to complete service ($1/\mu$)	5	(i.e., Mean time to repair)
Maximum # of customers in the system (M)	10	(i.e., Number of repairable units)
Specific time for delay distribution calculation (t)	20	

Solve

Results:

Combined effective overall arrival rate (λ_{eff})	0.191372
Service rate (μ)	0.2
System utilization (ρ_{eff})	0.956858
Fraction of time server is idle (p_0)	0.043142
Expected queue size (Lq)	3.301993
Expected system size (L)	4.258851
Expected waiting time in the queue (Wq)	17.254349
Expected waiting time in the system (W)	22.254349
Probability that an arrival's line delay exceeds t	0.36331

Probability Table:

Size	prob(n)	Cumulative Prob(n)
0	0.043142	0.043142
1	0.071903	0.115045
2	0.107855	0.222899
3	0.143806	0.366706
4	0.167774	0.534479
5	0.167774	0.702253
6	0.139812	0.842065
7	0.093208	0.935272
8	0.046604	0.981876
9	0.015535	0.997411
10	0.002589	1.000000

Figure E.3 Single-server finite-source model.

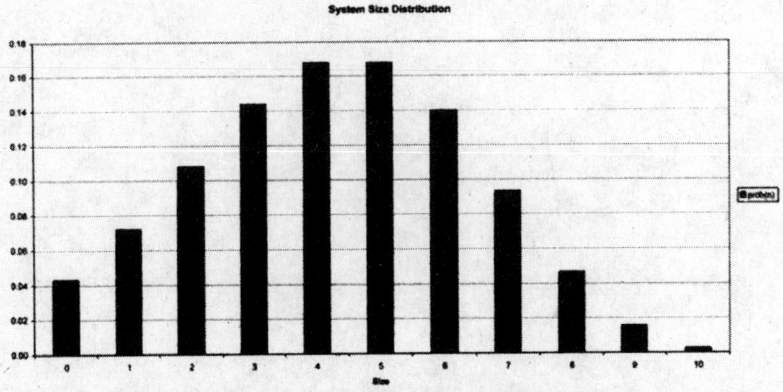

Figure E.4 System size distribution.

E.1 Instructions for Downloading

The Excel version is provided as a self-extracting compressed file and the OpenOffice version is distributed as a compressed zip file. Both distribution files can be obtained from the Wiley public ftp site. The installation instructions found there contain information on requirements to install the software. To access the files, type the following location address in your ftp or web browsing program:

`ftp://ftp.wiley.com/public/sci_tech_med/queueing_theory/`

From the queueing_theory area, you can download either of the two versions by double-clicking on the preferred file and saving the file to your hard drive.

- Excel Version: qtsplus-xcel.exe
- OpenOffice Version: QtsPlus4Calc.zip

Please consult the readme.txt file for the latest detailed instructions for downloading and installing. To learn more about titles from Wiley, visit the Publisher's web site, `http://www.wiley.com`.

INDEX

Fundamentals of Queueing Theory, Fourth Edition.
By D. Gross, J. F. Shortle, J. M. Thompson, and C. M. Harris
Copyright © 2009 John Wiley & Sons, Inc.

WILEY SERIES IN PROBABILITY AND STATISTICS

ESTABLISHED BY WALTER A. SHEWHART AND SAMUEL S. WILKS

Editors: *David J. Balding, Noel A. C. Cressie, Garrett M. Fitzmaurice, Iain M. Johnstone, Geert Molenberghs, David W. Scott, Adrian F. M. Smith, Ruey S. Tsay, Sanford Weisberg*

Editors Emeriti: *Vic Barnett, J. Stuart Hunter, Jozef L. Teugels*

The ***Wiley Series in Probability and Statistics*** is well established and authoritative. It covers many topics of current research interest in both pure and applied statistics and probability theory. Written by leading statisticians and institutions, the titles span both state-of-the-art developments in the field and classical methods.

Reflecting the wide range of current research in statistics, the series encompasses applied, methodological and theoretical statistics, ranging from applications and new techniques made possible by advances in computerized practice to rigorous treatment of theoretical approaches.

This series provides essential and invaluable reading for all statisticians, whether in academia, industry, government, or research.

† ABRAHAM and LEDOLTER · Statistical Methods for Forecasting
AGRESTI · Analysis of Ordinal Categorical Data
AGRESTI · An Introduction to Categorical Data Analysis, *Second Edition*
AGRESTI · Categorical Data Analysis, *Second Edition*
ALTMAN, GILL, and McDONALD · Numerical Issues in Statistical Computing for the Social Scientist
AMARATUNGA and CABRERA · Exploration and Analysis of DNA Microarray and Protein Array Data
ANDĚL · Mathematics of Chance
ANDERSON · An Introduction to Multivariate Statistical Analysis, *Third Edition*
* ANDERSON · The Statistical Analysis of Time Series
ANDERSON, AUQUIER, HAUCK, OAKES, VANDAELE, and WEISBERG · Statistical Methods for Comparative Studies
ANDERSON and LOYNES · The Teaching of Practical Statistics
ARMITAGE and DAVID (editors) · Advances in Biometry
ARNOLD, BALAKRISHNAN, and NAGARAJA · Records
* ARTHANARI and DODGE · Mathematical Programming in Statistics
* BAILEY · The Elements of Stochastic Processes with Applications to the Natural Sciences
BALAKRISHNAN and KOUTRAS · Runs and Scans with Applications
BALAKRISHNAN and NG · Precedence-Type Tests and Applications
BARNETT · Comparative Statistical Inference, *Third Edition*
BARNETT · Environmental Statistics
BARNETT and LEWIS · Outliers in Statistical Data, *Third Edition*
BARTOSZYNSKI and NIEWIADOMSKA-BUGAJ · Probability and Statistical Inference
BASILEVSKY · Statistical Factor Analysis and Related Methods: Theory and Applications
BASU and RIGDON · Statistical Methods for the Reliability of Repairable Systems
BATES and WATTS · Nonlinear Regression Analysis and Its Applications
BECHHOFER, SANTNER, and GOLDSMAN · Design and Analysis of Experiments for Statistical Selection, Screening, and Multiple Comparisons

*Now available in a lower priced paperback edition in the Wiley Classics Library.
†Now available in a lower priced paperback edition in the Wiley–Interscience Paperback Series.

BELSLEY · Conditioning Diagnostics: Collinearity and Weak Data in Regression
† BELSLEY, KUH, and WELSCH · Regression Diagnostics: Identifying Influential Data and Sources of Collinearity
BENDAT and PIERSOL · Random Data: Analysis and Measurement Procedures, *Third Edition*
BERRY, CHALONER, and GEWEKE · Bayesian Analysis in Statistics and Econometrics: Essays in Honor of Arnold Zellner
BERNARDO and SMITH · Bayesian Theory
BHAT and MILLER · Elements of Applied Stochastic Processes, *Third Edition*
BHATTACHARYA and WAYMIRE · Stochastic Processes with Applications
BILLINGSLEY · Convergence of Probability Measures, *Second Edition*
BILLINGSLEY · Probability and Measure, *Third Edition*
BIRKES and DODGE · Alternative Methods of Regression
BISWAS, DATTA, FINE, and SEGAL · Statistical Advances in the Biomedical Sciences: Clinical Trials, Epidemiology, Survival Analysis, and Bioinformatics
BLISCHKE AND MURTHY (editors) · Case Studies in Reliability and Maintenance
BLISCHKE AND MURTHY · Reliability: Modeling, Prediction, and Optimization
BLOOMFIELD · Fourier Analysis of Time Series: An Introduction, *Second Edition*
BOLLEN · Structural Equations with Latent Variables
BOLLEN and CURRAN · Latent Curve Models: A Structural Equation Perspective
BOROVKOV · Ergodicity and Stability of Stochastic Processes
BOULEAU · Numerical Methods for Stochastic Processes
BOX · Bayesian Inference in Statistical Analysis
BOX · R. A. Fisher, the Life of a Scientist
BOX and DRAPER · Response Surfaces, Mixtures, and Ridge Analyses, *Second Edition*
* BOX and DRAPER · Evolutionary Operation: A Statistical Method for Process Improvement
BOX and FRIENDS · Improving Almost Anything, *Revised Edition*
BOX, HUNTER, and HUNTER · Statistics for Experimenters: Design, Innovation, and Discovery, *Second Editon*
BOX, JENKINS, and REINSEL · Time Series Analysis: Forcasting and Control, *Fourth Edition*
BOX and LUCEÑO · Statistical Control by Monitoring and Feedback Adjustment
BRANDIMARTE · Numerical Methods in Finance: A MATLAB-Based Introduction
† BROWN and HOLLANDER · Statistics: A Biomedical Introduction
BRUNNER, DOMHOF, and LANGER · Nonparametric Analysis of Longitudinal Data in Factorial Experiments
BUCKLEW · Large Deviation Techniques in Decision, Simulation, and Estimation
CAIROLI and DALANG · Sequential Stochastic Optimization
CASTILLO, HADI, BALAKRISHNAN, and SARABIA · Extreme Value and Related Models with Applications in Engineering and Science
CHAN · Time Series: Applications to Finance
CHARALAMBIDES · Combinatorial Methods in Discrete Distributions
CHATTERJEE and HADI · Regression Analysis by Example, *Fourth Edition*
CHATTERJEE and HADI · Sensitivity Analysis in Linear Regression
CHERNICK · Bootstrap Methods: A Guide for Practitioners and Researchers, *Second Edition*
CHERNICK and FRIIS · Introductory Biostatistics for the Health Sciences
CHILÈS and DELFINER · Geostatistics: Modeling Spatial Uncertainty
CHOW and LIU · Design and Analysis of Clinical Trials: Concepts and Methodologies, *Second Edition*
CLARKE · Linear Models: The Theory and Application of Analysis of Variance
CLARKE and DISNEY · Probability and Random Processes: A First Course with Applications, *Second Edition*

*Now available in a lower priced paperback edition in the Wiley Classics Library.
†Now available in a lower priced paperback edition in the Wiley–Interscience Paperback Series.

* COCHRAN and COX · Experimental Designs, *Second Edition*
CONGDON · Applied Bayesian Modelling
CONGDON · Bayesian Models for Categorical Data
CONGDON · Bayesian Statistical Modelling
CONOVER · Practical Nonparametric Statistics, *Third Edition*
COOK · Regression Graphics
COOK and WEISBERG · Applied Regression Including Computing and Graphics
COOK and WEISBERG · An Introduction to Regression Graphics
CORNELL · Experiments with Mixtures, Designs, Models, and the Analysis of Mixture Data, *Third Edition*
COVER and THOMAS · Elements of Information Theory
COX · A Handbook of Introductory Statistical Methods
* COX · Planning of Experiments
CRESSIE · Statistics for Spatial Data, *Revised Edition*
CSÖRGŐ and HORVÁTH · Limit Theorems in Change Point Analysis
DANIEL · Applications of Statistics to Industrial Experimentation
DANIEL · Biostatistics: A Foundation for Analysis in the Health Sciences, *Eighth Edition*
* DANIEL · Fitting Equations to Data: Computer Analysis of Multifactor Data, *Second Edition*
DASU and JOHNSON · Exploratory Data Mining and Data Cleaning
DAVID and NAGARAJA · Order Statistics, *Third Edition*
* DEGROOT, FIENBERG, and KADANE · Statistics and the Law
DEL CASTILLO · Statistical Process Adjustment for Quality Control
DEMARIS · Regression with Social Data: Modeling Continuous and Limited Response Variables
DEMIDENKO · Mixed Models: Theory and Applications
DENISON, HOLMES, MALLICK and SMITH · Bayesian Methods for Nonlinear Classification and Regression
DETTE and STUDDEN · The Theory of Canonical Moments with Applications in Statistics, Probability, and Analysis
DEY and MUKERJEE · Fractional Factorial Plans
DILLON and GOLDSTEIN · Multivariate Analysis: Methods and Applications
DODGE · Alternative Methods of Regression
* DODGE and ROMIG · Sampling Inspection Tables, *Second Edition*
* DOOB · Stochastic Processes
DOWDY, WEARDEN, and CHILKO · Statistics for Research, *Third Edition*
DRAPER and SMITH · Applied Regression Analysis, *Third Edition*
DRYDEN and MARDIA · Statistical Shape Analysis
DUDEWICZ and MISHRA · Modern Mathematical Statistics
DUNN and CLARK · Basic Statistics: A Primer for the Biomedical Sciences, *Third Edition*
DUPUIS and ELLIS · A Weak Convergence Approach to the Theory of Large Deviations
EDLER and KITSOS · Recent Advances in Quantitative Methods in Cancer and Human Health Risk Assessment
* ELANDT-JOHNSON and JOHNSON · Survival Models and Data Analysis
ENDERS · Applied Econometric Time Series
† ETHIER and KURTZ · Markov Processes: Characterization and Convergence
EVANS, HASTINGS, and PEACOCK · Statistical Distributions, *Third Edition*
FELLER · An Introduction to Probability Theory and Its Applications, Volume I, *Third Edition,* Revised; Volume II, *Second Edition*
FISHER and VAN BELLE · Biostatistics: A Methodology for the Health Sciences
FITZMAURICE, LAIRD, and WARE · Applied Longitudinal Analysis
* FLEISS · The Design and Analysis of Clinical Experiments

*Now available in a lower priced paperback edition in the Wiley Classics Library.
†Now available in a lower priced paperback edition in the Wiley–Interscience Paperback Series.

FLEISS · Statistical Methods for Rates and Proportions, *Third Edition*
† FLEMING and HARRINGTON · Counting Processes and Survival Analysis
FULLER · Introduction to Statistical Time Series, *Second Edition*
† FULLER · Measurement Error Models
GALLANT · Nonlinear Statistical Models
GEISSER · Modes of Parametric Statistical Inference
GELMAN and MENG · Applied Bayesian Modeling and Causal Inference from Incomplete-Data Perspectives
GEWEKE · Contemporary Bayesian Econometrics and Statistics
GHOSH, MUKHOPADHYAY, and SEN · Sequential Estimation
GIESBRECHT and GUMPERTZ · Planning, Construction, and Statistical Analysis of Comparative Experiments
GIFI · Nonlinear Multivariate Analysis
GIVENS and HOETING · Computational Statistics
GLASSERMAN and YAO · Monotone Structure in Discrete-Event Systems
GNANADESIKAN · Methods for Statistical Data Analysis of Multivariate Observations, *Second Edition*
GOLDSTEIN and LEWIS · Assessment: Problems, Development, and Statistical Issues
GREENWOOD and NIKULIN · A Guide to Chi-Squared Testing
GROSS, SHORTLE, THOMPSON, and HARRIS · Fundamentals of Queueing Theory, *Fourth Edition*
GROSS, SHORTLE, THOMPSON, and HARRIS · Solutions Manual to Accompany Fundamentals of Queueing Theory, *Fourth Edition*
* HAHN and SHAPIRO · Statistical Models in Engineering
HAHN and MEEKER · Statistical Intervals: A Guide for Practitioners
HALD · A History of Probability and Statistics and their Applications Before 1750
HALD · A History of Mathematical Statistics from 1750 to 1930
† HAMPEL · Robust Statistics: The Approach Based on Influence Functions
HANNAN and DEISTLER · The Statistical Theory of Linear Systems
HARTUNG, KNAPP, and SINHA · Statistical Meta-Analysis with Applications
HEIBERGER · Computation for the Analysis of Designed Experiments
HEDAYAT and SINHA · Design and Inference in Finite Population Sampling
HEDEKER and GIBBONS · Longitudinal Data Analysis
HELLER · MACSYMA for Statisticians
HINKELMANN and KEMPTHORNE · Design and Analysis of Experiments, Volume 1: Introduction to Experimental Design, *Second Edition*
HINKELMANN and KEMPTHORNE · Design and Analysis of Experiments, Volume 2: Advanced Experimental Design
HOAGLIN, MOSTELLER, and TUKEY · Exploratory Approach to Analysis of Variance
* HOAGLIN, MOSTELLER, and TUKEY · Exploring Data Tables, Trends and Shapes
* HOAGLIN, MOSTELLER, and TUKEY · Understanding Robust and Exploratory Data Analysis
HOCHBERG and TAMHANE · Multiple Comparison Procedures
HOCKING · Methods and Applications of Linear Models: Regression and the Analysis of Variance, *Second Edition*
HOEL · Introduction to Mathematical Statistics, *Fifth Edition*
HOGG and KLUGMAN · Loss Distributions
HOLLANDER and WOLFE · Nonparametric Statistical Methods, *Second Edition*
HOSMER and LEMESHOW · Applied Logistic Regression, *Second Edition*
HOSMER, LEMESHOW, and MAY · Applied Survival Analysis: Regression Modeling of Time-to-Event Data, *Second Edition*
† HUBER · Robust Statistics

*Now available in a lower priced paperback edition in the Wiley Classics Library.
†Now available in a lower priced paperback edition in the Wiley–Interscience Paperback Series.

HUBERTY · Applied Discriminant Analysis
HUBERTY and OLEJNIK · Applied MANOVA and Discriminant Analysis, *Second Edition*
HUNT and KENNEDY · Financial Derivatives in Theory and Practice, *Revised Edition*
HURD and MIAMEE · Periodically Correlated Random Sequences: Spectral Theory and Practice
HUSKOVA, BERAN, and DUPAC · Collected Works of Jaroslav Hajek—with Commentary
HUZURBAZAR · Flowgraph Models for Multistate Time-to-Event Data
IMAN and CONOVER · A Modern Approach to Statistics
† JACKSON · A User's Guide to Principle Components
JOHN · Statistical Methods in Engineering and Quality Assurance
JOHNSON · Multivariate Statistical Simulation
JOHNSON and BALAKRISHNAN · Advances in the Theory and Practice of Statistics: A Volume in Honor of Samuel Kotz
JOHNSON and BHATTACHARYYA · Statistics: Principles and Methods, *Fifth Edition*
JOHNSON and KOTZ · Distributions in Statistics
JOHNSON and KOTZ (editors) · Leading Personalities in Statistical Sciences: From the Seventeenth Century to the Present
JOHNSON, KOTZ, and BALAKRISHNAN · Continuous Univariate Distributions, Volume 1, *Second Edition*
JOHNSON, KOTZ, and BALAKRISHNAN · Continuous Univariate Distributions, Volume 2, *Second Edition*
JOHNSON, KOTZ, and BALAKRISHNAN · Discrete Multivariate Distributions
JOHNSON, KEMP, and KOTZ · Univariate Discrete Distributions, *Third Edition*
JUDGE, GRIFFITHS, HILL, LÜTKEPOHL, and LEE · The Theory and Practice of Econometrics, *Second Edition*
JUREČKOVÁ and SEN · Robust Statistical Procedures: Aymptotics and Interrelations
JUREK and MASON · Operator-Limit Distributions in Probability Theory
KADANE · Bayesian Methods and Ethics in a Clinical Trial Design
KADANE AND SCHUM · A Probabilistic Analysis of the Sacco and Vanzetti Evidence
KALBFLEISCH and PRENTICE · The Statistical Analysis of Failure Time Data, *Second Edition*
KARIYA and KURATA · Generalized Least Squares
KASS and VOS · Geometrical Foundations of Asymptotic Inference
† KAUFMAN and ROUSSEEUW · Finding Groups in Data: An Introduction to Cluster Analysis
KEDEM and FOKIANOS · Regression Models for Time Series Analysis
KENDALL, BARDEN, CARNE, and LE · Shape and Shape Theory
KHURI · Advanced Calculus with Applications in Statistics, *Second Edition*
KHURI, MATHEW, and SINHA · Statistical Tests for Mixed Linear Models
KLEIBER and KOTZ · Statistical Size Distributions in Economics and Actuarial Sciences
KLUGMAN, PANJER, and WILLMOT · Loss Models: From Data to Decisions, *Third Edition*
KLUGMAN, PANJER, and WILLMOT · Solutions Manual to Accompany Loss Models: From Data to Decisions, *Third Edition*
KOTZ, BALAKRISHNAN, and JOHNSON · Continuous Multivariate Distributions, Volume 1, *Second Edition*
KOVALENKO, KUZNETZOV, and PEGG · Mathematical Theory of Reliability of Time-Dependent Systems with Practical Applications
KOWALSKI and TU · Modern Applied U-Statistics
KROONENBERG · Applied Multiway Data Analysis
KVAM and VIDAKOVIC · Nonparametric Statistics with Applications to Science and Engineering

*Now available in a lower priced paperback edition in the Wiley Classics Library.
†Now available in a lower priced paperback edition in the Wiley–Interscience Paperback Series.

LACHIN · Biostatistical Methods: The Assessment of Relative Risks
LAD · Operational Subjective Statistical Methods: A Mathematical, Philosophical, and Historical Introduction
LAMPERTI · Probability: A Survey of the Mathematical Theory, *Second Edition*
LANGE, RYAN, BILLARD, BRILLINGER, CONQUEST, and GREENHOUSE · Case Studies in Biometry
LARSON · Introduction to Probability Theory and Statistical Inference, *Third Edition*
LAWLESS · Statistical Models and Methods for Lifetime Data, *Second Edition*
LAWSON · Statistical Methods in Spatial Epidemiology
LE · Applied Categorical Data Analysis
LE · Applied Survival Analysis
LEE and WANG · Statistical Methods for Survival Data Analysis, *Third Edition*
LEPAGE and BILLARD · Exploring the Limits of Bootstrap
LEYLAND and GOLDSTEIN (editors) · Multilevel Modelling of Health Statistics
LIAO · Statistical Group Comparison
LINDVALL · Lectures on the Coupling Method
LIN · Introductory Stochastic Analysis for Finance and Insurance
LINHART and ZUCCHINI · Model Selection
LITTLE and RUBIN · Statistical Analysis with Missing Data, *Second Edition*
LLOYD · The Statistical Analysis of Categorical Data
LOWEN and TEICH · Fractal-Based Point Processes
MAGNUS and NEUDECKER · Matrix Differential Calculus with Applications in Statistics and Econometrics, *Revised Edition*
MALLER and ZHOU · Survival Analysis with Long Term Survivors
MALLOWS · Design, Data, and Analysis by Some Friends of Cuthbert Daniel
MANN, SCHAFER, and SINGPURWALLA · Methods for Statistical Analysis of Reliability and Life Data
MANTON, WOODBURY, and TOLLEY · Statistical Applications Using Fuzzy Sets
MARCHETTE · Random Graphs for Statistical Pattern Recognition
MARDIA and JUPP · Directional Statistics
MASON, GUNST, and HESS · Statistical Design and Analysis of Experiments with Applications to Engineering and Science, *Second Edition*
McCULLOCH, SEARLE, and NEUHAUS · Generalized, Linear, and Mixed Models, *Second Edition*
McFADDEN · Management of Data in Clinical Trials, *Second Edition*
* McLACHLAN · Discriminant Analysis and Statistical Pattern Recognition
McLACHLAN, DO, and AMBROISE · Analyzing Microarray Gene Expression Data
McLACHLAN and KRISHNAN · The EM Algorithm and Extensions, *Second Edition*
McLACHLAN and PEEL · Finite Mixture Models
McNEIL · Epidemiological Research Methods
MEEKER and ESCOBAR · Statistical Methods for Reliability Data
MEERSCHAERT and SCHEFFLER · Limit Distributions for Sums of Independent Random Vectors: Heavy Tails in Theory and Practice
MICKEY, DUNN, and CLARK · Applied Statistics: Analysis of Variance and Regression, *Third Edition*
* MILLER · Survival Analysis, *Second Edition*
MONTGOMERY, JENNINGS, and KULAHCI · Introduction to Time Series Analysis and Forecasting
MONTGOMERY, PECK, and VINING · Introduction to Linear Regression Analysis, *Fourth Edition*
MORGENTHALER and TUKEY · Configural Polysampling: A Route to Practical Robustness
MUIRHEAD · Aspects of Multivariate Statistical Theory

*Now available in a lower priced paperback edition in the Wiley Classics Library.
†Now available in a lower priced paperback edition in the Wiley–Interscience Paperback Series.

MULLER and STOYAN · Comparison Methods for Stochastic Models and Risks
MURRAY · X-STAT 2.0 Statistical Experimentation, Design Data Analysis, and Nonlinear Optimization
MURTHY, XIE, and JIANG · Weibull Models
MYERS and MONTGOMERY · Response Surface Methodology: Process and Product Optimization Using Designed Experiments, *Second Edition*
MYERS, MONTGOMERY, and VINING · Generalized Linear Models. With Applications in Engineering and the Sciences
† NELSON · Accelerated Testing, Statistical Models, Test Plans, and Data Analyses
† NELSON · Applied Life Data Analysis
NEWMAN · Biostatistical Methods in Epidemiology
OCHI · Applied Probability and Stochastic Processes in Engineering and Physical Sciences
OKABE, BOOTS, SUGIHARA, and CHIU · Spatial Tesselations: Concepts and Applications of Voronoi Diagrams, *Second Edition*
OLIVER and SMITH · Influence Diagrams, Belief Nets and Decision Analysis
PALTA · Quantitative Methods in Population Health: Extensions of Ordinary Regressions
PANJER · Operational Risk: Modeling and Analytics
PANKRATZ · Forecasting with Dynamic Regression Models
PANKRATZ · Forecasting with Univariate Box-Jenkins Models: Concepts and Cases
* PARZEN · Modern Probability Theory and Its Applications
PEÑA, TIAO, and TSAY · A Course in Time Series Analysis
PIANTADOSI · Clinical Trials: A Methodologic Perspective
PORT · Theoretical Probability for Applications
POURAHMADI · Foundations of Time Series Analysis and Prediction Theory
POWELL · Approximate Dynamic Programming: Solving the Curses of Dimensionality
PRESS · Bayesian Statistics: Principles, Models, and Applications
PRESS · Subjective and Objective Bayesian Statistics, *Second Edition*
PRESS and TANUR · The Subjectivity of Scientists and the Bayesian Approach
PUKELSHEIM · Optimal Experimental Design
PURI, VILAPLANA, and WERTZ · New Perspectives in Theoretical and Applied Statistics
† PUTERMAN · Markov Decision Processes: Discrete Stochastic Dynamic Programming
QIU · Image Processing and Jump Regression Analysis
* RAO · Linear Statistical Inference and Its Applications, *Second Edition*
RAUSAND and HØYLAND · System Reliability Theory: Models, Statistical Methods, and Applications, *Second Edition*
RENCHER · Linear Models in Statistics
RENCHER · Methods of Multivariate Analysis, *Second Edition*
RENCHER · Multivariate Statistical Inference with Applications
* RIPLEY · Spatial Statistics
* RIPLEY · Stochastic Simulation
ROBINSON · Practical Strategies for Experimenting
ROHATGI and SALEH · An Introduction to Probability and Statistics, *Second Edition*
ROLSKI, SCHMIDLI, SCHMIDT, and TEUGELS · Stochastic Processes for Insurance and Finance
ROSENBERGER and LACHIN · Randomization in Clinical Trials: Theory and Practice
ROSS · Introduction to Probability and Statistics for Engineers and Scientists
ROSSI, ALLENBY, and McCULLOCH · Bayesian Statistics and Marketing
† ROUSSEEUW and LEROY · Robust Regression and Outlier Detection
* RUBIN · Multiple Imputation for Nonresponse in Surveys
RUBINSTEIN and KROESE · Simulation and the Monte Carlo Method, *Second Edition*
RUBINSTEIN and MELAMED · Modern Simulation and Modeling
RYAN · Modern Engineering Statistics

*Now available in a lower priced paperback edition in the Wiley Classics Library.
†Now available in a lower priced paperback edition in the Wiley–Interscience Paperback Series.

RYAN · Modern Experimental Design
RYAN · Modern Regression Methods
RYAN · Statistical Methods for Quality Improvement, *Second Edition*
SALEH · Theory of Preliminary Test and Stein-Type Estimation with Applications
* SCHEFFE · The Analysis of Variance
SCHIMEK · Smoothing and Regression: Approaches, Computation, and Application
SCHOTT · Matrix Analysis for Statistics, *Second Edition*
SCHOUTENS · Levy Processes in Finance: Pricing Financial Derivatives
SCHUSS · Theory and Applications of Stochastic Differential Equations
SCOTT · Multivariate Density Estimation: Theory, Practice, and Visualization
† SEARLE · Linear Models for Unbalanced Data
† SEARLE · Matrix Algebra Useful for Statistics
† SEARLE, CASELLA, and McCULLOCH · Variance Components
SEARLE and WILLETT · Matrix Algebra for Applied Economics
SEBER · A Matrix Handbook For Statisticians
† SEBER · Multivariate Observations
SEBER and LEE · Linear Regression Analysis, *Second Edition*
† SEBER and WILD · Nonlinear Regression
SENNOTT · Stochastic Dynamic Programming and the Control of Queueing Systems
* SERFLING · Approximation Theorems of Mathematical Statistics
SHAFER and VOVK · Probability and Finance: It's Only a Game!
SILVAPULLE and SEN · Constrained Statistical Inference: Inequality, Order, and Shape Restrictions
SMALL and McLEISH · Hilbert Space Methods in Probability and Statistical Inference
SRIVASTAVA · Methods of Multivariate Statistics
STAPLETON · Linear Statistical Models
STAPLETON · Models for Probability and Statistical Inference: Theory and Applications
STAUDTE and SHEATHER · Robust Estimation and Testing
STOYAN, KENDALL, and MECKE · Stochastic Geometry and Its Applications, *Second Edition*
STOYAN and STOYAN · Fractals, Random Shapes and Point Fields: Methods of Geometrical Statistics
STREET and BURGESS · The Construction of Optimal Stated Choice Experiments: Theory and Methods
STYAN · The Collected Papers of T. W. Anderson: 1943–1985
SUTTON, ABRAMS, JONES, SHELDON, and SONG · Methods for Meta-Analysis in Medical Research
TAKEZAWA · Introduction to Nonparametric Regression
TANAKA · Time Series Analysis: Nonstationary and Noninvertible Distribution Theory
THOMPSON · Empirical Model Building
THOMPSON · Sampling, *Second Edition*
THOMPSON · Simulation: A Modeler's Approach
THOMPSON and SEBER · Adaptive Sampling
THOMPSON, WILLIAMS, and FINDLAY · Models for Investors in Real World Markets
TIAO, BISGAARD, HILL, PEÑA, and STIGLER (editors) · Box on Quality and Discovery: with Design, Control, and Robustness
TIERNEY · LISP-STAT: An Object-Oriented Environment for Statistical Computing and Dynamic Graphics
TSAY · Analysis of Financial Time Series, *Second Edition*
UPTON and FINGLETON · Spatial Data Analysis by Example, Volume II: Categorical and Directional Data
† VAN BELLE · Statistical Rules of Thumb, *Second Edition*
VAN BELLE, FISHER, HEAGERTY, and LUMLEY · Biostatistics: A Methodology for the Health Sciences, *Second Edition*

*Now available in a lower priced paperback edition in the Wiley Classics Library.
†Now available in a lower priced paperback edition in the Wiley–Interscience Paperback Series.

VESTRUP · The Theory of Measures and Integration
VIDAKOVIC · Statistical Modeling by Wavelets
VINOD and REAGLE · Preparing for the Worst: Incorporating Downside Risk in Stock Market Investments
WALLER and GOTWAY · Applied Spatial Statistics for Public Health Data
WEERAHANDI · Generalized Inference in Repeated Measures: Exact Methods in MANOVA and Mixed Models
WEISBERG · Applied Linear Regression, *Third Edition*
WELSH · Aspects of Statistical Inference
WESTFALL and YOUNG · Resampling-Based Multiple Testing: Examples and Methods for *p*-Value Adjustment
WHITTAKER · Graphical Models in Applied Multivariate Statistics
WINKER · Optimization Heuristics in Economics: Applications of Threshold Accepting
WONNACOTT and WONNACOTT · Econometrics, *Second Edition*
WOODING · Planning Pharmaceutical Clinical Trials: Basic Statistical Principles
WOODWORTH · Biostatistics: A Bayesian Introduction
WOOLSON and CLARKE · Statistical Methods for the Analysis of Biomedical Data, *Second Edition*
WU and HAMADA · Experiments: Planning, Analysis, and Parameter Design Optimization
WU and ZHANG · Nonparametric Regression Methods for Longitudinal Data Analysis
YANG · The Construction Theory of Denumerable Markov Processes
YOUNG, VALERO-MORA, and FRIENDLY · Visual Statistics: Seeing Data with Dynamic Interactive Graphics
ZELTERMAN · Discrete Distributions—Applications in the Health Sciences
* ZELLNER · An Introduction to Bayesian Inference in Econometrics
ZHOU, OBUCHOWSKI, and McCLISH · Statistical Methods in Diagnostic Medicine

*Now available in a lower priced paperback edition in the Wiley Classics Library.
†Now available in a lower priced paperback edition in the Wiley–Interscience Paperback Series.